- Ideal PAT. P Reamer
- No-Dog - for conduit
- Klein tape measurer

Electrical
Level Two

Trainee Guide
Tenth Edition

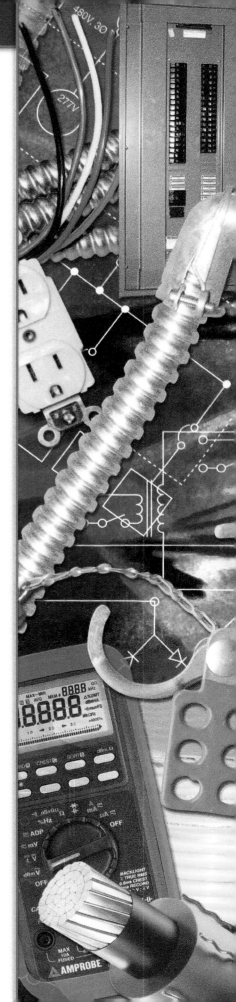

NCCER

Chief Executive Officer: Don Whyte
President: Boyd Worsham
Chief Operations Officer: Katrina Kersch
Electrical Curriculum Project Manager: Veronica Westfall
Director, Product Development: Tim Davis
Senior Production Manager: Erin O'Nora
Senior Manager of Projects: Chris Wilson
Testing/Assessment Project Manager: Elizabeth Schlaupitz
Project Assistant: Lauren Corley
Lead Technical Writer: Veronica Westfall
Technical Writers: Gary Ferguson, Veronica Westfall, John Mueller, Chris Wilson

Managing Editor: Graham Hack
Desktop Publishing Manager: James McKay
Art Manager: Kelly Sadler/Karyn Payne
Multimedia Project Manager: Alan Youngblood
Digital Content Coordinator: Rachael Downs
Production Specialists: Gene Page, Eric Caraballoso
Production Assistance: Adrienne Payne, David Gregoire, Edward Fortman, Joanne Hart, Olga Trofymenko
Editors: Jordan Hutchinson, Karina Kuchta, Hannah Murray
Product Development Program Specialist: Tim Douglas

Composition: NCCER
Printer/Binder: LSC Communications
Cover Printer: LSC Communications
Text Fonts: Palatino and Univers
Comtent Technologies: Gnostyx

Credits and acknowledgments for content borrowed from other sources and reproduced, with permission, in this textbook appear at the end of each module.

Copyright © 2020, 2017, 2014, 2011, 2008, 2005, 2002, 1999, 1996, 1992 by NCCER, Alachua, FL 32615, and published by by Pearson Education, Inc. or its affiliates, 221 River Street, Hoboken, NJ 07030. All rights reserved. Printed in the United States of America. This publication is protected by Copyright and permission should be obtained from NCCER prior to any prohibited reproduction, storage in a retrieval system, or transmission in any form or by any means, electronic, mechanical, photocopying, recording, or likewise. For information regarding permission(s), write to: NCCER Product Development, 13614 Progress Blvd., Alachua, FL 32615.

Pearson

Perfect Bound: 978-0-13-689782-8
Case Bound: 978-0-13-689774-3

Preface

To the Trainee

Electricity powers the applications that make our daily lives more productive and efficient. The demand for electricity has led to vast job opportunities in the electrical field. Electricians constitute one of the largest construction occupations in the United States, and they are among the highest-paid workers in the construction industry. According to the U.S. Bureau of Labor Statistics, the demand for trained electricians is projected to increase, creating more job opportunities for skilled craftspeople in the electrical industry.

Electricians install electrical systems in structures such as homes, office buildings, and factories. These systems include wiring and other electrical components, such as circuit breaker panels, switches, and lighting. Electricians follow blueprints, the *National Electrical Code®*, and state and local codes. They use specialized tools and testing equipment, such as ammeters, ohmmeters, and voltmeters. Electricians learn their trade through craft and apprenticeship programs. These programs provide classroom instruction and on-the-job learning with experienced electricians.

We wish you success as you embark on your first year of training in the electrical craft, and hope that you will continue your training beyond this textbook. There are more than 700,000 people employed in electrical work in the United States, and there are many opportunities awaiting those with the skills and desire to move forward in the construction industry.

New with *Electrical Level Two*

NCCER and Pearson are pleased to present *Electrical Level Two*, which has been updated to meet the 2020 *National Electrical Code®* and includes revisions to the Module Examinations.

In addition to the 2020 *NEC®* changes, this edition of *Electrical Level Two* features several updated modules. *Electric Lighting* (Module ID 26203-20) has been revised to reflect current lighting technology and terminology being used in the industry. The module has been refocused to reflect the growth of LED and other modern lighting. Both *Conductor Installations* (Module ID 26206-20) and *Grounding and Bonding* (Module ID 26209-20) have had extensive updates to the NEC requirements. Figures and tables throughout the modules have been updated to reflect the NEC and new technology.

Our website, **www.nccer.org**, has information on the latest product releases and training.

Your feedback is welcome. You may email your comments to **curriculum@nccer.org** or send general comments and inquiries to **info@nccer.org**.

NCCER Standardized Curricula

NCCER is a not-for-profit 501(c)(3) education foundation established in 1996 by the world's largest and most progressive construction companies and national construction associations. It was founded to address the severe workforce shortage facing the industry and to develop a standardized training process and curricula. Today, NCCER is supported by hundreds of leading construction and maintenance companies, manufacturers, and national associations. The NCCER Standardized Curricula was developed by NCCER in partnership with Pearson, the world's largest educational publisher.

Some features of the NCCER Standardized Curricula are as follows:

- An industry-proven record of success
- Curricula developed by the industry, for the industry
- National standardization providing portability of learned job skills and educational credits
- Compliance with the Office of Apprenticeship requirements for related classroom training. (*CFR 29:29*)
- Well-illustrated, up-to-date, and practical information

NCCER also maintains the NCCER Registry, which provides transcripts, certificates, and wallet cards to individuals who have successfully completed a level of training within a craft in NCCER's Curricula. *Training programs must be delivered by an NCCER Accredited Training Sponsor in order to receive these credentials.*

Special Features

In an effort to provide a comprehensive and user-friendly training resource, this curriculum showcases several informative features. Whether you are a visual or hands-on learner, these features are intended to enhance your knowledge of the construction industry as you progress in your training. Some of the features you may find in the curriculum are explained below.

Introduction

This introductory page, found at the beginning of each module, lists the module Objectives, Performance Tasks, and Trade Terms. The Objectives list the knowledge you will acquire after successfully completing the module. The Performance Tasks give you an opportunity to apply your knowledge to real-world tasks. The Trade Terms are industry-specific vocabulary that you will learn as you study this module.

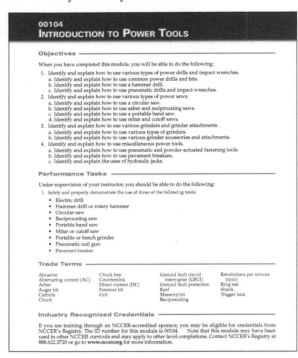

Figures and Tables

Photographs, drawings, diagrams, and tables are used throughout each module to illustrate important concepts and provide clarity for complex instructions. Text references to figures and tables are emphasized with *italic* type.

Notes, Cautions, and Warnings

Safety features are set off from the main text in highlighted boxes and categorized according to the potential danger involved. Notes simply provide additional information. Cautions flag a hazardous issue that could cause damage to materials or equipment. Warnings stress a potentially dangerous situation that could result in injury or death to workers.

Trade Features

Trade features present technical tips and professional practices based on real-life scenarios similar to those you might encounter on the job site.

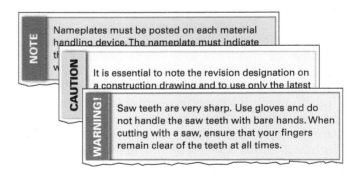

Bowline Trivia

Some people use this saying to help them remember how to tie a bowline: "The rabbit comes out of his hole, around a tree, and back into the hole."

Case History

Case History features emphasize the importance of safety by citing examples of the costly (and often devastating) consequences of ignoring best practices or OSHA regulations.

Going Green

Going Green features present steps being taken within the construction industry to protect the environment and save energy, emphasizing choices that can be made on the job to preserve the health of the planet.

Did You Know

Did You Know features introduce historical tidbits or interesting and sometimes surprising facts about the trade.

Step-by-Step Instructions

Step-by-step instructions are used throughout to guide you through technical procedures and tasks from start to finish. These steps show you how to perform a task safely and efficiently.

> Perform the following steps to erect this system area scaffold:
>
> *Step 1* Gather and inspect all scaffold equipment for the scaffold arrangement.
>
> *Step 2* Place appropriate mudsills in their approximate locations.
>
> *Step 3* Attach the screw jacks to the mudsills.

Trade Terms

Each module presents a list of Trade Terms that are discussed within the text and defined in the Glossary at the end of the module. These terms are presented in the text with **bold, blue** type upon their first occurrence. To make searches for key information easier, a comprehensive Glossary of Trade Terms from all modules is located at the back of this book.

Section Review

Each section of the module wraps up with a list of Additional Resources for further study and Section Review questions designed to test your knowledge of the Objectives for that section.

Review Questions

The end-of-module Review Questions can be used to measure and reinforce your knowledge of the module's content.

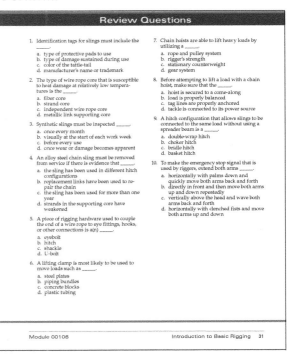

NCCER Standardized Curricula

NCCER's training programs comprise more than 80 construction, maintenance, pipeline, and utility areas and include skills assessments, safety training, and management education.

Boilermaking
Cabinetmaking
Carpentry
Concrete Finishing
Construction Craft Laborer
Construction Technology
Core Curriculum: Introductory Craft Skills
Drywall
Electrical
Electronic Systems Technician
Heating, Ventilating, and Air Conditioning
Heavy Equipment Operations
Heavy Highway Construction
Hydroblasting
Industrial Coating and Lining Application Specialist
Industrial Maintenance Electrical and Instrumentation Technician
Industrial Maintenance Mechanic
Instrumentation
Ironworking
Manufactured Construction Technology
Masonry
Mechanical Insulating
Millwright
Mobile Crane Operations
Painting
Painting, Industrial
Pipefitting
Pipelayer
Plumbing
Reinforcing Ironwork
Rigging
Scaffolding
Sheet Metal
Signal Person
Site Layout
Sprinkler Fitting
Tower Crane Operator
Welding

Maritime

Maritime Industry Fundamentals
Maritime Electrical
Maritime Pipefitting
Maritime Structural Fitter
Maritime Welding
Maritime Aluminum Welding

Green/Sustainable Construction

Building Auditor
Fundamentals of Weatherization
Introduction to Weatherization
Sustainable Construction Supervisor
Weatherization Crew Chief
Weatherization Technician
Your Role in the Green Environment

Energy

Alternative Energy
Introduction to the Power Industry
Introduction to Solar Photovoltaics
Power Generation Maintenance Electrician
Power Generation I&C Maintenance Technician
Power Generation Maintenance Mechanic
Power Line Worker
Power Line Worker: Distribution
Power Line Worker: Substation
Power Line Worker: Transmission
Solar Photovoltaic Systems Installer
Wind Energy
Wind Turbine Maintenance Technician

Pipeline

Abnormal Operating Conditions, Control Center
Abnormal Operating Conditions, Field and Gas
Corrosion Control
Electrical and Instrumentation
Field and Control Center Operations
Introduction to the Pipeline Industry
Maintenance
Mechanical

Safety

Field Safety
Safety Orientation
Safety Technology

Supplemental Titles

Applied Construction Math
Tools for Success

Management

Construction Workforce Development Professional
Fundamentals of Crew Leadership
Mentoring for Craft Professionals
Project Management
Project Supervision

Spanish Titles

Acabado de concreto: nivel uno (*Concrete Finishing Level One*)
Aislamiento: nivel uno (*Insulating Level One*)
Albañilería: nivel uno (*Masonry Level One*)
Andamios (*Scaffolding*)
Carpintería: Formas para carpintería, nivel tres (*Carpentry: Carpentry Forms, Level Three*)
Currículo básico: habilidades introductorias del oficio (*Core Curriculum: Introductory Craft Skills*)
Electricidad: nivel uno (*Electrical Level One*)
Herrería: nivel uno (*Ironworking Level One*)
Herrería de refuerzo: nivel uno (*Reinforcing Ironwork Level One*)
Instalación de rociadores: nivel uno (*Sprinkler Fitting Level One*)
Instalación de tuberías: nivel uno (*Pipefitting Level One*)
Instrumentación: nivel uno, nivel dos, nivel tres, nivel cuatro (*Instrumentation Levels One through Four*)
Orientación de seguridad (*Safety Orientation*)
Paneles de yeso: nivel uno (*Drywall Level One*)
Seguridad de campo (*Field Safety*)

Acknowledgments

This curriculum was revised as a result of the farsightedness and leadership of the following sponsors:

ABC of Iowa
ABC of Western Pennsylvania
Beacon Electrical Contractors
Cianbro Corporation
Elm Electrical, Inc.
Gaylor Electric, Inc.
Gould Construction Institute

Industrial Management and Training Institute
National Field Services
Madison Comprehensive High School
Putnam Career and Technical Center
Tri-City Electrical Contractors

This curriculum would not exist were it not for the dedication and unselfish energy of those volunteers who served on the Authoring Team. A sincere thanks is extended to the following:

Paul Asselin
David Coelho
Tim Dean
Tim Ely
Mark Kozloski
Dan Lamphear

David Lewis
John Lupacchino
Gerard McDonald
Scott Mitchell
John Mueller
Steve Newton

Ron Otts
Mike Powers
Josiha Schuh
Wayne Stratton
James Westfall

NCCER Partners

American Fire Sprinkler Association
Associated Builders and Contractors, Inc.
Associated General Contractors of America
Association for Career and Technical Education
Construction Industry Institute
Construction Users Roundtable
Gulf States Shipbuilders Consortium
ISN Software Corporation
Manufacturing Institute
Mason Contractors Association of America
Merit Contractors Association of Canada
NACE International
National Association of Women in Construction
National Insulation Association
National Technical Honor Society
NAWIC Education Foundation
North American Crane Bureau
North American Technician Excellence
Pearson
Prov

SkillsUSA®
Steel Erectors Association of America
University of Florida, M. E. Rinker Sr., School of Construction Management

Contents

Module One
Alternating Current
The foundation for safe and successful electrical installations is a sound understanding of DC and AC electrical principles. AC electricity also has a frequency component, so knowledge of AC waveforms and the effects of reactive and inductive components in a circuit is essential. This module describes AC circuits and explains how to apply Ohm's law to solve for unknown circuit values. (Module ID 26201-20; 17.5 Hours)

Module Two
Motors: Theory and Application
The electric motor is the workhorse of modern industry. Its functions are almost unlimited. To control the motors that drive machinery and equipment, we must have electrical supply circuits that perform certain functions. They must provide electrical current to cause the motor to operate in the manner needed to make it perform its intended function. This module describes AC and DC motors, including their components, circuits, and connections. (Module ID 26202-20; 20 Hours)

Module Three
Electric Lighting
Electric lighting is used extensively throughout residential structures, commercial businesses, industrial plants, and outdoor sites. It provides illumination for the performance of visual tasks with a maximum of comfort and a minimum of eyestrain and fatigue, allowing individuals to perform their daily living and work-related tasks more easily. This module introduces the principles of human vision and the characteristics of light. It also covers different types of light sources and describes the operating characteristics and installation requirements of various lighting fixtures (luminaires). (Module ID 26203-20; 15 Hours)

Module Four
Conduit Bending
The normal installation of intermediate metal conduit (IMC), rigid metal conduit (RMC), and electrical metallic tubing (EMT) requires many changes of direction in the conduit runs, ranging from simple offsets at the point of termination at outlet boxes and cabinets to complicated angular offsets at columns, beams, cornices, and so forth. This module describes how to make conduit bends using mechanical, hydraulic, and electric benders. (Module ID 26204-20; 15 Hours)

Module Five
Pull and Junction Boxes
Pull boxes and junction boxes are provided in an electrical installation to facilitate the installation of conductors, or to provide a junction point for the connection of conductors, or both. This module describes how to size and install pull and junction boxes. It also identifies various specialty enclosures, including conduit bodies, FS and FD boxes, and handholes. (Module ID 26205-20; 12.5 Hours)

Module Six
Conductor Installations
In most cases, the installation of conductors in raceway systems is merely routine. However, there are certain practices that can reduce labor and materials and help prevent damage to the conductors. This module describes how to prepare conduit for conductors. It also explains how to set up and complete a cable-pulling operation. (Module ID 26206-20; 10 Hours)

Module Seven
Cable Tray

Cable trays are the usual means of supporting cable systems in industrial applications. This module covers various types of cable tray, supports, and associated fittings. It also explains how to determine the loads on a cable tray and calculate fill per *NEC®* requirements. (Module ID 26207-20; 7.5 Hours)

Module Eight
Conductor Terminations and Splices

Anyone involved with electrical systems of any type must be familiar with wire connectors and splicing, as they are both necessary to make the numerous electrical joints required during the course of an electrical installation. This module explains how to prepare cable ends for terminations and splices. It also describes how to train cable at termination points and describes crimping techniques. (Module ID 26208-20; 7.5 Hours)

Module Nine
Grounding and Bonding

The grounding system is a major part of the electrical system. Its purpose is to protect life and equipment against the various electrical faults that can occur. This module explains the grounding and bonding requirements of *NEC Article 250*. It also explains how to size the main and system bonding jumpers as well as the grounding electrode conductor for various AC systems. (Module ID 26209-20; 15 Hours)

(continued)

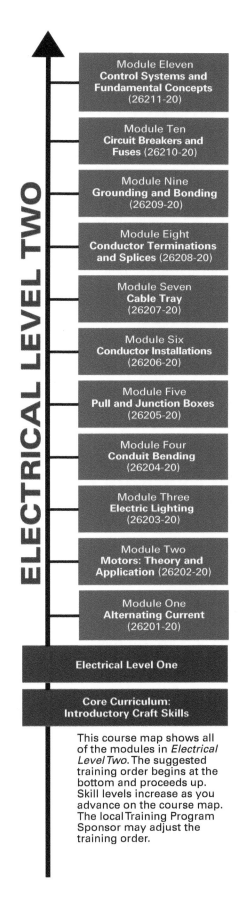

Module Ten
Circuit Breakers and Fuses

All electrical circuits and their related components are subject to destructive overcurrents, due to harsh environments, general deterioration, accidental damage, damage from natural causes, excessive expansion, and overloading of the electrical system. Reliable protective devices prevent or minimize costly damage to transformers, conductors, motors, equipment, and the many other components and loads that make up the complete electrical system. Fuses and circuit breakers are two types of automatic overload devices that are normally used in electrical circuits to prevent fires and the destruction of the circuit and its associated equipment. This module describes the operating principles of circuit breakers and fuses, and explains how to select and install overcurrent devices. (Module ID 26210-20; 12.5 Hours)

Module Eleven
Control Systems and Fundamental Concepts

Contactors are used to make or break the circuit to high-current, non-motor loads, or are used in motor circuits if overload protection is separately provided. Meanwhile, a relay is an electromagnetic device whose contacts are used in the control circuits of magnetic starters, contactors, solenoids, timers, and other relays. This module describes the operating principles of contactors and relays, including both mechanical and solid-state devices. It also explains how to select and install relays and troubleshoot control circuits. (Module ID 26211-20; 12.5 Hours)

Glossary

Alternating Current

Overview

The foundation for safe and successful electrical installations is a sound understanding of DC and AC electrical principles. AC electricity also has a frequency component, so knowledge of AC waveforms and the effects of reactive and inductive components in a circuit is essential. This module describes AC circuits and explains how to apply Ohm's law to solve for unknown circuit values.

Module 26201-20

Trainees with successful module completions may be eligible for credentialing through the NCCER Registry. To learn more, go to **www.nccer.org** or contact us at 1.888.622.3720. Our website, **www.nccer.org**, has information on the latest product releases and training.

Your feedback is welcome. You may email your comments to **curriculum@nccer.org**, send general comments and inquiries to **info@nccer.org**, or fill in the User Update form at the back of this module.

This information is general in nature and intended for training purposes only. Actual performance of activities described in this manual requires compliance with all applicable operating, service, maintenance, and safety procedures under the direction of qualified personnel. References in this manual to patented or proprietary devices do not constitute a recommendation of their use.

Copyright © 2020 by NCCER, Alachua, FL 32615, and published by Pearson, New York, NY 10013. All rights reserved. Printed in the United States of America. This publication is protected by Copyright, and permission should be obtained from NCCER prior to any prohibited reproduction, storage in a retrieval system, or transmission in any form or by any means, electronic, mechanical, photocopying, recording, or likewise. To obtain permission(s) to use material from this work, please submit a written request to NCCER Product Development, 13614 Progress Blvd., Alachua, FL 32615.

26201-20 V10.0

From *Electrical, Trainee Guide*. NCCER.
Copyright © 2020 by NCCER. Published by Pearson. All rights reserved.

26201-20
ALTERNATING CURRENT

Objectives

When you have completed this module, you will be able to do the following:

1. Identify AC waveforms.
   a. Define the terminology of sine waves.
   b. Define AC phase relationships.
   c. Identify nonsinusoidal waveforms.
2. Determine unknown values in AC circuits.
   a. Find unknown values in purely resistive AC circuits.
   b. Find unknown values in inductive AC circuits.
   c. Find unknown values in capacitive AC circuits.
   d. Find unknown values in combination circuits.
3. Make power calculations in AC circuits.
   a. Calculate true power.
   b. Calculate apparent power.
   c. Calculate reactive power.
   d. Calculate power factor.
   e. Use the power triangle to determine unknown values.
4. Identify transformers and explain how they operate.
   a. Identify the basic components in a transformer.
   b. Identify transformer operating characteristics.
   c. Calculate turns and voltage ratios.
   d. Identify various types of transformers and their applications.

Performance Tasks

This is a knowledge-based module. There are no Performance Tasks.

Trade Terms

Capacitance
Frequency
Hertz (Hz)
Impedance
Inductance
Micro (μ)

Peak voltage
Radian
Reactance
Root-mean-square (rms)
Self-inductance

Industry Recognized Credentials

If you are training through an NCCER-accredited sponsor, you may be eligible for credentials from NCCER's Registry. The ID number for this module is 26201-20. Note that this module may have been used in other NCCER curricula and may apply to other level completions. Contact NCCER's Registry at 888.622.3720 or go to **www.nccer.org** for more information.

Contents

- **1.0.0** AC Waveforms 1
  - **1.1.0** Sine Wave Terminology 3
    - 1.1.1 Frequency 4
    - 1.1.2 Period 4
    - 1.1.3 Wavelength 4
    - 1.1.4 Peak Value 4
    - 1.1.5 Average Value 4
    - 1.1.6 Root-Mean-Square or Effective Value 5
  - **1.2.0** AC Phase Relationships 6
    - 1.2.1 Phase Angle 6
    - 1.2.2 Phase Angle Diagrams 6
  - **1.3.0** Nonsinusoidal Waveforms 7
- 2.0.0 Unknown Values in AC Circuits 12
  - **2.1.0** Purely Resistive AC Circuits 12
  - **2.2.0** Inductive AC Circuits 13
    - 2.2.1 Factors Affecting Inductance 13
    - 2.2.2 Voltage and Current in an Inductive AC Circuit 14
    - 2.2.3 Inductive Reactance 16
  - **2.3.0** Capacitive AC Circuits 16
    - 2.3.1 Factors Affecting Capacitance 18
    - 2.3.2 Equivalent Capacitance 19
    - 2.3.3 Capacitor Specifications 19
    - 2.3.4 Voltage and Current in a Capacitive AC Circuit 20
    - 2.3.5 Capacitive Reactance 21
  - **2.4.0** Combination Circuits 22
    - 2.4.1 RL Circuits 24
    - 2.4.2 RC Circuits 27
    - 2.4.3 LC Circuits 28
    - 2.4.4 RLC Circuits 30
- 3.0.0 Power in AC Circuits 34
  - **3.1.0** True Power 34
  - **3.2.0** Apparent Power 34
  - **3.3.0** Reactive Power 34
  - **3.4.0** Power Factor 35
  - **3.5.0** Power Triangle 36
- 4.0.0 Transformers 38
  - **4.1.0** Components of a Transformer 38
    - 4.1.1 Core Characteristics 38
    - 4.1.2 Transformer Windings 38
  - **4.2.0** Operating Characteristics 39
    - 4.2.1 Energized with No Load 39
    - 4.2.2 Phase Relationship 40
  - **4.3.0** Turns and Voltage Ratios 40

4.4.0	Types of Transformers		42
4.4.1	Isolation Transformer		42
4.4.2	Autotransformer		42
4.4.3	Current Transformer		43
4.4.4	Potential Transformer		45

Figures and Tables

Figure 1	Conductor moving across a magnetic field	1
Figure 2	Angle versus rate of cutting lines of flux	2
Figure 3	One cycle of alternating voltage	3
Figure 4	Frequency measurement	4
Figure 5	Amplitude values for a sine wave	5
Figure 6	Voltage waveforms 90° out of phase	6
Figure 7	Waves in phase	7
Figure 8	AC waveforms	9
Figure 9	Resistive AC circuit	13
Figure 10	Voltage and current in a resistive AC circuit	13
Figure 11	Factors affecting the inductance of a coil	14
Figure 12	Inductor voltage and current relationship	15
Figure 13	Capacitors	17
Figure 14	Charging and discharging a capacitor	17
Figure 15	Capacitors in parallel	19
Figure 16	Capacitors in series	19
Figure 17	Voltage and current in a capacitive AC circuit	20
Figure 18	Summary of AC circuit phase relationships	23
Figure 19	Series RL circuit and vector diagram	24
Figure 20	Series RL circuit with waveforms and vector diagram	25
Figure 21	Parallel RL circuit with waveforms and vector diagram	26
Figure 22	Series RC circuit with vector diagrams	27
Figure 23	Parallel RC circuit with vector diagrams	28
Figure 24	Series LC circuit with vector diagram	29
Figure 25	Parallel LC circuit with vector diagram	29
Figure 26	Series RLC circuit and vector diagram	30
Figure 27	Parallel RLC circuit and vector diagram	31
Figure 28	Power calculations in an AC circuit	35
Figure 29	RLC circuit calculation	36
Figure 30	Power triangle	37
Figure 31	Basic components of a transformer	38
Figure 32	Transformer action	38
Figure 33	Steel laminated core	39
Figure 34	Cutaway view of a transformer core	39
Figure 35	Transformer winding polarity	41
Figure 36	Transformer turns ratio	42
Figure 37	Tapped transformers	43
Figure 38	Importance of an isolation transformer	44
Figure 39	Autotransformer schematic diagram	44
Figure 40	Current transformer schematic diagram	44
Figure 41	Potential transformer	45
Table 1	Series R and XL Combinations	25
Table 2	Parallel R and XL Combinations	27
Table 3	Series R and XC Combinations	27
Table 4	Parallel R and XC Combinations	29

Section One

1.0.0 AC Waveforms

Objective

Identify AC waveforms.
a. Define the terminology of sine waves.
b. Define AC phase relationships.
c. Identify nonsinusoidal waveforms.

Trade Terms

Frequency: The number of cycles an alternating electric current, sound wave, or vibrating object undergoes per second.

Hertz (Hz): A unit of frequency; one hertz equals one cycle per second.

Peak voltage: The peak value of a sinusoidally varying (cyclical) voltage or current is equal to the root-mean-square (rms) value multiplied by the square root of two (1.414). AC voltages are usually expressed as rms values; that is, 120 volts, 208 volts, 240 volts, 277 volts, 480 volts, etc., are all rms values. The peak voltage, however, differs. For example, the peak value of 120 volts (rms) is actually 120 × 1.414 = 170 volts.

Radian: An angle at the center of a circle, subtending (opposite to) an arc of the circle that is equal in length to the radius.

Root-mean-square (rms): The square root of the average of the square of the function taken throughout the period. The rms value of a sinusoidally varying voltage or current is the effective value of the voltage or current.

Alternating current (AC) and its associated voltage reverses between positive and negative polarities and varies in amplitude with time. One complete waveform or cycle includes a complete set of variations, with two alternations in polarity. Many sources of voltage change direction with time and produce a resultant waveform. The most common AC waveform is the sine wave.

To understand how the alternating current sine wave is generated, some of the basic principles learned in magnetism should be reviewed. Two principles form the basis of all electromagnetic phenomena:

- An electric current in a conductor creates a magnetic field that surrounds the conductor.
- Relative motion between a conductor and a magnetic field, when at least one component of that relative motion is in a direction that is perpendicular to the direction of the field, creates a voltage in the conductor.

Figure 1 shows how these principles are applied to generate an AC waveform in a simple one-loop rotary generator. The conductor loop rotates through the magnetic field to generate the induced AC voltage across its open terminals. The magnetic flux shown here is vertical.

There are three factors affecting the magnitude of voltage developed by a conductor through a magnetic field: the strength of the magnetic field; the length of the conductor; and the rate at which the conductor cuts directly across or perpendicular to the magnetic field.

Assuming that the strength of the magnetic field and the length of the conductor making the loop are both constant, the voltage produced will vary depending on the rate at which the loop cuts directly across the magnetic field.

The rate at which the conductor cuts the magnetic field depends on the speed of the generator in revolutions per minute (rpm) and the angle at which the conductor is traveling through the field. If the generator is operated at a constant rpm, the voltage produced at any moment will depend on the angle at which the conductor is cutting the field at that instant.

In *Figure 2*, the magnetic field is shown as parallel lines called lines of flux. These lines always go from the north to south poles in a generator. The motion of the conductor is shown by the large arrow.

Assuming the speed of the conductor is constant, as the angle between the flux and the conductor motion increases, the number of flux lines cut in a given time (the rate) increases. When the conductor is moving parallel to the lines of flux (angle of 0°), it is not cutting any of them, and the voltage will be zero.

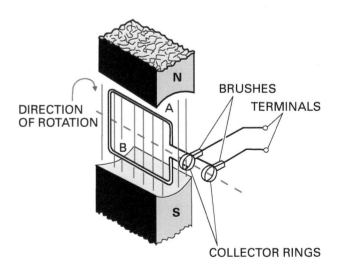

Figure 1 Conductor moving across a magnetic field.

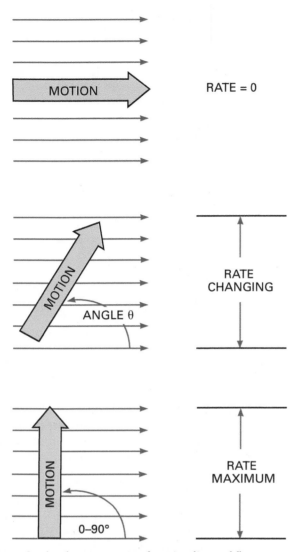

Figure 2 Angle versus rate of cutting lines of flux.

The angle between the lines of flux and the motion of the conductor is called θ (theta). The magnitude of the voltage produced will be proportional to the sine of the angle. Sine is a trigonometric function. Each angle has a sine value that never changes.

The sine of 0° is 0. It increases to a maximum of 1 at 90°. From 90° to 180°, the sine decreases back to 0. From 180° to 270°, the sine decreases to –1. Then from 270° to 360° (back to 0°), the sine increases to its original 0.

Because voltage is proportional to the sine of the angle, as the loop goes 360° around the circle the voltage will increase from 0 to its maximum at 90°, back to 0 at 180°, down to its maximum negative value at 270°, and back up to 0 at 360°, as shown in *Figure 3*.

Notice that at 180° the polarity reverses. This is because the conductor has turned completely around and is now cutting the lines of flux in the opposite direction. This can be shown using the left-hand rule for generators. The curve shown in *Figure 3* is called a sine wave because its shape is generated by the trigonometric function sine. The value of voltage at any point along the sine wave can be calculated if the angle and the maximum obtainable voltage (E_{max}) are known. The formula used is:

$$E = E_{max} \times \sin(\theta)$$

Where:

E = voltage induced
E_{max} = maximum induced voltage
θ = angle at which the voltage is induced

Why Do Power Companies Generate and Distribute AC Power Instead of DC Power?

The transformer is the key. Power plants generate and distribute AC power because it permits the use of transformers, which makes power delivery more economical. Transformers used at generation plants step the AC voltage up, which decreases the current. Decreased current allows smaller-sized wires to be used for the power transmission lines. Smaller wire is less expensive and easier to support over the long distances that the power must travel from the generation plant to remotely located substations. At the substations, transformers are again used to step AC voltages back down to a level suitable for distribution to homes and businesses.

There is no such thing as a DC transformer. This means DC power would have to be transmitted at low voltages and high currents over very large-sized wires, making the process very uneconomical. When DC is required for special applications, the AC voltage may be converted to DC voltage by using rectifiers, which make the change electrically, or by using AC motor–DC generator sets, which make the change mechanically.

Using this formula, the values of voltage anywhere along the sine wave in *Figure 3* can be calculated. Sine values can be found using either a scientific calculator or trigonometric tables. As an example, the following values can be calculated along a sine wave with an E_{max} of 10 volts (V):

$\theta = 0°$, $\sin(0°) = 0$
$E = E_{max} \times \sin(\theta)$
$E = (10V)(0)$
$E = 0V$

$\theta = 45°$, $\sin(45°) = 0.707$
$E = E_{max} \times \sin(\theta)$
$E = (10V)(0.707)$
$E = 7.07V$

$\theta = 90°$, $\sin(90°) = 1.0$
$E = E_{max} \times \sin(\theta)$
$E = (10V)(1.0)$
$E = 10V$

$\theta = 135°$, $\sin(135°) = 0.707$
$E = E_{max} \times \sin(\theta)$
$E = (10V)(0.707)$
$E = 7.07V$

$\theta = 180°$, $\sin(180°) = 0$
$E = E_{max} \times \sin(\theta)$
$E = (10V)(0)$
$E = 0V$

$\theta = 225°$, $\sin(225°) = -0.707$
$E = E_{max} \times \sin(\theta)$
$E = (10V)(-0.707)$
$E = -7.07V$

$\theta = 270°$, $\sin(270°) = -1.0$
$E = E_{max} \times \sin(\theta)$
$E = (10V)(-1.0)$
$E = -10V$

$\theta = 315°$, $\sin(315°) = -0.707$
$E = E_{max} \times \sin(\theta)$
$E = (10V)(-0.707)$
$E = -7.07V$

1.1.0 Sine Wave Terminology

There are a number of AC voltage terms are specific to sine waves. It is important to understand this terminology when working with AC. The following sections discuss some of these terms, including frequency, period, wavelength, peak value, average value, and effective value.

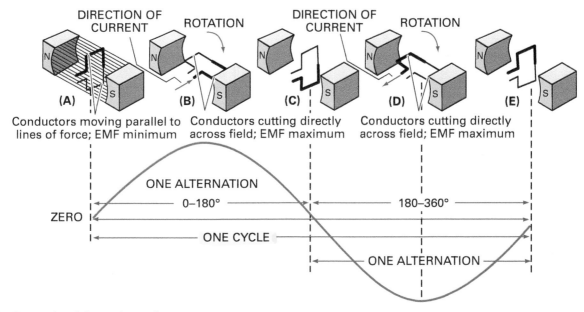

Figure 3 One cycle of alternating voltage.

1.1.1 Frequency

The frequency of a waveform is the number of times per second an identical pattern repeats itself. Each time the waveform changes from zero to a peak value and back to zero is called an alternation. Two alternations form one cycle. The number of cycles per second is the frequency. The unit of frequency is **hertz (Hz)**. One hertz equals one cycle per second (cps).

For example, let us determine the frequency of the waveform shown in *Figure 4*.

In one-half second, the basic sine wave is repeated five times. Therefore, the frequency (f) is:

$$f = \frac{5 \text{ cycles}}{0.5 \text{ seconds}} = 10 \text{ cycles per second (Hz)}$$

1.1.2 Period

The period of a waveform is the time (t) required to complete one cycle. The period is the inverse of frequency:

$$t = \frac{1}{f}$$

Where:

t = period (seconds)
f = frequency (Hz or cps)

For example, let us determine the period of the waveform in *Figure 4*. If there are five cycles in one-half second, then the frequency for one cycle is 10 cps (5 ÷ 0.5 = 10). Therefore, the period is:

$$t = \frac{1}{\text{cps}} = \frac{1}{10} = 0.1 \text{ seconds}$$

1.1.3 Wavelength

The wavelength or λ (lambda) is the distance traveled by a waveform during one period. Because electromagnetic radiation travels at the speed of light (186,000 miles/second or 300,000 kilometers/second), the wavelength of electrical waveforms equals the product of the period and the speed of light (c):

$$\lambda = tc \quad or \quad \lambda = \frac{c}{f}$$

Where:

λ = wavelength (meters)
t = period (seconds)
c = speed of light (meters/second)
f = frequency (Hz or cps)

1.1.4 Peak Value

A peak value is a maximum value of voltage (V_M) or current (I_M). For example, specifying that a sine wave has a **peak voltage** of 170V applies to either the positive or the negative peak. To include both peak amplitudes, the peak-to-peak (p–p) value may be specified. In the previous example, the peak-to-peak value is 340V, double the peak value of 170V, because the positive and negative peaks are symmetrical. However, the two opposite peak values cannot occur at the same time. Furthermore, in some waveforms the two peaks are not equal. The positive peak value and peak-to-peak value of a sine wave are shown in *Figure 5*.

1.1.5 Average Value

The average value is calculated from all the values in a sine wave for one alternation or half cycle. The half cycle is used for the average because over a full cycle the average value is zero, which is useless for comparison purposes. If the sine values for all angles up to 180° in one alternation are added and then divided by the number of values, the average in this example equals 0.637.

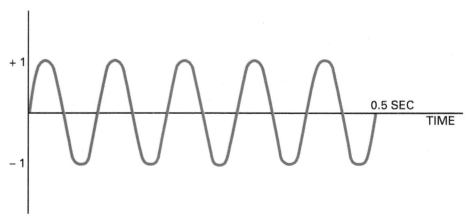

Figure 4 Frequency measurement.

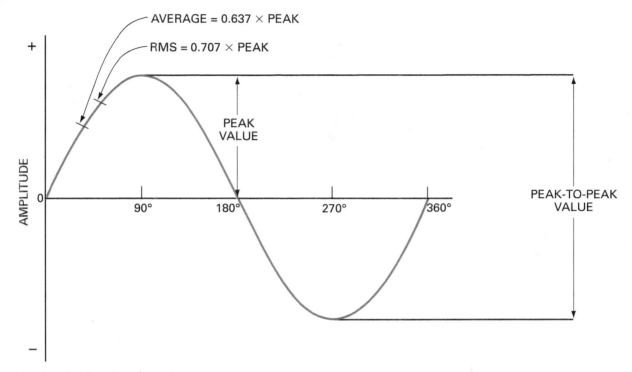

Figure 5 Amplitude values for a sine wave.

Since the peak value of the sine is 1 and the average equals 0.637, the average value can be calculated as follows:

Average value = 0.637 × peak value

For example, with a peak of 170V, the average value is 0.637 × 170V, which equals approximately 108V. *Figure 5* shows where the average value would fall on a sine wave.

1.1.6 Root-Mean-Square or Effective Value

Meters used in AC circuits indicate a value called the effective value. The effective value is the value of the AC current or voltage wave that indicates the same energy transfer as an equivalent direct current (DC) or voltage.

Think About It

Frequency

The frequency of the utility power generated in the United States is normally 60Hz. In some European countries and elsewhere, utility power is often generated at a frequency of 50Hz. Which of these frequencies (60Hz or 50Hz) has the shortest period?

The direct comparison between DC and AC is in the heating effect of the two currents. Heat produced by current is a function of current amplitude only and is independent of current direction. Thus, heat is produced by both alternations of the AC wave, although the current changes direction during each alternation.

In a DC circuit, the current maintains a steady amplitude. Therefore, the heat produced is steady and is equal to I^2R. In an AC circuit, the current is continuously changing; periodically high, periodically low, and periodically zero. To produce the same amount of heat from AC as from an equivalent amount of DC, the instantaneous value of the AC must at times exceed the DC value.

By averaging the heating effects of all the instantaneous values during one cycle of alternating current, it is possible to find the average heat produced by the AC current during the cycle. The amount of DC required to produce that heat will be equal to the effective value of the AC.

The most common method of specifying the amount of a sine wave of voltage or current is by stating its value at 45°, which is 70.7% of the peak. This is its root-mean-square (rms) value. Therefore:

Value of rms = 0.707 × peak value

For example, with a peak of 170V, the rms value is 0.707 × 170, or approximately 120V. This is the voltage of the commercial AC power line, which is always given in rms value.

1.2.0 AC Phase Relationships

In AC systems, phase is involved in two ways: the location of a point on a voltage or current wave with respect to the starting point of the wave, or with respect to some corresponding point on the same wave. In the case of two waves of the same frequency, it is the time at which an event of one takes place with respect to a similar event of the other.

Often, the event is the starting of the waves at zero or the points at which the waves reach their maximum values. When two waves are compared in this manner, there is a phase lead or lag of one with respect to the other unless they are alternating in unison, in which case they are said to be in phase.

1.2.1 Phase Angle

Suppose that a generator started its cycle at 90° where maximum voltage output is produced instead of starting at the point of zero output. The two output voltage waves are shown in *Figure 6*. Each is the same waveform of alternating voltage, but wave B starts at the maximum value while wave A starts at zero. The complete cycle of wave B through 360° takes it back to the maximum value from which it started.

Wave A starts and finishes its cycle at zero. With respect to time, wave B is ahead of wave A in its values of generated voltage. The amount it leads in time equals one quarter revolution, which is 90°. This angular difference is the phase angle between waves B and A. Wave B leads wave A by the phase angle of 90°.

The 90° phase angle between waves B and A is maintained throughout the complete cycle and in all successive cycles as long as they both have the same frequency. At any instant in time, wave B has the value that A will have 90° later. For instance, at 180°, wave A is at zero, but B is already at its negative maximum value, the point where wave A will be later at 270°.

To compare the phase angle between two waves, both waves must have the same frequency. Otherwise, the relative phase keeps changing. Both waves must also have sine wave variations, because this is the only kind of waveform that is measured in angular units of time. The amplitudes can be different for the two waves. The phases of two voltages, two currents, or a current with a voltage can be compared.

1.2.2 Phase Angle Diagrams

To compare AC phases, it is much more convenient to use vector diagrams corresponding to the voltage and current waveforms, as shown in *Figure 6*. V_A and V_B represent the vector quantities corresponding to the generator voltage.

A vector is a quantity that has magnitude and direction. The length of the arrow indicates the magnitude of the alternating voltage in rms, peak, or any AC value as long as the same measure is used for all the vectors. The angle of the arrow with respect to the horizontal axis indicates the phase angle.

In *Figure 6*, the vector V_A represents the voltage wave A, with a phase angle of 0°. This angle can be considered as the plane of the loop in the rotary generator where it starts with zero

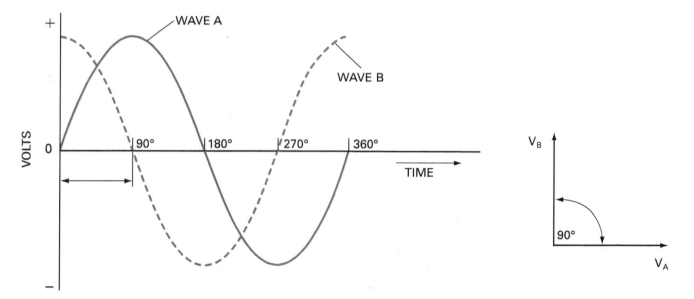

Figure 6 Voltage waveforms 90° out of phase.

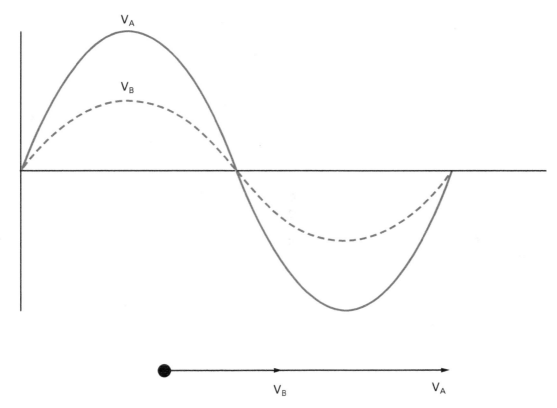

Figure 7 Waves in phase.

output voltage. The vector V_B is vertical to show the phase angle of 90° for this voltage wave, corresponding to the vertical generator loop at the start of its cycle. The angle between the two vectors is the phase angle.

The symbol for a phase angle is θ (theta). In *Figure 7*, θ = 0°. *Figure 7* shows the waveforms and vector diagram of two waves that are in phase but have different amplitudes.

1.3.0 Nonsinusoidal Waveforms

The sine wave is the basic waveform for AC variations for several reasons. This waveform is produced by a rotary generator, as the output is proportional to the angle of rotation. Because of its derivation from circular motion, any sine wave can be analyzed in angular measure, expressed either as a degree from 0° to 360° or as a radian from 0 to 2π radians.

In many electronic applications, however, other waveshapes are important. Any waveform that is not a sine (or cosine) wave is a nonsinusoidal waveform. Common examples are the square wave and sawtooth wave in *Figure 8*. One device that generates a sawtooth waveform is an oscilloscope.

With nonsinusoidal waveforms for either voltage or current, there are important differences and similarities to consider. Note the following comparisons with sine waves:

- In all cases, the cycle is measured between two points having the same amplitude and varying in the same direction. The period is the time for one cycle.
- Peak amplitude is measured from the zero axis to the maximum positive or negative value. However, peak-to-peak amplitude is better for measuring nonsinusoidal waveshapes because they can have asymmetrical peaks, as with the rectangular wave in *Figure 8*.
- The rms value 0.707 of peak applies only to sine waves, as this factor is derived from the sine values in the angular measure used only for the sine waveform.
- Phase angles apply only to sine waves, as angular measure is used only for sine waves. Note that the phase angle is indicated only on the sine wave of *Figure 8*.

Think About It
Phase Angles

Why is the phase angle 90° in *Figure 6* and 0° in *Figure 7*? Why is the vector diagram in *Figure 7* shown as a straight line?

Left-Hand Rule for Generators

Hand rules for generators and motors give direction to the basic principles of induction. For a generator, if you move a conductor through a magnetic field made up of flux lines, you will induce an EMF, which drives current through a conductor. The left-hand rule for generators will help you determine which direction the current will flow in the conductor. It states that if you hold the thumb, first, and middle fingers of the left hand at right angles to one another with the first finger pointing in the flux direction (from the north pole to the south pole), and the thumb pointing in the direction of motion of the conductor, the middle finger will point in the direction of the induced voltage (EMF). The polarity of the EMF determines the direction in which current will flow as a result of this induced EMF. The left-hand rule for generators is also called Fleming's first rule.

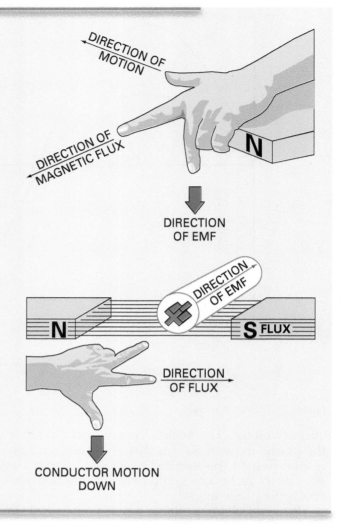

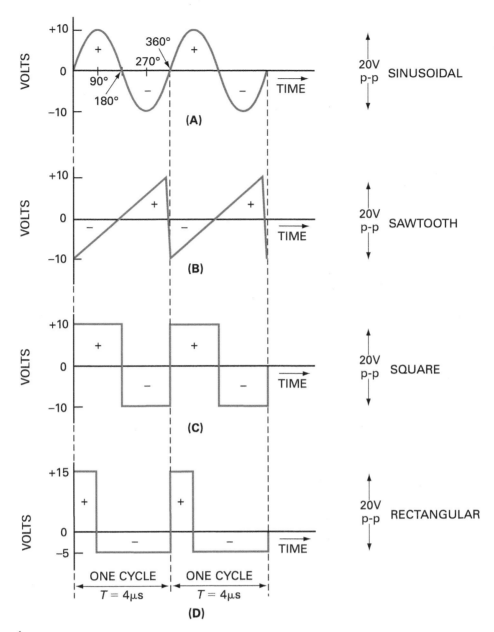

Figure 8 AC waveforms.

RMS Amplitude

The root-mean-square (rms) value, also called the effective value, is the value assigned to an alternating voltage or current that results in the same power dissipation in a given resistance as DC voltage or current of the same numerical value. This is illustrated below. As shown, 120 (peak) VAC will not produce the same light (350 lumens versus 500 lumens) as 120VDC from a 60W lamp. In order to produce the same light (500 lumens), 120V rms must be applied to the lamp. This requires that the applied sinusoidal AC waveform have a peak voltage of about 170V (170V × 0.707V = 120V).

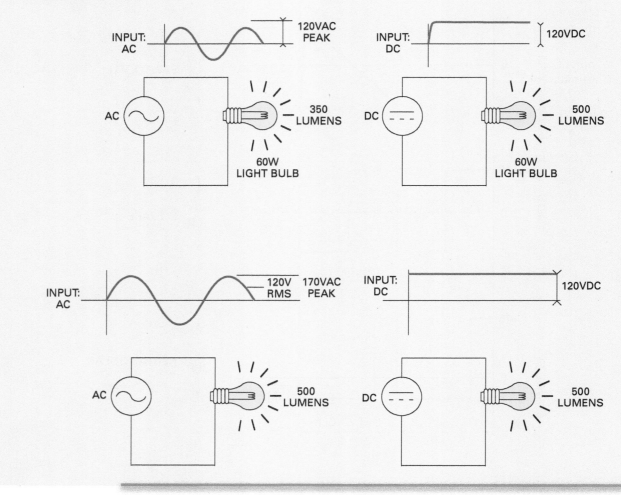

1.0.0 Section Review

1. If a sine wave is repeated ten times per half-second, its frequency is _____.
   a. 10Hz
   b. 20Hz
   c. 40Hz
   d. 60Hz

   10/0.5

2. To compare the phase angle between two waves, both waves must have the same _____.
   a. voltage
   b. current
   c. potential
   d. frequency

3. One device that requires a sawtooth waveform is a(n) _____.
   a. refrigerator
   b. compressor
   c. electric drill
   d. oscilloscope

Section Two

2.0.0 Unknown Values in AC Circuits

Objective

Determine unknown values in AC circuits.
a. Find unknown values in purely resistive AC circuits.
b. Find unknown values in inductive AC circuits.
c. Find unknown values in capacitive AC circuits.
d. Find unknown values in combination circuits.

Trade Terms

Capacitance: The storage of electricity in a capacitor; capacitance produces an opposition to voltage change. The unit of measurement for capacitance is the farad (F) or microfarad (µF).

Impedance: The opposition to current flow in an AC circuit; impedance includes resistance (R), capacitive reactance (X_C), and inductive reactance (X_L). Impedance is measured in ohms (Ω).

Inductance: The creation of a voltage due to a time-varying current; also, the opposition to current change, causing current changes to lag behind voltage changes. The unit of measure for inductance is the henry (H).

Micro (µ): Prefix designating one-millionth of a unit. For example, one microfarad is one-millionth of a farad.

Reactance: The opposition to alternating current (AC) due to capacitance (X_C) and/or inductance (X_L).

Self-inductance: A magnetic field induced in the conductor carrying the current.

An AC circuit has an AC voltage source. Note the circular symbol with the sine wave inside it shown in *Figure 9*. It is used for any source of sine wave alternating voltage. This voltage connected across an external load resistance produces alternating current of the same waveform, frequency, and phase as the applied voltage.

According to Ohm's law, current (I) equals voltage (E) divided by resistance (R). When E is an rms value, I is also an rms value. For any instantaneous value of E during the cycle, the value of I is for the corresponding instant of time.

2.1.0 Purely Resistive AC Circuits

In an AC circuit with only resistance, the current variations are in phase with the applied voltage, as shown in *Figure 9*. This in-phase relationship between E and I means that such an AC circuit can be analyzed by the same methods used for DC circuits since there is no phase angle to consider. Components that have only resistance include resistors, the filaments for incandescent light bulbs, and vacuum tube heaters.

In purely resistive AC circuits, the voltage, current, and resistance are related by Ohm's law because the voltage and current are in phase.

$$I = \frac{E}{R}$$

This equation can be rearranged to solve for the desired value (E = IR and R = E/I). Unless otherwise noted, the calculations in AC circuits are generally in rms values. For example, in *Figure 9*, the 120V applied across the 10Ω resistance R_L produces an rms current of 12A. This is determined as follows:

$$I = \frac{E}{R_L} = \frac{120V}{10\Omega} = 12A$$

Furthermore, the rms power (true power) dissipation is I^2R or:

$$P = I^2R$$
$$P = 12A^2 \times 10\Omega$$
$$P = 144A \times 10\Omega$$
$$P = 1,440W$$

Figure 10 shows the relationship between voltage and current in purely resistive AC circuits. The voltage and current are in phase, their cycles begin and end at the same time, and their peaks occur at the same time.

What's wrong with this picture?

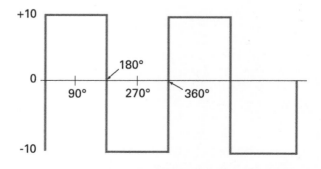

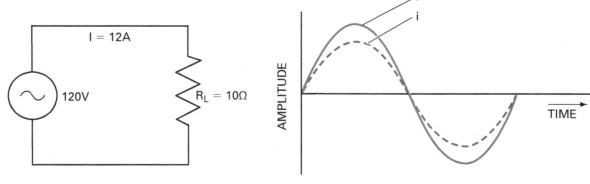

Figure 9 Resistive AC circuit.

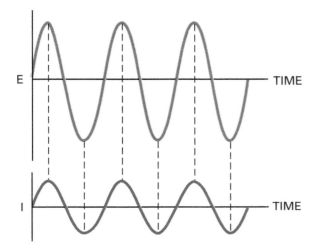

Figure 10 Voltage and current in a resistive AC circuit.

The value of the voltage shown in *Figure 10* depends on the applied voltage to the circuit. The value of the current depends on the applied voltage and the amount of resistance. If resistance is changed, it will affect only the magnitude of the current.

The total resistance in any AC circuit, whether it is a series, parallel, or series-parallel circuit, is calculated using the same rules that were learned and applied to DC circuits with resistance. Power computations are discussed later in this module.

2.2.0 Inductive AC Circuits

The characteristic of an electrical circuit that opposes the change of current flow is called inductance. It is the result of the expanding and collapsing field caused by the changing current. This moving flux cuts across the conductor that is providing the current, producing induced voltage in the wire itself. Furthermore, any other conductor in the field, whether carrying current or not, is also cut by the varying flux and has induced voltage. This induced current opposes the current flow that generated it.

In DC circuits, a change must be initiated in the circuit to cause inductance. The current must change to provide motion of the flux. A steady DC of 10A cannot produce any induced voltage as long as the current value is constant. A current of 1A changing to 2A does induce voltage. Also, the faster the current changes, the higher the induced voltage becomes, because when the flux moves at a higher speed it can induce more voltage.

However, in an AC circuit the current is continuously changing and producing induced voltage. Lower frequencies of AC require more inductance to produce the same amount of induced voltage as a higher frequency current. The current can have any waveform as long as the amplitude is changing.

The ability of a conductor to induce voltage in itself when the current changes is its self-inductance, or simply *inductance*. The symbol for inductance is L and its unit is the henry (H). One henry is the amount of inductance that allows one volt to be induced when the current changes at the rate of one ampere per second.

2.2.1 Factors Affecting Inductance

An inductor is a coil of wire that may be wound on a core of metal or paper, or it may be self-supporting. It may consist of turns of wire placed side by side to form a layer of wire over the core or coil form. The inductance of a coil or inductor depends on its physical construction. Some of the factors affecting inductance are:

- *Number of turns* – The greater the number of turns, the greater the inductance. In addition, the spacing of the turns on a coil also affects inductance. A coil that has widely-spaced turns has a lower inductance than one that has the same number of more closely-spaced turns. The reason for this higher inductance is that the closely-wound turns produce a more concentrated magnetic field, causing the coil to exhibit a greater inductance.

- *Coil diameter* – The inductance increases directly as the cross-sectional area of the coil increases.
- *Length of the core* – When the length of the core is decreased, the turn spacing is decreased, increasing the inductance of the coil.
- *Core material* – The core of the coil can be either a magnetic material (such as iron) or a non-magnetic material (such as paper or air). Coils wound on a magnetic core produce a stronger magnetic field than those with non-magnetic cores, giving them higher values of inductance. Air-core coils are used where small values of inductance are required.
- *Winding the coil in layers* – The more layers used to form a coil, the greater the effect the magnetic field has on the conductor. Layering a coil can increase the inductance.

Factors affecting the inductance of a coil can be seen in *Figure 11*.

2.2.2 Voltage and Current in an Inductive AC Circuit

The self-induced voltage across an inductance L is produced by a change in current with respect to time ($\Delta i/\Delta t$) and can be stated as:

$$V_L = L\frac{\Delta i}{\Delta t}$$

Where:

V_L = self-induced voltage in volts
L = inductance in henrys
Δ = change (Δi is change in current, and Δt is change in time)
$\Delta i/\Delta t$ = amperes per second

This gives the voltage in terms of how much magnetic flux is cut per second. When the magnetic flux associated with the current varies the same as I, this formula gives the same results for calculating induced voltage. Remember that the induced voltage across the coil is actually the result of inducing electrons to move in the conductor, so there is also an induced current.

For example, what is the self-induced voltage V_L across a 4H inductance produced by a current change of 12A per second?

$$V_L = L\frac{\Delta i}{\Delta t}$$

$$V_L = 4H \times \frac{12A}{1\text{sec}}$$

$$V_L = 4 \times 12 = 48V$$

The current through a 200 microhenry (µH) inductor changes from 0 to 200 milliamps (mA) in 2 microseconds (µsec). Note that the prefix micro (µ) means one-millionth, so a microhenry is equal to one millionth of a henry (1µH = 1 × 10^{-6} henrys). What is the V_L?

$$V_L = L\frac{\Delta i}{\Delta t}$$

$$V_L = (200 \times 10^{-6})\frac{200 \times 10^{-3}}{2 \times 10^{-6}}$$

$$V_L = 20V$$

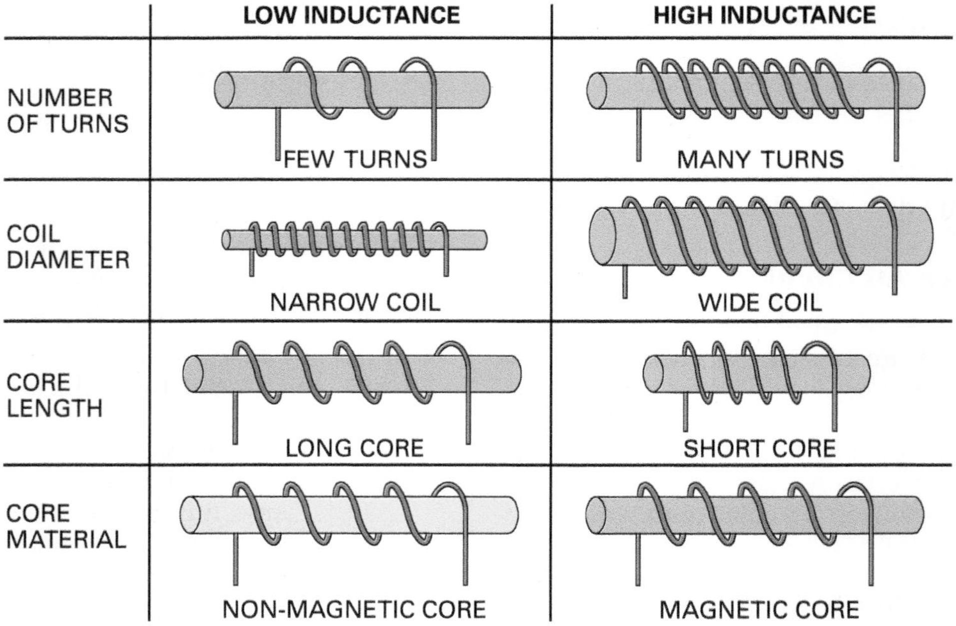

	LOW INDUCTANCE	HIGH INDUCTANCE
NUMBER OF TURNS	FEW TURNS	MANY TURNS
COIL DIAMETER	NARROW COIL	WIDE COIL
CORE LENGTH	LONG CORE	SHORT CORE
CORE MATERIAL	NON-MAGNETIC CORE	MAGNETIC CORE

Figure 11 Factors affecting the inductance of a coil.

Think About It

Inductance in an AC Circuit

Can you name three commonly used electrical devices that insert inductance into an AC circuit?

The induced voltage is an actual voltage that can be measured, although V_L is produced only while the current is changing. When $\Delta i/\Delta t$ is present for only a short time, V_L is in the form of a voltage pulse. With a sine wave current that is always changing, V_L is a sinusoidal voltage that is 90° out of phase with I_L.

The current that flows in an inductor is induced by the changing magnetic field that surrounds the inductor. This changing magnetic field is produced by an AC voltage source that is applied to the inductor. The magnitude and polarity of the induced current depend on the field strength, direction, and rate at which the field cuts the inductor windings. The overall effect is that the current is out of phase and lags the applied voltage by 90°.

At 270° in *Figure 12*, the applied electromotive force (EMF) is zero, but it is increasing in the positive direction at its greatest rate of change. Likewise, electron flow due to the applied EMF is also increasing at its greatest rate. As the electron flow increases, it produces a magnetic field that is building with it. The lines of flux cut the conductor as they move outward from it with the expanding field.

As the lines of flux cut the conductor, they induce a current into it. The induced current is at its maximum value because the lines of flux are expanding outward through the conductor at their greatest rate. The direction of the induced current is in opposition to the force that generated it. Therefore, at 270° the applied voltage is zero and is increasing to a positive value, while the current is at its maximum negative value.

At 0° in *Figure 12*, the applied voltage is at its maximum positive value, but its rate of change is zero. Therefore, the field it produces is no longer expanding and is not cutting the conductor. Because there is no relative motion between the field and conductor, no current is induced. Therefore, at 0° voltage is at its maximum positive value, while current is zero.

At 90° in *Figure 12*, voltage is once again zero, but this time it is decreasing toward negative at its greatest rate of change. Because the applied

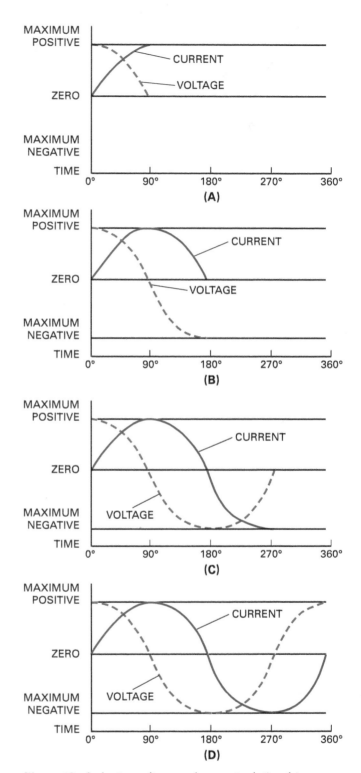

Figure 12 Inductor voltage and current relationship.

voltage is decreasing, the magnetic field is collapsing inward on the conductor. This has the effect of reversing the direction of motion between the field and conductor that existed at 0°.

Therefore, the current will flow in a direction opposite of what it was at 0°. Also, because the applied voltage is decreasing at its greatest rate, the field is collapsing at its greatest rate. This causes the flux to cut the conductor at the greatest rate, causing the induced current magnitude to be maximum. At 90°, the applied voltage is zero decreasing toward negative, while the current is maximum positive.

At 180° in *Figure 12*, the applied voltage is at its maximum negative value, but just as at 0°, its rate of change is zero. At 180°, therefore, current will be zero. This explanation shows that the voltage peaks positive first, then 90° later the current peaks positive. Current thus lags the applied voltage in an inductor by 90°. This can easily be remembered using the phrase "ELI the ICE man." ELI represents voltage (E), inductance (L), and current (I). In an inductor, the voltage leads the current just as the letter E leads or comes before the letter I. The word ICE will be explained in the section on capacitance.

2.2.3 Inductive Reactance

The opposing force that an inductor presents to the flow of alternating current cannot be called resistance because it is not the result of friction within a conductor. The name given to this force is inductive reactance because it is the reaction of the inductor to alternating current. Inductive reactance is measured in ohms and its symbol is X_L.

Remember that the induced voltage in a conductor is proportional to the rate at which magnetic lines of force cut the conductor. The greater the rate or higher the frequency, the greater the counter-electromotive force (CEMF). Also, the induced voltage increases with an increase in inductance; the more turns, the greater the CEMF. Reactance then increases with an increase of frequency and with an increase in inductance. The formula for inductive reactance is as follows:

$$X_L = 2\pi fL$$

Where:

X_L = inductive reactance in ohms
2π = 6.28 (This is a constant in which the Greek letter pi [π] represents 3.14, so 2 × pi = 6.28.)
f = frequency of the alternating current in hertz
L = inductance in henrys

For example, if f is equal to 60Hz and L is equal to 20H, find X_L:

$X_L = 2\pi fL$
$X_L = 6.28 \times 60Hz \times 20H$
$X_L = 7,536\Omega$

Once calculated, the value of X_L is used like resistance in a form of Ohm's law:

$$I = \frac{E}{X_L}$$

Where:

I = effective current (amps)
E = effective voltage (volts)
X_L = inductive reactance (ohms)

Unlike a resistor, there is no power dissipation in an ideal inductor. An inductor limits current, but it uses no net energy because the energy required to build up the field in the inductor is given back to the circuit when the field collapses.

2.3.0 Capacitive AC Circuits

A capacitor is a device that stores an electric charge in a dielectric material. **Capacitance** is the ability to store a charge. In storing a charge, a capacitor opposes a change in voltage. *Figure 13* shows a simple capacitor in a circuit, schematic representations of two types of capacitors, and a photo of common capacitors.

Figure 14 (A) shows a capacitor in a DC circuit. When voltage is applied, the capacitor begins to charge, as shown in *Figure 14 (B)*. The charging continues until the potential difference across the capacitor is equal to the applied voltage. This charging current is transient or temporary since it flows only until the capacitor is charged to the applied voltage. Then there is no current in the circuit. *Figure 14 (C)* shows this with the voltage across the capacitor equal to the battery voltage or 10V.

The capacitor can be discharged by connecting a conducting path across the dielectric. The stored charge across the dielectric provides the potential difference to produce a discharge current, as shown in *Figure 14 (D)*. Once the capacitor is completely discharged, the voltage across it equals zero, and there is no discharge current.

ELI in ELI the ICE Man

Remembering the phrase "ELI" as in "ELI the ICE man" is an easy way to remember the phase relationships that always exist between voltage and current in an inductive circuit. An inductive circuit is a circuit that has more inductive reactance than capacitive reactance. The L in ELI indicates inductance. The E (voltage) is stated before the I (current) in ELI, meaning that the voltage leads the current in an inductive circuit.

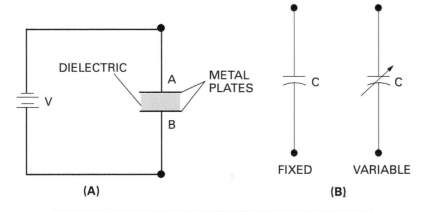

Figure 13 Capacitors.

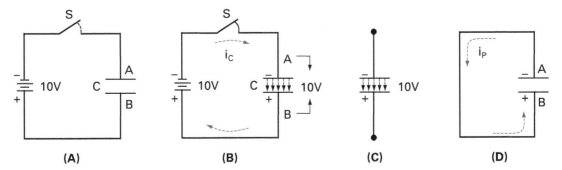

Figure 14 Charging and discharging a capacitor.

Module 26201-20 Alternating Current 17

In a capacitive circuit, the charge and discharge current must always be in opposite directions. Current flows in one direction to charge the capacitor and in the opposite direction when the capacitor is allowed to discharge.

Current will flow in a capacitive circuit with AC voltage applied because of the capacitor charge and discharge current. There is no current through the dielectric, which is an insulator. While the capacitor is being charged by increasing applied voltage, the charging current flows in one direction to the plates. While the capacitor is discharging as the applied voltage decreases, the discharge current flows in the reverse direction. With alternating voltage applied, the capacitor alternately charges and discharges.

First, the capacitor is charged in one polarity, and then it discharges; next, the capacitor is charged in the opposite polarity, and then it discharges again. The cycles of charge and discharge current provide alternating current in the circuit at the same frequency as the applied voltage. The amount of capacitance in the circuit will determine how much current is allowed to flow.

Capacitance is measured in farads (F), where one farad is the capacitance when one coulomb is stored in the dielectric with a potential difference of one volt. Smaller values are measured in microfarads (µF). A small capacitance will allow less charge and discharge current to flow than a larger capacitance. The smaller capacitor has more opposition to alternating current, because less current flows with the same applied voltage.

In summary, capacitance exhibits the following characteristics:

- DC is blocked by a capacitor. Once charged, no current will flow in the circuit.
- AC flows in a capacitive circuit with AC voltage applied.
- A smaller capacitance allows less current.

2.3.1 Factors Affecting Capacitance

A capacitor consists of two conductors separated by an insulating material called a dielectric. There are many types and sizes of capacitors with different dielectric materials. The capacitance of a capacitor is determined by three factors:

- *Area of the plates* – The initial charge displacement on a set of capacitor plates is related to the number of free electrons in each plate. Larger plates will produce a greater capacitance than smaller ones. Therefore, the capacitance of a capacitor varies directly with the area of the plates. For example, if the area of the plates is doubled, the capacitance is doubled. If the size of the plates is reduced by 50%, the capacitance would also be reduced by 50%.
- *Distance between plates* – As two capacitor plates are brought closer together, more electrons will move away from the positively charged plate and move into the negatively charged plate. This is because the mutual attraction between the opposite charges on the plates increases as the plates move closer together. This added movement of charge is an increase in the capacitance of the capacitor. In a capacitor composed of two plates of equal area, the capacitance varies inversely with the distance between the plates. For example, if the distance between the plates is decreased by one-half, the capacitance will be doubled. If the distance between the plates is doubled, the capacitance would be one-half as great.
- *Dielectric permittivity* – Another factor that determines the value of capacitance is the permittivity of the dielectric. The dielectric is the material between the capacitor plates in which the electric field appears. Relative permittivity expresses the ratio of the electric field strength in a dielectric to that in a vacuum. Permittivity has nothing to do with the dielectric strength of the medium or the breakdown voltage. An insulating material that will withstand a higher applied voltage than some other substance does not always have a higher dielectric permittivity. Many insulating materials have a greater dielectric permittivity than air. For a given applied voltage, a greater attraction exists between the opposite charges on the capacitor plates, and an electric field can be set up more easily than when the dielectric is air. The capacitance of the capacitor is increased when the permittivity of the dielectric is increased if all the other parameters remain unchanged.

Capacitance

The concept of capacitance, like many electrical quantities, is often hard to visualize or understand. A comparison with a balloon may help to make this concept clearer. Electrical capacitance has a charging effect similar to blowing up a balloon and holding it closed. The expansion capacity of the balloon can be changed by changing the thickness of the balloon walls. A balloon with thick walls will expand less (have less capacity) than one with thin walls. This is like a small 10µF capacitor that has less capacity and will charge less than a larger 100µF capacitor.

2.3.2 Equivalent Capacitance

Connecting capacitors in parallel is equivalent to adding the plate areas. Therefore, the total capacitance is the sum of the individual capacitances, as illustrated in *Figure 15*.

A 10µF capacitor in parallel with a 5µF capacitor, for example, provides a 15µF capacitance for the parallel combination. The voltage is the same across the parallel capacitors. Note that adding parallel capacitance is just like combining resistances in series.

Connecting capacitances in series is equivalent to increasing the thickness of the dielectric. Therefore, the combined capacitance is less than the smallest individual value. The combined equivalent capacitance is calculated by the reciprocal formula, as shown in *Figure 16*.

Capacitors connected in series are combined like resistors in parallel; any of the formulas (and "shortcut" calculations) used to calculate parallel resistance values can be used. For example, the combined capacitance of two equal capacitors of 10µF in series can be calculated using the product-over-sum method, as follows:

$$C_T = \frac{C_1 \times C_2}{C_1 + C_2}$$

$$C_T = \frac{10\mu F \times 10\mu F}{10\mu F + 10\mu F}$$

$$C_T = \frac{100\mu F}{20\mu F}$$

$$C_T = 5\mu F$$

> **NOTE**
> General circuit calculations, Ohm's law, and methods used to calculate resistance are covered in more detail in *Electrical Level One*, modules 26103 ("Introduction to Electrical Circuits") and 26104 ("Electrical Theory").

Capacitors are used in series to provide a higher voltage breakdown rating for the combination. For instance, each of three equal capacitances in series has one-third the applied voltage.

In series, the voltage across each capacitor is inversely proportional to its capacitance. The smaller capacitance has the larger proportion of the applied voltage. The reason is that the series capacitances all have the same charge because they are in one current path. With equal charge, a smaller capacitance has a greater potential difference.

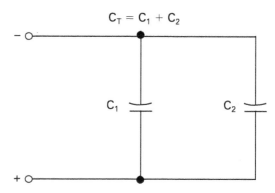

Figure 15 Capacitors in parallel.

$$C_T = \frac{1}{\frac{1}{C_1} + \frac{1}{C_2}}$$

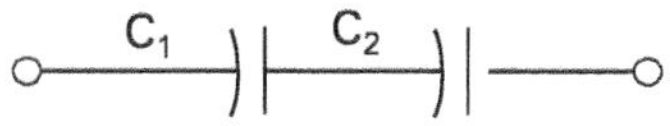

Figure 16 Capacitors in series.

2.3.3 Capacitor Specifications

The capacitor specifications list the maximum potential difference that can be applied across the plates without puncturing the dielectric. Usually, the voltage rating is for temperatures up to about 60°C. High temperatures result in a lower voltage rating. Voltage ratings for general-purpose paper, mica, and ceramic capacitors are typically 200V to 500V. Ceramic capacitors with ratings of 1 to 5kV are also available.

Electrolytic capacitors are commonly used in 25V, 150V, and 450V ratings. In addition, 6V and 10V electrolytic capacitors are often used in transistor circuits. For applications where a lower voltage rating is permissible, more capacitance can be obtained in a smaller physical size.

> **Think About It**
> ## Capacitance
> Suppose you had a motor with a bad 30µF starting capacitor and no 30µF direct replacement capacitor was available. As a temporary measure, you are authorized to substitute two equal-value capacitors in its place. What size capacitors (µF) should be used if you are connecting them in parallel?

The potential difference across the capacitor depends on the applied voltage and is not necessarily equal to the voltage rating. A voltage rating higher than the potential difference applied across the capacitor provides a safety factor for long life in service. With electrolytic capacitors, however, the actual capacitor voltage should be close to the rated voltage to produce the oxide film that provides the specified capacitance.

The voltage ratings are for applied DC voltage. The breakdown rating is lower for AC voltage because of the internal heat produced by continuous charge and discharge.

Capacitors are also rated by leakage resistance. After the charging voltage is removed, a perfect capacitor would keep its charge indefinitely. After a long period of time, however, the charge will be neutralized by a small leakage current through the dielectric and across the insulated case between terminals, because there is no perfect insulator. For paper, ceramic, and mica capacitors, the leakage current is very slight, or inversely, the leakage resistance is very high. For paper, ceramic, or mica capacitors, the resistance is 100MΩ or more. Electrolytic capacitors may have a leakage resistance of 0.5MΩ or less.

2.3.4 Voltage and Current in a Capacitive AC Circuit

In a capacitive circuit driven by an AC voltage source, the voltage is continuously changing. Thus, the charge on the capacitor is also continuously changing. The four parts of *Figure 17* show the variation of the alternating voltage and current in a capacitive circuit for each quarter of one cycle.

The solid line represents the voltage across the capacitor, and the dotted line represents the current. The line running through the center is the zero or reference point for both the voltage and the current. The bottom line marks off the time of the cycle in terms of electric degrees. Assume that the AC voltage has been acting on the capacitor for some time before the time represented by the starting point of the sine wave.

At the beginning of the first quarter-cycle (0° to 90°), the voltage has just passed through zero and is increasing in the positive direction. Since the zero point is the steepest part of the sine wave, the voltage is changing at its greatest rate.

The charge on a capacitor varies directly with the voltage; therefore, the charge on the capacitor is also changing at its greatest rate at the beginning of the first quarter-cycle. In other words, the greatest number of electrons are moving off one plate and onto the other plate. Thus, the capacitor current is at its maximum value.

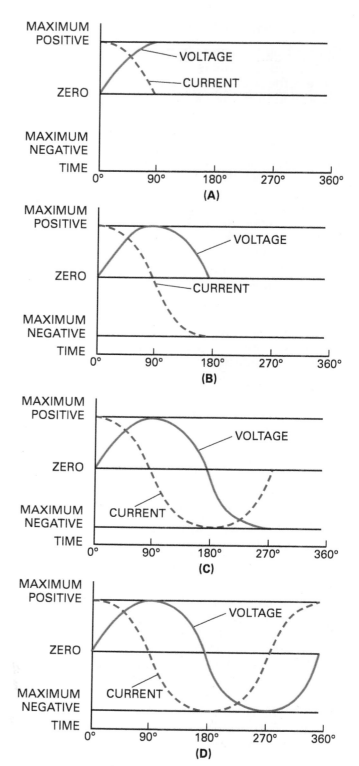

Figure 17 Voltage and current in a capacitive AC circuit.

As the voltage proceeds toward maximum at 90°, its rate of change becomes lower and lower, making the current decrease toward zero. At 90°, the voltage across the capacitor is maximum, the capacitor is fully charged, and there is no further movement of electrons from plate to plate. That is why the current at 90° is zero.

At the end of the first quarter-cycle, the alternating voltage stops increasing in the positive direction and starts to decrease. It is still a positive voltage; but to the capacitor, the decrease in voltage means that the plate that has just accumulated an excess of electrons must lose some electrons. The current flow must reverse its direction. Part (B) of the figure shows the current curve to be below the zero line (negative current direction) during the second quarter-cycle (90° to 180°).

At 180°, the voltage has dropped to zero. This means that, for a brief instant, the electrons are equally distributed between the two plates; the current is maximum because the rate of change of voltage is maximum.

Just after 180°, the voltage has reversed polarity and starts building to its maximum negative peak, which is reached at the end of the third quarter-cycle (180° to 270°). During the third quarter-cycle, the rate of voltage change gradually decreases as the charge builds to a maximum at 270°. At this point, the capacitor is fully charged and carries the full impressed voltage. Because the capacitor is fully charged, there is no further exchange of electrons and the current flow is zero at this point. The conditions are exactly the same as at the end of the first quarter-cycle (90°), but the polarity is reversed.

Just after 270°, the impressed voltage once again starts to decrease, and the capacitor must lose electrons from the negative plate. It must discharge, starting at a minimum rate of flow and rising to a maximum. This discharging action continues through the last quarter-cycle (270° to 360°) until the impressed voltage has reached zero. The beginning of the entire cycle is 360°, and everything starts over again.

In *Figure 17*, note that the current always arrives at a certain point in the cycle 90° ahead of the voltage because of the charging and discharging action. This voltage-current phase relationship in a capacitive circuit is exactly opposite to that in an inductive circuit. The current through a capacitor leads the voltage across the capacitor by 90°. A convenient way to remember this is the phrase "ELI the ICE man" (ELI refers to inductors, as previously explained). ICE pertains to capacitors as follows:

- I = current
- C = capacitor
- E = voltage

In capacitors (C), current (I) leads voltage (E) by 90°.

It is important to realize that the current and voltage are both going through their individual cycles at the same time during the period the AC voltage is impressed. The current does not go through part of its cycle (charging or discharging) and then stop and wait for the voltage to catch up. The amplitude and polarity of the voltage and the amplitude and direction of the current are continually changing.

Their positions, with respect to each other and to the zero line at any electrical instant or any degree between 0° and 360°, can be seen by reading upward from the time-degree line. The current swing from the positive peak at 0° to the negative peak at 180° is not a measure of the number of electrons or the charge on the plates. It is a picture of the direction and strength of the current in relation to the polarity and strength of the voltage appearing across the plates.

2.3.5 Capacitive Reactance

Capacitors offer a very real opposition to current flow. This opposition arises from the fact that, at a given voltage and frequency, the number of electrons that go back and forth from plate to plate is limited by the storage ability (capacitance) of the capacitor. As the capacitance is increased, a greater number of electrons changes plates every cycle. Since current is a measure of the number of electrons passing a given point in a given time, the current is increased.

Increasing the frequency will also decrease the opposition offered by a capacitor. This occurs because the capacitor has less time to charge to full potential. As a result, more electrons will pass a given point in a given time (greater current flow). The opposition that a capacitor offers to AC is therefore inversely proportional to frequency and capacitance. This opposition is called capacitive reactance. Capacitive reactance decreases with increasing frequency or, for a given frequency, the capacitive reactance decreases with increasing capacitance. The symbol for capacitive reactance is X_C. The formula is:

$$X_C = \frac{1}{2\pi f C}$$

Where:

X_C = capacitive reactance in ohms
f = frequency in hertz (Hz)
C = capacitance in farads (F)
2π = 6.28 (2 × 3.14)

For example, what is the capacitive reactance of a 0.05µF capacitor in a circuit whose frequency is 1 megahertz?

$$X_C = \frac{1}{2\pi f C}$$

$$X_C = \frac{1}{(6.28)(10^6 Hz)(5 \times 10^{-8} F)}$$

$$X_C = \frac{1}{3.14 \times 10^{-1}} = \frac{1}{0.314} = 3.18\Omega$$

The capacitive reactance of a 0.05µF capacitor operated at a frequency of 1 megahertz is 3.18 ohms. Suppose this same capacitor is operated at a lower frequency of 1,500 hertz instead of 1 megahertz. What is the capacitive reactance now? Substituting where $1,500 = 1.5 \times 10^3$ hertz:

$$X_C = \frac{1}{2\pi f C}$$

$$X_C = \frac{1}{(6.28)(1.5 \times 10^3 Hz)(5 \times 10^{-8} F)}$$

$$X_C = \frac{1}{4.71 \times 10^{-4}} = \frac{1}{0.000471} = 2,123\Omega$$

Note a very interesting point from these two examples. As frequency is decreased from 1 megahertz to 1,500 hertz, the capacitive reactance increases from 3.18 ohms to 2,123 ohms. Capacitive reactance increases as the frequency decreases.

2.4.0 Combination Circuits

AC circuits often contain inductors, capacitors, and/or resistors connected in series or parallel combinations. When this is done, it is important to determine the resulting phase relationship between the applied voltage and the current in the circuit. The simplest method of combining factors that have different phase relationships is vector addition with the trigonometric functions. Each quantity is represented as a vector, and the resultant vector and phase angle are then calculated.

ICE in ELI the ICE Man

Remembering the phrase "ICE" as in "ELI the ICE man" is an easy way to remember the phase relationships that always exist between voltage and current in a capacitive circuit. A capacitive circuit is a circuit in which there is more capacitive reactance than inductive reactance. This is indicated by the C in ICE. The I (current) is stated before the E (voltage) in ICE, meaning that the current leads the voltage in a capacitive circuit.

Think About It

Frequency and Capacitive Reactance

A variable capacitor is used in the tuner of an AM radio to tune the radio to the desired station. Will its capacitive reactance value be higher or lower when it is tuned to the low end of the frequency band (550kHz) than it would be when tuned to the high end of the band (1,440kHz)?

In purely resistive circuits, the voltage and current are in phase. In inductive circuits, the voltage leads the current by 90°. In capacitive circuits, the current leads the voltage by 90°. *Figure 18* shows the phase relationships of these components used in AC circuits. Recall that these characteristics are summarized by the phrase "ELI the ICE man":

ELI = **E** Leads **I** (inductive)
ICE = **I** Capacitive (leads) **E**

The impedance Z of a circuit is defined as the total opposition to current flow. The magnitude of the impedance Z is given by the following equation in a series circuit:

$$Z = \sqrt{R^2 + X^2}$$

Where:

Z = impedance (ohms)
R = resistance (ohms)
X = net reactance (ohms)

The current through a resistance is always in phase with the voltage applied to it; thus resistance is shown along the 0° axis. The voltage across an inductor leads the current by 90°; thus inductive reactance is shown along the 90° axis. The voltage across a capacitor lags the current by 90°; thus capacitive reactance is shown along the –90° axis. The net reactance is the difference between the inductive reactance and the capacitive reactance:

- X = net reactance (ohms)
- X_L = inductive reactance (ohms)
- X_C = capacitive reactance (ohms)

The impedance Z is the vector sum of the resistance R and the net reactance X. The angle, called the phase angle, gives the phase relationship between the applied voltage and current.

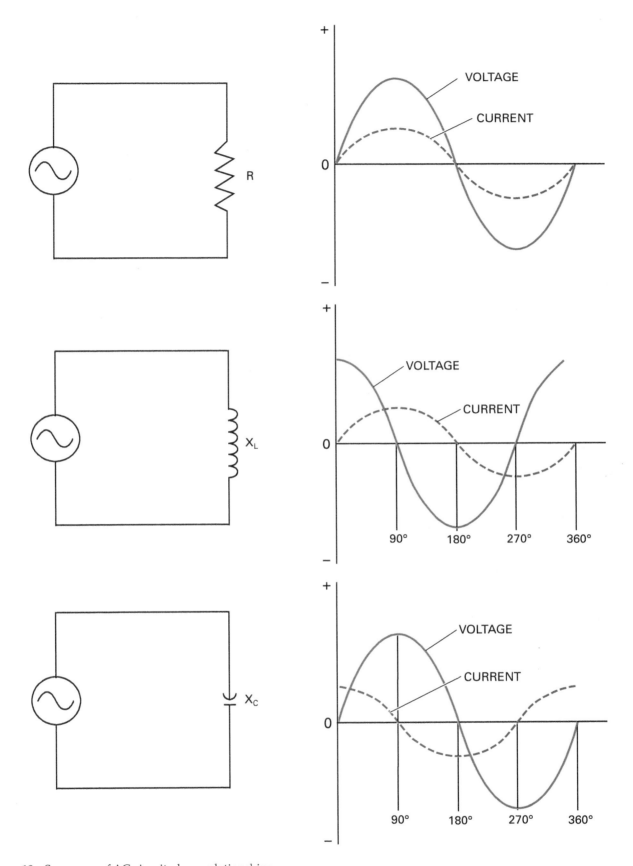

Figure 18 Summary of AC circuit phase relationships.

2.4.1 RL Circuits

RL circuits combine resistors and inductors in a series, parallel, or series-parallel configuration. In a pure inductive circuit, the current lags the voltage by an angle of 90°. In a circuit containing both resistance and inductance, the current will lag the voltage by some angle between zero and 90°.

Figure 19 shows a series RL circuit. Because it is a series circuit, the current is the same in all portions of the loop. Using the values shown, the circuit will be analyzed for unknown values such as X_L, Z, I, E_L, and E_R.

The solution would be worked as follows:

Step 1 Compute the value of X_L.

$X_L = 2\pi fL$
$X_L = 6.28 \times 100\text{Hz} \times 4\text{H}$
$X_L = 2,512\Omega$

Step 2 Draw vectors R and X_L as shown in *Figure 19*. R is drawn horizontally because the circuit current and voltage across R are in phase. It therefore becomes the reference line from which other angles are measured. X_L is drawn upward at 90° from R because voltage across X_L leads circuit current through R.

Step 3 Compute the value of circuit impedance Z, which is equal to the vector sum of X_L and R.

$\tan = \dfrac{X_L}{R} = \dfrac{2,512}{1,500} = 1.67$

arctan 1.67 = 59.1°

cos 59.1° = 0.5135

Find Z using the cosine function:

$\cos = \dfrac{R}{Z}$ so $Z = \dfrac{R}{\cos}$

$Z = \dfrac{1,500}{0.5135} = 2,921\Omega$

Step 4 Compute the circuit current using Ohm's law for AC circuits.

$I = \dfrac{E}{Z} = \dfrac{100\text{V}}{2,921\Omega} = 0.034\text{A}$

Step 5 Compute voltage drops in the circuit.

$E_L = IX_L = 0.034 \times 2,512 = 85\text{V}$
$E_R = IR = 0.034 \times 1,500 = 51\text{V}$

Note that the voltage drops across the resistor and inductor do not equal the supply voltage because they must be added vectorially. This would be done as follows (because of rounding, numbers are not exact):

$\tan = \dfrac{E_L}{E_R} = \dfrac{85}{51} = 1.67$

arctan 1.67 = 59.1°

$\cos = \dfrac{E_R}{E_Z}$ $E_Z = \dfrac{E_R}{\cos}$

$E_Z = \dfrac{51}{0.5135}$ = approx. 100V = E_S

In this inductive circuit, the current lags the applied voltage by an angle equal to 59.1°.

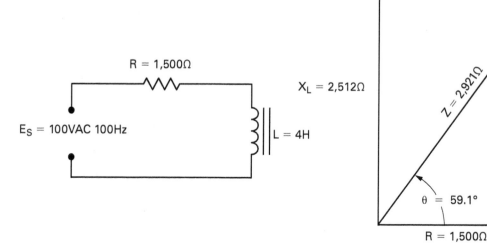

Figure 19 Series RL circuit and vector diagram.

> **NOTE:** You could also use the Pythagorean theorem (discussed later in this module) to find this answer.

Figure 20 shows another series RL circuit, its associated waveforms, and vector diagrams. This circuit is used to summarize the characteristics of a series RL circuit:

- The current I flows through all the series components.
- The voltage across X_L, labeled V_L, can be considered an IX_L voltage drop, just as V_R is used for an IR voltage drop.
- The current I through X_L must lag V_L by 90°, as this is the angle between current through an inductance and its self-induced voltage.
- The current I through R and its IR voltage drop have the same phase. There is no reactance to sine wave current in any resistance. Therefore, I and IR have the same phase, or this phase angle is 0°.
- V_T is the vector sum of the two out-of-phase voltages V_R and V_L.
- Circuit current I lags V_T by the phase angle.
- Circuit impedance is the vector sum of R and X_L.

In a series circuit, the higher the value of X_L compared with R, the more inductive the circuit is. This means there is more voltage drop across the inductive reactance, and the phase angle increases toward 90°. The series current lags the applied generator voltage.

Vector Analysis

When using vector analysis, the horizontal line is the in-phase value and the vertical line pointing up represents the leading value. The vertical line pointing down represents the lagging value.

Several combinations of X_L and R in series are listed in *Table 1* with their resultant impedance and phase angle. Note that a ratio of 10:1 or more for X_L/R means that the circuit is practically all inductive. The phase angle of 84.3° is only slightly less than 90° for the ratio of 10:1, and the total impedance Z is approximately equal to X_L. The voltage drop across X_L in the series circuit will be equal to the applied voltage, with almost none across R.

At the opposite extreme, when R is 10 times as large as X_L, the series circuit is mainly resistive. The phase angle of 5.7° means the current has almost the same phase as the applied voltage, the total impedance Z is approximately equal to R, and the voltage drop across R is practically equal to the applied voltage, with almost none across X_L.

Table 1 Series R and X_L Combinations

R (Ω)	X_L (Ω)	Z (Ω) (Approx.)	Phase Angle (θ) (°)
1	10	$\sqrt{101}$ = 10	84.3°
10	10	$\sqrt{200}$ = 14	45°
10	1	$\sqrt{101}$ = 10	5.7°

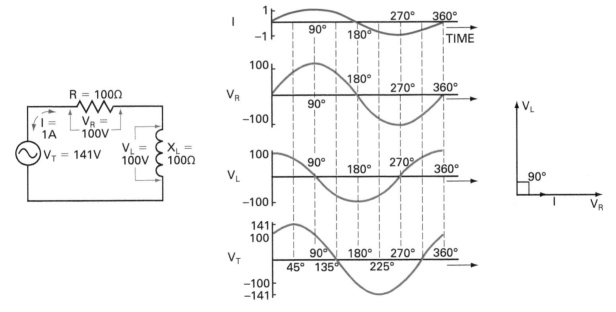

Figure 20 Series RL circuit with waveforms and vector diagram.

In a parallel RL circuit, the resistance and inductance are connected in parallel across a voltage source. Such a circuit thus has a resistive branch and an inductive branch.

The 90° phase angle must be considered for each of the branch currents, instead of voltage drops in a series circuit. Remember that any series circuit has different voltage drops, but one common current. A parallel circuit has different branch currents, but one common voltage.

In the parallel circuit in *Figure 21*, the applied voltage V_A is the same across X_L, R, and the generator, since they are all in parallel. There cannot be any phase difference between these voltages. Each branch, however, has its individual current. For the resistive branch $I_R = V_A/R$; in the inductive branch $I_L = V_A/X_L$.

The resistive branch current I_R has the same phase as the generator voltage V_A. The inductive branch current I_L lags V_A, however, because the current in an inductance lags the voltage across it by 90°.

The total line current, therefore, consists of I_R and I_L, which are 90° out of phase with each other. The phasor sum of I_R and I_L equals the total line current I_T. These phase relations are shown by the waveforms and vectors in *Figure 21*. I_T will lag V_A by some phase angle that results from the vector addition of I_R and I_L.

The impedance of a parallel RL circuit is the total opposition to current flow by the R of the resistive branch and the X_L of the inductive branch. Since X_L and R are vector quantities, they must be added vectorially.

If the line current and the applied voltage are known, Z can also be calculated by the equation:

$$Z = \frac{V_A}{I_{Line}}$$

The Z of a parallel RL circuit is always less than the R or X_L of any one branch. The branch of a parallel RL circuit that offers the most opposition to current flow has the lesser effect on the phase angle of the current.

Several combinations of X_L and R in parallel are listed in *Table 2*. When X_L is 10 times R, the parallel circuit is practically resistive because there is little inductive current in the line. The small value of I_L results from the high X_L. The total impedance of the parallel circuit is approximately equal to the resistance then, since the high value of X_L in a parallel branch has little effect. The phase angle of –5.7° is practically 0° because almost all the line current is resistive.

As X_L becomes smaller, it provides more inductive current in the main line. When X_L is 1/10R, practically all the line current is the I_L component. Then, the parallel circuit is practically all inductive, with a total impedance practically equal to X_L. The phase angle of –84.3° is almost –90° because the line current is mostly inductive. Note

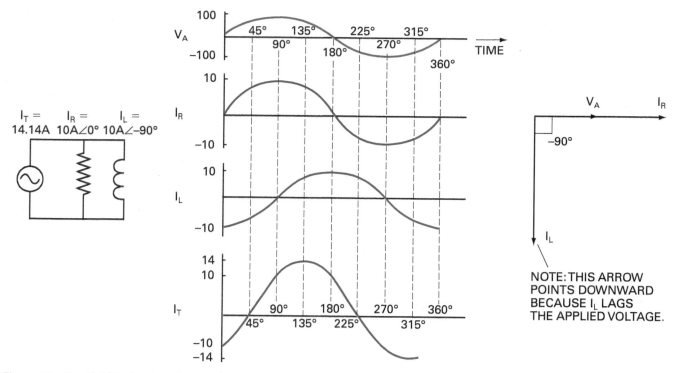

Figure 21 Parallel RL circuit with waveforms and vector diagram.

Table 2 Parallel R and XL Combinations

R (Ω)	X_L (Ω)	I_R (A)	I_L (A)	I_T (A) (Approx.)	$Z_T = V_A/I_T$ (Ω)	Phase Angle (θ)I (°)
1	10	10	1	$\sqrt{101} = 10$	1	−5.7°
10	10	1	1	$\sqrt{2} = 1.4$	7.07	−45°
10	1	1	10	$\sqrt{101} = 10$	1	−84.3°

that these conditions are opposite from the case of X_L and R in series.

2.4.2 RC Circuits

In a circuit containing resistance only, the current and voltage are in phase. In a circuit of pure capacitance, the current leads the voltage by an angle of 90°. In a circuit that has both resistance and capacitance, the current will lead the voltage by some angle between 0° and 90°. This is known as an RC circuit.

In a series RC circuit, the current (I) is the same in X_C and R since they are in series. *Figure 22 (A)* shows a series RC circuit with resistance R in series with capacitive reactance X_C. Each has its own series voltage drop, equal to IR for the resistance and IX_C for the reactance.

In *Figure 22 (B)*, the phasor is shown horizontal as the reference phase, because I is the same throughout the series circuit. The resistive voltage drop IR has the same phase as I. The capacitor voltage IX_C must be 90° clockwise from I and IR, as the capacitive voltage lags. Note that the IX_C phasor is downward, exactly opposite from an IX_L phasor, because of the opposite phase angle.

If the capacitive reactance alone is considered, its voltage drop lags the series current I by 90°. The IR voltage has the same phase as I, however, because resistance provides no phase shift. Therefore, R and X_C combined in series must be added by vectors because they are 90° out of phase with each other, as shown in *Figure 22 (C)*.

As with inductive reactance, θ (theta) is the phase angle between the generator voltage and its series current. As shown in *Figure 22 (B)* and *Figure 22 (C)*, θ can be calculated from the voltage or impedance triangle.

With series X_C the phase angle is negative, clockwise from the zero reference angle of I because the X_C voltage lags its current. To indicate the negative phase angle, this 90° phasor points downward from the horizontal reference, instead of upward as with the series inductive reactance.

In series, the higher the X_C compared with R, the more capacitive the circuit. There is more voltage drop across the capacitive reactance, and the phase angle increases toward −90°. The series X_C always makes the current lead the applied voltage. With all X_C and no R, the entire applied voltage is across X_C and equals −90°. Several combinations of X_C and R in series are listed in *Table 3*.

In a parallel RC circuit, as shown in *Figure 23 (A)*, a capacitive branch as well as a resistive branch are connected across a voltage source. The current that leaves the voltage source divides among the

Table 3 Series R and X_C Combinations

R (Ω)	X_C (Ω)	Z (Ω) (Approx.)	Phase Angle (θ)z (°)
1	10	$\sqrt{101} = 10$	84.3°
10	10	$\sqrt{200} = 14$	45°
10	1	$\sqrt{101} = 10$	5.7°

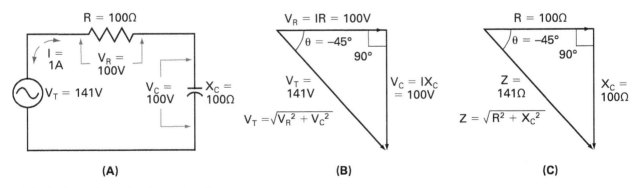

Figure 22 Series RC circuit with vector diagrams.

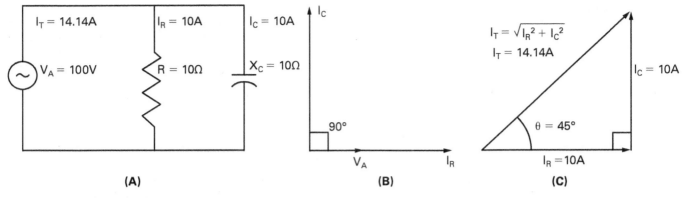

Figure 23 Parallel RC circuit with vector diagrams.

branches, so there are different currents in each branch. The current is therefore not a common quantity, as it is in the series RC circuit.

In a parallel RC circuit, the applied voltage is directly across each branch. Therefore, the branch voltages are equal in value to the applied voltage and all voltages are in phase. Since the voltage is common throughout the parallel RC circuit, it serves as the common quantity in any vector representation of parallel RC circuits. This means the reference vector will have the same phase relationship or direction as the circuit voltage. Note in *Figure 23* (B) that V_A and I_R are both shown as the 0° reference.

Current within an individual branch of an RC parallel circuit is dependent on the voltage across the branch and on the R or X_C contained in the branch. The current in the resistive branch is in phase with the branch voltage, which is the applied voltage. The current in the capacitive branch leads V_A by 90°. Since the branch voltages are the same, I_C leads I_R by 90°, as shown in *Figure 23* (B). Since the branch currents are out of phase, they have to be added vectorially to find the line current.

The phase angle, θ, is 45° because R and X_C are equal, resulting in equal branch currents. The phase angle is between the total current I_T and the generator voltage V_A. However, the phase of V_A is the same as the phase of I_R. Therefore, θ is also between I_T and I_R.

The impedance of a parallel RC circuit represents the total opposition to current flow offered by the resistance and capacitive reactance of the circuit. The equation for calculating the impedance of a parallel RC circuit is:

$$Z = \frac{RX_C}{\sqrt{R^2 + X_C^2}} \quad or \quad Z = \frac{V_A}{I_T}$$

For the example shown in *Figure 23*, Z is:

$$Z = \frac{V_A}{I_T} = \frac{100V}{14.14A} = 7.07\Omega$$

This is the opposition in ohms across the generator. This Z of 7.07Ω is equal to the resistance of 10Ω in parallel with the reactance of 10Ω. Notice that the impedance of equal values of R and X_C is not one-half, but equals 70.7% of either one.

When X_C is high relative to R, the parallel circuit is practically resistive because there is little leading capacitive current in the main line. The small value of I_C results from the high reactance of shunt X_C. The total impedance of the parallel circuit is approximately equal to the resistance, because the high value of X_C in a parallel branch has little effect.

As X_C becomes smaller, it provides more leading capacitive current in the main line. When X_C is very small relative to R, practically all the line current is the I_C component. The parallel circuit is practically all capacitive, with a total impedance practically equal to X_C.

The characteristics of different circuit arrangements are shown in *Table 4*.

2.4.3 LC Circuits

An LC circuit consists of an inductance and a capacitance connected in series or in parallel with a voltage source. There is no resistor physically in an LC circuit, but every circuit contains some resistance. Because the circuit resistance of the wiring and voltage source is usually so small, it has little or no effect on circuit operation.

In a circuit with both X_L and X_C, the opposite phase angles enable one to cancel the effect of the other. For X_L and X_C in series, the net reactance is the difference between the two series reactances, resulting in less reactance than either one. In parallel circuits, the I_L and I_C branch currents cancel. The net line current is then the difference

Table 4 Parallel R and X_C Combinations

R (Ω)	X_C (Ω)	I_R (A)	I_C (A)	I_T (A) (Approx.)	Z_T (Ω) (Approx.)	Phase Angle (θ) (°)
1	10	10	1	$\sqrt{101}$ = 10	1	5.7°
10	10	1	1	$\sqrt{2}$ = 1.4	7.07	45°
10	1	1	10	$\sqrt{101}$ = 10	1	84.3°

between the two branch currents, resulting in less total line current than either branch current.

As in all series circuits, the current in a series LC circuit is the same at all points. Therefore, the current in the inductor is the same as, and in phase with, the current in the capacitor. Because of this, on the vector diagram for a series LC circuit, the direction of the current vector is the reference or in the 0° direction, as shown in *Figure 24*.

When there is current flow in a series LC circuit, the voltage drops across the inductor and capacitor depend on the circuit current and the values of X_L and X_C. The voltage drop across the inductor leads the circuit current by 90°, and the voltage drop across the capacitor lags the circuit current by 90°. Using Kirchhoff's voltage law, the source voltage equals the sum of the voltage drops across the inductor and capacitor, with respect to the polarity of each.

Since the current through both is the same, the voltage across the inductor leads that across the capacitor by 180°. The method used to add the two voltage vectors is to subtract the smaller vector from the larger, and assign the resultant the direction of the larger. When applied to a series LC circuit, this means the applied voltage is equal to the difference between the voltage drops (E_L and E_C), with the phase angle between the applied voltage (E_T) and the circuit current determined by the larger voltage drop.

In a series LC circuit, one or both of the voltage drops are always greater than the applied voltage, but remember that they are 180° out of phase. One of them effectively cancels a portion of the other so that the total voltage drop is always equal to the applied voltage.

Recall that X_L is 180° out of phase with X_C. The impedance is then the vector sum of the two reactances. The reactances are 180° apart, so their vector sum is found by subtracting the smaller one from the larger.

Unlike RL and RC circuits, the impedance in an LC circuit is either purely inductive or purely capacitive.

In a parallel LC circuit there is an inductance and a capacitance connected in parallel across a

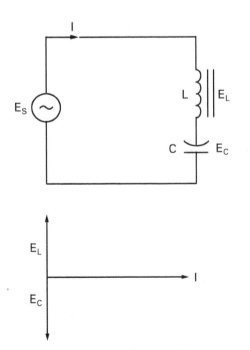

Figure 24 Series LC circuit with vector diagram.

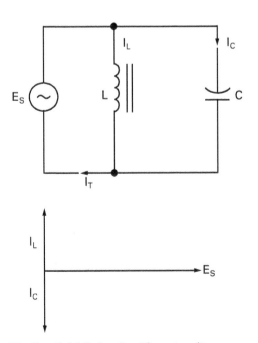

Figure 25 Parallel LC circuit with vector diagram.

voltage source. *Figure 25* shows a parallel LC circuit with its vector diagram.

As in any parallel circuit, the voltage across the branches is the same as the applied voltage. Since they are actually the same voltage, the branch voltages and applied voltage are in phase. Because of this, the voltage is used as the 0° phase reference and the phases of the other circuit quantities are expressed in relation to the voltage.

The currents in the branches of a parallel LC circuit are both out of phase with the circuit voltage. The current in the inductive branch (I_L) lags the voltage by 90°, while the current in the capacitive branch (I_C) leads the voltage by 90°. Since the voltage is the same for both branches, currents I_L and I_C are therefore 180° out of phase. The amplitudes of the branch currents depend on the value of the reactance in the respective branches.

With the branch currents being 180° out of phase, the line current is equal to their vector sum. This vector addition is done by subtracting the smaller branch current from the larger.

The line current for a parallel LC circuit, therefore, has the phase characteristics of the larger branch current. Thus, if the inductive branch current is the larger, the line current is inductive and lags the applied voltage by 90°; if the capacitive branch current is the larger, the line current is capacitive, and leads the applied voltage by 90°.

The line current in a parallel LC circuit is always less than one of the branch currents and sometimes less than both. The line current is less than the branch currents because the two branch currents are 180° out of phase. As a result of the phase difference, some cancellation takes place between the two currents when they combine to produce the line current. The impedance of a parallel LC circuit can be found using the following equations:

$$Z = \frac{X_L \times X_C}{X_L - X_C} \text{ (for } X_L \text{ larger than } X_C\text{)}$$

or:

$$Z = \frac{X_L \times X_C}{X_C - X_L} \text{ (for } X_C \text{ larger than } X_L\text{)}$$

When using these equations, the impedance will have the phase characteristics of the smaller reactance.

2.4.4 RLC Circuits

Circuits in which the inductance, capacitance, and resistance are all connected in series are called series RLC circuits. The fundamental properties of series RLC circuits are similar to those for series LC circuits. The differences are caused by the effects of the resistance. Any practical series LC circuit contains some resistance. When the resistance is very small compared to the circuit reactance, it has almost no effect on the circuit and can be considered as zero. When the resistance is appreciable, though, it has a significant effect on the circuit operation and therefore must be considered in any circuit analysis. In a series RLC circuit, the same current flows through each component. The phase relationships between the voltage drops are the same as they were in series RC, RL, and LC circuits. The voltage drops across the inductance and capacitance are 180° out of phase. With current the same throughout the circuit as a reference, the inductive voltage drop (E_L) leads the resistive voltage drop (E_R) by 90°, and the capacitive voltage drop (E_C) lags the resistive voltage drop by 90°.

Figure 26 shows a series RLC circuit and the vector diagram used to determine the applied voltage. The vector sum of the three voltage drops is equal to the applied voltage. However, to calculate this vector sum, a combination of the methods learned for LC, RL, and RC circuits must be used. First, calculate the combined voltage drop of the two reactances. This value is designated E_X and is found as in pure LC circuits by subtracting the smaller reactive voltage drop from the larger. This is shown in *Figure 26* as E_X. The result of this calculation is the net reactive voltage drop and is either inductive or capacitive, depending on which of the individual voltage drops is larger. In *Figure 26*, the net reactive voltage drop is inductive because $E_L > E_C$. Once the net reactive voltage drop is known, it is added vectorially to the voltage drop across the resistance.

The angle between the applied voltage E_A and the voltage across the resistance E_R is the same as

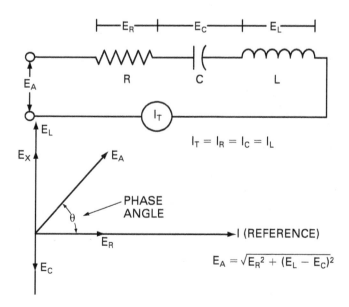

Figure 26 Series RLC circuit and vector diagram.

the phase angle between E_A and the circuit current. The reason for this is that E_R and I are in phase.

The impedance of a series RLC circuit is the vector sum of the inductive reactance, the capacitive reactance, and the resistance. This is done using the same method as for voltage drop calculations.

When X_L is greater than X_C, the net reactance is inductive, and the circuit acts essentially as an RL circuit. Similarly, when X_C is greater than X_L, the net reactance is capacitive, and the circuit acts as an RC circuit.

The same current flows in every part of a series RLC circuit. The current always leads the voltage across the capacitance by 90° and is in phase with the voltage across the resistance. The phase relationship between the current and the applied voltage, however, depends on the circuit impedance. If the impedance is inductive (X_L greater than X_C), the current is inductive and lags the applied voltage by some phase angle less than 90°. If the impedance is capacitive (X_C greater than X_L), the current is capacitive and leads the applied voltage by some phase angle also less than 90°. The angle of the lead or lag is determined by the relative values of the net reactance and the resistance.

The greater the value of X or the smaller the value of R, the larger the phase angle, and the more reactive (or less resistive) the current. Similarly, the smaller the value of X or the larger the value of R, the more resistive (or less reactive) the current. If either R or X is 10 or more times greater than the other, the circuit will essentially act as though it is purely resistive or reactive, as the case may be.

A parallel RLC circuit is basically a parallel LC circuit with an added parallel branch of resistance. The solution of a parallel circuit involves the solution of a parallel LC circuit, and then the solution of either a parallel RL circuit or a parallel RC circuit. The reason for this is that a parallel combination of L and C appears to the source as a pure L or a pure C. So by solving the LC portion of a parallel RLC circuit first, the circuit is reduced to an equivalent RL or RC circuit.

The distribution of the voltage in a parallel RLC circuit is no different from what it is in a parallel LC circuit, or in any parallel circuit. The branch voltages are all equal and in phase, since they are the same as the applied voltage. The resistance is simply another branch across which the applied voltage appears. Because the voltages throughout the circuit are the same, the applied voltage is again used as the θ phase reference. The current relationship in a parallel RLC circuit is shown in *Figure 27*.

The three branch currents in a parallel RLC circuit are an inductive current I_L, a capacitive current I_C, and a resistive current I_R. Each is independent of the others, and depends only on the applied voltage and the branch resistance or reactance.

The three branch currents all have different phases with respect to the branch voltages. I_L lags the voltage by 90°, I_C leads the voltage by 90°, and I is in phase with the voltage. Since the voltages are the same, I_L and I_C are 180° out of phase with each other, and both are 90° out of phase with I_R. Because I_R is in phase with the voltage, it has the same zero-reference direction as the voltage. So I_C leads I_R by 90°, and I_L lags I_R by 90°.

The line current (I_T), or total current, is the vector sum of the three branch currents, and can be calculated by adding I_L, I_C, and I_R vectorially. Whether the line current leads or lags the applied

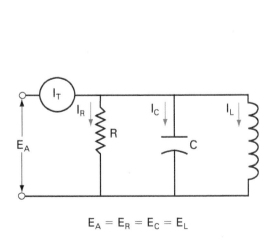
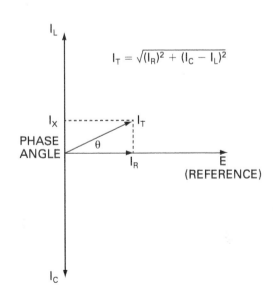

Figure 27 Parallel RLC circuit and vector diagram.

voltage depends on which of the reactive branch currents (I_L or I_C) is the larger. If I_L is larger, I_T lags the applied voltage. If I_C is larger, I_T leads the applied voltage.

To determine the impedance of a parallel RLC circuit, first determine the net reactance X of the inductive and capacitive branches. Then use X to determine the impedance Z, the same as in a parallel RL or RC circuit.

Whenever Z is inductive, the line current will lag the applied voltage. Similarly, when Z is capacitive, the line current will lead the applied voltage.

Think About It

AC Circuits

The photo below shows a simple series circuit composed of an On/Off switch, small lamp, motor, and capacitor. How would you classify this circuit? When energized, which components insert resistance, inductive reactance, and capacitive reactance into the circuit?

2.0.0 Section Review

1. What is the voltage (E) in a purely resistive circuit with a current draw of 22A and a 10Ω resistance?
   a. 32V
   b. 120V
   c. 220V
   d. 440V

2. Which of the following coils is likely to show the highest inductance?
   a. A coil with few turns
   b. A coil with a non-magnetic core
   c. A coil with a narrow coil
   d. A coil with a short core

3. The total capacitance of three 10μF capacitors and one 5μF capacitor connected in parallel is _____.
   a. 5μF
   b. 10μF
   c. 15μF
   d. 35μF

4. A circuit's impedance is represented by the letter _____.
   a. I
   b. R
   c. X
   d. Z

SECTION THREE

3.0.0 POWER IN AC CIRCUITS

Objective

Make power calculations in AC circuits.
a. Calculate true power.
b. Calculate apparent power.
c. Calculate reactive power.
d. Calculate power factor.
e. Use the power triangle to determine unknown values.

In DC circuits, the power consumed is the sum of all the I^2R heating in the resistors. It is also equal to the power produced by the source, which is the product of the source voltage and current. In AC circuits containing only resistors, the above relationship also holds true.

3.1.0 True Power

The power consumed by resistance is called true power and is measured in units of watts. True power is the product of the resistor current squared and the resistance:

$P_T = I^2R$

This formula applies because current and voltage have the same phase across a resistance. To find the corresponding value of power as a product of voltage and current, this product must be multiplied by the cosine of the phase angle θ:

$P_T = I^2R$ *or* $P_T = E \times I \times \cos(\theta)$

Where E and I are in rms values to calculate the true power in watts, multiplying I by the cosine of the phase angle provides the resistive component for true power equal to I^2R.

For example, a series RL circuit has 2A through a 100Ω resistor in series with the X_L of 173Ω. Therefore:

$P_T = I^2R$
$P_T = 4A \times 100Ω$
$P_T = 400W$

Furthermore, in this circuit the phase angle is 60° with a cosine of 0.5. The applied voltage is 400V. Therefore:

$P_T = E \times I \times \cos(\theta)$
$P_T = 400V \times 2A \times 0.5$
$P_T = 400W$

In both cases, the true power is the same (400W) because this is the amount of power supplied by the generator and dissipated in the resistance. Either formula can be used for calculating the true power.

3.2.0 Apparent Power

In ideal AC circuits containing resistors, capacitors, and inductors, the only mechanism for power consumption is $I^2_{eff}R$ heating in the resistors. Inductors and capacitors consume no power. The only function of inductors and capacitors is to store and release energy. However, because of the phase shifts that are introduced by these elements, the power consumed by the resistors is not equal to the product of the source voltage and current. The product of the source voltage and current is called apparent power and has units of volt-amperes (VA).

The apparent power is the product of the source voltage and the total current. Therefore, apparent power is actual power delivered by the source. The formula for apparent power is:

$P_A = (E_A)(I)$

Figure 28 shows a series RL circuit and its associated vector diagram.

This circuit is used to calculate the apparent power and compare it to the circuit's true power:

$P_A = (E_A)(I)$ $P_T = E \times I \times \cos(\theta)$
$P_A = (400V)(2A)$ $\theta = \dfrac{R}{Z} = \dfrac{100}{200} = \cos(60°)$
$P_A = 800VA$ $P_T = 400V \times 2A \times \cos(60°)$
                           $P_T = 400V \times 2A \times 0.5$
                           $P_T = 400W$

Note that the apparent power formula is the product of EI alone without considering the cosine of the phase angle.

3.3.0 Reactive Power

Reactive power is that portion of the apparent power that is caused by inductors and capacitors in the circuit. Inductance and capacitance are always present in real AC circuits. No work is performed by reactive power; the power is stored in the inductors and capacitors, then returned to the circuit. Therefore, reactive power is always out of phase with true power. The units for reactive power are volt-amperes-reactive (VARs).

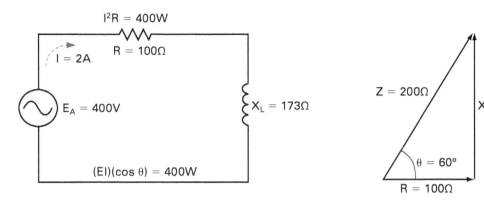

Figure 28 Power calculations in an AC circuit.

In general, for any phase angle θ between E and I, multiplying EI by sine θ gives the vertical component at 90° for the value of the VARs. In *Figure 28*, the value of sine 60° is 800 × 0.866 = 692.8 VARs.

Note that the factor sine θ for the VARs gives the vertical or reactive component of the apparent power EI. However, multiplying EI by cosine θ as the power factor gives the horizontal or resistive component for the real power.

3.4.0 Power Factor

Because it indicates the resistive component, cosine θ is the power factor (pf) of the circuit, converting the EI product to real power. For series circuits, use the formula:

$$pf = \cos(\theta) = \frac{R}{Z}$$

For parallel circuits, use the following formula:

$$pf = \cos(\theta) = \frac{I_R}{I_T}$$

In *Figure 28* as an example of a series circuit, R and Z are used for the calculations:

$$pf = \cos(\theta) = \frac{R}{Z} = \frac{100\Omega}{200\Omega} = 0.5$$

The power factor is not an angular measure but a numerical ratio with a value between 0 and 1, equal to the cosine of the phase angle.

With all resistance and zero reactance, R and Z are the same for a series circuit of I_R and I_T and are the same for a parallel circuit. The ratio is 1. Therefore, unity power factor means a resistive circuit. At the opposite extreme, all reactance with zero resistance makes the power factor zero, meaning that the circuit is all reactive.

The power factor gives the relationship between apparent power and true power. The power factor can thus be defined as the ratio of true power to apparent power:

$$pf = \frac{P_T}{P_A}$$

For example, calculate the power factor of the circuit shown in *Figure 29*.

The true power is the product of the resistor current squared and the resistance:

$P_T = I^2R$
$P_T = 10A^2 \times 10\Omega$
$P_T = 100A \times 10\Omega$
$P_T = 1,000W$

The apparent power is the product of the source voltage and total current:

$P_A = (I_T)(E)$
$P_A = 10.2A \times 100V$
$P_A = 1,020VA$

Third-Order Harmonics

Harmonics are frequencies that are multiples of the basic frequency. For example, 180Hz is three times the frequency of 60Hz and is therefore known as a third-order harmonic. Harmonics are caused by a variety of devices, such as personal computers, uninterruptible power supplies, adjustable speed drives, and electronic ballasts. Third-order harmonics, also known as triplen harmonics or triplens, may cause transformer overheating and should be a consideration when sizing transformers.

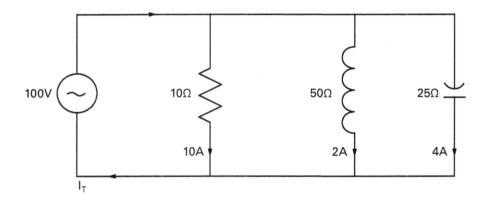
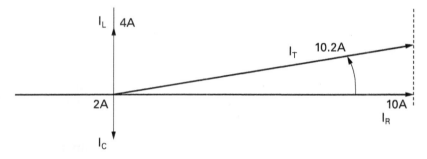

Figure 29 RLC circuit calculation.

Calculating total current:

$$I_T = \sqrt{I_R^2 + (I_C - I_L)^2} = \sqrt{10A^2 + (4A - 2A)^2}$$
$$I_T = \sqrt{100A + (2A)^2} = \sqrt{100A + 4A}$$
$$I_T = \sqrt{104A} = 10.2A$$

The power factor is the ratio of true power to apparent power:

$$pf = \frac{P_T}{P_A}$$
$$pf = \frac{1,000}{1,020}$$
$$pf = 0.98$$

As illustrated in the previous example, the power factor is determined by the system load. If the load contained only resistance, the apparent power would equal the true power and the power factor would be at its maximum value of one. Purely resistive circuits have a power factor of unity or one. If the load is more inductive than capacitive, the apparent power will lag the true power and the power factor will be lagging. If the load is more capacitive than inductive, the apparent power will lead the true power and the power factor will be leading. If there is any reactive load on the system, the apparent power will be greater than the true power and the power factor will be less than one.

3.5.0 Power Triangle

The phase relationships among the three types of AC power are easily visualized on the power triangle shown in *Figure 30*. The true power (W) is the horizontal leg, the apparent power (VA) is the hypotenuse, and the cosine of the phase angle between them is the power factor. The vertical leg of the triangle is the reactive power and has units of volt-amperes-reactive (VARs).

As illustrated on the power triangle (*Figure 30*), the apparent power will always be greater than the true power or reactive power. Also, the apparent power is the result of the vector addition of true and reactive power. The power magnitude relationships shown in *Figure 30* can be derived from the Pythagorean theorem for right triangles:

$$c^2 = a^2 + b^2$$

Therefore, c also equals the square root of $a^2 + b^2$, as shown below:

$$c = \sqrt{a^2 + b^2}$$

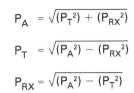

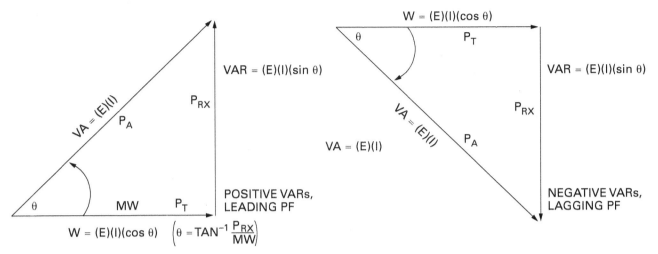

Figure 30 Power triangle.

3.0.0 Section Review

1. True power is the product of the resistor current squared and the _____.

   a. resistance
   b. voltage
   c. frequency
   d. reactance

2. Apparent power is the product of the source voltage and the _____.

   a. resistance
   b. frequency
   c. current
   d. reactance

3. Which of the following is *true* regarding reactive power?

   a. Reactive power performs no work.
   b. Reactive power is always in phase with true power.
   c. Reactive power is measured in watts.
   d. Reactive power is caused by harmonics in the circuit.

4. The power factor of a circuit with a resistance of 75Ω and an impedance of 150Ω is _____.

   a. 1.00
   b. 0.75
   c. 0.50
   d. 0.25

5. The apparent power is the result of the vector addition of the true power and the _____.

   a. power factor
   b. reactive power
   c. inductance
   d. phase angle

SECTION FOUR

4.0.0 TRANSFORMERS

Objective

Identify transformers and explain how they operate.
  a. Identify the basic components in a transformer.
  b. Identify transformer operating characteristics.
  c. Calculate turns and voltage ratios.
  d. Identify various types of transformers and their applications.

A transformer is a device that transfers electrical energy from one circuit to another by electromagnetic induction (transformer action). The electrical energy is always transferred without a change in frequency, but may involve changes in the magnitudes of voltage and current. Because a transformer works on the principle of electromagnetic induction, it must be used with an input source voltage that varies in amplitude.

4.1.0 Components of a Transformer

The basic components of a transformer are shown in *Figure 31*. In its most basic form, a transformer consists of:

- A primary coil or winding
- A secondary coil or winding
- A core that supports the coils or windings

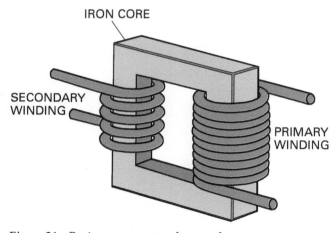

Figure 31 Basic components of a transformer.

A simple transformer action is shown in *Figure 32*. The primary winding is connected to a 60Hz AC voltage source. The magnetic field or flux builds up (expands) and collapses (contracts) around the primary winding. The expanding and contracting magnetic field around the primary winding cuts the secondary winding and induces an alternating voltage into the winding. This voltage causes AC to flow through the load. The voltage may be stepped up or down depending on the design of the primary and secondary windings.

4.1.1 Core Characteristics

Commonly used core materials are air, soft iron, and steel. Each of these materials is suitable for particular applications and unsuitable for others. Generally, air-core transformers are used when the voltage source has a high frequency (above 20kHz). Iron-core transformers are usually used when the source frequency is low (below 20kHz). A soft-iron transformer is very useful where the transformer must be physically small yet efficient. The iron-core transformer provides better power transfer than the air-core transformer. Laminated sheets of steel are often used in a transformer to reduce one type of power loss known as eddy currents. These are undesirable currents, induced into the core, which circulate around the core. Laminating the core reduces these currents to smaller levels. These steel laminations are insulated with a non-conducting material, such as varnish, and then formed into a core as shown in *Figure 33*. It takes about 50 such laminations to make a core one inch (25 mm) thick. The most efficient transformer core is one that offers the best path for the most lines of flux, with the least loss in magnetic and electrical energy.

4.1.2 Transformer Windings

A transformer consists of two coils called windings, which are wrapped around a core. The transformer operates when a source of AC voltage is connected to one of the windings and a load

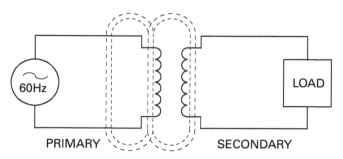

Figure 32 Transformer action.

Think About It

Power Factor

In power distribution circuits, it is desirable to achieve a power factor approaching a value of 1 in order to obtain the most efficient transfer of power. In AC circuits where there are large inductive loads such as in motors and transformers, the power factor can be considerably less than 1. For example, in a highly inductive motor circuit, if the voltage is 120V, the current is 12A, and the current lags the voltage by 60°, the power factor is 0.5 or 50% (cosine of 60° = 0.5). The apparent power is 1,440VA (120V × 12A), but the true power is only 720W [120V × (0.5 × 12A) = 720W]. This is a very inefficient circuit. What would you do to this circuit in order to achieve a circuit having a power factor as close to 1 as possible?

---

device is connected to the other. The winding that is connected to the source is called the primary winding. The winding that is connected to the load is called the secondary winding. *Figure 34* shows a cutaway view of a typical transformer.

The wire is coated with varnish so that each turn of the winding is insulated from every other turn. In a transformer designed for high-voltage applications, sheets of insulating material such as paper are placed between the layers of windings to provide additional insulation.

When the primary winding is completely wound, it is wrapped in insulating paper or cloth. The secondary winding is then wound on top of the primary winding. After the secondary winding is complete, it too is covered with insulating paper. Next, the iron core is inserted into and around the windings as shown.

Sometimes, terminals may be provided on the enclosure for connections to the windings. The figure shows four leads, two from the primary and two from the secondary. These leads are to be connected to the source and load, respectively.

4.2.0 Operating Characteristics

The operating characteristics of transformers are determined by the applied voltage and the winding design. These affect both the exciting current (no-load condition) and the phase relationship between the windings during transformer operation.

4.2.1 Energized with No Load

A no-load condition is said to exist when a voltage is applied to the primary, but no load is connected to the secondary. Assume the output of the secondary is connected to a load by a switch that is open. Because of the open switch, there is no current flowing in the secondary winding. With the switch open and an AC voltage applied to the primary, there is, however, a very small amount of current, called exciting current, flowing in the

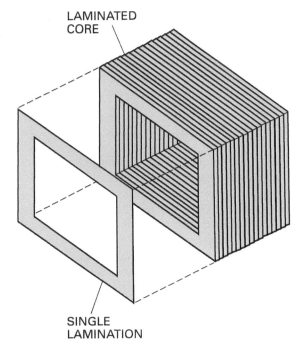

Figure 33 Steel laminated core.

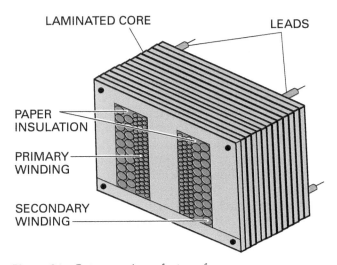

Figure 34 Cutaway view of a transformer core.

Transformers

Transformers are essential to all electrical systems and all types of electronic equipment. They are especially essential to the operation of AC high-voltage power distribution systems. Transformers are used to both step up voltage and step down voltage throughout the distribution process. For example, a typical power generation plant might generate AC power at 13,800V, step it up to 230,000V for distribution over long transmission lines, step it down to 13,800V again at substations located at different points for local distribution, and finally step it down again to 240V and 120V for lighting and local power use.

---

primary. Essentially, what this current does is excite the coil of the primary to create a magnetic field. The amount of exciting current is determined by three factors: the amount of voltage applied (E_A); the resistance (R) of the primary coil's wire and core losses; and the X_L, which is dependent on the frequency of the exciting current. These factors are all controlled by transformer design.

This very small amount of exciting current serves two functions:

- Most of the exciting energy is used to support the magnetic field of the primary.
- A small amount of energy is used to overcome the resistance of the wire and core. This is dissipated in the form of heat (power loss).

Exciting current will flow in the primary winding at all times to maintain this magnetic field, but no transfer of energy will take place as long as the secondary circuit is open.

4.2.2 Phase Relationship

The secondary voltage of a simple transformer may be either in phase or out of phase with the primary voltage. This depends on the direction in which the windings are wound and the arrangement of the connection to the external circuit (load). Simply, this means that the two voltages may rise and fall together, or one may rise while the other is falling. Transformers in which the secondary voltage is in phase with the primary are referred to as like-wound transformers, while those in which the voltages are 180° out of phase are called unlike-wound transformers.

Dots are used to indicate points on a transformer schematic symbol that have the same instantaneous polarity (points that are in phase). The use of phase-indicating dots is illustrated in *Figure 35*. In the first part of the figure, both the primary and secondary windings are wound from top to bottom in a clockwise direction, as viewed from above the windings. When constructed in this manner, the top lead of the primary and the top lead of the secondary have the same polarity. This is indicated by the dots on the transformer symbol.

The second part of the figure illustrates a transformer in which the primary and secondary are wound in opposite directions. As viewed from above the windings, the primary is wound in a clockwise direction from top to bottom, while the secondary is wound in a counterclockwise direction. Notice that the top leads of the primary and secondary have opposite polarities. This is indicated by the dots being placed on opposite ends of the transformer symbol. Thus, the polarity of voltage at the terminals of the transformer secondary depends on the direction in which the secondary is wound with respect to the primary.

4.3.0 Turns and Voltage Ratios

To understand how a transformer can be used to step up or step down voltage, the term *turns ratio* must be understood. The total voltage induced into the secondary winding of a transformer is determined mainly by the ratio of the number of turns in the primary to the number of turns in the secondary, and by the amount of voltage applied to the primary. Therefore, to set up a formula:

$$\text{Turns ratio} = \frac{\text{number of turns in the primary}}{\text{number of turns in the secondary}}$$

The first transformer in *Figure 36* shows a transformer whose primary consists of 10 turns of wire, and whose secondary consists of a single turn of wire. As lines of flux generated by the primary expand and collapse, they cut both the 10 turns of the primary and the single turn of the secondary. Since the length of the wire in the secondary is approximately the same as the length of the wire in each turn of the primary, the EMF induced into the secondary will be the same as the EMF induced into each turn of the primary.

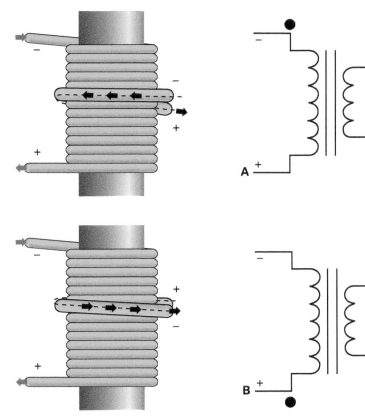

Figure 35 Transformer winding polarity.

This means that if the voltage applied to the primary winding is 10 volts, the CEMF in the primary is almost 10 volts. Thus, each turn in the primary will have an induced CEMF of approximately $\frac{1}{10}$ of the total applied voltage, or one volt. Since the same flux lines cut the turns in both the secondary and the primary, each turn will have an EMF of one volt induced into it. The first transformer in *Figure 36* has only one turn in the secondary, thus, the EMF across the secondary is one volt.

The second transformer represented in *Figure 36* has a 10-turn primary and a two-turn secondary. Since the flux induces one volt per turn, the total voltage across the secondary is two volts. Notice that the volts per turn are the same for both primary and secondary windings. Since the CEMF in the primary is equal (or almost equal) to the applied voltage, a proportion may be set up to express the value of the voltage induced in terms of the voltage applied to the primary and the number of turns in each winding. This proportion also shows the relationship between the number of turns in each winding and the voltage across each winding, and is expressed by the equation:

$$\frac{E_S}{E_P} = \frac{N_S}{N_P}$$

Where:

N_P = number of turns in the primary
E_P = voltage applied to the primary
E_S = voltage induced in the secondary
N_S = number of turns in the secondary

The equation shows that the ratio of secondary voltage to primary voltage is equal to the ratio of secondary turns to primary turns. The equation can be written as:

$$E_P N_S = E_S N_P$$

For example, a transformer has 100 turns in the primary, 50 turns in the secondary, and 120VAC applied to the primary (E_P). What is the voltage across the secondary (E_S)?

$$\frac{E_S}{E_P} = \frac{N_S}{N_P} \quad or \quad E_S = \frac{E_P N_S}{N_P}$$

$$E_S = \frac{120V \times 50 \text{ turns}}{100 \text{ turns}} = 60VAC$$

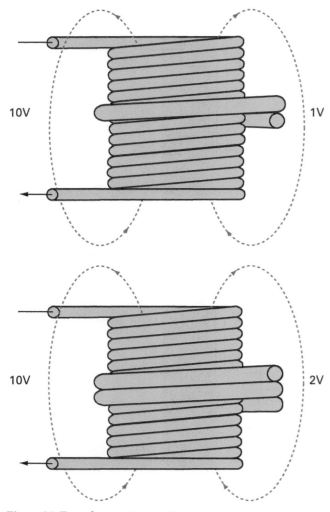

Figure 36 Transformer turns ratio.

The transformers in *Figure 36* have fewer turns in the secondary than in the primary. As a result, there is less voltage across the secondary than across the primary. A transformer in which the voltage across the secondary is less than the voltage across the primary is called a step-down transformer. The ratio of a 10-to-1 step-down transformer is written as 10:1.

A transformer that has fewer turns in the primary than in the secondary will produce a greater voltage across the secondary than the voltage applied to the primary. A transformer in which the voltage across the secondary is greater than the voltage applied to the primary is called a step-up transformer. The ratio of a 1-to-4 step-up transformer should be written 1:4. Notice in the two ratios that the value of the primary winding is always stated first.

4.4.0 Types of Transformers

Transformers are widely used to permit the use of trip coils and instruments of moderate current and voltage capacities and to measure the characteristics of high-voltage and high-current circuits. Since secondary voltage and current are directly related to primary voltage and current, measurements can be made under the low-voltage or low-current conditions of the secondary circuit and still determine primary characteristics. Tripping transformers and instrument transformers are examples of this use of transformers.

The primary or secondary coils of a transformer can be tapped to permit multiple input and output voltages. *Figure 37* shows several tapped transformers. The center-tapped transformer is particularly important because it can be used in conjunction with other components to convert an AC input to a DC output.

4.4.1 Isolation Transformer

Isolation transformers are wound so that their primary and secondary voltages are equal. Their purpose is to electrically isolate a piece of electrical equipment from the power distribution system.

Many pieces of electronic equipment use the metal chassis on which the components are mounted as part of the circuit (*Figure 38*). Personnel working with this equipment may accidentally come in contact with the chassis, completing the circuit to ground, and receive a shock. If the resistances of their body and the ground path are low, the shock can be fatal. Placing an isolation transformer in the circuit breaks the ground current path that includes the worker. Current can no longer flow from the power supply through the chassis and worker to ground; however, the equipment is still supplied with the normal operating voltage and current.

4.4.2 Autotransformer

In a transformer, it is not necessary for the primary and secondary to be separate and distinct windings. *Figure 39* is a schematic diagram of what is known as an autotransformer. Note that a single coil of wire is tapped to produce what is electrically both a primary and a secondary winding.

The voltage across the secondary winding has the same relationship to the voltage across the primary that it would have if they were two distinct windings. The movable tap in the secondary is used to select a value of output voltage either higher or lower than E_P, within the range of the transformer. When the tap is at Point A, E_S is less than E_P; when the tap is at Point B, E_S is greater than E_P.

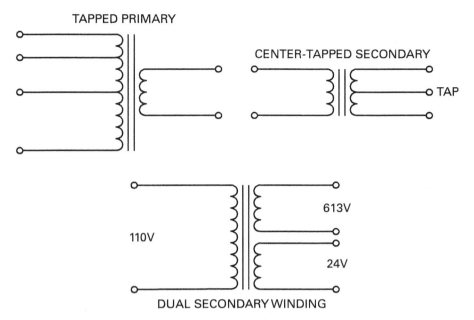

Figure 37 Tapped transformers.

Autotransformers rely on self-induction to induce their secondary voltage. The term *autotransformer* can be broken down into two words: auto, meaning self; and transformer, meaning to change potential. The autotransformer is made of one winding that acts as both a primary and a secondary winding. It may be used as either a step-up or step-down transformer. Some common uses of autotransformers are as variable AC voltage supplies and fluorescent light ballast transformers, and to reduce the line voltage for various types of low-voltage motor starters.

4.4.3 Current Transformer

A current transformer differs from other transformers in that the primary is a conductor to the load and the secondary is a coil wrapped around the wire to the load. Just as any ammeter is connected in line with a circuit, the current transformer is connected in series with the current to be measured. *Figure 40* is a diagram of a current transformer.

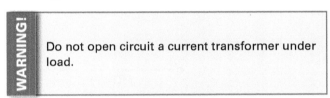

WARNING! Do not open circuit a current transformer under load.

Think About It
Turns and Voltage Ratios

What is the magnitude of the voltage and current supplied by the secondary of the transformer in the circuit shown below?

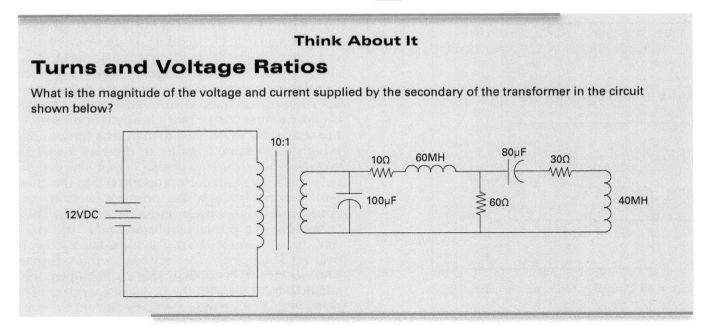

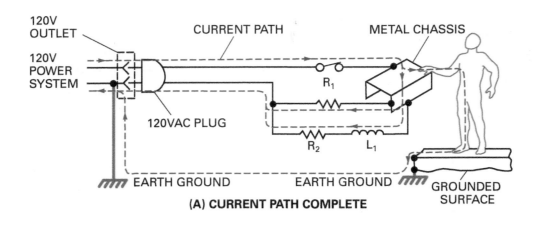

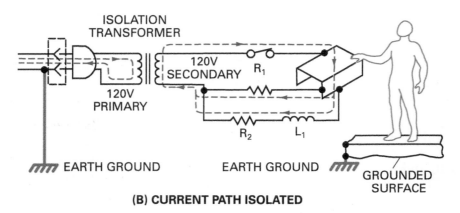

Figure 38 Importance of an isolation transformer.

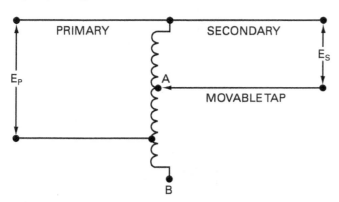

Figure 39 Autotransformer schematic diagram.

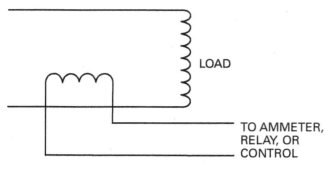

Figure 40 Current transformer schematic diagram.

Because current transformers are series transformers, the usual voltage and current relationships do not apply. Current transformers vary considerably in rated primary current, but are usually designed with ampere-turn ratios such that the secondary delivers five amperes at full primary load.

Current transformers are generally constructed with only a few turns or no turns in the primary. The voltage in the secondary is induced by the changing magnetic field that exists around a single conductor. The secondary is wound on a circular core, and the large conductor that makes up the primary passes through the hole in its center. Because the primary has few or no turns, the secondary must have many turns (providing a high turns ratio) in order to produce a usable voltage. The advantage of this is that you get an output off the secondary proportional to the current flowing through the primary, without an appreciable voltage drop across the primary. This is because the primary voltage equals the current times the impedance. The impedance is kept near zero by using no or very few primary turns. The disadvantage is that you cannot open the secondary circuit with the primary energized. To do so would cause the secondary current to drop rapidly to zero. This would cause the magnetic field generated by the secondary current to collapse rapidly. The rapid collapse of the secondary field through the many turns of the secondary winding would induce a dangerously high voltage in the secondary, creating an equipment and personnel hazard.

Because the output of current transformers is proportional to the current in the primary, they are most often used to power current-sensing meters and relays. This allows the instruments to respond to primary current without having to handle extreme magnitudes of current.

4.4.4 Potential Transformer

The primary of a potential transformer is connected across or in parallel with the voltage to be measured, just as a voltmeter is connected across a circuit. *Figure 41* shows the schematic diagram for a potential transformer.

Potential transformers are basically the same as any other single-phase transformer. Although primary voltage ratings vary widely according to the specific application, secondary voltage ratings are usually 120V, a convenient voltage for meters and relays.

Isolation Transformers

In addition to being used to protect personnel from receiving electrical shocks, shielded isolation transformers are widely used to prevent electrical disturbances on power lines from being transmitted into related load circuits. The shielded isolation transformer has a grounded electrostatic shield between the primary and secondary windings that acts to direct unwanted signals to ground.

Because the output of potential transformers is proportional to the phase-to-phase voltage of the primary, they are often used to power voltage-sensing meters and relays. This allows the instruments to respond to primary voltage while having to handle only 120V. Also, potential transformers are essentially single-phase step-down transformers. Therefore, power to operate low-voltage auxiliary equipment associated with high-voltage switchgear can be supplied off the high-voltage lines that the equipment serves via potential transformers.

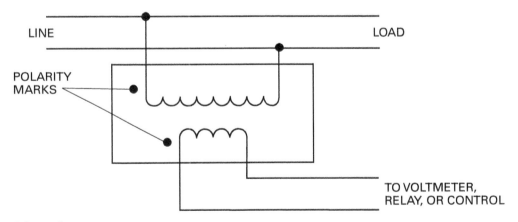

Figure 41 Potential transformer.

Case History

AC Power

Working with AC power can be dangerous unless proper safety methods and procedures are followed. The National Institute for Occupational Safety and Health (NIOSH) investigated 224 incidents of electrocutions that resulted in occupational fatalities. One hundred twenty-one of the victims were employed in the construction industry. Two hundred twenty-one of the incidents (99%) involved AC. Of the 221 AC electrocutions, 74 (33%) involved AC voltages less than 600V and 147 (66%) involved 600V or more. Forty of the lower-voltage electrocutions involved 120/240V.

Factors relating to the causes of these electrocutions included the lack of enforcement of existing employer policies including the use of personal protective equipment and the lack of supervisory intervention when existing policies were being violated. Of the 224 victims, 194 (80%) had some type of electrical safety training. Thirty-nine victims had no training at all. It is notable that 100 of the victims had been on the job less than one year.

The Bottom Line: Never assume that you are safe when working at lower voltages. All voltage levels must be considered potentially lethal. Also, safety training does no good if you don't put it into practice every day. Always put safety first.

Think About It
Putting It All Together

A power company's distribution system has capacitor banks that are automatically switched into the system by a temperature switch during hot weather. Why?

Current Transformers

Although the use of a current transformer completely isolates the secondary and the related ammeter from the high-voltage lines, the secondary of a current transformer should never be left open circuited. To do so may result in dangerously high voltage being induced in the secondary.

Potential Transformers

In addition to being used to step down high voltages for the purpose of safe metering, potential transformers are widely used in all kinds of control devices where the condition of high voltages must be monitored. One such example involves the use of a potential transformer-operated contactor in emergency lighting standby generator circuits. Under normal conditions with utility power applied, the contactor is energized and its normally closed contacts are open. If the power fails, the contactor de-energizes, causing its contacts to close and activating the standby generator circuit.

4.0.0 Section Review

1. Which of the following is *true* regarding transformer cores?

    a. Air-core transformers are used when the voltage source has a frequency below 20kHz.
    b. An iron-core transformer provides better power transfer than an air-core transformer.
    c. Laminated sheets of steel are used to increase eddy currents.
    d. The most efficient transformers have the fewest laminations.

2. On a transformer schematic, dots are used to indicate _____.

    a. switches
    b. polarity
    c. number of turns
    d. core type

3. A transformer has 200 turns in the primary, 100 turns in the secondary, and 120VAC applied to the primary (E_P). The voltage across the secondary (E_S) is _____.

    a. 480VAC
    b. 240VAC
    c. 120VAC
    d. 60VAC

4. A voltage-sensing relay is likely to be powered by a(n) _____.

    a. isolation transformer
    b. potential transformer
    c. autotransformer
    d. current transformer

Review Questions

1. An electric current always produces $\underline{1.0.0}$.
   a. mutual inductance
   b. a magnetic field
   c. capacitive reactance
   d. high voltage

2. The number of cycles an alternating electric current undergoes per second is known as $\underline{1.1.1}$.
   a. amperage
   b. frequency
   c. voltage
   d. resistance

3. What is the peak voltage in a circuit with an rms voltage of 120VAC?
   a. 117 volts
   b. 120 volts
   c. 150 volts
   d. 170 volts

   1.414(120)
   ↑
   rms

4. Which of the following conditions exist in a circuit of pure resistance?
   a. The voltage and current are in phase.
   b. The voltage and current are 90° out of phase.
   c. The voltage and current are 120° out of phase.
   d. The voltage and current are 180° out of phase.

5. In a purely resistive AC circuit where 240V is applied across a 10-ohm resistor, the amperage is $\underline{2.1.0}$.
   a. 10A
   b. 24A
   c. 60A
   d. 120A

   240/10

6. When the current increases in an AC circuit, what role does inductance play?
   a. It increases the current.
   b. It plays no role at all.
   c. It causes the overcurrent protection to open.
   d. It reduces the current.

7. Which of the following coils is likely to show the lowest inductance?
   a. A coil with many turns
   b. A coil with a magnetic core
   c. A coil with a narrow coil
   d. A coil with a short core

8. Which of the following conditions exist in a circuit of pure inductance?
   a. The voltage and current are in phase.
   b. The voltage and current are 90° out of phase.
   c. The voltage and current are 120° out of phase.
   d. The voltage and current are 180° out of phase.

9. Reactance increases with an increase in $\underline{2.2.3}$.
   a. capacitance
   b. inductance
   c. magnetism
   d. resistance

10. Capacitance is measured in $\underline{2.3.0}$
    a. farads
    b. joules
    c. henrys
    d. amps

11. The total capacitance of two 15µF capacitors connected in parallel is _____.
    a. 5µF
    b. 10µF
    c. 15µF
    d. 30µF

    15+15

12. The total capacitance of two 15µF capacitors in series is $\underline{2.3.2}$.
    a. 5µF
    b. 7.5µF
    c. 15µF
    d. 30µF

    15/2

13. The total capacitance of two 5µF capacitors and one 10µF capacitor in parallel is $\underline{2.3.2}$
    a. 5µF
    b. 15µF
    c. 20µF
    d. 25µF

14. The opposition to current flow offered by the capacitance of a circuit is known as 2.3.5.
    a. mutual inductance
    b. pure resistance
    c. inductive reactance
    d. capacitive reactance

15. The total opposition to current flow in an AC circuit is known as 2.4.0.
    a. resistance
    b. capacitive reactance
    c. inductive reactance
    d. impedance

16. The true power in a circuit with 4A through a 50Ω resistor is _____. $4^2(50)$
    a. 200W
    b. 400W
    c. 600W
    d. 800W

17. The power factor in a circuit with a resistance of 50Ω and an impedance of 100Ω is 3.4.0.
    a. 0.25
    b. 0.5   50/100
    c. 0.75
    d. 1.0

18. A power factor is not an angular measure, but a numerical ratio with a value between 0 and 1, equal to the 3.4.0.
    a. sine of the phase angle
    b. tangent of the phase angle
    c. cosine of the phase angle
    d. cotangent of the phase angle

19. The two windings of a conventional transformer are known as the 4.1.2.
    a. mutual and inductive windings
    b. high and low voltage windings
    c. primary and secondary windings
    d. step-up and step-down windings

20. A transformer with 10 turns in the primary and two turns in the secondary has a turns ratio of 4.3.0.
    a. 5:1   10:2
    b. 8:1   5:1
    c. 10:1
    d. 20:1

Supplemental Exercises

1. In a circuit containing resistance only, the voltage and current are _____ with one another.
2. The power consumed by _____ is called true power.
3. The power factor is the ratio of _____ power to _____ power.
4. LC circuits consist of a(n) _____ and a(n) _____ connected in series or in parallel with a voltage source.
5. As in all series circuits, the _____ in a series LC circuit is the same at all points.
6. On a transformer with 100 volts applied to the primary, 200 turns in the primary, and 100 turns in the secondary, the secondary voltage is _____ volts.
7. A(n) _____ can be used to electrically isolate a piece of equipment from the power distribution system.
8. Reactive power is the portion of apparent power that is caused by _____ and _____ in the circuit.
9. The primary winding in a transformer is connected to the _____.
10. Apparent power is the product of the source _____ and the _____.
11. The secondary winding in a transformer is connected to the _____.
12. True power is measured in _____.
13. VAR is the unit of measurement for _____ power.
14. True or False? The average value is calculated from all the values in a sine wave for only half a cycle.
15. The average value is = _____ x peak value.
16. In purely resistive circuits, the current and voltage are related by Ohm's law, which is E = IR. Define the variables.
    E = _____
    I = _____
    R = _____

Trade Terms Introduced in This Module

Capacitance: The storage of electricity in a capacitor; capacitance produces an opposition to voltage change. The unit of measurement for capacitance is the farad (F) or microfarad (µF).

Frequency: The number of cycles an alternating electric current, sound wave, or vibrating object undergoes per second.

Hertz (Hz): A unit of frequency; one hertz equals one cycle per second.

Impedance: The opposition to current flow in an AC circuit; impedance includes resistance (R), capacitive reactance (X_C), and inductive reactance (X_L). Impedance is measured in ohms (Ω).

Inductance: The creation of a voltage due to a time-varying current; also, the opposition to current change, causing current changes to lag behind voltage changes. The unit of measure for inductance is the henry (H).

Micro (µ): Prefix designating one-millionth of a unit. For example, one microfarad is one-millionth of a farad.

Peak voltage: The peak value of a sinusoidally varying (cyclical) voltage or current is equal to the root-mean-square (rms) value multiplied by the square root of two (1.414). AC voltages are usually expressed as rms values; that is, 120 volts, 208 volts, 240 volts, 277 volts, 480 volts, etc., are all rms values. The peak voltage, however, differs. For example, the peak value of 120 volts (rms) is actually $120 \times 1.414 = 170$ volts.

Radian: An angle at the center of a circle, subtending (opposite to) an arc of the circle that is equal in length to the radius.

Reactance: The opposition to alternating current (AC) due to capacitance (X_C) and/or inductance (X_L).

Root-mean-square (rms): The square root of the average of the square of the function taken throughout the period. The rms value of a sinusoidally varying voltage or current is the effective value of the voltage or current.

Self-inductance: A magnetic field induced in the conductor carrying the current.

Additional Resources

This module presents thorough resources for task training. The following resource material is suggested for further study.

Principles of Electric Circuits, Thomas L. Floyd. Latest Edition. New York, NY: Pearson Education, Inc.

Figure Credits

Greenleee / A Textron Company, Module Opener

Section Review Answer Key

Section 1.0.0

Answer	Section Reference	Objective
1. b*	1.1.1	1a
2. d	1.2.1	1b
3. d	1.3.0	1c

Section 2.0.0

Answer	Section Reference	Objective
1. c*	2.1.0	2a
2. d	2.2.1	2b
3. d*	2.3.2	2c
4. d	2.4.0	2d

Section 3.0.0

Answer	Section Reference	Objective
1. a	3.1.0	3a
2. c	3.2.0	3b
3. a	3.3.0	3c
4. c*	3.4.0	3d
5. b	3.5.0	3e

Section 4.0.0

Answer	Section Reference	Objective
1. b	4.1.1	4a
2. b	4.2.2	4b
3. d*	4.3.0	4c
4. b	4.4.4	4d

*Calculations for these answers are provided on the following page(s).

Section Review Calculations

1.0.0 SECTION REVIEW

Question 1

Divide the number of cycles (times the sine wave is repeated) by the amount of time in seconds to determine the frequency in Hz (cycles per second):

$$f = \frac{10 \text{ cycles}}{0.5 \text{ seconds}} = 20 \text{ cycles per second (Hz)}$$

The frequency is **20Hz**.

2.0.0 SECTION REVIEW

Question 1

Use Ohm's law to find the voltage (solve for E):

$E = IR$
$E = 22A \times 10\Omega$
$E = 220V$

The voltage is **220V**.

Question 3

Add the capacitance values to find their combined capacitance in series:

$C_T = C_1 + C_2 + C_3 + C_4$
$C_T = 10\mu F + 10\mu F + 10\mu F + 5\mu F$
$C_T = 35\mu F$

The total capacitance is **35μF**.

3.0.0 SECTION REVIEW

Question 4

Divide the resistance by the impedance to find the power factor:

$$pf = \frac{R}{Z} = \frac{75\Omega}{150\Omega} = 0.5$$

The power factor is **0.50**.

4.0.0 Section Review

Question 3

Use the formula for finding the voltage across the secondary:

$N_S = 100$ turns

$$\frac{E_S}{E_P} = \frac{N_S}{N_P} \quad or \quad E_S = \frac{E_P N_S}{N_P}$$

$$E_S = \frac{120V \times 100 \text{ turns}}{200 \text{ turns}} = 60VAC$$

The voltage across the secondary is **60VAC**.

This page is intentionally left blank.

NCCER CURRICULA — USER UPDATE

NCCER makes every effort to keep its textbooks up-to-date and free of technical errors. We appreciate your help in this process. If you find an error, a typographical mistake, or an inaccuracy in NCCER's curricula, please fill out this form (or a photocopy), or complete the online form at **www.nccer.org/olf**. Be sure to include the exact module ID number, page number, a detailed description, and your recommended correction. Your input will be brought to the attention of the Authoring Team. Thank you for your assistance.

Instructors – If you have an idea for improving this textbook, or have found that additional materials were necessary to teach this module effectively, please let us know so that we may present your suggestions to the Authoring Team.

NCCER Product Development and Revision
13614 Progress Blvd., Alachua, FL 32615

Email: curriculum@nccer.org
Online: www.nccer.org/olf

❏ Trainee Guide ❏ Lesson Plans ❏ Exam ❏ PowerPoints Other _____

Craft / Level: _____ Copyright Date: _____

Module ID Number / Title: _____

Section Number(s): _____

Description: _____

_____

_____

_____

_____

Recommended Correction: _____

_____

_____

_____

_____

Your Name: _____

Address: _____

_____

Email: _____ Phone: _____

This page is intentionally left blank.

Motors: Theory and Application

Overview

The electric motor is the workhorse of modern industry. Its functions are almost unlimited. To control the motors that drive machinery and equipment, we must have electrical supply circuits that perform certain functions. They must provide electrical current to cause the motor to operate in the manner needed to make it perform its intended function. This module describes AC and DC motors, including their components, circuits, and connections.

Module 26202-20

Trainees with successful module completions may be eligible for credentialing through the NCCER Registry. To learn more, go to **www.nccer.org** or contact us at 1.888.622.3720. Our website, **www.nccer.org**, has information on the latest product releases and training.

Your feedback is welcome. You may email your comments to **curriculum@nccer.org**, send general comments and inquiries to **info@nccer.org**, or fill in the User Update form at the back of this module.

This information is general in nature and intended for training purposes only. Actual performance of activities described in this manual requires compliance with all applicable operating, service, maintenance, and safety procedures under the direction of qualified personnel. References in this manual to patented or proprietary devices do not constitute a recommendation of their use.

Copyright © 2020 by NCCER, Alachua, FL 32615, and published by Pearson, New York, NY 10013. All rights reserved. Printed in the United States of America. This publication is protected by Copyright, and permission should be obtained from NCCER prior to any prohibited reproduction, storage in a retrieval system, or transmission in any form or by any means, electronic, mechanical, photocopying, recording, or likewise. To obtain permission(s) to use material from this work, please submit a written request to NCCER Product Development, 13614 Progress Blvd., Alachua, FL 32615.

26202-20 V10.0

From *Electrical, Trainee Guide*. NCCER.
Copyright © 2020 by NCCER. Published by Pearson. All rights reserved.

26202-20
MOTORS: THEORY AND APPLICATION

Objectives

Upon completion of this module, you will be able to do the following:

1. Identify direct current (DC) motors and describe their operating characteristics.
   a. Understand how DC motors operate.
   b. Identify types of DC motors.
2. Identify alternating current (AC) motors and describe their operating characteristics.
   a. Understand how AC motors operate.
   b. Identify three-phase induction motors.
   c. Identify synchronous motors.
   d. Identify single-phase induction motors.
3. Identify variable-speed drives and describe their operating characteristics.
   a. Identify types of adjustable speed loads.
   b. Identify types of motor speed control.
   c. Identify braking methods.
4. Identify motor enclosures, frame designations, and operating characteristics.
   a. Identify types of motor enclosures.
   b. Identify NEMA frame designations.
   c. Identify motor operating characteristics using nameplate data.
5. Identify the connections and terminal markings for AC motors.
   a. Identify the terminals of wye-connected motors.
   b. Identify the terminals of delta-connected motors.
6. Identify the *NEC*® requirements for motors.
   a. Identify *NEC*® installation requirements.
   b. Identify *NEC*® motor protection requirements.

Performance Tasks

Under the supervision of the instructor, you should be able to do the following:

1. Identify various types of motors and their application(s).
2. Collect data from a motor nameplate.
3. Connect the terminals for a dual-voltage motor.

Trade Terms

Armature
Branch circuits
Brush
Circuit breaker
Commutator
Continuous duty
Controller
Delta-connected motor
Duty
Equipment
Field poles
Horsepower

Hours
Intermittent duty
Overcurrent
Overload
Periodic duty
Revolutions per minute (rpm)
Rotation
Synchronous speed
Thermal protector
Varying duty
Wye-connected motor

Industry Recognized Credentials

If you are training through an NCCER-accredited sponsor, you may be eligible for credentials from NCCER's Registry. The ID number for this module is 26202-20. Note that this module may have been used in other NCCER curricula and may apply to other level completions. Contact NCCER's Registry at 888.622.3720 or go to **www.nccer.org** for more information.

> **NOTE**
> NFPA 70®, *National Electrical Code*® and *NEC*® are registered trademarks of the National Fire Protection Association, Quincy, MA.

Contents

- 1.0.0 DC Motor Operation 1
  - 1.1.0 How DC Motors Operate ....
    - 1.1.1 DC Motor Components 2
    - 1.1.2 The Neutral Plane 5
    - 1.1.3 Armature Reaction 7
    - 1.1.4 Counter-Electromotive Force (CEMF) 7
    - 1.1.5 Starting Resistance 9
  - 1.2.0 DC Motor Types 10
    - 1.2.1 Shunt DC Motors 11
    - 1.2.2 Series DC Motors 12
    - 1.2.3 Compound DC Motors 13
- 2.0.0 AC Motor Operation 15
  - 2.1.0 How AC Motors Operate 15
    - 2.1.1 Rotating Fields 15
    - 2.1.2 Rotor Behavior in a Rotating Field 15
    - 2.1.3 Induction 17
    - 2.1.4 Torque 17
    - 2.1.5 Slip 17
    - 2.1.6 Starting Current 18
    - 2.1.7 Loaded Torque 19
    - 2.1.8 Overload Condition 19
    - 2.1.9 Power Factor 19
    - 2.1.10 Reversing Rotation 19
  - 2.2.0 Three-Phase Induction Motors 21
    - 2.2.1 Squirrel Cage Induction Motor 21
    - 2.2.2 Wound-Rotor Induction Motor 23
  - 2.3.0 Synchronous Motors 24
    - 2.3.1 Operating Characteristics 24
    - 2.3.2 Construction 24
    - 2.3.3 Principles of Operation 25
    - 2.3.4 Rotor Field Excitation 25
    - 2.3.5 Synchronous Motor Pullout 27
    - 2.3.6 Synchronous Motor Torque Angle 27
  - 2.4.0 Single-Phase Induction Motors 28
    - 2.4.1 Single-Phase Induction Motors 28
    - 2.4.2 Split-Phase Induction Motor 28
    - 2.4.3 Capacitor-Type Induction Motor 29
    - 2.4.4 Shaded-Pole Induction Motor 33
    - 2.4.5 Single-Phase Synchronous Motor 33
    - 2.4.6 Multiple-Speed Induction Motors 35

3.0.0	Variable-Speed Drives	39
3.1.0	Adjustable Speed Loads	39
3.1.1	Typical Torque-Speed Curves	40
3.1.2	Motor Heating	40
3.2.0	Motor Speed Control	40
3.2.1	Varying the Speed of a DC Shunt Motor	42
3.2.2	Varying the Speed of an AC Motor	43
3.3.0	Braking Methods	44
3.3.1	DC Injection Braking	44
3.3.2	Dynamic Braking	44
4.0.0	Motor Enclosures, Frame Designations, and Operating Characteristics	46
4.1.0	Motor Enclosures	46
4.1.1	Open Motor	46
4.1.2	Enclosed Motor	47
4.2.0	NEMA Frame Designations	48
4.2.1	Small Machines	48
4.2.2	Medium Machines	49
4.3.0	Motor Ratings and Nameplate Data	50
4.3.1	Voltage	50
4.3.2	Amperage	51
4.3.3	Speed	51
4.3.4	Horsepower	52
4.3.5	NEMA Design Letters	52
4.3.6	Insulation Class	53
4.3.7	Frequency	54
4.3.8	Service Factor	54
4.3.9	NEMA kVA Code Letters	54
4.3.10	Bearings	54
4.3.11	Power Factor	55
4.3.12	Duty Rating	55
5.0.0	Connections and Terminal Markings for AC Motors	57
5.1.0	Wye-Connected Motor Terminals	58
5.2.0	Delta-Connected Motor Terminals	58
6.0.0	*NEC*® Requirements for Motors	61
6.1.0	Motor Installation	61
6.2.0	Motor Protection	64
6.2.1	Thermal Protectors	65
6.2.2	Branch Considerations	66

Figures and Tables

Figure 1 Magnetic field 2
Figure 2 Basic motor action 2
Figure 3 Motor action 2
Figure 4 Torque 3
Figure 5 Single-loop armature DC motor 4
Figure 6 Brushes, brush rigging, and commutator connections 4
Figure 7 Self-starting motor 5
Figure 8 Neutral plane 5
Figure 9 Neutral plane in a single-loop DC motor 6
Figure 10 Two-loop armature DC motor 7
Figure 11 Armature reaction 8
Figure 12 Interpoles 8
Figure 13 Counter-electromotive force (CEMF) 9
Figure 14 No CEMF 10
Figure 15 Shunt motor 10
Figure 16 Typical DC motor 11
Figure 17 Shunt DC motor 11
Figure 18 Operating characteristics of a typical DC shunt motor ... 12
Figure 19 Series DC motor 12
Figure 20 Operating characteristics of a typical series motor 13
Figure 21 Long and short shunts 13
Figure 22 Operating characteristics of a typical DC compound motor 14
Figure 23 AC generation 16
Figure 24 Typical torque-speed curve 18
Figure 25 Torque and current curves 20
Figure 26 Power factor versus load for an induction motor 20
Figure 27 Three-phase induction motor rotational direction change 21
Figure 28 Producing torque 21
Figure 29 Main components of an induction motor 22
Figure 30 Squirrel cage rotor 22
Figure 31 Wound rotor 23
Figure 32 Wound-rotor motor circuit 23
Figure 33 Synchronous motor operation at start 25
Figure 34 Simplification of a synchronous motor 25
Figure 35 Pole assembly 26
Figure 36 Simplified synchronous motor excitation circuit 26
Figure 37 Torque angle 27
Figure 38 AC induction motor 28
Figure 39 Split-phase motor 29
Figure 40 Capacitor motor 30
Figure 41 Capacitor-start motor schematic 30
Figure 42 Torque-slip curves 32
Figure 43 Capacitor-start, capacitor-run motor schematic 32
Figure 44 Four-pole shaded-pole motor 33
Figure 45 Two-pole shaded-pole motor 34

Figure 46	Two-winding, two-speed motor	36
Figure 47	High-speed consequent-pole motor	36
Figure 48	Low-speed consequent-pole motor	36
Figure 49	Single-phase consequent-pole motor	37
Figure 50	Types of adjustable speed loads	39
Figure 51	Electric drive operation in four quadrants	40
Figure 52	Four-quadrant operation for a squirrel cage motor	41
Figure 53	Four-quadrant operation for a DC motor	41
Figure 54	DC shunt motor schematic	42
Figure 55	End view of a foot-mounted motor	48
Figure 56	Lettering of dimension sheets for foot-mounted machines (side view)	48
Figure 57	Nameplate data	50
Figure 58	High-voltage and low-voltage connection diagrams shown on motor nameplate	51
Figure 59	Nameplate showing amperage, speed, horsepower, and bearings	52
Figure 60	Nameplate showing NEMA design letter and insulation class	53
Figure 61	Dual-voltage, three-phase wye connection	57
Figure 62	Dual-voltage, three-phase delta connection	57
Figure 63	Coil arrangement in a wye-connected motor	58
Figure 64	Battery hookup for wye-connected motor lead identification	58
Figure 65	Coil arrangement in a delta-connected motor	59
Figure 66	Battery hookup for delta-connected motor lead identification	59
Figure 67	Summary of requirements for motors, motor circuits, and controllers	63

Table 1	Motor Enclosure Types	47
Table 2	Frame Dimension Chart	49
Table 3	Induction Motor Voltages	51
Table 4	Motor Operation	52
Table 5	Typical Currents for 220V, 60-Cycle Squirrel Cage Motors	53
Table 6	Locked-Rotor Indicating Code Letters [Data from *NEC Table 430.7(B)*]	55
Table 7	Summary of *NEC®* Requirements for Motor Installations	62
Table 8	Motor Protection Devices (Data from *NEC Table 430.52*)	65
Table 9	Duty Cycle Service [Data from *NEC Table 430.22(E)*]	66

SECTION ONE

1.0.0 DC Motor Operation

Objective

Identify direct current (DC) motors and describe their operating characteristics.
a. Understand how DC motors operate.
b. Identify types of DC motors.

Performance Task

1. Identify various types of motors and their application(s).

Trade Terms

Armature: The rotating windings of a DC motor.

Brush: A conductor between the stationary and rotating parts of a machine. It is usually made of carbon.

Commutator: A device used on electric motors or generators to maintain a unidirectional current.

Controller: A device that serves to govern, in some predetermined manner, the electric power delivered to the apparatus to which it is connected.

Equipment: A general term including material, fittings, devices, appliances, fixtures, apparatus, and the like used as a part of, or in connection with, an electrical installation.

Field poles: The stationary portion of a DC motor that produces the magnetic field.

Horsepower: The rated output capacity of the motor. It is based on breakdown torque, which is the maximum torque a motor will develop without an abrupt drop in speed.

Revolutions per minute (rpm): The approximate full-load speed at the rated power line frequency. The speed of a motor is determined by the number of poles in the winding.

Rotation: For single-phase motors, the standard rotation, unless otherwise noted, is counterclockwise facing the lead or opposite shaft end. All motors can be reconnected at the terminal board for opposite rotation unless otherwise indicated.

The electric motor is the workhorse of modern industry. Its functions are almost unlimited. To control the motors that drive machinery and equipment, we must have electrical supply circuits that perform certain functions. They must provide electrical current to cause the motor to operate in the manner needed to make it perform its intended function. They must also provide protection for the motor from adverse mechanical and electrical conditions. These functions are frequently combined within electrical equipment that we classify as motor control centers.

A thorough understanding of the functions of the various components of a motor control center is desirable from both a maintenance and a troubleshooting standpoint. Properly maintained motor control centers ensure a minimum of downtime for unscheduled repairs, increase productivity, and contribute to a safer working environment.

1.1.0 How DC Motors Operate

When a bar of magnetic material is given an induced magnetic charge, a field of magnetic force is developed around the bar. Picture this field as consisting of magnetic lines of force (flux) that exist in the space surrounding the bar. These lines of force appear to leave one end of the bar and extend outside the bar to the opposite end.

The end of the bar magnet where magnetic lines appear to start is called the north pole of the magnet, while the end where these lines reenter the magnet is called the south pole. Actually, the lines extend inside the magnet from the south pole to the north pole, completing a closed loop. This principle is shown in *Figure 1*.

There are several characteristics of these magnetic lines of force that must be remembered when dealing with electric motors. They are the following:

- Magnetic lines of force are continuous and always form closed loops.
- Magnetic lines of force do not cross.
- Magnetic lines of force with polarities in the same direction repel each other. In other words, a north pole will repel a north pole and a south pole will repel a south pole.
- Magnetic lines of force having polarities in opposite directions tend to attract each other and combine. In other words, a south pole will be attracted by a north pole and vice versa.

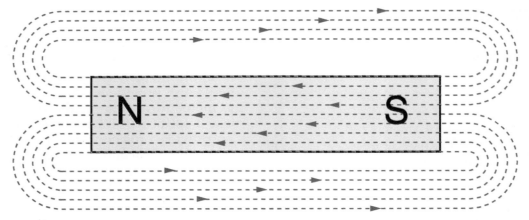

Figure 1 Magnetic field.

- Magnetic lines of force tend to shorten themselves. Therefore, the magnetic lines of force existing between two unlike poles cause the poles to tend to pull together.
- Magnetic lines of force pass through all known materials, magnetic or nonmagnetic. Some materials provide a much easier path for these lines than others. These materials have high permeability and low reluctance. (Reluctance is discussed later in this module.)

1.1.1 DC Motor Components

A DC motor consists of a few major components, each with a specific purpose in the motor's operation.

The armature is a movable electromagnet located between the poles of another fixed permanent (field) magnet, as shown in *Figure 2*. The magnetic field from the armature conductors interacts with the magnetic field from the field magnet. The result of the field interaction is motor action.

Current in a conductor also has its associated magnetic field. When a conductor is placed in another magnetic field from a separate source, the

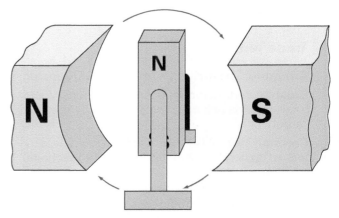

Figure 2 Basic motor action.

two fields can react to produce motor action. The conductor must be perpendicular to the magnetic field, as illustrated in *Figure 3*. This way, the perpendicular magnetic field of the current is in the same plane as the external magnetic field.

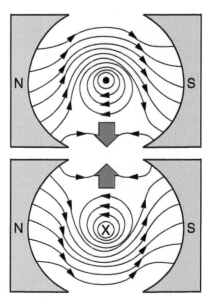

Figure 3 Motor action.

Did You Know?
Early Electric Motors

The first US patent for a motor was issued to Thomas Davenport in 1837. He reported that he used silk from his wife's wedding gown as insulation for the conductors, but despite this sacrifice, his motor was not commercially successful. Practical electric motors, like the practical light bulb, did not appear until the late 19th century.

Unless the two fields are in the same plane, they cannot affect each other. In the same plane, however, lines of force in the same direction reinforce to make a stronger field, while lines in the opposite direction cancel and result in a weaker field. The stronger field tends to move the conductor toward the weaker field, as illustrated in *Figure 3*. These directions are summarized as follows:

- With the conductor at 90°, or perpendicular to the external field, the reaction between the two magnetic fields is at its maximum.
- With the conductor at 0°, or parallel to the external field, there is no effect between them.
- When the conductor rise is at an angle between 0° and 90°, only the perpendicular component is effective.

In motor action, the wire moves only in a straight line and it stops moving when out of the field, even though current still exists. A practical motor must develop continuous rotary motion. To produce this, a twisting force called *torque* must be developed.

Torque is produced by mounting a loop in a fixed magnetic field. Current is applied and the flux lines along both sides of the loop interact, causing the loop to act like a lever with a force pushing on its two sides in opposite directions. This is shown in *Figure 4*.

The combined forces result in a turning force or torque because the rotor or armature is arranged to pivot on its axis. The overall turning force on the armature depends on several factors, including field strength, armature current strength, and the physical construction of the armature, especially the distance from the loop sides to the axis lines.

Because of the lever action, the forces on the sides of the armature loop will increase as the loop sides are farther from the axis; therefore, larger armatures will produce greater torques.

In the practical motor, the torque determines the energy available for doing useful work. The greater the torque, the greater the energy. If a motor does not develop enough torque to turn its load, it stalls.

To get continuous rotation, the armature must be kept moving in the same direction. This requires reversing the direction of current through the armature for every 180° of revolution. A commutator is used to provide this switching action. This is shown in *Figure 5*.

The commutator on a DC motor is a conducting ring that is split into two segments, with each segment connected to an end of the armature loop. Current enters the side of the armature closest to the south pole of the field and leaves the side closest to the north pole of the field. The interaction of the two fields produces a torque, and the armature rotates in that direction.

A brush makes contact with each segment of the commutator, providing a connection between the movable commutator and the stationary DC power source. *Figure 6* shows various brushes and commutator connections used in DC motors.

The problem of switching commutator segments in a simple single-loop motor is that when the motor stops, there is no way of predicting the position of the armature at rest. If the armature stops in a position where the commutator is in the middle of switching, the motor will not start unless you physically turn the armature.

This problem can be overcome by winding more coils on the armature and by using more commutator segments. This will produce a self-starting motor. The motor shown in *Figure 7* uses three armature coils and three commutator segments. Regardless of where the armature comes to rest, there is always a path for current that will produce torque to rotate the armature.

> **NOTE**
> The brushless DC motor was developed to eliminate commutator problems in missiles and spacecraft operating above the Earth's atmosphere. Two general types of brushless motors are in use: the inverter-induction motor and a DC motor with an electronic commutator.

Brushless DC Motors

Brushless DC motors do not produce carbon dust and are ideal in sensitive electronic applications. They are commonly used in personal computers and CD/DVD players.

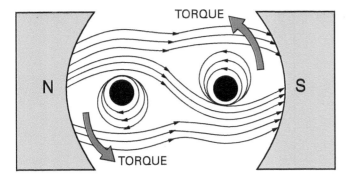

Figure 4 Torque.

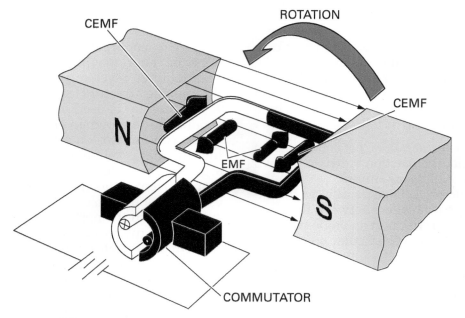

Figure 5 Single-loop armature DC motor.

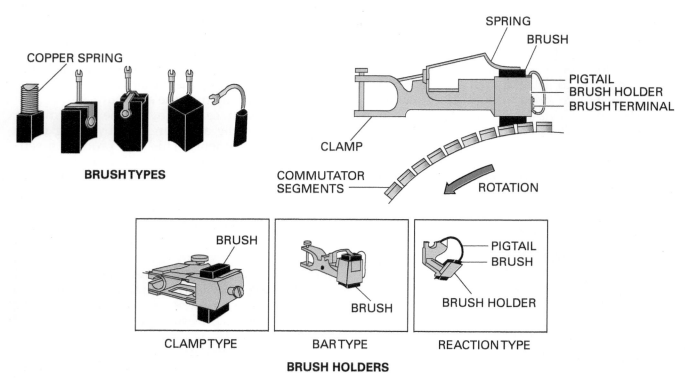

Figure 6 Brushes, brush rigging, and commutator connections.

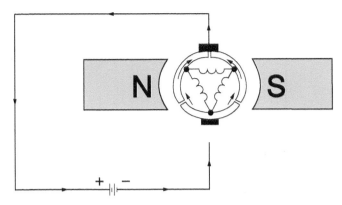

Figure 7 Self-starting motor.

1.1.2 The Neutral Plane

The armature turns when torque is produced, and torque is produced as long as the fields of the magnet and armature interact. When the loop reaches a position perpendicular to the field, the interaction of the magnetic fields stops. This position is the neutral plane, shown in *Figure 8*.

In the neutral plane, no torque is produced and the rotation of the armature should stop. However, inertia tends to keep the armature in motion even after the prime moving force is removed; thus the armature tends to rotate past the neutral plane. At the neutral position, the commutator disconnects from the brushes. When the armature goes past neutral, the sides of the loop reverse positions. The switching action of the commutator maintains the direction of current through the armature. Current still enters the armature side that is closest to the south pole.

Because the magnet's field direction remains the same throughout, the interaction of fields after commutation keeps the torque going in the original direction; thus, continuous rotation is maintained (*Figure 9*).

Although such an elementary DC motor can be built and operated, it has two serious shortcomings that prevent it from being useful: first, such a motor cannot always start by itself; second, once started, it operates very irregularly.

When the elementary DC motor runs, its operation is erratic because it produces torque irregularly. Maximum torque is produced only when the plane of the single-loop armature is parallel with the plane of the field. This is the position at right angles to the neutral plane. When the armature passes this plane of maximum torque, less and less torque is developed until it arrives at the neutral plane again. Inertia carries the armature past the neutral plane, and the motor continues to turn. Its irregularity in producing torque, however, prevents the single-loop elementary DC motor from being used for practical jobs.

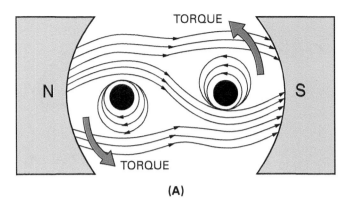
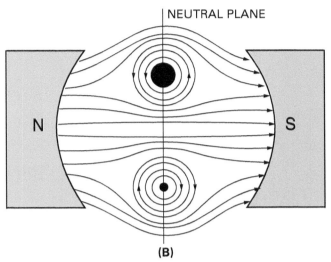

Figure 8 Neutral plane.

The basic DC motor is improved by building the armature with two or more loops. The loops are placed at right angles to each other; when one loop lies in the neutral plane, the other is in the plane of maximum torque (*Figure 10*).

In this case, the commutator is split into two pairs or four segments, with one segment associated with each end of each armature loop. This sets up two parallel loop circuits. Only one loop at a time is ever connected if power is supplied through one pair of fixed brushes to one set of ring segments.

In this multi-loop armature, the commutator serves two functions: it maintains current through the armature in the same direction at all times, and it switches power to the armature loop nearing the maximum torque position.

This motor is self-starting because at least one winding will have interaction with the main field. With this two-loop system, the torque developed is steadier and stronger but still somewhat erratic because only one loop at a time provides the torque that drives the motor.

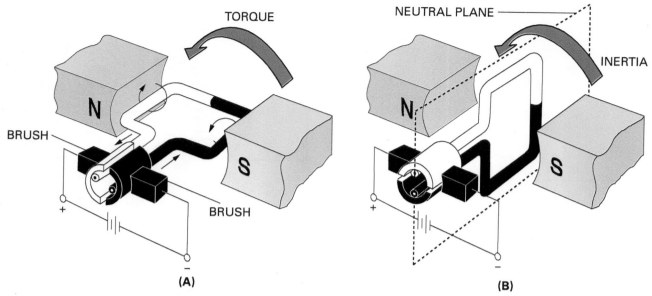

Figure 9 Neutral plane in a single-loop DC motor.

Think About It

Theory of Torque

A motor provides torque (a turning force) similar to that needed to turn a nut. In the illustration shown here, which wrench should be able to turn the nut more easily?

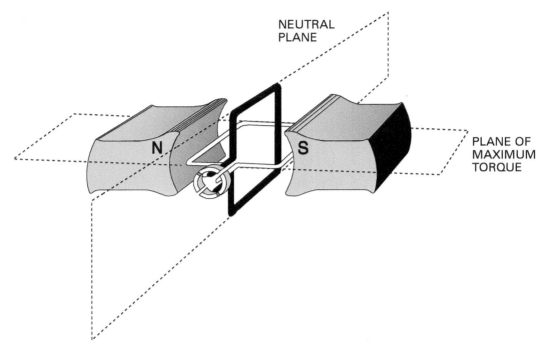

Figure 10 Two-loop armature DC motor.

1.1.3 Armature Reaction

When a motor armature is supplied with current, a magnetic flux is built up around the conductors of the armature windings. Armature reaction is caused by two magnetic fields: the main magnetic field from the field magnets and the magnetic field produced by the armature. These two fields combine to produce a new resultant magnetic field.

The resultant field is distorted and shifts opposite the main field and opposite the direction of armature rotation. This distortion shifts the neutral plane of the motor. See *Figure 11* for an illustration of armature reaction.

The amount of armature reaction determines how far the neutral plane is shifted. The amount of armature reaction depends on the amount and direction of the armature current. The concern over the neutral plane shift occurs because of the need for commutation.

Commutation, or the switching of the armature polarity, must take place at the neutral plane in order to allow the output current from the machine to remain in the same direction without arcing. When commutation takes place anywhere other than the neutral plane, it is like a switch that is opened during high current—it will draw an arc.

This armature reaction can be overcome by installing interpole windings (*Figure 12*). Interpoles are special electromagnetic pole pieces that are connected in series with the armature winding. The armature current causes a magnetic field to form around the windings. Their action is self-regulating, and the interpole field will apply the proper amount of cancellation field for any set of conditions. For a high armature reaction, the canceling field is strong. For a low armature reaction, the canceling field is weaker.

1.1.4 Counter-Electromotive Force (CEMF)

When a DC motor is in operation, it acts much like a DC generator. A magnetic field is produced by the **field poles**, and a loop of wire in the armature turns and cuts this magnetic field. To understand counter-electromotive force (CEMF), first disregard the fact that external current is being applied to the rotor via the carbon brushes on the commutator segments. As the armature wires rotate and cut the magnetic field of the field poles, a voltage is induced in them similar to that which was discussed in induction motors. This induced voltage (or electromotive force [EMF]) causes a current to flow in them and a resulting magnetic field is created.

Think About It

Two-Loop DC Motors

Using the basic principles of motor action, explain why the two-loop DC motor is self-starting and why its torque is erratic.

MAIN FIELD WITH NO ARMATURE CURRENT FLOW

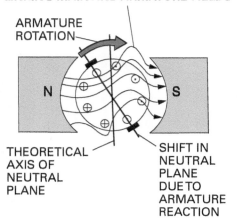

There are two magnetic fields in the gap between the pole pieces of the electric motor. One is the main magnetic field. The second is the magnetic field of the armature.

DISTORTED FIELD FORMED BY LINKING MAIN AND ARMATURE FIELDS

Combining the two fields results in a distorted main field whose perpendicular neutral plane is shifted backward against the direction of rotation.

ARMATURE FIELD, ASSUMING THERE IS NO MAIN FIELD

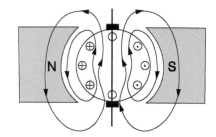

Figure 11 Armature reaction.

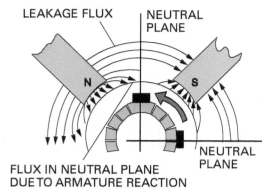

Interpoles produce a local field at the neutral plane that opposes the flux produced by armature reaction to restore the original neutral plane.

Figure 12 Interpoles.

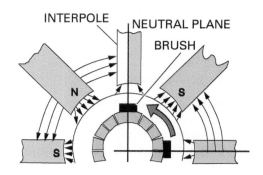

Interpoles are used on practical DC motors to counteract armature reaction.

Electromotive force is another word for *voltage* (also called *electrical pressure*, or *difference of potential*).

Before analyzing the relative direction between the current induced in the armature windings and the current that caused it in the field poles, first remember the left-hand rule. Using your left hand, hold it so your index finger points in the direction of the magnetic field (north to south) and your thumb points in the direction of rotational force on a given conductor. Your middle finger will now point in the direction of current flow for that conductor. This current would be in opposition to the current that is flowing from the battery. Because this induced voltage and induced current are opposite to those of the battery, they are called *CEMF*. The two currents are flowing in opposite directions. This would mean that the battery voltage and the CEMF are opposite in polarity (*Figure 13*).

When first discussing CEMF, we disregarded the fact that external DC was being applied to the armature via the brushes. The induced voltage and resulting current flow was then shown to flow opposite to the externally applied current. This was an oversimplification because only one current flows. Because the CEMF can never become as large as the external applied voltage and because they are opposite in polarity, the CEMF works to cancel only a part of the applied voltage. The single current that flows is smaller because of the CEMF.

Because the CEMF of a motor is generated by the action of the armature windings cutting the lines of force set up by the field poles, its value will depend on the field strength and the armature speed. The effective voltage acting in the armature is equal to the terminal voltage (V_T) minus the CEMF. Ohm's law can be used to determine the value of armature current, as follows:

$$\text{Armature } (I_A) = \frac{V_T - \text{CEMF}}{\text{armature resistance } (R_A)}$$

Where:

$$\text{CEMF} = V_T - (I_A \times R_A)$$

Example:
Find the value of CEMF of a DC motor when the terminal voltage is 240V and the armature current is 60A. The armature resistance has been measured at 0.08Ω.

Solution:

$$\text{CEMF} = V_T - (I_A \times R_A)$$
$$\text{CEMF} = 240 - (60 \times 0.08)$$
$$\text{CEMF} = 240 - 4.8$$
$$\text{CEMF} = 235.2V$$

CEMF acts as an automatic current limiter that reduces armature current to a level adequate to drive the motor but not so great as to risk burning out the armature. CEMF acts as a load for the DC power supply feeding the motor, so that the low-resistance motor windings do not draw excessive amounts of current.

If the armature stalls and no CEMF is produced, the motor draws so much current that it heats up. This reaction is shown in *Figure 14*. CEMF is present in all motors and is necessary for a motor's operation.

1.1.5 Starting Resistance

Large DC motors require that a starting resistance be inserted in series with the motor armature. The current drawn by the armature is governed by CEMF and the armature resistance. When starting, CEMF will be zero because the rotor is at a standstill. There is also no inductive reactance, as in AC induction motors. This means that the starting current will be abnormally high unless limited by external starting resistance.

Figure 15 shows a shunt motor that is connected directly across a 250V line. The armature resistance is known to be 0.5Ω. The full-load current of the motor is known to be 25A, and the shunt field current is 1A. The resulting armature current under full-load conditions would therefore be 24A.

If starting resistance is not used, the value of the armature current (I_A) can be found using the following equation:

$$I_A = \frac{V_T - \text{CEMF}}{R_A}$$
$$I_A = \frac{250V - 0V}{0.5\Omega}$$
$$I_A = 500A$$

This amount of starting current is too high and may result in excessive torque and heat that may cause damage to the motor. When starting resistance is added in series with the armature, the starting current can be limited to 1.5 times the full-load current value:

Starting I_A = 1.5 × full-load current
Starting I_A = 1.5 × 24A
Starting I_A = 36A

After starting, this external resistance can be removed from service. If we want to limit the starting armature current to 1.5 times the full-load value, we can solve for the size of resistance that would be required using the previous equations. At the moment of motor start, when the rotor is at a standstill and the CEMF is zero, the series resistance will be:

$$R_{starting} = \frac{(V_T - \text{ECMF}) - (I_A \times R_A)}{I_A}$$

$$R_{starting} = \frac{(250V - 0V) - (36A \times 0.5\Omega)}{36A} = 6.44\Omega$$

To find the wattage required in the starting resistance, take the square of the current multiplied by the resistance, where watt loss is calculated by the I^2R method.

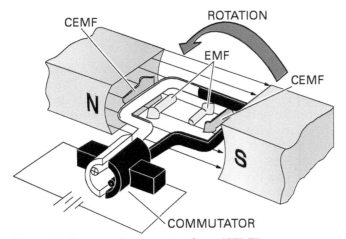

Figure 13 Counter-electromotive force (CEMF).

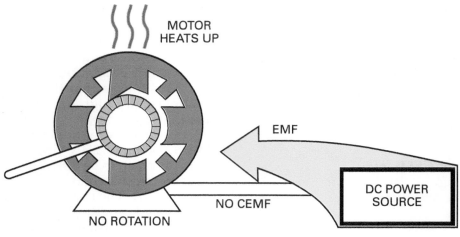

Figure 14 No CEMF.

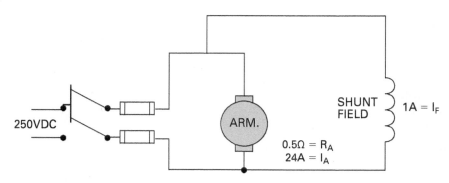

Figure 15 Shunt motor.

Example:
Find the power developed in both watts and horsepower in a DC motor that has a terminal voltage of 240V and an armature current of 60A. The armature resistance is known to be 0.08Ω.

Solution:
First, find the CEMF. Then use the CEMF value with the power equation to determine the power in watts, as follows:

$$CEMF = V_T - (I_A \times R_A)$$
$$CEMF = 240 - (60 \times 0.08)$$
$$CEMF = 240 - 4.8$$
$$CEMF = 235.2V$$

$$Power = IE$$
$$Power = 60A \times 235.2V$$
$$Power = 14,112W$$

Now convert to horsepower. This means you are converting an electrical term (watts) into a mechanical term (horsepower). To convert to horsepower:

$$Horsepower = \frac{watts}{746}$$
$$Horsepower = \frac{14,112W}{746}$$
$$Horsepower = 18.92hp$$

1.2.0 DC Motor Types

Two basic types of motor connections are in common use: series motor and shunt motor. The series motor is so called because the field coils are connected in series with the armature winding. The shunt motor has the field coils connected in parallel with the armature (rotor) winding. An additional type of motor is a compound motor. This motor has both a series- and a shunt-connected field. *Figure 16* shows a typical DC motor.

Think About It

CEMF

Can a motor's CEMF equal the applied terminal voltage? If not, why not?

1.2.1 Shunt DC Motors

The field circuit of a shunt motor is connected across the supply line and is in parallel with the armature. A shunt motor connection is shown in *Figure 17*.

When an external load is applied to the shunt motor, it tends to slow down. The slight decrease in speed causes a corresponding decrease in CEMF. Because the armature resistance is low, the resulting increases in armature current and torque are relatively large. Therefore, the torque is increased until it matches the opposing torque of the load. The speed of the motor then remains constant at the new value as long as the load is constant.

If the load on the shunt motor is reduced, the motor tends to speed up. The increased speed causes a corresponding increase in CEMF and a relatively large decrease in armature current and torque.

The amount of current through the armature of a shunt motor depends largely on the load. The larger the load, the larger the armature current; the smaller the load, the smaller the armature current. The change in speed causes a change in CEMF and armature current in each case.

The main advantage of a shunt-wound motor is that its speed is fairly constant, changing only a few revolutions per minute (rpm) when the amount of load changes. The main disadvantage is that the motor does not develop much torque when it is first started. If a motor is to be started with a large load, it is generally series connected.

> **WARNING!** Never open the shunt field circuit of a DC motor when the motor is operating, especially when unloaded. An open field may cause the motor to rotate at dangerously high speeds. Large DC shunt motors have a field rheostat with a no-field release feature that disconnects the motor from the power source if the field circuit opens.

The operating characteristics of a shunt-wound motor are as follows:

- *Torque* – A DC shunt motor has high torque at any rated speed. At startup, a DC shunt motor can develop up to 150% of its normal running torque as long as the resistors in the starting circuit can withstand the heating effect of the current.
- *Speed control* – DC motors have excellent speed control. To operate the motor above rated speed, a field rheostat is used to reduce the field current and field flux. To operate below rated speed, resistors are used to reduce the armature voltage.
- *Speed regulation* – The speed regulation of a shunt motor drops from 5% to 10% from no-load to full-load. As a result, a shunt motor is superior to the series DC motor but is inferior to a differential compound-wound DC motor. (Differential compound-wound DC motors are discussed in more detail later in this module.)

The operating characteristics of a motor should be matched to the type of load. *Figure 18* shows the operating characteristics of a typical DC shunt motor.

Notice that the motor speed is relatively independent of the torque (load applied) from 0 to 150% of the rated capacity of the motor. This motor is suited to applications that require a relatively constant speed over a wide load range.

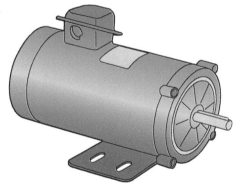

Figure 16 Typical DC motor.

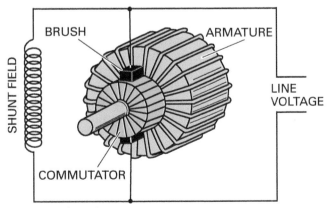

Figure 17 Shunt DC motor.

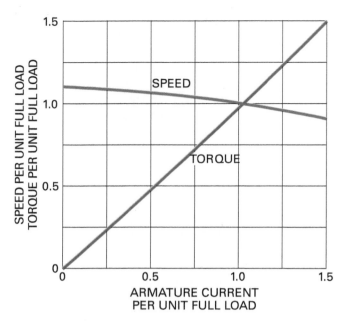

Figure 18 Operating characteristics of a typical DC shunt motor.

1.2.2 Series DC Motors

The field coils of a series motor are connected in series with the armature (*Figure 19*). The value of current through the armature and the field is the same. Therefore, if the armature current changes, the field current must also change.

As the motor speeds up, the armature current and field current decrease. With a weaker field, the armature speed will increase still more. The limiting factor on the speed is the load.

If there is no load on the motor, the armature will speed up to such an extent that the windings might be thrown from the slots and the commutator destroyed by the excessive centrifugal forces. For this reason, series motors are not belt-connected to their loads. The belt might break, allowing the motor to destroy itself. Series motors are usually connected to their loads directly or through gears.

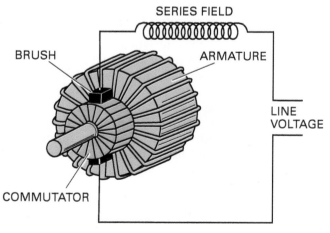

Figure 19 Series DC motor.

DC Motor Applications

DC motors were developed before AC motors, and one of their first uses was in electric trolleys. The DC motor is still widely used in applications that require accurate speed control or high starting torque. For example, in an elevator, the motor must start under a heavy load and accelerate smoothly. It must also stop precisely and reverse direction easily. A DC motor is a good choice for this application.

The series motor is used where there is a wide variation in both torque and speed requirements, such as traction equipment, blowers, hoists, cranes, and so forth. The operating characteristics of a series motor are as follows (the operating characteristics of a typical DC series motor are shown in *Figure 20*):

- *Torque* – A DC series motor develops 500% of its full-load torque at starting. Therefore, this type of motor is used in applications where large amounts of starting torque are needed, such as cranes, railway applications, and other equipment with high starting torque demands. With a series motor, any increase in load causes an increase in both the armature current and the field current. Because torque depends on the interaction of these two flux fields, the torque increases as the square of the value of the current increases. Therefore, series motors produce greater torque than shunt motors for the same increase in current. The series motor shows a greater reduction in speed for an equal change in load.
- *Speed control* – The speed control of a series motor is poorer than that of a shunt motor because if the load is reduced, a simultaneous reduction of current occurs in both the armature and field windings, and therefore, there is a greater increase in speed than there would be in a shunt-wound motor. If the mechanical load were to be disconnected completely from a series motor, the motor would continue to accelerate until the motor armature self-destructed. For this reason, series-wound motors are always permanently connected to their loads.
- *Speed regulation* – The speed of a series DC motor is controlled by varying the applied voltage. A series motor controller is usually designed to start, stop, reverse, and regulate speed. The direction of rotation of a series motor is changed by reversing either the armature or field winding current flow.

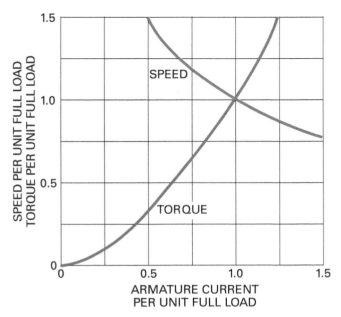

Figure 20 Operating characteristics of a typical series motor.

Notice that the motor speed varies greatly with respect to the torque (load applied). With less than half of its rated load applied, the motor operates at more than 150% of its rated speed. When 150% of the rated load is applied to the motor, it drops to 75% of its rated speed. Such motors find application where a constant heavy load exists or where great speed variations are tolerable.

1.2.3 Compound DC Motors

Compound DC motors are used whenever it is necessary to obtain speed regulation characteristics not obtainable with either the shunt- or series-wound motor. Because many applications require high starting torque and constant speed under load, the compound motor is used. Some industrial applications include drives for elevators, stamping presses, rolling mills, and metal shears. The compound motor has a normal shunt winding and a series winding on each field pole. They may be connected as a long shunt, as shown in *Figure 21* (A), or a short shunt, as shown in *Figure 21* (B). When the series winding is connected to aid the shunt winding, the machine is known as a cumulative compound motor. When the series field opposes the shunt field, the machine is known as a differential compound motor.

The operating characteristics of a compound DC motor are as follows:

- *Torque* – The operating characteristics of a cumulative compound-wound motor are a combination of the series motor and the shunt motor. A cumulative compound-wound motor develops high torque for sudden increases in load.

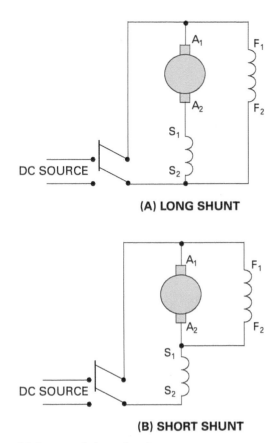

Figure 21 Long and short shunts.

- *Speed control* – Unlike the series motor, the cumulative compound-wound motor has definite no-load speeds and will not build up self-destructive speeds if the load is removed. Speed control of a cumulative compound-wound motor can be controlled by inserting resistors in the armature circuit to reduce the applied voltage. When the motor is to be used for installations where the rotation must be reversed frequently, such as in elevators, hoists, and railways, the controller should have voltage dropping resistors and switching arrangements to accomplish reversal.

Compound DC Motors

The compound motor avoids some of the limitations of the series-wound and shunt-wound motors. The shunt field has a constant current, so the motor will not self-destruct like a series motor. The series winding, on the other hand, provides strong torque.

- *Speed regulation* – The speed regulation of a cumulative compound-wound motor is inferior to that of a shunt motor and superior to that of a series motor.

Figure 22 shows the operating characteristics of a DC compound motor. Notice that the motor speed is relatively constant over the operating range. Its speed does vary with the torque somewhat more than the shunt motor, but will not run away or markedly decrease, as with the series motor. Such motors find application where the load is not known exactly or where some speed variation is tolerable with load variation.

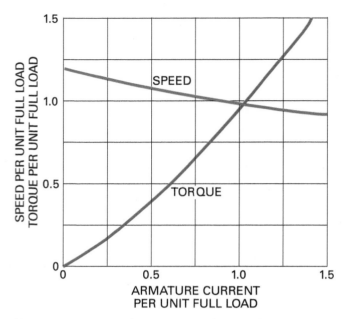

Figure 22 Operating characteristics of a typical DC compound motor.

Permanent Magnet DC Motors

Many ¼hp to 3hp variable-speed DC motors available for constant or diminishing torque applications use permanent magnets for the field poles instead of shunt or series windings. They employ variable DC armature voltages up to 90V or 180V for speed control. However, they are inefficient if they use only rheostat control of the armature voltage.

Think About It

DC Motors

Given the speed and torque characteristics of shunt and DC motors, which one would be better suited to the varying loads of an escalator?

1.0.0 Section Review

1. When studying the operation of electric motors, it is important to remember that magnetic lines of force _____.

    a. form infinite, straight lines
    b. do not cross
    c. attract materials with the same polarity
    d. pass through conductive materials only

2. A motor with the field coils connected in parallel with the rotor winding is a _____.

    a. shielded motor
    b. series motor
    c. shunt motor
    d. sealed motor

Section Two

2.0.0 AC Motor Operation

Objective

Identify alternating current (AC) motors and describe their operating characteristics.
  a. Understand how AC motors operate.
  b. Identify three-phase induction motors.
  c. Identify synchronous motors.
  d. Identify single-phase induction motors.

Performance Task

1. Identify various types of motors and their application(s).

Trade Terms

Overload: Operation of equipment in excess of the normal, full-load rating, or of a conductor in excess of rated ampacity, which, after a sufficient length of time, will cause damage or dangerous overheating. (A fault, such as a short circuit or ground fault, is not an overload.)

Synchronous speed: The speed of the revolving field of the stator, which is dependent on the supply frequency and number of poles in a stator winding.

Alternating current motors can be divided into two major types: single-phase motors and polyphase motors. The single-phase motor is normally limited to fractional horsepower ratings up to about 5hp They are commonly used to power fans, small pumps, appliances, and other devices not requiring a great amount of power. Single-phase motors are not likely to be connected to complicated motor control circuitry.

Polyphase motors make up the majority of motors needed to drive large machinery such as pumps, large fans, and compressors. These motors have several advantages over single-phase motors in that they do not require a separate winding or other device to start the motor. They have relatively high starting torque and good speed regulation for most applications.

There are two classes of polyphase motors: induction and synchronous. The rotor of a synchronous motor revolves at synchronous speed, or the speed of the revolving magnetic field in the stator. The rotor of an induction motor revolves at a speed somewhat less than synchronous speed. The differences in rotor speed are due to differences in construction and operation. Both will be discussed in depth after a review of motor theory.

2.1.0 How AC Motors Operate

AC motors consist of two parts: the stator (the stationary part) and the rotor (the revolving part). The stator is connected to the incoming three-phase AC power. The rotor in an induction motor is not connected to the power supply, whereas the rotor of a synchronous motor is connected to external power. Both induction and synchronous motors operate on the principle of a rotating magnetic field.

2.1.1 Rotating Fields

This section shows how the stator windings can be connected to a three-phase AC input to create a magnetic field that rotates. Another magnetic field in the rotor can be made to chase it by being attracted and repelled by the stator field. Because the rotor is free to turn, it follows the rotating magnetic field in the stator.

Polyphase AC is brought into the stator and connected to windings that are physically displaced 120° apart. These windings are connected to form north and south magnetic poles, as shown in *Figure 23*. An analysis of the electromagnetic polarity of the poles at points 1 through 7 in *Figure 23* shows how the three-phase AC creates magnetic fields that rotate.

At point 1, the magnetic field in coil (pole) 1–1A is at its maximum. Negative voltages are shown in 1–2A and 3–3A. The negative voltages in these windings create smaller magnetic fields that will tend to aid the field set up in 1–1A.

At point 2, phase 3 creates a maximum negative flux in 3–3A windings. This strong negative field is aided by the weaker magnetic fields in 1–1A and 1–2A.

The three-phase AC input rises and falls with each cycle. Analyzing each point on the voltage graph shows that the resultant magnetic field rotates clockwise. When the three-phase input completes a full cycle at point 7, the magnetic field has completed an entire revolution of 360°.

2.1.2 Rotor Behavior in a Rotating Field

An oversimplification of rotor behavior shows how the magnetic field of the stator influences the rotor. Assume that a simple bar magnet is placed in the center of the stator diagrams shown in *Figure 23*. Also assume that the bar magnet is free to

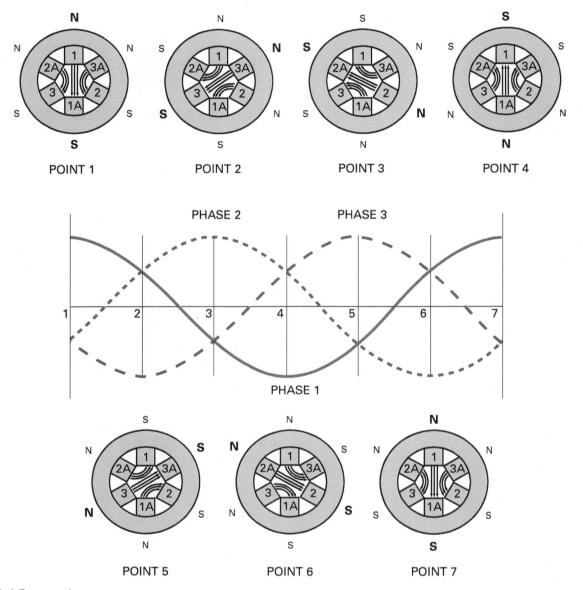

Figure 23 AC generation.

rotate. It has been aligned such that at point 1, its south pole is opposite the large north of the stator field.

Unlike poles attract, and like poles repel. As the AC completes a cycle, going from point 1 to point 7, the stator field rotates and pulls the bar magnet with it because of the attraction of unlike poles and the repulsion of like poles. The bar magnet is rotating at the same speed as the revolving flux of the stator. The speed of the revolving flux is known as synchronous speed. The synchronous speed of a motor is given by the equation:

$$N_S = \frac{120f}{P}$$

Where:

- N_S = synchronous speed in rpm
- f = frequency in cycles per second (also known as *hertz*, or *Hz*)
- P = number of magnetic poles
- 120 = a constant derived by multiplying 60 seconds/minute (to convert speed in rpm to seconds) by 2 (for one pair of poles)

2.1.3 Induction

Current flowing through a conductor sets up a magnetic field around the length of the conductor. Conversely, a conductor in a magnetic field will produce a current when the magnetic lines of flux cut across the conductor. This action is called induction because there is no physical connection between the magnetic field and the conductor. Current is induced in the conductor.

2.1.4 Torque

The torque on the rotor of an induction motor tends to turn the rotor in the same direction as the rotating field. If the motor is not driving a load, it will accelerate to nearly the same speed as the rotating field. As the rotor accelerates, the magnitude of the induced voltage in the rotor decreases. This is because the relative motion between the rotating field and the rotor conductors is reduced. It is impossible for an induction motor to operate at synchronous speed because there would be no relative motion between the rotating field and the rotor. Therefore, there would be no induced voltage, no rotor current, no rotor magnetic field, and no torque.

2.1.5 Slip

In an induction motor, the rotor always rotates at a speed less than the synchronous speed. The rotor speed is such that sufficient torque is produced to balance the restraining torque caused by motor friction and mechanical load. The difference between the synchronous speed and the rotor speed is known as *slip*. Slip is expressed mathematically as follows:

$$S = \frac{N_S - N_R}{N_S} \times 100$$

Where:

S = slip
N_S = synchronous speed
N_R = rotor speed

In this formula, the quantity is multiplied by 100 in order to be expressed as a percentage.

Example:

A four-pole, 208V, 2hp, 60Hz, three-phase induction motor has a no-load speed of 1,790 rpm and a full-load speed of 1,650 rpm. Find the percent slip for each case below:

- No-load condition
- Full-load condition
- Locked-rotor condition (standstill)

Solution:

First, determine synchronous speed:

$$N_S = \frac{120f}{P}$$

$$N_S = \frac{120 \times 60}{4}$$

$$N_S = 1{,}800 \text{ rpm}$$

Using the formula for finding slip, insert N_S and the rotor speed to calculate the slip at no-load condition:

$$S = \frac{N_S - N_R}{N_S} \times 100$$

$$S = \frac{1{,}800 - 1{,}790}{1{,}800} \times 100$$

$$S = 0.556\%$$

At full-load condition:

$$S = \frac{N_S - N_R}{N_S} \times 100$$

$$S = \frac{1{,}800 - 1{,}650}{1{,}800} \times 100$$

$$S = 8.33\%$$

At locked-rotor condition:

$$S = \frac{N_S - N_R}{N_S} \times 100$$

$$S = \frac{1{,}800 - 0}{1{,}800} \times 100$$

$$S = 100\%$$

Figure 24 shows how torque relates to speed over the operating range of a motor. Note that speed is proportional to torque on the left side up to pullout torque. Beyond this point, however, torque decreases as speed increases.

Wound-Rotor Motor Applications

Like DC motors, wound-rotor motors are used where high inertia loads must be started easily or often. Wound-rotor motors have starting torques in the range of 225% of full-load torque. They are used for hoists, hydraulic gates, yard locomotives, and cranes.

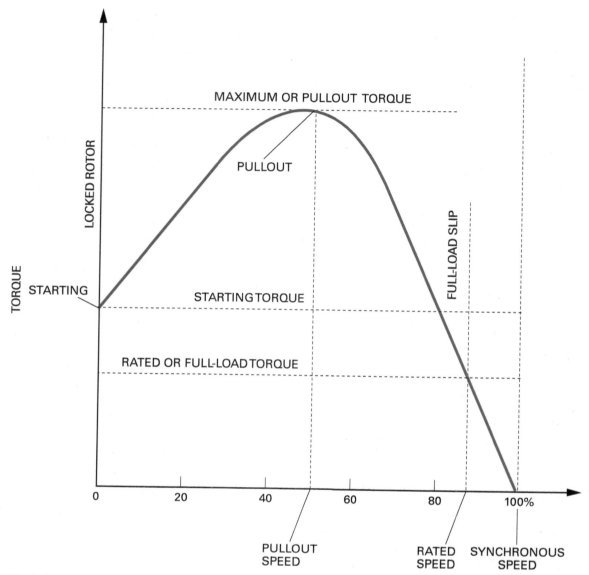

Figure 24 Typical torque-speed curve.

Slip is the difference between the synchronous speed and the actual speed of the rotor in an induction motor. Slip is necessary to permit motor action to occur. Under increasing load, the rotor torque increases. Because percent slip is proportional to torque, the amount of slip will increase. This increase means a higher current draw by the motor due to the greater difference between the rotor and the magnetic field. Motor supply voltages, current, torque, speed, and rotor impedance are closely related. By changing the resistance and reactance of the rotor, the characteristics of the motor can be changed; however, for any particular rotor design these characteristics are fixed.

2.1.6 Starting Current

At the moment a three-phase induction motor is started, the current supplied to the motor stator terminals may be as high as six times the motor full-load current. This is because at starting, the rotor is at rest; therefore, the rotating magnetic field of the stator cuts the rotor at the maximum rate, inducing large amounts of EMF in the rotor. This results in proportionally high currents at the input terminals of the motor, as previously discussed. Because of this high inrush, current starting protection as high as 300% of full-load current must be provided to allow the motor to start and come up to speed.

Because 100% slip exists at the instant the motor is energized (as shown in *Figure 24*), the rotor current lags the rotor EMF by a large angle. This means that the maximum current flow occurs in a rotor conductor at a time after the maximum amount of stator flux has passed by. This results in a high starting current at a low power factor, which results in a low value of starting torque.

As the rotor speeds up, the rotor frequency and reactance decrease, causing the torque to increase up to its maximum value then decrease to the value needed to carry the load.

2.1.7 Loaded Torque

If a load is now placed on the shaft, the rotor will tend to slow down. As it slows down, more flux lines are cut until enough torque is developed to overcome the load placed on the shaft.

The motor now runs under load at a slower speed than before the load was placed on the shaft. This normal range of operation is shown in the lower right corner of *Figure 24* as the rated or full-load torque.

In this range, the slip will vary from 2% to 10%, depending on the load applied and the motor. Rated slip will occur at the point at which 100% rated load is applied. Increased load means increased slip, which means the rotor is now rotating slower. An induction motor is considered to be a constant speed motor. Now, examine how much speed fluctuates from no-load speed to full-load speed.

Example:

A two-pole induction motor has a no-load slip of 2% and a full-load slip of 8%. What are the no-load speed, full-load speed, and percent speed change?

Solution:

$$\text{Nominal} = \frac{120 \times 60}{2} = 3{,}600 \text{ rpm}$$

$$\text{No-load speed} = \frac{100\% - 2\%}{100\%} \times 3{,}600 \text{ rpm} = 3{,}528 \text{ rpm}$$

$$\text{Full-load speed} = \frac{100\% - 8\%}{100\%} \times 3{,}600 \text{ rpm} = 3{,}312 \text{ rpm}$$

Percent speed change

$$= \frac{\text{no-load speed} - \text{full-load speed}}{\text{no-load speed}} \times 100\%$$

$$= \frac{3{,}528 - 3{,}312}{3{,}528} \times 100\% = 6.12\%$$

2.1.8 Overload Condition

If the load is increased above full-rated load, everything happens as stated before to increase torque up to a certain point. *Figure 25* shows typical torque and current curves. Note how the torque climbs as the load is increased.

This will continue as load is increased until the pullout torque point is reached. Beyond this point, the torque decreases and the motor will quickly stall. A typical situation is when a bench circular saw or a lathe stalls on a heavy cut. The machine will slow down as its cutting load is increased until it suddenly stalls and hums or growls loudly. The condition will persist until the load is relieved or a fuse blows or a breaker trips. The motor has simply reached a point (**overload** condition) where it cannot continue to increase its torque. Any further increase in load will cause a stall.

2.1.9 Power Factor

The power factor of an induction motor is poor at no-load and low-load conditions. At no-load conditions, the power factor can be as low as 15% lagging. However, as load is increased, the power factor increases. At high-rated load, the power factor may be as high as 85% to 90% lagging.

The power factor at no-load speed is low because the magnetizing component of input current is a large part of the total input current of the motor. When the load on the motor is increased, the in-phase current supplied to the motor increases, but the magnetizing component of current remains practically the same. This means that the resultant line current is more nearly in phase with the voltage, and the power factor is improved when the motor is loaded compared with an unloaded motor, which chiefly draws its magnetizing current.

Figure 26 shows the increase in power factor from no-load conditions to full-load conditions. In the no-load diagram, the in-phase current (I_{ENERGY}) is small when compared to the magnetizing current (I_M); thus, the power factor is poor at no-load conditions. In the full-load diagram, the in-phase current has increased, while the magnetizing current remains the same. As a result, the angle of lag of the line current decreases, and the power factor increases.

2.1.10 Reversing Rotation

The direction of rotation of a three-phase induction motor can be readily reversed. The motor will rotate in the opposite direction if any two of the three incoming leads are reversed, as shown in *Figure 27*.

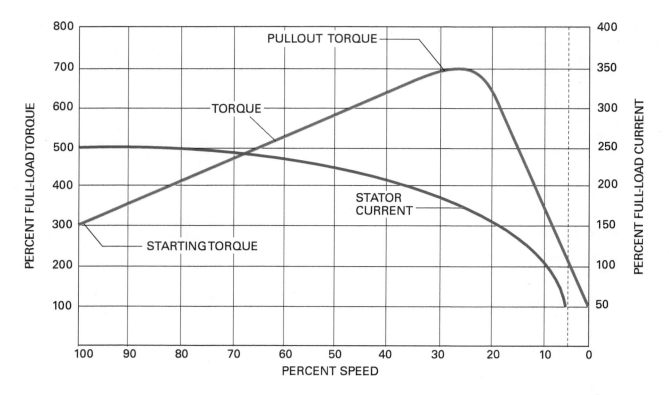

Figure 25 Torque and current curves.

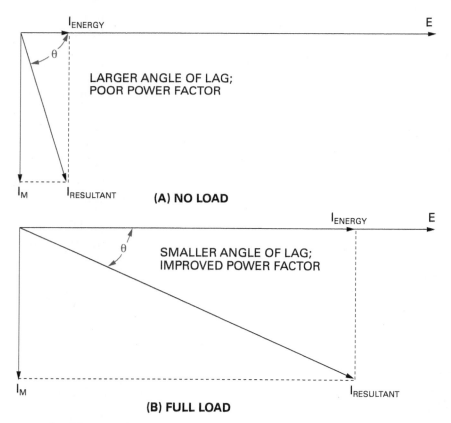

Figure 26 Power factor versus load for an induction motor.

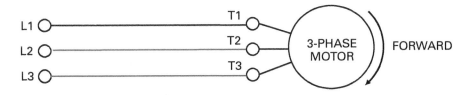

(A) ROTATION BEFORE CONNECTIONS ARE CHANGED

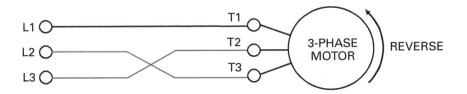

(B) ROTATION AFTER CONNECTIONS ARE CHANGED

Figure 27 Three-phase induction motor rotational direction change.

2.2.0 Three-Phase Induction Motors

In a three-phase induction motor, the driving torque is caused by the reaction of a current-carrying conductor in a magnetic field. In induction motors, the rotor currents are supplied by electromagnetic induction. The stator windings are supplied with three-phase power and produce a rotating magnetic field.

The rotor is not electrically connected to the power supply. The induction motor derives its name from the mutual inductance taking place between the stator and the rotor under operating conditions. The rotating field produced by the stator cuts the rotor conductors, inducing a voltage into the conductors. The induced voltage causes rotor current. This develops motor torque due to the reaction of a current-carrying conductor in a magnetic field. This torque causes the rotor to rotate. This principle is shown in *Figure 28*.

The three-phase (3φ) induction motor has a frame or stationary part—the stator. The stator is made of laminated steel rings with slots on the inside circumference. The motor stator windings are the phase windings. The windings are symmetrically placed on the stator and may be either wye- or delta-connected. Depending on how the stator is wound, it may have two, four, or any even number of poles.

There are two varieties of three-phase induction motors: the squirrel cage rotor motor and the wound-rotor motor.

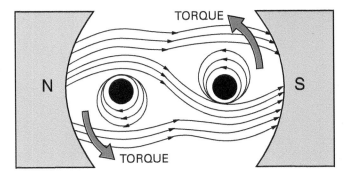

Figure 28 Producing torque.

2.2.1 Squirrel Cage Induction Motor

The squirrel cage is the rotor most commonly used for three-phase motors. Three-phase squirrel cage induction motors consist of a stator, a rotor, and two end shields that house the bearings that support the rotor shaft. The frame is usually made of cast steel. The stator core is pressed into the frame. In this rotor, the bars are connected together at the ends by shorting rings made of similar material. The conductor bars carry large currents at low voltages. The bearings can be either sleeve or ball type. *Figure 29* shows the main components of an induction motor.

It is not necessary to insulate the bars from the core because the current will follow the path of least resistance and is confined to the cage windings. *Figure 30* shows how a squirrel cage rotor is constructed.

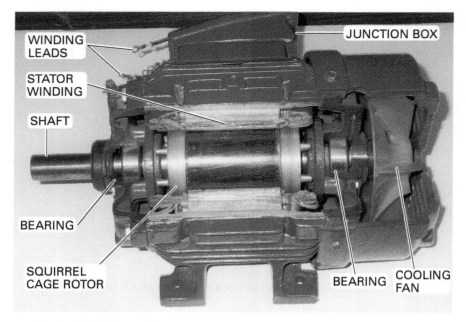

Figure 29 Main components of an induction motor.

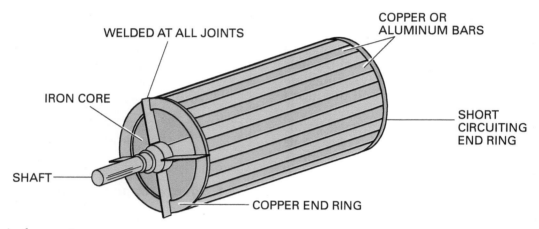

Figure 30 Squirrel cage rotor.

The speed of a three-phase squirrel cage induction motor depends on the frequency of the applied voltage and the number of poles. As a result, these motors are used in applications where the speed either remains constant or can be controlled by other means, such as variable frequency drives.

The squirrel cage rotor motor is a general-purpose motor. It is used to drive loads that require variable torque at relatively constant speed with high full-load efficiency. Uses include powering blowers, centrifugal pumps, and fans. Because of the absence of any moving electrical contacts, they are suitable for use where they are exposed to flammable dust or gas.

If the load requires special operating characteristics, such as high starting torque, the squirrel cage rotor can be designed with high resistance bars for a starting circuit and low resistance bars for running operation. A rotor of this type is called a double squirrel cage rotor.

Squirrel Cage Motor Applications

Squirrel cage induction motors have many favorable attributes. For example, maintenance costs are low because these motors have no brushes or slip rings but work entirely through induction. They also have a high starting torque, so they are useful in common applications such as overhead doors, large compressors, fans, and printing presses.

2.2.2 Wound-Rotor Induction Motor

A wound rotor (*Figure 31*) has a winding that is similar to the three-phase stator windings. The rotor windings are usually wye-connected with the free ends of the windings connected to three slip rings mounted on the rotor shaft. The slip rings are shown physically mounted on the end of the rotor shaft in *Figure 31*. They are used with brushes to form an electromechanical connection to the rotor.

Slip rings are contact surfaces mounted on the shaft of a motor or generator to which the rotor windings are connected and against which the brushes ride. The brushes are sliding contacts, usually made of carbon, that make continuous electrical connection to the rotating part of a motor or generator.

The wound-rotor motor often uses an external wye-connected resistor connected to the rotor through slip rings. The resistor provides a means of varying the rotor resistance. This can be used when the motor is started to produce a high starting torque. As the motor accelerates, the resistance is reduced. When the motor has reached full speed, the slip rings are short circuited, and the operation is similar to that of a squirrel cage rotor induction motor. A schematic representation of a wound-rotor motor is shown in *Figure 32*.

In a wound-rotor induction motor, the insertion of resistance in the rotor circuit not only limits the starting surge of current but also produces a high starting torque and provides a means of adjusting the speed. Speed can be varied by as much as 50% to 75%; the greater the resistance inserted in the rotor circuit, the lower the speed will be below synchronous speed. When the motor is operating below full speed, the percent slip is increased and the motor is operating at reduced efficiency and horsepower. When all resistance is cut completely out, the speed is somewhat less than that obtained with squirrel cage rotors. If the full resistance of the speed controller is cut into the rotor circuit when the motor is running, the rotor current decreases and the motor slows down. As the rotor speed decreases, more voltage is induced in the rotor windings and more rotor current is developed to create the necessary torque at the reduced rotor speed.

If all the resistance is removed from the rotor circuit, both the current and motor speed will increase. However, the rotor speed will always be less than the synchronous speed of the field developed by the stator windings. This is also true of a squirrel cage induction motor. The speed of a wound-rotor motor can be controlled manually or automatically with timing relays, contactors, and push-button speed selection.

The wound-rotor induction motor is used when it is necessary to vary the rotor resistance, to limit starting current, or to vary the motor speed. Because the rotor circuit heat generation is largely external to the rotor windings, the wound-rotor motor is used for applications that require frequent starts without overheating the motor. The advantages of the wound-rotor motor are high starting torque with moderate starting current, smooth acceleration under heavy load, no excessive heating during starting, good running characteristics, and adjustable speed control. The chief disadvantage is that both initial and maintenance costs are greater than those of the squirrel cage rotor motor.

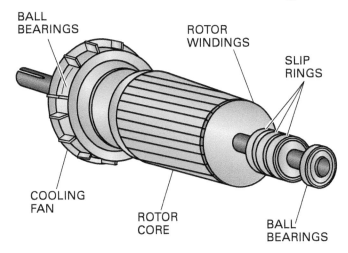

Figure 31 Wound rotor.

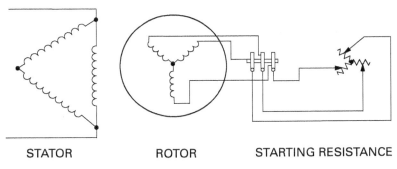

Figure 32 Wound-rotor motor circuit.

> **Think About It**
> ### Wound-Rotor Motors
> When might you want a wound-rotor motor to run like a squirrel cage motor at full speed?

2.3.0 Synchronous Motors

The synchronous motor is a three-phase motor that operates at synchronous speed (0% slip) from no-load conditions to full-load conditions. These motors are used in precision applications.

2.3.1 Operating Characteristics

This type of motor has a revolving field that is energized by a source separate from the stator winding. The rotor is excited by a DC source. The magnetic field set up by the direct current on the rotor then locks in with the rotating magnetic field of the stator and causes the rotor to revolve at synchronous speed. By changing the magnitude of DC excitation, the power factor of the motor can be changed over a wide variety of power factors from leading to lagging. Because of the unique ability of synchronous motors to change power factors, they are often used as power-factor correctors. They are most often used in applications that require constant speed from no-load conditions to full-load conditions.

2.3.2 Construction

The construction of synchronous motors is essentially the same as the construction of three-phase generators. They have three stator windings that are 120° apart and a wound rotor that is connected to slip rings where the rotor excitation current is applied.

When three-phase AC is applied to the stator, a revolving magnetic field is created just as it is in induction motors. The rotor is energized with DC, which creates a magnetic field around the rotor. The strong rotating magnetic field of the stator attracts the rotor field. This results in a strong turning force on the rotor shaft.

This is how the synchronous motor works once it is started. However, one of the disadvantages of this type of motor is that it cannot be started just by applying AC to the stator. When AC is applied to the stator, the high-speed rotating magnetic field rushes past the rotor poles so quickly that the rotor does not have a chance to get started. The rotor is locked; it is repelled in one direction and then in another direction. In its purest form, the synchronous motor has no starting torque.

This is more easily understood using *Figure 33*. When the stator and rotor fields are energized, the poles of the rotating field approach the rotor poles of opposite polarity. The attracting force will tend to turn the rotor in a direction opposite the rotating field. As the rotor starts to move in that direction, the rotating field moves past the rotor poles and tends to pull the rotor in the same direction as the rotating field. The result is no starting torque.

To allow this type of motor to start, a squirrel cage winding is added to the rotor to cause it to start like an induction motor. This winding is called an *amortisseur winding*. The rotor windings are constructed so that definite north and south poles are created, and these poles, when excited by DC, will lock in with the revolving field. The rotor windings are wound about the salient field poles, which are connected in series for opposite polarity.

The number of field poles must equal the number of stator poles. The rotor field windings are brought out to slip rings that are mounted on the rotor shaft. The field current is supplied through carbon brushes to the field windings. *Figure 34* shows a simplification of a synchronous motor. *Figure 35* shows the construction of the rotor pole assembly.

GOING GREEN

Variable-Speed Drives

Variable-speed drives, known as VSDs or ASDs (adjustable-speed drives), are powerful electronic devices that are available for virtually any size motor in all types of applications. Of the various types of VSDs available, the most efficient versions for AC motors are VFDs (variable-frequency drives), which control both the frequency and the voltage applied to the motor. By changing the frequency of the rotating stator field, you change the speed of the rotor, thus changing the speed of the motor. VSDs are often used in energy management systems to conserve energy by supporting variable loads, such as those that occur in heating, ventilating, and air conditioning systems. In addition to controlling the speed of a motor, VSDs are available to control both motor starting and stopping functions, as well as to provide controlled acceleration and deceleration.

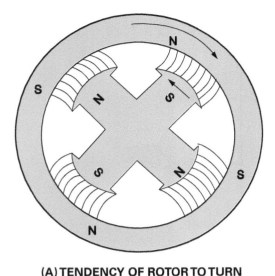
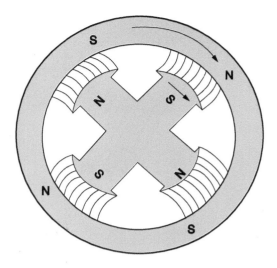

(A) TENDENCY OF ROTOR TO TURN COUNTERCLOCKWISE

(B) TENDENCY OF ROTOR TO TURN CLOCKWISE

Figure 33 Synchronous motor operation at start.

2.3.3 Principles of Operation

When a synchronous motor is started, current is first applied to the stator windings. Current is induced in the amortisseur winding, and the motor starts as an induction motor. The motor then comes up to near-synchronous speed (about 5% to 10% slip). At that point, the field is excited, and the motor, turning at high speed, pulls into synchronism. When this occurs, the rotor is turning at synchronous speed. The squirrel cage winding will not be generating any current and therefore will not affect the synchronous motor's operation.

The amortisseur windings serve an additional purpose. When the load changes frequently, the motor speed is not steady because the torque angle (discussed later) oscillates (or hunts) back and forth, trying to settle at its required value. This momentary change in speed creates a current due to induction, and there will be torque in the amortisseur winding. This momentary torque serves to dampen or stabilize the oscillating torque angle. That is why amortisseur windings are sometimes referred to as damper windings.

2.3.4 Rotor Field Excitation

The rotor must be excited from an external DC source. *Figure 36* shows a simplified synchronous motor excitation circuit. Notice that the DC field current can be varied by the rheostat; however, this does not change the speed of the motor. It only changes the power factor of the motor stator circuit. If full resistance is applied to the rotor

---

GOING GREEN

Synchronous Motors

Three-phase synchronous motors can be used in industrial applications to correct the low power factor of a number of induction motors or other inductive devices that are operating at less than their rated load levels. Synchronous motors can accomplish power factor correction while driving their own mechanical loads. Correcting a low power factor created by inductive loads through the use of synchronous motors reduces energy costs by making efficient use of the power supplied to the industrial facility. The use of synchronous motors can eliminate the need for dedicated capacitor banks or switched capacitor banks and avoid the surges caused by them.

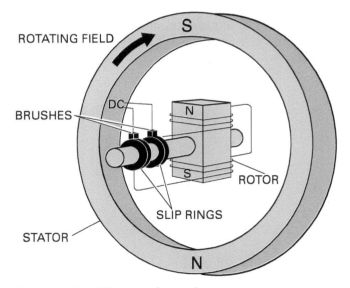

Figure 34 Simplification of a synchronous motor.

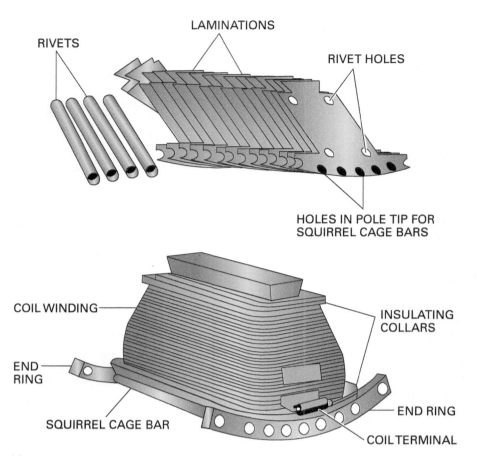

Figure 35 Pole assembly.

field circuit, then the field strength of the rotor is at its minimum and the power factor is extremely lagging. As the DC field strength is increased, the power factor improves. If current is increased sufficiently, the power factor can be increased to near unity or 100%. This value of field current is referred to as *normal excitation*. By increasing the rotor field strength further, the power factor decreases but in a leading direction; that is, the stator circuit becomes capacitive and the motor is said to be overexcited. The synchronous motor can be used to counteract the lagging power factor in circuits by adding capacitive reactance to the circuit, thereby bringing the overall power factor closer to unity.

If the rotor DC field windings of a synchronous motor are open when the stator is energized, a high AC voltage will be induced in it because the rotating field sweeps through the large number of turns at synchronous speed.

It is therefore necessary to connect a resistor of low resistance across the rotor DC field winding during the starting period. During the starting period, the DC field winding is disconnected

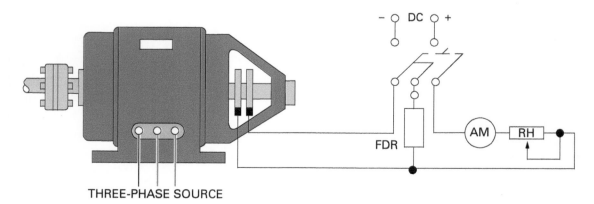

Figure 36 Simplified synchronous motor excitation circuit.

from the source and the resistor is connected across the field terminals. This permits alternating current to flow in the DC field winding. Because the impedance of the winding is high compared with the inserted external resistance, the internal voltage drop limits the terminal voltage to a safe value.

2.3.5 Synchronous Motor Pullout

When a synchronous motor loses synchronism with the system to which it is connected, it is said to be out of step. This occurs when the following take place singly or in combination:

- Excessive load applied to the shaft
- Supply voltage reduced excessively
- Motor excitation lost or too low

Torque pulsations applied to the shaft of a synchronous motor are also a possible cause of loss of synchronism if the pulsations occur at an unfavorable period relative to the natural frequency of the rotor with respect to the power system.

A prevalent cause of loss of synchronism is a fault occurring on the supply system. Underexcitation of the rotor is also a distinct possibility.

Synchronous motor pullout is significant in that the squirrel cage or amortisseur winding is designed for starting only. Therefore, it is not as hardy as those found in induction motors. The amortisseur winding will not overheat if the motor starts, accelerates, and reaches synchronous speed within a time interval determined to be normal for the motor. However, the motor must continue to operate at synchronous speed. If the motor operates at a speed less than synchronous, the amortisseur winding may overheat and suffer damage.

Protection against a synchronous motor losing synchronism can be provided by polarized field frequency relays and out-of-step relays as well as various digital methods.

2.3.6 Synchronous Motor Torque Angle

When the rotor is brought up to high speed (close to synchronous speed), it will lock onto the rotating magnetic field. Under these conditions, a running torque will be developed. The rotor will rotate at synchronous speed in a direction and at a speed determined by synchronous speed.

While the motor is running, the two rotating fields will line up perfectly. The rotor pole will always lag behind the stator pole by some angle. This angle is called the torque angle and is shown in *Figure 37*.

As the load on the shaft increases, the torque angle increases even though the rotor continues to turn at synchronous speed. This behavior continues until the torque angle is approximately 90°. At that point, the motor is developing a maximum torque. Any further increase in load will cause either of the following to occur:

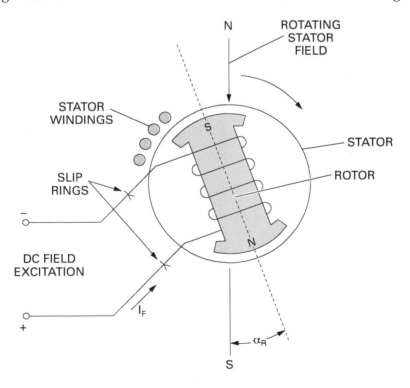

Figure 37 Torque angle.

- If the increase in load is momentary or very small, the rotor will slip a pole. In other words, the stator field will lose hold of the rotor and grab onto it again the next time around.
- If the increase in load is large enough and is not momentary, the motor will lose synchronism and will either stall or cause the rotor to suffer thermal damage.

In both cases, a noticeable straining sound will be heard.

The synchronous motor should not be used in applications where fluctuations in torque are violent. As a rule, the synchronous motor is also not used in small sizes (under 50hp) because it requires DC excitation. It is more difficult to start than induction motors and falls out of step quite readily when system disturbances occur. Common applications for synchronous motors are in motor generator sets, air compressors, and compressors in refrigerating plants.

2.4.0 Single-Phase Induction Motors

Single-phase motors operate on a single-phase power supply. This is important because in the typical home or office and in many areas of industrial plants the only power source available is single-phase AC. Not only do single-phase AC motors eliminate the need for three-phase AC lines, but they are also easier to manufacture in small sizes and are, therefore, less expensive.

Examples of the many applications of single-phase AC motors today are refrigerators, freezers, washers, dryers, power tools, copy machines, heating systems, water pumps, computer peripherals, and various small appliances.

2.4.1 Single-Phase Induction Motors

Single-phase AC induction motors are extremely common. Unlike polyphase induction motors, the stator field in the single-phase motor does not rotate. Instead, it simply alternates polarity between poles as the AC voltage changes polarity.

Voltage is induced in the rotor, and a magnetic field is produced around the rotor. This field will always be in opposition to the stator field. However, the interaction between the rotor and stator fields will not produce rotation (*Figure 38*). Because this force is across the rotor and through the pole pieces, there is no rotary motion, just a push and/or pull along this line.

If the rotor is rotated by some outside force (a twist of your hand, for example), the push-pull along the line is disturbed. Look at the fields shown in *Figure 38* as the motor begins to rotate.

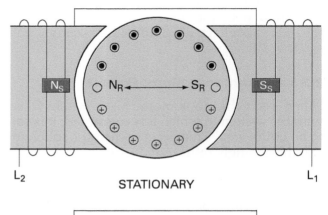
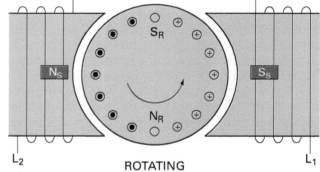

Figure 38 AC induction motor.

At that instant, the south pole on the rotor is being attracted to the left-hand pole. The north rotor pole is being attracted to the right-hand pole. All of this is a result of the rotor being rotated 90° by the outside force.

The pull that now exists between the two fields becomes a rotary force, turning the rotor toward magnetic correspondence with the stator. Because the two fields continuously alternate, they will never actually line up and the rotor will continue to turn once started.

Because a single-phase rotor will rotate if it has a rotating magnetic field present, all that remains is to find a means of generating a rotating field at the start. There are a number of practical means for generating a rotating field. All the methods used for single-phase induction motors involve the simulation of a second phase for a starting circuit.

2.4.2 Split-Phase Induction Motor

The split-phase motor, shown schematically in *Figure 39*, has a stator composed of slotted laminations that contain an auxiliary (starting) winding and a running (main) winding. The axes of these

Single-Phase Induction Motors

Outside large industrial and commercial facilities, single-phase induction motors are the most common type of motor used. While they are initially less expensive than polyphase motors, they are also less efficient and more costly to maintain. They are typically available in small sizes from ⅛hp to 1hp (previously referred to as *fractional horsepower sizes*) and in sizes up to 10hp.

two windings are displaced by an angle of 90 electrical degrees. The starting winding has fewer turns and smaller wire than the running winding and, therefore, has different electrical characteristics. The main winding occupies the lower half of the slots, and the starting winding occupies the upper half. The two windings are connected in parallel across the single-phase line supplying the motor. The motor derives its name from the action of the stator during the starting period.

When energized with single-phase AC, the two windings are physically different enough in position and construction to produce a magnetic revolving field that rotates around the stator air gap at synchronous speed. As the rotating field moves around the air gap, it cuts across the rotor conductors and induces a voltage in them. The interaction between the rotor and stator causes the rotor to accelerate in the direction in which the stator field is rotating.

When the rotor has come up to about 75% of synchronous speed, a centrifugally operated switch disconnects the starting winding from the line supply and the motor continues to run on the main winding alone. As the motor ages, the centrifugal switch contacts pit and corrode. When this happens, they may get stuck in the closed position. To safeguard against the winding burning up, a thermal relay is also used. If the motor draws the high starting current for more than 5 or 10 seconds, the relay will de-energize.

In a split-phase motor, the starting torque is 150% to 200% of the full-load torque and the starting current is six to eight times the full-load current. Fractional-horsepower split-phase motors are used in a variety of devices, such as washers, oil burners, and ventilating fans. The direction of rotation of the split-phase motor can be reversed by interchanging the starting winding leads.

2.4.3 Capacitor-Type Induction Motor

The capacitor-type motor is a modified form of split-phase motor. A typical capacitor-type motor is shown in *Figure 40*. The capacitor is located on top of the motor.

To develop a larger starting torque than that available with a standard split-phase motor, a capacitor is placed in series with the auxiliary winding of a split-phase motor, as shown in *Figure 41*. This is called a capacitor-start motor. The capacitor tends to create a greater electrical phase separation of the two windings. Also, because the reactance of a capacitor is 180° out of phase with the inductive reactance of the motor

Figure 39 Split-phase motor.

Figure 40 Capacitor motor.

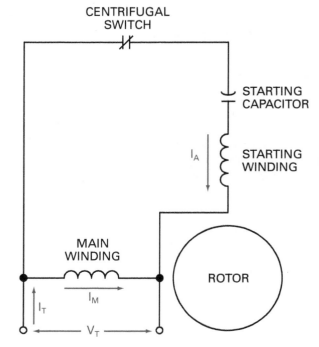

Figure 41 Capacitor-start motor schematic.

windings when they are combined, they yield a lower total impedance. This allows a larger current to produce a greater magnetic field.

The net effect of the capacitor is to give its motor a starting torque of about four times its rated torque. The split-phase motor, on the other hand, produces a starting torque of about one to two times its rated torque. When the capacitor motor has come up to speed and the starting winding has been disconnected, it will have the same running characteristics as the split-phase motor.

To reverse the direction of rotation of the capacitor-start motor and split-phase motor, the connection of either winding would have to be reversed. Because the starting winding is disconnected at a high speed, this reversal can be accomplished only at standstill or at low speeds when the centrifugal switch is still closed.

The capacitor-start motor is made in sizes from ¼hp to 10hp (150W to 7.5kW). The starting capacitor is the dry-type electrolytic capacitor made for AC use. Typical values are from 200 to 600 microfarads (μF). *Figure 42* shows a comparison of torque slip curves for a split-phase and capacitor-start motor. It also shows the typical effect of the starting capacitor.

A variation of the capacitor-start motor is one in which the capacitor and auxiliary winding are not disconnected. The centrifugal switch in *Figure 41* is eliminated, and the auxiliary winding is left in all the time. This motor is called a capacitor-run motor.

The capacity used for running under load is not the same as that needed for starting. Furthermore, the capacitors used for starting cannot be used for continuous operation. Because the capacitor used in this motor is in all the time, it must be of a different type; that is, it must be capable of operating continuously. The net result is that the motor has improved running characteristics; however, it does not provide a starting torque as large as that of the capacitor-start motor.

Among the improvements are higher efficiency and power factor at rated load, lower line current, and very quiet operation. It should also be pointed out that the start winding must be designed for continuous operation. This makes the motor somewhat more costly.

Another variation is the capacitor-start, capacitor-run motor. This motor combines the useful features of the capacitor-start and the capacitor-run motors by using two different capacitors, as shown in *Figure 43*.

Modern Split-Phase Induction Motor

Photo (A) shows a centrifugally-actuated start winding switch that is closed when the motor is at rest. The start winding switch is opened and closed by the movement of an actuator disk against a contact lever. The actuator disk is moved back and forth on the rotor shaft by a centrifugal weight assembly. When the motor is at rest, springs retract the weights and cause the actuator disk to move toward the bearing at the end of the shaft. This pushes on the contact lever, closing the switch contacts. After the motor starts and reaches about 75% of its rated speed, the weights swing out against the spring tension and retract the contact disk from the contact lever. This opens the switch contacts, removing the starting winding from the circuit. If the switch does not open after starting, the motor will operate at a reduced speed until it overheats the starting winding and activates the thermal relay. Photo (B) shows the starting winding (green) and the running winding (copper).

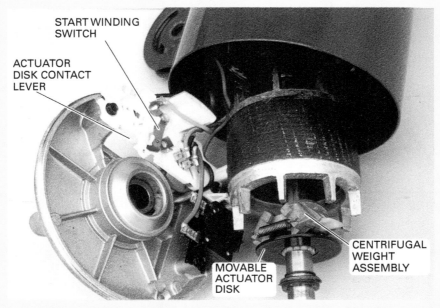

(A) CENTRIFUGALLY-ACTUATED START WINDING SWITCH

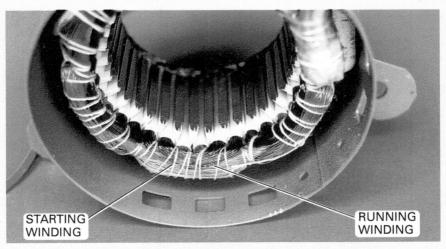

(B) STARTING WINDING AND RUNNING WINDING

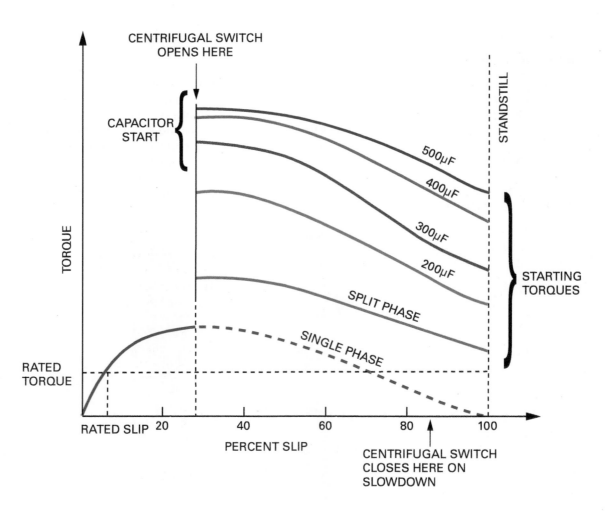

Figure 42 Torque-slip curves.

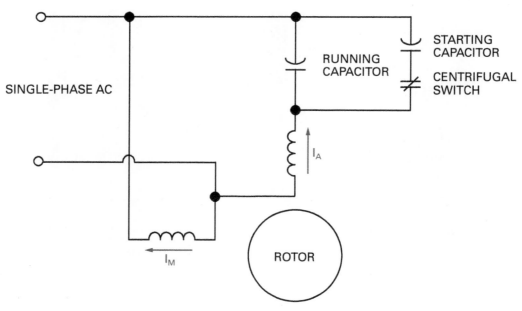

Figure 43 Capacitor-start, capacitor-run motor schematic.

2.4.4 Shaded-Pole Induction Motor

The shaded-pole motor employs a salient-pole stator and a cage rotor. The projecting poles on the stator resemble those of DC machines, except that the entire magnetic circuit is laminated and a portion of each pole is split to accommodate a short-circuited copper strap called a shading coil. This motor is generally manufactured in very small sizes and runs up to $\frac{1}{20}$hp. A four-pole motor of this type is illustrated in *Figure 44*.

The shading coils are placed around the leading pole tip, and the main pole winding is concentrated and wound around the entire pole. The four coils that make up the main winding are connected in series across the motor terminals. An inexpensive type of two-pole motor that uses shading coils is illustrated in *Figure 45*.

Figure 45 shows that during part of the cycle when the main pole flux ($\varphi 1$) is increasing, the shading coil is cut by the flux, and the resulting induced EMF and current in the shading coil tend to prevent the flux from rising readily through it. Thus, the greater portion of the flux rises in the portion of the pole that is not in the vicinity of the shading coil ($\varphi 1 > \varphi 2$). When the flux reaches its maximum value, the rate of change of flux is zero, and the voltage and current in the shading coil are also at zero. At this time, the flux is distributed more uniformly over the entire pole face ($\varphi 1 = \varphi 2$).

As the main flux decreases toward zero, the induced voltage and current in the shading coil reverse their polarity, and the resulting force tends to prevent the flux from collapsing through the iron in the region of the shading coil ($\varphi 2 > \varphi 1$).

The result is that the main flux rises first in the unshaded portion of the pole and later in the shaded portion. This action is equivalent to a sweeping movement of the field across the pole face in the direction of the shaded pole. The cage rotor conductors are cut by this moving field, and the force exerted on them causes the rotor to turn in the direction of the sweeping field.

Most shaded-pole motors have only one edge of the pole split; therefore, the direction of rotation is not reversible. However, some shaded-pole motors have both leading and trailing pole tips split to accommodate shading coils. The leading pole tip shading coils form one series group, and the trailing pole tip shading coils form another series group. Only the shading coils in one group are simultaneously active, while those in the other group are on an open circuit.

The shaded-pole motor is similar in operating characteristics to the split-phase motor. The former has the advantages of simpler construction and lower cost. It has no sliding electrical contacts and is reliable in operation. However, it has low starting torque, low efficiency, and a high noise level. It is normally used to operate small fans. The shading coil and split pole are also used in timers to make them self-starting.

2.4.5 Single-Phase Synchronous Motor

The single-phase synchronous motor, as its name implies, runs at synchronous speed. It finds use where a constant speed is needed, such as in tape recorders and clocks. It is started in the same way

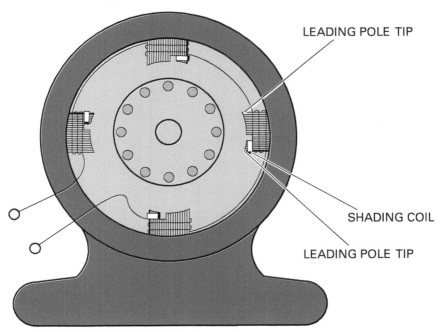

Figure 44 Four-pole shaded-pole motor.

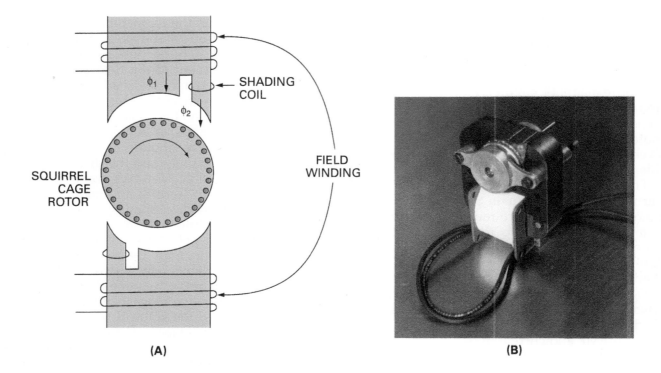

Figure 45 Two-pole shaded-pole motor.

Capacitor-Start, Capacitor-Run Motors

These motors run quietly and smoothly and have a high starting torque. Their construction is similar to a split-phase motor in that they use a centrifugal switch to remove only the start capacitor from the starting winding while leaving the run capacitor connected to the winding. These motors have a higher power factor than ordinary split-phase motors.

Shaded-Pole Motors

The efficiency of this type of motor can be as low as 5%. However, this low efficiency is rarely significant because these motors use very little power to begin with ($\frac{1}{20}$hp or less).

as any of the single-phase induction motors and therefore has a rotating field. By having a modified rotor, the motor pulls into synchronism and runs at synchronous speed.

2.4.6 Multiple-Speed Induction Motors

The speed of an induction motor depends on the power supply frequency and the number of pairs of poles used in the motor. Obviously, to alter motor speed it is merely necessary to change one of these two factors. By far the most common method used involves changing the number of poles, generally at some type of external controller.

There are two types of multiple-speed squirrel cage induction motors in common use: the multiple-winding motor and the consequent-pole motor. Both feature poles that may be changed, as required, by shifting key external connections, and in this way they provide for operating the motor at a limited number of different speeds. They are described as follows:

- *Multiple-winding motor* – In a multiple-winding motor, two or more separate windings are placed in the stator core slots, one over the other, as shown in *Figure 46*. For example, a four-pole winding can be positioned in the core slots and have a two-pole winding placed on top of it. The windings are insulated from each other and arranged so that only one winding at a time can be energized. Switching speeds is normally accomplished by switching contacts that are in the motor controller external to the motor itself.
- *Consequent-pole motor* – Consequent-pole motors have a single winding with two speeds. The motor is constructed to have a certain number of poles for high-speed operation and then, by a switching action, double this number of poles to give low-speed operation. The switching action is illustrated by the use of the motor in *Figure 47*. If you trace the wiring in *Figure 47*, you can see how the system is constructed so that both magnetic north and south poles are produced at the winding projections. With power applied to the two-pole motor, a rotating magnetic field of 3,600 rpm is produced.

In *Figure 48*, the connections are changed so that the system is phased to produce four magnetic north poles at the winding projections. Because every north pole must have a south pole, consequent south poles are produced between the projecting north poles as a consequence of having formed north poles. Accordingly, in *Figure 48*, there are twice as many pole groups as in *Figure 47*. Therefore, a four-pole rotating magnetic field of 1,800 rpm is produced.

Figure 49 shows the short-jumpering arrangement of the consequent-pole motor. In this, all windings are in series, and alternate north and south poles are produced. To produce consequent poles, the series connection is replaced with a parallel connection accomplished by the long-jumpering arrangement. By connecting the motor in this manner, four salient monopoles are produced and, as a result, create four opposite consequent poles. In the practical consequent-pole motor, all necessary internal connection rearrangements are accomplished at an external control panel.

Consequent-pole motor characteristics depend on the intended application. In a constant-horsepower motor, torque varies inversely with speed. It is used for driving machine tools.

In a constant-torque motor, horsepower varies directly with speed. It is used to drive pumps and air compressors, as well as in constant-pressure blowers. In a variable-torque, variable-horsepower motor, both torque and horsepower change with changes in speed. This is the type of motor found in household fans and air conditioners.

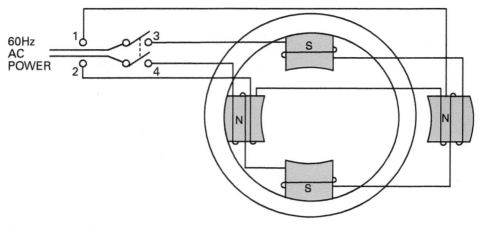

Figure 46 Two-winding, two-speed motor.

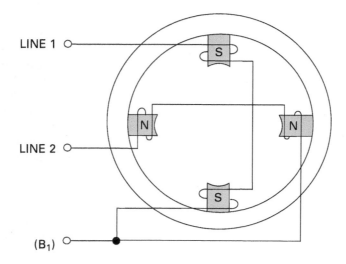

Figure 47 High-speed consequent-pole motor.

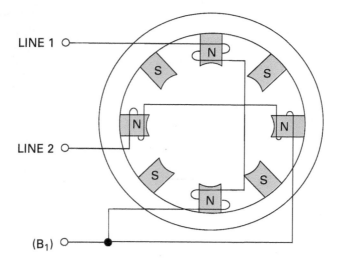

Figure 48 Low-speed consequent-pole motor.

Figure 49 Single-phase consequent-pole motor.

Hysteresis Synchronous Motor

This unusual synchronous motor has an external rotor. These motors have low noise levels, high efficiency, and constant speed. They are used for applications such as tape recorders and clocks.

2.0.0 Section Review

1. If a motor has 4 poles and is operating at a frequency of 60Hz, its synchronous speed is _____.
   a. 240 rpm
   b. 480 rpm
   c. 1,200 rpm
   d. 1,800 rpm

2. The type of motor most likely to be used in a location with flammable dust and torque fluctuations is a _____.
   a. single-phase induction motor
   b. wound-rotor induction motor
   c. squirrel cage induction motor
   d. synchronous motor

3. A type of motor used in precision applications that require constant speed with zero slip is a _____.
   a. single-phase induction motor
   b. wound-rotor induction motor
   c. squirrel cage induction motor
   d. synchronous motor

4. A heating system is most likely to use a _____.
   a. single-phase induction motor
   b. wound-rotor induction motor
   c. squirrel cage induction motor
   d. synchronous motor

Section Three

3.0.0 Variable-Speed Drives

Objective

Identify variable-speed drives and describe their operating characteristics.

a. Identify types of adjustable speed loads.
b. Identify types of motor speed control.
c. Identify braking methods.

The use of adjustable speed in industrial equipment is increasing due to the need for better equipment control and for energy savings where partial power is required. AC drives compete with DC drives, eddy current drives, and mechanical and hydraulic systems as methods to control speed. Reliability, cost, and control capabilities are the major factors in system selection.

A drive system includes both the drive controller and the motor being driven. This module focuses specifically on the electronic drive components and covers various types of control for both DC and AC drives. This section provides a basic review of some fundamental principles that are important to understand when starting up, operating, or troubleshooting a variable-speed drive system.

3.1.0 Adjustable Speed Loads

Most drive controllers can be adjusted or modified to optimize performance and provide the most efficient and cost-effective drive, depending on the load characteristics of the application.

It is important to understand the speed and torque characteristics as well as the maximum horsepower requirements for the type of load to be considered. Based on this, either a constant-torque controller or a variable-torque controller is selected. The most common types of loads are shown in *Figure 50*. Note that a load requires the same amount of torque at low speed as at high speed.

For a constant-torque load, the torque remains constant throughout the speed range and the horsepower increases and decreases in direct proportion to the speed. This applies to applications such as conveyors, as well as applications in which shock loads, overloads, or high inertia loads are encountered.

A variable-torque load requires much lower torque at lower speeds. This applies to applications such as centrifugal fans, pumps, and blowers.

A constant-horsepower load requires high torque at low speeds, low torque at high speeds, and thus constant horsepower at any speed. It applies to applications such as lathes requiring slow speeds for deep cuts and high speeds for finishing. Usually, very high starting torques are required.

For industrial applications, motors are required to function at varying torques and speeds and in forward and reverse directions. Besides operating as a motor, the machine may also function as a brake or a generator for short periods. The various operating modes for industrial drives are shown in *Figure 51*. Positive speed and negative speed (rotation) are plotted on the horizontal axis, and torque is plotted on the vertical axis. The quadrants of operation are labeled 1, 2, 3, and 4.

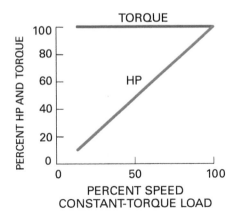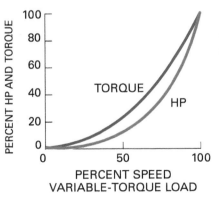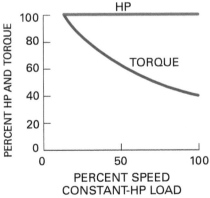

Figure 50 Types of adjustable speed loads.

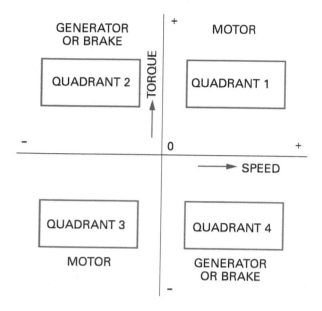

Figure 51 Electric drive operation in four quadrants.

A machine operating in quadrant 1 has positive torque and speed, which means that they both act in the same direction (in this case, clockwise). A machine in this quadrant functions as a motor. It delivers mechanical power to a load. The machine will also act as a motor in quadrant 3, but torque and speed are reversed from quadrant 1 (counterclockwise).

While operating in quadrant 2, a machine will develop a positive torque and a negative speed. The torque is acting clockwise, and the speed is counterclockwise. In this quadrant, the machine absorbs mechanical power from the load and functions as a generator. This mechanical power is converted into electric power and is generally transmitted back into the line. The electric power may also be dissipated in an external resistor, which is known as dynamic braking.

Depending on its connections, a machine may also be used as a brake while operating in quadrant 2. Absorbed mechanical power is converted to electric power, then converted into heat. If the machine absorbs electric line power as it converts mechanical power into electric power, it functions as a brake. Both power inputs are dissipated as heat. Large power drives seldom use the brake mode of operation because it is very inefficient. The circuitry is generally chosen so that the machine will function as a generator when it operates in quadrant 2. Quadrant 4 operation is identical to quadrant 2 except that speed and torque are reversed.

3.1.1 Typical Torque-Speed Curves

A three-phase motor has a torque-speed curve that is a good example of an electrical machine's behavior as a generator brake. The solid curve in *Figure 52* is the torque-speed curve for a machine acting as a motor in quadrant 1, a brake in quadrant 2, and a generator in quadrant 4.

If the stator leads are reversed, the torque-speed curve is shown by the dotted curve. Now the motor operates as a motor in quadrant 3, a generator in quadrant 2, and a brake in quadrant 4. The machine functions as a brake or a generator in quadrants 2 and 4, but it always runs as a motor in quadrants 1 and 3.

Figure 53 shows the torque-speed curve of a DC shunt motor. Motor, generator, and brake modes are apparent. The dotted curve represents reversed armature leads.

Variable-speed electric drives are designed to vary speed and torque in a smooth and continuous manner so as to satisfy load requirements. Typically, this is accomplished by shifting the torque-speed characteristic back and forth along the horizontal axis. The torque-speed characteristic of the motor is shifted by varying the armature voltage. Also, the curve of an induction motor can be shifted by varying the voltage and frequency applied to the stator.

In describing the various methods of motor control, only the behavior of power circuits will be discussed. The many ways of shaping and controlling triggering pulses will not be covered. They constitute a complex subject that involves sophisticated electronics, logic circuits, integrated circuits, and microprocessors.

3.1.2 Motor Heating

Because a variable-speed drive system includes both the drive controller and the motor, the design engineer should always consider the capabilities of the motor to perform acceptably under the desired operating conditions. A factor to observe is motor heating. When operating a motor at reduced speeds, the ability to dissipate heat is also reduced due to the slower cooling fan speed. This factor should be considered when maintaining the motor, modifying its enclosure or surrounding area, or troubleshooting the drive system.

3.2.0 Motor Speed Control

It is important to understand how DC or AC motor speed can be varied in order to understand how a drive controller accomplishes that task.

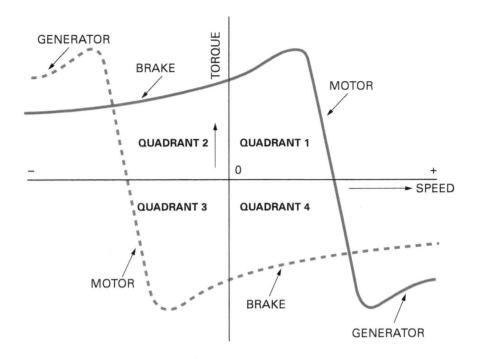

Figure 52 Four-quadrant operation for a squirrel cage motor.

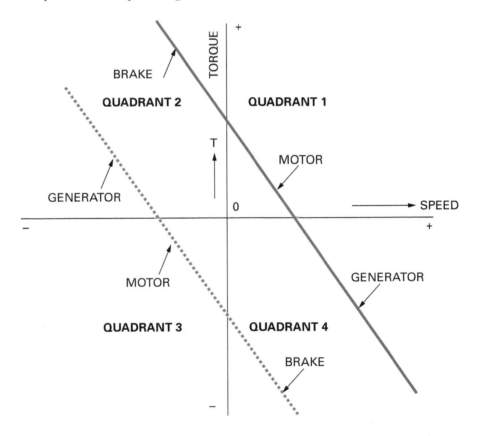

Figure 53 Four-quadrant operation for a DC motor.

This section reviews the fundamentals of DC and AC motor speed control.

3.2.1 Varying the Speed of a DC Shunt Motor

A DC shunt motor is shown in *Figure 54* (A). Basically, there are two ways of varying the running speed of a DC shunt motor:

- Adjusting the voltage (and current) applied to the field winding. As the field voltage is increased, the motor slows down. This method is shown in *Figure 54* (B).
- Adjusting the voltage (and current) applied to the armature. As the armature voltage is increased, the motor speeds up. This method is shown in *Figure 54* (C).

As the field voltage is increased by reducing RV in *Figure 54* (B), the field current is increased. This results in a stronger magnetic field, which induces a greater CEMF in the armature winding. The greater CEMF tends to oppose the applied DC voltage and reduces the armature current, I_A. Therefore, an increased field current causes the motor to slow down until the induced CEMF has returned to near its normal value.

Going in the other direction, if the field current is reduced, the magnetic field gets weaker. This causes a reduction in CEMF created by the rotating armature winding. The armature current increases, forcing the motor to spin faster, until the CEMF is once again approximately equal to what it was before. An increase in armature speed compensates for the reduction in magnetic field strength.

This method of speed control has certain positive features. It can be accomplished by a small, inexpensive rheostat because the current in the field winding is fairly low. Also, because of the low value of the field current, I_F, the rheostat R_V does not dissipate very much energy. Therefore, this method is energy efficient.

However, there is a major drawback to speed control from the field winding: to increase the speed, you must reduce I_F and weaken the magnetic field, thereby lessening the motor's torque-producing ability. The ability of a motor to create torque depends on two things: the current in the armature conductors and the strength of the magnetic field. As I_F is reduced, the magnetic field is weakened, and the motor's torque-producing ability declines. Unfortunately, it is at this point that the motor needs all the torque-producing ability it can get because it probably requires greater torque to drive the load at a faster speed.

From the torque-producing point of view, armature control is much better, as shown in *Figure 54* (C). As the armature voltage and current are increased by reducing RV, the motor starts running faster, which normally requires more torque.

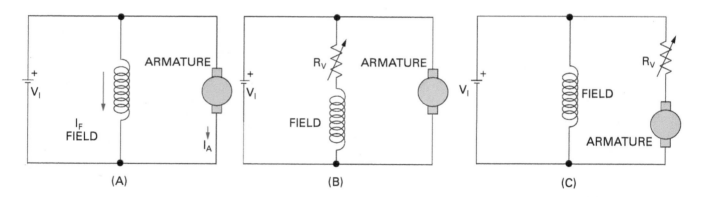

Figure 54 DC shunt motor schematic.

Using a Motor as Both a Generator and a Brake

GOING GREEN

DC motors that power subway cars are also used for regenerative braking. When driven by the train's momentum, such as when slowing down upon approaching a station, the motor acts as a generator and puts current back into the system. More precisely, the train's motors are not consuming power to produce motion. Instead, the train's great forward inertia turns the motors, producing current. At the same time, the resistive torque resulting from the power generation overcomes the train's forward momentum and slows the train. The same regenerative braking process is used in many hybrid cars.

The reason for the rise in speed is that the increased armature voltage demands an increased CEMF to limit the increase in armature current to a reasonable amount. The only way the CEMF can increase is for the armature winding to spin faster because the magnetic field strength is fixed. In this instance, the elements are all present for increased torque production because the magnetic field strength is kept constant and I_A is increased.

The problem with the armature control method of *Figure 54* (C) is that R_V, the rheostat, must handle the armature current, which is relatively large. Therefore, the rheostat must be physically large and expensive, and it will waste a considerable amount of energy.

3.2.2 Varying the Speed of an AC Motor

The principle of speed control for adjustable-frequency drives is based on the following fundamental formula for a standard AC motor:

$$N_s = \frac{120f}{P}$$

Where:

N_s = synchronous speed (rpm)
f = frequency
P = number of poles
120 = constant

The number of poles of a particular motor is set in its design and manufacture.

The adjustable-frequency system controls the frequency (f) applied to the motor. The speed (N_s) of the motor is then proportional to this applied frequency. Control frequency is adjusted by means of a potentiometer or external signal, depending on the application.

The frequency output of the controller is adjustable over its design speed range. Therefore, the speed of the motor is adjustable over this same range. Because an electronic means of generating variable frequencies is being used, the speed range often exceeds the 60 hertz (Hz) rated speed of the motor.

When variable-frequency speed control is employed, the motor supply voltage cannot be allowed to remain at a steady value. The magnitude of the motor voltage must be increased or decreased in proportion to the frequency. That is, the voltage-to-frequency ratio, V/f, must remain approximately constant. (The voltage-to-frequency ratio is sometimes referred to as the *volts-per-hertz ratio*.)

For instance, if the motor has a nameplate rating of 240V at 60Hz, the voltage-to-frequency ratio is 4 (because 240 ÷ 60 = 4). If the motor is speeded up by adjusting its variable-frequency inverter to 90Hz, the voltage magnitude must be increased to 360V because 4 × 90 = 360. If the motor is slowed down by adjusting the inverter frequency to 45Hz, the voltage magnitude must be decreased to 180V because 4 × 45 = 180.

The stator's magnetic field strength must remain constant under all operating conditions. If the stator field strength should happen to rise much above the design value, the motor's core material would go into magnetic saturation. This would effectively lower the core's permeability, thereby inhibiting proper induction of voltage and current in the rotor loops (or bars), and detracting from the torque-producing capability of the motor. On the other hand, if the stator field strength should happen to fall much below the design value, the weakened magnetic field would simply induce lower values of voltage and current in the rotor loops. This would also detract from the torque-producing ability of the motor.

Therefore, the magnetic field produced by the stator windings must hold a constant rms value, regardless of frequency. The magnetizing current of an induction motor is the current that flows through the stator winding when the rotor is spinning at steady-state speed with no torque load. The magnetizing current for an induction motor is given by Ohm's law:

$$I_{mag} = \frac{V}{X_L}$$

Where:

V = rms value of the applied stator voltage
X_L = inductive reactance of the stator winding

In the equation, X_L does not remain constant as the supply frequency is adjusted; it varies in proportion to the frequency ($X_L = 2\pi fL$). Therefore, V must also be varied in proportion to the frequency, so that the Ohm's law division operation yields an unvarying value of magnetizing current.

Alternatively, using $X_L = 2\pi fL$, the equation can be rewritten as follows:

$$I_{mag} = \frac{V}{X_L}$$

$$I_{mag} = \frac{V}{2\pi fL}$$

$$I_{mag} = \frac{1}{2\pi L} \times \frac{V}{f}$$

Because 1 ÷ 2πL is a constant determined by the motor's construction, the magnetizing current is kept constant by maintaining the V/f ratio.

The controller can automatically maintain the required volts/cycle (V/Hz) ratio to the motor at any speed. This provides maximum motor capability throughout the speed range.

The V/Hz setting is typically preset at the factory. However, on many controllers it can be adjusted or changed to fine-tune controller operation.

3.3.0 Braking Methods

Most people think of braking in terms of mechanical friction braking, such as that used in cars. However, friction braking requires a physical connection between moving parts, which results in wear and maintenance. For many motors, it is often more effective to employ injection or dynamic braking.

3.3.1 DC Injection Braking

DC injection braking is a method of braking in which direct current (DC) is applied to the stationary windings of an AC motor after the AC voltage is removed. This is an efficient and effective method of stopping most AC motors. DC injection braking provides a quick and smooth braking action on all types of loads, including high-speed and high-inertia loads.

In an AC induction motor, when the AC voltage is removed, the motor will coast to a standstill over a period of time because there is no induced field to keep it rotating. Because the coasting time may be unacceptable, particularly in an emergency situation, electric braking can be used to provide a more immediate stop.

By applying a DC voltage to the stationary windings when the AC is removed, a magnetic field is created in the stator that will not change polarity. This constant magnetic field in the stator creates a magnetic field in the rotor. Because the magnetic field of the stator is not changing in polarity, it will attempt to stop the rotor when the magnetic fields are aligned. When the fields are aligned north and south, the two magnetic fields are attracted to each other. The resistive pull of the fields is used to stop the motor's rotational inertia. The same is true when the magnetic poles are the same, but instead of attraction, the magnetic forces repel one another, causing the motor to slow.

The stopping time is based on the amount of DC current applied. Normally, the applied current is three times the full load current of the motor. The higher the current level, the faster the motor will stop. For instance, in paper mills where sudden motor stopping time is an issue, the applied current is less than normal to avoid tearing the paper.

The only thing that can keep the rotor from stopping with the first alignment is the rotational inertia of the load connected to the motor shaft. However, because the braking action of the stator is present at all times, the motor is stopped quickly and smoothly to a standstill.

Because there are no parts that come in physical contact during braking, maintenance is kept to a minimum.

3.3.2 Dynamic Braking

Dynamic braking is another method for stopping a motor. It is achieved by reconnecting a running motor to act as a generator immediately after it is turned off, rapidly stopping the motor. The generator action converts the mechanical energy of rotation to electrical energy that can be dissipated as heat in a resistor or used as a load as in a generator. Using the motor as a generator is not very practical due to the fact that the motor stopping time is very limited.

Dynamic braking of a DC motor may be needed because DC motors are often used for lifting and moving heavy loads that may be difficult to stop. For example, forklifts often use dynamic braking.

There must be access to the rotor windings in order to reconnect the motor to act as a generator. On a DC motor, access is accomplished through the brushes on the commutator. The armature terminals of the DC motor are disconnected from the power supply and immediately connected across a resistor, which acts as a load. The smaller the resistance of the resistor, the greater the rate of energy dissipation and the faster the motor slows down.

The field windings of the DC motor are left connected to the power supply. The armature generates CEMF, which causes current to flow through the resistor and armature. The current causes heat to be dissipated in the resistor, removing energy from the system and slowing the motor rotation.

The generated CEMF decreases as the speed of the motor decreases. As the motor speed approaches zero, the generated voltage also approaches zero. This means that the braking action lessens as the speed of the motor decreases. As a result, a motor cannot be braked to a complete stop using dynamic braking. Dynamic braking also cannot hold a load once it is stopped because without the generated CEMF there is no more braking action.

For this reason, electromechanical friction brakes are sometimes used along with dynamic braking in applications that require the load to be held, or in applications where a large, heavy load is to be stopped.

3.0.0 Section Review

1. With a constant-torque load, the horsepower _____.
    a. varies in direct proportion to the speed
    b. varies inversely with the speed
    c. remains constant
    d. lags the speed by 10 percent

2. When variable-frequency speed control is used on a motor with a nameplate rating of 480V at 60Hz, the voltage-to-frequency ratio is _____.
    a. 1
    b. 2
    c. 4
    d. 8

3. When using a motor with dynamic braking, _____.
    a. the field windings are disconnected during braking action
    b. electromechanical friction brakes may be required with large loads
    c. the braking action can be used to hold a load after it has stopped
    d. a large load can be braked to an immediate stop

Section Four

4.0.0 Motor Enclosures, Frame Designations, and Operating Characteristics

Objective

Identify motor enclosures, frame designations, and operating characteristics.
a. Identify types of motor enclosures.
b. Identify NEMA frame designations.
c. Identify motor operating characteristics using nameplate data.

Performance Task

2. Collect data from a motor nameplate.

Trade Terms

Circuit breaker: A device designed to open and close a circuit by nonautomatic means and to open the circuit automatically on a predetermined overcurrent without injury to itself when properly applied within its rating.

Continuous duty: Operation at a substantially constant load for an indefinitely long time.

Duty: Describes the length of operation. There are four designations for circuit duty: continuous, periodic, intermittent, and varying.

Hours: The duty cycle of a motor. Most fractional horsepower motors are marked continuous for around-the-clock operation at the nameplate rating in the rated ambient conditions. Motors marked one-half are for ½-hour ratings, and those marked one are for 1-hour ratings.

Intermittent duty: Operation for alternate intervals of (1) load and no load, (2) load and rest, or (3) load, no load, and rest.

Overcurrent: Any current in excess of the rated current of equipment or the ampacity of a conductor. It may result from an overload, short circuit, or ground fault.

Periodic duty: Intermittent operation at a substantially constant load for a short and definitely specified time.

Varying duty: Operation at varying loads and/or intervals of time.

The National Electrical Manufacturers Association (NEMA) classifies motors by enclosure size and frame designation. These classifications were developed to insure interchangeability of motors among manufacturers.

4.1.0 Motor Enclosures

Motors are usually designed with covers over the moving parts. These covers, called enclosures, are classified by NEMA according to the degree of environmental protection provided and the method of cooling. If the cover has openings, the motor is classified as an open motor; if the enclosure is complete, the motor is classified as an enclosed motor. Each of these types of motors has many varieties. *Table 1* lists the various types for both open and totally enclosed motors.

4.1.1 Open Motor

The most common type of motor is the open motor. It has ventilating openings that permit the passage of external cooling air over and around its windings. If these are limited in size and shape, the motor is called a protected motor because it is protected from any large pieces of material that may somehow enter the motor, thus damaging its internal parts. A protected motor also prevents a person from touching the rotating or electrically energized parts of the motor. Drip-proof and splash-proof motors are constructed to prevent entry of drops of liquid.

The different types of open motors as defined by NEMA are as follows:

- *General purpose* – This type has ventilating openings that permit the passage of external cooling air over and around the windings of the machine.
- *Drip-proof* – Ventilating openings are constructed so that successful operation is not affected by drops of liquid or solid particles that strike or enter the enclosure at any angle from 0° to 15° downward from the vertical.
- *Splash-proof* – Ventilating openings are constructed so that successful operation is not affected by drops of liquid or solid particles that strike or enter the enclosure at any angle not greater than 100° downward from the vertical.
- *Guarded* – Openings giving direct access to live metal or rotating parts (except smooth surfaces) are limited in size by the structural parts or by screens, baffles, grills, expanded metal, or other means to prevent accidental contact with hazardous parts.

Table 1 Motor Enclosure Types

Open	Totally Enclosed
General purpose	Nonventilated
Drip-proof	Fan-cooled
Splash-proof	Fan-cooled guarded
Guarded	Explosion-proof
Semi-guarded	Dust- and ignition-proof
Drip-proof guarded	Pipe-ventilated
Externally ventilated	Water-cooled
Pipe-ventilated	Water-to-air-cooled
Weather-protected (Type I & Type II)	
Encapsulated windings	
Sealed windings	

- *Semi-guarded* – Some of the ventilating openings, usually in the top half, are guarded as in the case of a guarded machine, but the others are left open.
- *Drip-proof guarded* – This type of machine has ventilating openings as in a guarded machine.
- *Externally ventilated* – This designates a machine that is ventilated by a separate motor-driven blower mounted on the machine enclosure. Mechanical protection may be as defined above. This machine is sometimes known as a blower-ventilated or force-ventilated machine.
- *Pipe-ventilated* – Openings for the admission of ventilating air are arranged so that inlet ducts or pipes can be connected to them.
- *Weather-protected* – Type I: Ventilation passages are designed to minimize the entrance of rain, snow, and airborne particles to the electrical parts. Type II: In addition to the enclosure described for a Type I machine, ventilating passages at both intake and discharge are arranged so that high-velocity air and airborne particles blown into the machine by storms or high winds can be discharged without entering the internal ventilating passages leading directly to the electric parts.
- *Encapsulated windings* – An AC squirrel cage machine having random windings filled with an insulating resin, which also forms a protective coating.
- *Sealed windings* – An AC squirrel cage machine that makes use of form-wound coils and an insulation system that, through the use of materials, processes, or a combination of materials and processes, results in a sealing of the windings and connections against contaminants.

4.1.2 Enclosed Motor

The totally enclosed motor is designed to prevent the free exchange of air between the inside and outside of the actual motor housing. It is used where hostile environmental conditions and the motor application require maximum protection of the internal parts of the motor.

The different types of totally enclosed motors as defined by NEMA are as follows:

- *Nonventilated* – Not equipped for cooling by means external to the enclosing parts.
- *Fan-cooled* – Equipped for exterior cooling by means of a fan or fans that are integral with the machine but external to the enclosing parts.
- *Fan-cooled guarded* – All openings giving direct access to the fan are limited in size by design of the structural parts or by screens, grills, expanded metal, etc., to prevent accidental contact with the fan.
- *Explosion-proof* – Designed and constructed to withstand an explosion of a specified gas or vapor that may occur within it and to prevent the ignition of the specified gas or vapor surrounding the machine by sparks, flashes, or explosions of the specified gas or vapor that may occur within the machine casing.
- *Dust- and ignition-proof* – Designed and constructed in a manner that will exclude ignitable amounts of dust or amounts that might affect performance or rating, and that will not permit arcs, sparks, or heat otherwise generated or liberated inside the enclosure to cause ignition of exterior accumulations or atmospheric suspensions of a specific dust on or in the vicinity of the enclosure.
- *Pipe-ventilated* – Openings are so arranged that when inlet and outlet ducts or pipes are connected to them there is no free exchange of the internal air and the air outside the case.
- *Water-cooled* – Cooled by circulating water, with the water or water conductors coming in direct contact with the machine parts.
- *Water-to-air-cooled* – Cooled by circulating air, which in turn is cooled by circulating water.

Self-Cooling Motors

Conventional squirrel cage fan-cooled motors may overheat when operated at reduced speeds. Many manufacturers now offer inverter duty-rated motors with increased self-cooling capability.

4.2.0 NEMA Frame Designations

Frame sizes were developed by NEMA to ensure interchangeability of motors among manufacturers. Frame sizes are listed on motor nameplates and provide information about the machine's physical dimensions. Key dimensions are shown in *Figure 55* and *Figure 56*. A few of these are:

- Distance from motor feet to shaft center line, known as the D dimension
- Bolt-hole center-to-center distance between front and back feet, known as the 2F dimension
- Exposed shaft distance from shaft end to shaft shoulder, known as the N-W dimension

Manufacturer tables are available to correlate frame size to dimensions. An example is shown in *Table 2*. The system for designating the frames of motors and generators consists of a series of numbers in combination with letters. Note that metric sizes vary, but also use an alphanumeric system with dimensions given in millimeters.

More compact design, better ventilation, and insulation systems with higher temperature ratings have enabled manufacturers to house motors in increasingly smaller frame sizes. NEMA re-rates occurred in 1952 and 1964. Motors manufactured before 1952 are generally referred to as pre-U-frame motors. Those manufactured between 1952 and 1964 are called U-frame motors; those manufactured since 1964 are called T-frame motors.

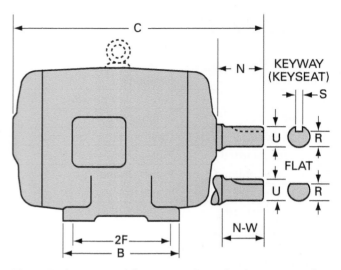

Figure 56 Lettering of dimension sheets for foot-mounted machines (side view).

4.2.1 Small Machines

The frame number for small machines is the D dimension in inches multiplied by 16. The following letters shall immediately follow the frame number to denote variations:

- **B** – Carbonator pump motors
- **C** – Type C face-mounting motors
- **G** – Gasoline pump motors
- **H** – A frame having an F dimension larger than that of the same frame without the suffix H
- **J** – Jet pump motors
- **K** – Sump pump motors
- **M** – Oil burner motors
- **N** – Oil burner motors

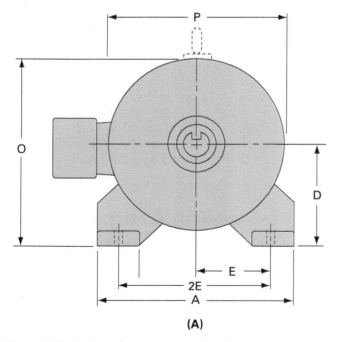

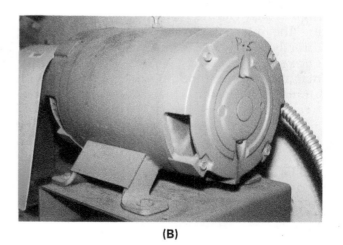

Figure 55 End view of a foot-mounted motor.

- **Y** – Special mounting dimensions (must obtain dimensional diagram from manufacturer)
- **Z** – All mounting dimensions are standard except the shaft extension

4.2.2 Medium Machines

The system for numbering frames of medium machines is as follows:

- The first two digits of the frame number are equal to four times the D dimension in inches. (If this product is not a whole number, the first two digits of the frame number shall be the next higher whole number.)
- The third and, when required, the fourth digit of the frame number are obtained from the value of 2F in inches.

Figure 55 shows a typical end view of a foot-mounted machine. The many different dimensions can be found on a dimension sheet for that machine. The NEMA frame designation will provide information relating to both the D and 2F dimensions.

Table 2 may be used to determine the D dimension and 2F dimension for medium-size motors. The D dimension is the distance from the center line of the shaft to the bottom of the feet. The 2F dimension is the distance between the center lines of the mounting holes in the feet or in the base of the machine.

Medium machines also use letters that denote variations. These letters follow the frame number. Because there are many more varieties of medium-size machines, the letter relates to the different aspects of mounting and shaft orientation.

Table 2 Frame Dimension Chart

Frame Number Series	D	Third/Fourth Digit in Frame Number						
		1	2	3	4	5	6	7
		2F Dimensions						
140	3.50		3.00	3.50	4.00	4.50	5.00	5.50	6.25
160	4.00		3.50	4.00	4.50	5.00	5.50	6.25	7.00
180	4.50		4.00	4.50	5.00	5.50	6.25	7.00	8.00
200	4.50		4.50	5.00	5.50	6.50	7.00	8.00	9.00
210	5.00		4.50	5.00	5.50	6.50	7.00	8.00	9.00
220	5.50		5.00	5.50	6.25	6.75	7.50	9.00	10.00
250	6.25		5.50	6.25	7.00	8.25	9.00	10.00	11.00
280	7.00		6.25	7.00	8.00	9.50	10.00	11.00	12.50
320	8.00		7.00	8.00	9.00	10.50	11.00	12.00	14.00
360	9.00		8.00	9.00	10.00	11.25	12.25	14.00	16.00
400	10.00		9.00	10.00	11.00	12.25	13.75	16.00	18.00
440	11.00		10.00	11.00	12.50	14.50	16.50	18.00	20.00
500	12.50		11.00	12.50	14.00	16.00	18.00	20.00	22.00
580	14.50		12.50	14.00	16.00	18.00	20.00	22.00	25.00
680	17.00		16.00	18.00	20.00	22.00	25.00	28.00	32.00

Frame Number Series	D	Third/Fourth Digit in Frame Number						
		8	9	10	11	12	13	14	15
		2F Dimensions						
140	3.50	7.00	8.00	9.00	10.00	11.00	12.50	14.00	16.00
160	4.00	8.00	9.00	10.00	11.00	12.50	14.00	16.00	18.00
180	4.50	9.00	10.00	11.00	12.50	14.00	16.00	18.00	20.00
200	5.00	10.00	11.00	...	...	...	...	...	...
210	5.25	10.00	11.00	12.50	14.00	16.00	18.00	20.00	22.00
220	5.50	11.00	12.50	...	...	...	...	...	...
250	6.25	12.50	14.00	16.00	18.00	20.00	22.00	25.00	28.00
280	7.00	14.00	16.00	18.00	20.00	22.00	25.00	28.00	32.00
320	8.00	16.00	18.00	20.00	22.00	25.00	28.00	32.00	36.00
360	9.00	18.00	20.00	22.00	25.00	28.00	32.00	36.00	40.00
400	10.00	20.00	22.00	25.00	28.00	32.00	36.00	40.00	45.00
440	11.00	22.00	25.00	28.00	32.00	36.00	40.00	45.00	50.00
500	12.50	25.00	28.00	32.00	36.00	40.00	45.00	50.00	56.00
580	14.50	28.00	32.00	36.00	40.00	45.00	50.00	56.00	63.00
680	17.00	36.00	40.00	45.00	50.00	56.00	63.00	71.00	80.00

For example, to understand the NEMA frame designation, take a typical motor frame designation and determine the D and 2F dimensions. Then use these dimensions to determine the frame designation number.

Example 1:

A typical medium-size frame number is a 256T. Because this is a medium frame, divide the first two digits by 4:

$$25 \div 4 = 6.25$$

Therefore, the D dimension is 6.25 inches.

To determine the 2F dimension, use *Table 2* and the third digit in the frame number. The third digit is 6, and the frame is a 250 series. Using the table, the 2F dimension is 10 inches. The T in the frame number is included as part of a frame designation for which standard dimensions have been established.

Medium-size frames can have multiple letters that denote a variety of different applications and arrangements. A 256AT has the same dimensions, with the A added to denote an industrial DC machine.

Example 2:

A frame has a D dimension of 3.5 inches and a 2F dimension of 4 inches, and all standard dimensions have been established. Multiplying the D dimension by 4 will give the first two digits of the frame designation:

$$3.5 \times 4 = 14$$

Using the table, a 140 frame series and a 2F dimension of 4 inches provides a third digit of 3. Because it is a standard dimension frame, the letter will be the suffix. This frame has a designation of 143T.

Full-load torque (rather than horsepower) determines the frame size required to house the motor. Thus, a motor developing a large amount of horsepower at high speed will have the same frame size as a machine developing less horsepower at a slower speed.

4.3.0 Motor Ratings and Nameplate Data

Specified information that must be listed on a motor nameplate is regulated in *NEC Section 430.7*. Requirements can also be found in *NEMA Standards MG-1* and *MG-2*. A typical motor nameplate is shown in *Figure 57*. The ratings of an electric motor include:

- Voltage
- Amperage
- Speed
- Horsepower
- NEMA design letters
- Insulation class
- Frequency
- Service factor
- NEMA kVA code letters
- Bearings
- Power factor
- Duty rating

4.3.1 Voltage

There are two voltage ratings to be considered: the nominal voltage and the minimum starting voltage. The nominal rated voltage is defined as the voltage rating at which the motor is designed to operate. The minimum starting voltage may be defined as the lowest voltage at which a motor will start without drawing excessive current and tripping protective devices.

Power plant induction motors are designed to operate with a balanced three-phase voltage source applied at the terminals. The rated voltage on the nameplate is usually lower than the voltage of the electrical system. For example, a 460V motor is designed to operate in a 480V system. Here,

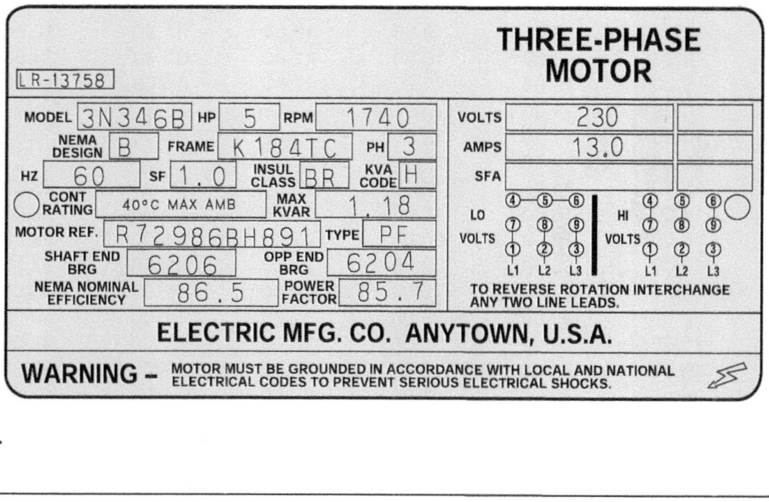

Figure 57 Nameplate data.

an assumption is made by motor manufacturers that there will be a voltage drop of 20V from the transformer down to the motor terminals (*Table 3*). The rated or nameplate voltage is the voltage at which the motor will operate most effectively. When other than rated voltage is applied, performance will change and motor life may be reduced.

Many three-phase motors have two voltages listed on the nameplate. For example, 230/460V means the motor can be connected for either 230V or 460V operation. In these cases, a connection diagram is usually found on the nameplate, as shown in *Figure 58*. These diagrams refer to low-voltage and high-voltage connections.

4.3.2 Amperage

The amperage rating appearing on a nameplate indicates the current the motor will draw at nameplate horsepower, frequency, and voltage. This is typically referred to as the *full-load amps* (*FLA*). *Figure 59* shows the amperage for a typical motor. Most manufacturers test to determine this value on a periodic basis during production, ensuring reasonable accuracy. The *NEC®* requires that the rated full-load current be the basis for determining the proper sizing of cable, overload protective devices, and other overcurrent protection in the motor circuit. Because many motors can be connected for one of two voltage ratings, they have two FLA ratings.

The FLA ratings are guaranteed if the induction motor is operating at full-load conditions and the applied voltage and frequency are the same as stated on the nameplate. When voltage and frequency are not the same, however, the current drawn by the motor at full-load conditions will be different from the nameplate indication (*Table 4*). It is possible to damage a motor operated below its rated voltage or frequency because the current the motor draws at full-load conditions increases in both cases. If the overload protective device is not sized properly, motor life may be shortened by this overcurrent condition.

4.3.3 Speed

The rated full-load speed is the value indicated in rpm on the nameplate. It is the speed at which the shaft will turn at the nameplate horsepower when supplied with power at the nameplate voltage and frequency. If the driven load is less than the nameplate horsepower, the shaft will turn faster than full-load speed.

If the motor is operating unloaded, the shaft will turn very close to synchronous speed. For example, with a full-load speed of 1,740 rpm, it can be inferred that the motor's synchronous speed is 1,800 rpm. The machine will operate from close to 1,800 rpm down to 1,740 rpm, from no-load to full-load conditions.

Table 3 Induction Motor Voltages

System Voltage	Rated Voltage
216	208
240	230
480	460
600	575
2,400	2,300
4,160	4,000
4,800	4,600
6,900 and 7,200	6,600
13,200 and 13,800	13,200

Starting Current

The total instantaneous starting current comprises the locked-rotor current plus the transient inrush that flows until the motor magnetic circuit stabilizes.

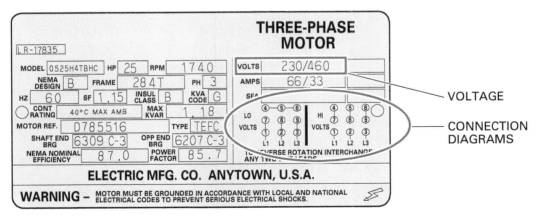

Figure 58 High-voltage and low-voltage connection diagrams shown on motor nameplate.

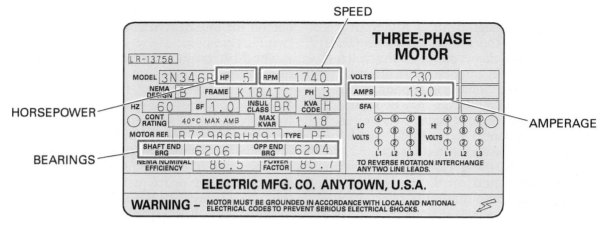

Figure 59 Nameplate showing amperage, speed, horsepower, and bearings.

Table 4 Motor Operation

Mode	Full-Load Current
110% of rated volts	7% decrease
90% of rated volts	11% increase
105% of rated frequency	5%–6% decrease
95% of rated frequency	5%–6% increase

Common synchronous speeds are 3,600, 1,800, 1,200, 900, and 600 rpm. Synchronous speed is rarely found on the motor nameplate unless the machine has been retrofitted and has not yet been tested for new full-load speed.

4.3.4 Horsepower

An induction motor is essentially a torque generator. It delivers a needed torque to a driven machine at a certain speed. Although *torque* and *horsepower* are closely related, they are not the same thing. Torque is the twisting force of a motor, and horsepower specifies the rate, or speed, of this force. This is illustrated in the following equation:

$$hp = \frac{T \times rpm}{5{,}250}$$

Where:

hp = horsepower
T = load torque in foot-pounds
rpm = shaft speed
5,250 = a mathematical constant

Horsepower is a rating used to specify the capability of an electric motor to produce mechanical power to drive a specific piece of equipment (*Figure 59*). For induction motors that are built to NEMA standards, the ratings will range from $\frac{1}{2}$ hp to 400hp, with 24 categories in all. If horsepower requirements fall between any two ratings, the larger motor size should be selected.

Remember, an induction motor will try to deliver any amount of horsepower the load requires. If properly sized, most motors operate at something less than the motor nameplate horsepower. Standard motors are designed to operate at nameplate values from sea level up to an altitude of 3,300' (1,006 m) if the ambient temperature does not exceed 104°F (40°C). Above this altitude, the nameplate horsepower no longer applies.

NEMA standards provide a method for determining the proper temperature rise, or the new maximum ambient temperature, at higher elevations. However, the standards do not provide a direct method for deriving the horsepower. Several methods are available to estimate true motor horsepower output.

4.3.5 NEMA Design Letters

The NEMA design letter defines the starting torque characteristics of an induction motor. It is one of the most important pieces of information on the nameplate; unfortunately, when a motor is replaced, the NEMA design letter is usually ignored, often leading to misapplication of the new machine. For fans or centrifugal pumps, starting torque requirements increase with the square of the change in speed. For mixers or loaded conveyor belts, however, starting torque requirements change very little with speed.

> **Think About It**
>
> **Motor Horsepower and Speed**
>
> Can you replace a fan motor rated at a specific horsepower and speed with a motor rated at the same horsepower but a higher speed to increase airflow? If not, why not?

To account for these differences, NEMA has formulated design letters A, B, C, D, and F. The difference among motors with these letters is mainly in the design of the rotor, although there are also a few external differences. Design A and B motors are intended to drive conventional loads such as fans, blowers, and centrifugal pumps. About 80% of industrial motors are NEMA Design B (*Figure 60*).

Generally, the starting current is about five to seven times the rated full-load current. From *Table 5*, it can be seen that for larger motors, the starting current can be very significant, and across-the-line starting of larger motors could result in objectionable line-voltage dips. These voltage dips could result in other control equipment dropping out on low voltage and could even cause lights to dim.

4.3.6 Insulation Class

The electrical insulation system in a motor determines the machine's ultimate life span more than any other component. By some estimates, over 60% of all motors brought to repair shops are there because of premature failure of the insulation system.

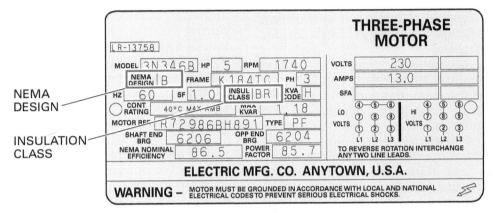

Figure 60 Nameplate showing NEMA design letter and insulation class.

Table 5 Typical Currents for 220V, 60-Cycle Squirrel Cage Motors

HP	Rated Full-Load Current	Starting (Maximum) Current	
		Classes B, C, D	Class F
½	2.0	12	—
1	3.5	24	—
1½	5.0	35	—
2	6.5	45	—
3	9	60	—
5	15	90	—
7½	22	120	—
10	27	150	—
15	40	220	—
20	52	290	—
25	64	365	—
30	78	435	270
40	104	580	360
50	125	725	450
60	150	870	540
75	185	1,085	675
100	246	1,450	900
125	310	1,815	1,125
150	360	2,170	1,350
200	480	2,900	1,800

The insulation class is a NEMA designation that identifies the class of material used to insulate the windings. Four letters designate the four classifications. They are A, B, F, and H. The insulation class defines the temperature that the insulation can be subjected to without suffering damage. The insulation class is shown on the nameplate in *Figure 60*.

Class A is now obsolete insofar as industrial motors are concerned. Class A was once the most common classification for motor insulation, especially for small motors. Class A comprises materials or combinations of materials such as cotton or paper, when suitably impregnated or coated, or other materials capable of operation at the temperature rise assigned for Class A insulation for the particular machine.

Class B is the predominant class of insulation used in motor manufacturing and rewinding today. This class is the basic standard of the industry. It includes mica, glass fiber, polyester, aramid laminates, and other materials with suitable bonding substances, or other materials, not necessarily inorganic, capable of operation at the temperature rise assigned for Class B insulation for the particular machine. (The insulation class may be designated more specifically by the use of additional letters, such as the BR shown in *Figure 60*.)

Class F incorporates materials that are similar to those in Class B but are capable of operation at the temperature rise assigned for Class F for the particular machine.

Class H insulation systems comprise, alone or combination, silicone elastomer, mica, glass fiber, polyester, aramid laminates and other materials with suitable bonding substances such as silicone resins, or other materials capable of operation at the temperature rise assigned for Class H insulation for the particular machine.

When replacing motors, ensure that the insulation class is equal to or better than that of the motor removed from service.

4.3.7 Frequency

Frequency is given for AC motors in hertz, or cycles per second. Standard frequencies for AC motors are 50Hz and 60Hz. Alternating current in the US is 60Hz.

4.3.8 Service Factor

The service factor is a multiplier for the nameplate horsepower rating that determines the amount of overload the motor can withstand. This extra horsepower is available if the motor is already operating at rated voltage and frequency and is in an environment that does not exceed the ambient temperature rating. The most common service factor appearing on a motor nameplate is 1.15.

4.3.9 NEMA kVA Code Letters

The high current draw of the motor during the first moments of startup is called the in-rush or locked-rotor current. This is the steady-state current of a motor with the rotor locked and with rated voltage applied at rated frequency. It can be derived from the kVA code letter on the motor nameplate. The letter corresponds to the kilovolt-amperes per rated horsepower (kVA/hp) required during the first moments of motor startup. NEMA has designated a set of code letters [*NEC Table 430.7(B)*] to define locked-rotor kilovolt-amperes (kVA) per horsepower. This code letter appears on the nameplate of all AC squirrel cage induction motors. *Table 6* provides the kVA/hp value for each kVA code—a letter from A to V, excluding I, O, and Q.

The locked-rotor current is required when sizing fuses or determining a **circuit breaker** setting in an induction motor circuit. The kVA rating is an indication of the current draw and, indirectly, the impedance of the locked rotor. The current drawn by the motor under stall conditions can be calculated using the values given in *Table 6*.

4.3.10 Bearings

Polyphase induction motors require either antifriction or sleeve bearings. Antifriction bearings are standard in medium (integral) horsepower motor sizes through 125hp/1,800 rpm. They are optional in 150 to 600hp/1,800 rpm sizes. Sleeve bearings are standard in 500hp/3,600 rpm and larger sizes.

Because radial loads are higher at the drive end of the motor, the drive-end bearing has a higher load rating than the bearing at the opposite end. A typical nameplate might depict both bearing duties as:

- Shaft end brg: 6,206
- Opp end brg: 6,204

Bearing internal clearances are: C1 and C2 (smaller-than-normal clearance); standard clearance (normal); and C3, C4, and C5 (larger-than-normal clearance). Electric motors usually require a C3 internal clearance. Some bearing manufacturers have a different designation for motor bearings that have a larger-than-normal internal clearance.

Table 6 Locked-Rotor Indicating Code Letters [Data from *NEC Table 430.7(B)*]

Code Letter	kVA Per Horsepower with Locked Rotor
A	0–3.14
B	3.15–3.54
C	3.55–3.99
D	4.0–4.49
E	4.5–4.99
F	5.0–5.59
G	5.6–6.29
H	6.3–7.09
J	7.1–7.99
K	8.0–8.99
L	9.0–9.99
M	10.0–11.19
N	11.2–12.49
P	12.5–13.99
R	14.0–15.99
S	16.0–17.99
T	18.0–19.99
U	20.0–22.39
V	22.4–AND UP

Reprinted with permission from NFPA 70®-2020, *National Electrical Code*®, Copyright © 2019, National Fire Protection Association, Quincy, MA. This reprinted material is not the complete and official position of the NFPA on the referenced subject, which is represented only by the standard in its entirety which may be obtained through the NFPA website at **www.nfpa.org**.

4.3.11 Power Factor

The power factor (pf) is the ratio of active power of an alternating or pulsating current (measured with a wattmeter) to the apparent power indicated by an ammeter and voltmeter. It is also referred to as the phase factor. The power factor is the measure of the system or equipment efficiency.

4.3.12 Duty Rating

All polyphase induction motors have either a duty or a time rating, which is the elapsed time the motor can operate at nameplate horsepower without shortening its life. The time rating of a motor is determined by operating the machine at full-load conditions and measuring the time it takes for the windings to heat up to the temperature rating of the insulation.

Standard time ratings are 5, 15, 30, and 60 minutes, and continuous or 24 hours. A motor with a time (or duty) rating other than continuous is a smaller motor that is given a higher horsepower rating for a shorter period of time, thus reducing size and cost. In power plant uses, most motor ratings are continuous at 104°F (40°C).

The most common machine rating is the continuous duty rating defining the output (in kilowatts for DC generators, kilovolt-amperes at a specified power factor for AC generators, and horsepower for motors) that can be carried indefinitely without exceeding established limitations. For intermittent duty, periodic duty, or varying duty, a machine may be given a short-time rating defining the load that can be carried for a specific time. Standard periods for short-time ratings are 5, 15, 30, and 60 minutes. Speeds, voltages, and frequencies are also specified in ratings, and provision is made for possible variations in voltage and frequency.

For example, motors must operate successfully at voltages 10% above and below rated voltage and, for AC motors, at frequencies 5% above and below rated frequency; the combined variation of voltage and frequency may not exceed 10%. Other performance conditions are so established that reasonable short-time overloads can be carried. The user of a motor can expect to be able to apply an overload of 25% for a short time at 90% of normal voltage with an ample margin of safety.

Service Classifications

Electric motor service classification depends on the type of service for which the motor is designed. General-purpose motors are those motors designed for use without restriction to a particular application. They meet certain specifications as standardized by NEMA. A definite-purpose motor is one that is designed in standard ratings and with standard operating characteristics for use under service conditions other than usual or for use on a particular type of application. A special-purpose motor is one with special operating characteristics or special mechanical construction, or both, that is designed for a particular application and that does not meet the definition of a general-purpose or a definite-purpose motor.

4.0.0 Section Review

1. A NEMA designation for a type of enclosed motor is _____.
    a. splash-proof
    b. drip-proof
    c. explosion-proof
    d. guarded

2. Using NEMA frame designations, the distance from the motor feet to the shaft center line is known as the _____.
    a. C dimension
    b. D dimension
    c. F dimension
    d. S dimension

3. A motor with a NEMA kVA code letter of M has a locked-rotor value of _____.
    a. 4.0–4.49 kVA/hp
    b. 4.5–4.99 kVA/hp
    c. 5.0–5.59 kVA/hp
    d. 10.0–11.19 kVA/hp

Section Five

5.0.0 Connections and Terminal Markings for AC Motors

Objective

Identify connections and terminal markings for AC motors.
a. Identify the terminals of wye-connected motors.
b. Identify the terminals of delta-connected motors.

Performance Task

3. Connect the terminals for a dual-voltage motor.

Trade Terms

Delta-connected motor: A type of motor terminal arrangement in which there are three sets of two coils connected together.

Wye-connected motor: A type of motor terminal arrangement in which three coils are connected together and three coils are isolated.

The markings on the external leads of an induction motor are sometimes missing or illegible, and proper identification must be made before the motor can be connected to the line. This section describes the procedures for identifying leads in either a wye-connected or delta-connected, three-phase, nine-lead motor.

The required materials for this procedure are:

- Appropriate personal protective equipment
- 12V battery such as an automotive battery
- Analog meter with a large scale and low range (digital meters may not clearly capture the voltage kick)
- Test leads and jumpers
- Momentary contact, normally open (N.O.) push-button switch
- Labels to mark leads as they are identified
- Three-phase, nine-lead induction motor

Before starting, identify whether the motor to be tagged is wye-connected or delta-connected. Compare the two connection diagrams in *Figure 61* and *Figure 62*. You will see that both types of motors have nine leads and six coils. In a **wye-connected motor**, three coils are connected together and three coils are isolated. A **delta-connected motor** has three sets of two coils connected together. Using an ohmmeter, insulate all motor leads from one another and check for continuity between each lead. Start by placing one probe on one lead, and check through the remaining leads. As you identify which leads show continuity, group them together. Continue this procedure with each lead until all leads are grouped. When you have completed this procedure, you should have either three wires in one group and three sets of two wires grouped together, which would indicate a wye-connected motor, or three sets of three wires grouped together, indicating a delta-connected motor.

> **NOTE**
> If the leads are partially tied together, mark or identify them in a way that will allow you to reconnect them after the testing procedure is completed.

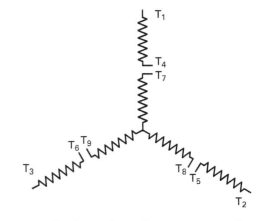

Figure 61 Dual-voltage, three-phase wye connection.

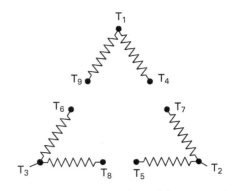

Figure 62 Dual-voltage, three-phase delta connection.

5.1.0 Wye-Connected Motor Terminals

The coil arrangement for a three-phase, wye-connected motor is shown in *Figure 63*. To identify the terminals of a wye-connected motor, proceed as follows:

Step 1 Taking the group of three common leads, arbitrarily identify them as leads 7, 8, and 9.

Step 2 Using the diagram in *Figure 64* as a guide, connect the positive lead from the battery to lead 7. Connect the lead from the battery negative through the N.O. switch to leads 8 and 9 simultaneously.

Step 3 Connect one of the three remaining lead pairs to the voltmeter terminals.

Step 4 Close the N.O. switch while observing the DC voltmeter. Use the lowest scale practical without over-ranging the meter. If the meter deflection is upward, note the voltage reading. Observe that the reading occurs only on the initial energization of the windings and then decays. Note the peak reading only and ignore the deflection in the opposite direction that occurs when the switch is opened. If the meter initially deflects downward, reverse the test lead connections.

Step 5 Continue with the remaining two lead pairs. The pair with the highest voltage reading is the winding associated with lead 7. The lead with positive polarity is identified as lead 4, and the lead with negative polarity is lead 1.

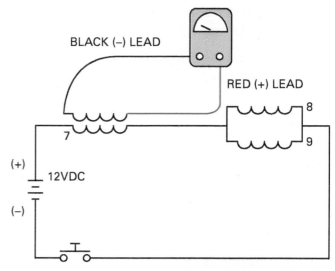

Figure 64 Battery hookup for wye-connected motor lead identification.

Step 6 Repeat Step 3, but apply the positive lead of the battery to lead 8 and the negative lead of the battery to leads 7 and 9. The positive lead of the pair with the highest voltage is identified as lead 5, and the lead with negative polarity is lead 2.

Step 7 Repeat Step 3, but apply the positive lead of the battery to lead 9 and the negative lead of the battery to leads 7 and 8. The positive lead of the pair with the highest voltage is marked lead 6, and the negative lead is lead 3.

Step 8 To confirm that all leads are correctly identified, connect the motor to the circuit. Be sure to observe proper connection procedures for the applied voltage. After connecting the motor to the circuit, start the motor and take current readings on all three lines. If the motor starts correctly and the current readings are approximately equal, the procedure was a success.

5.2.0 Delta-Connected Motor Terminals

A delta-connected motor has three sets of three leads. *Figure 65* shows how the coils are arranged in a delta-connected, three-phase motor. In this figure, the coils that are side by side are actually wound on the same poles on the motor. As discussed previously, this will allow some transformer interaction between adjacent coils. To identify the terminals of a delta-connected motor, proceed as follows:

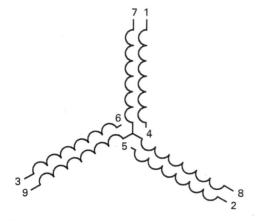

Figure 63 Coil arrangement in a wye-connected motor.

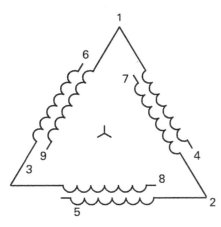

Figure 65 Coil arrangement in a delta-connected motor.

Step 1 Using an ohmmeter on a low scale, measure the resistance between each of the three leads in one group. When performing this measurement, you should see that the resistance between two of the leads is about twice that between either of those two and the third. The lead that shows the least resistance to the other two will be lead 1. Refer to all three wires in this set as set 1.

Step 2 Repeat Step 1 with the second set to identify the lead with the least resistance as lead 2. Refer to all three wires in this set as set 2.

Step 3 Repeat Step 1 with the final set of leads to identify the lead with the least resistance as lead 3. Refer to all three wires in this set as set 3.

Step 4 Using the diagram in *Figure 66* as a guide, connect lead 1 to the positive terminal of a DC voltage source and one of the remaining leads in that set (set 1) to the negative terminal. Attach the red lead of the voltmeter to lead 2 and the black lead to one of the two unknown leads in set 2. Press the push button and observe the meter needle. If the correct leads have been selected, a voltage will be induced into this coil. If not, connect the second lead to the voltmeter and repeat the test. If there is still no induced voltage, disconnect the unknown lead in set 1 from the DC voltage and connect the remaining unknown lead to the negative source. Repeat the test until the leads with an induced voltage have been identified. When these leads are located, identify the lead connected to the negative terminal of the DC voltage source as lead 4 and the lead connected to the negative voltmeter probe as lead 7. Identify the other lead in set 2 as lead 5.

Step 5 The remaining lead in set 1 will be lead 9. Leaving lead 1 on the positive DC terminal, connect the negative terminal to lead 9. Attach the red lead of the voltmeter to lead 3 and the black lead to one of the two unknown leads in set 3. Press the push button and observe the meter needle. If the correct leads have been selected, a voltage will be induced into this coil. If not, connect the second lead to the voltmeter and repeat the test. When these leads are located, identify the lead connected to the negative voltmeter probe as lead 6.

Step 6 The remaining lead in set 3 is lead 8. To verify this, connect lead 3 to the positive terminal of the DC voltage source and lead 8 to the negative terminal. Attach the red lead of the voltmeter to lead 2 and the black lead to lead 5. Press the push button and observe the meter needle. If the results are correct, a voltage will be induced into this coil, resulting in meter needle deflection.

Step 7 To confirm that all leads are correctly identified, connect the motor to the circuit. Be sure to observe proper connection procedures for the applied voltage. When the motor is connected to the circuit, start the motor and take current readings on all three lines. If the motor starts correctly and the current readings are approximately equal, the procedure was a success.

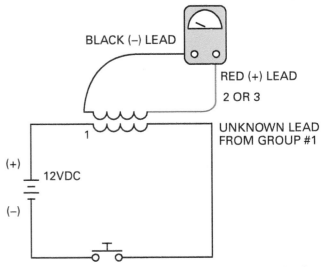

Figure 66 Battery hookup for delta-connected motor lead identification.

5.0.0 Section Review

1. A wye-connected motor has _____.
    a. no common point
    b. the coils connected in series
    c. the coils connected in parallel
    d. four coils

2. A delta-connected motor has _____.
    a. a common point
    b. the coils connected in series
    c. the coils connected in parallel
    d. four coils

Section Six

6.0.0 NEC® Requirements for Motors

Objective

Identify the *NEC®* requirements for motors.
a. Identify *NEC®* installation requirements.
b. Identify *NEC®* motor protection requirements.

Trade Terms

Branch circuits: The circuit conductors between the final overcurrent device protecting the circuit and the outlet(s).

Thermal protector: A protective device for assembly as an integral part of a motor or motor compressor that, when properly applied, protects the motor against dangerous overheating due to overload or failure to start.

The best motors on the market will operate improperly if they are installed incorrectly. Therefore, all personnel involved with the installation of electric motors should understand the procedures for installing the various types of motors that will be used.

6.1.0 Motor Installation

NEC Article 430 covers the application and installation of motors, motor circuits, and motor control connections, including conductors, short circuit and ground fault protection, starters, disconnects, and overload protection.

NEC Article 440 contains provisions for motor-driven equipment and for branch circuits and controllers for HVAC equipment.

All motors must be installed in a location that allows adequate ventilation to cool the motors. Furthermore, the motors should be located so that maintenance, troubleshooting, and repairs can be readily performed. Such work could consist of lubricating the motor bearings or perhaps replacing worn brushes. Testing the motor for open circuits and ground faults is also necessary from time to time.

When motors must be installed in locations where combustible material, dust, or similar material may be present, special precautions must be taken in selecting and installing the motors.

Any exposed live parts of motors operating at 50V or more between terminals must be guarded; that is, they must be installed in a room, enclosure, or location that allows access only by qualified persons (electrical maintenance personnel). If such a room, enclosure, or location is not feasible, an alternative is to elevate the motors not less than 8' (2.5 m) above the floor per *NEC Section 110.27(A)(4)* for 50V to 300V between ungrounded conductors, 8'-6" (2.6 m) for 301V to 600V, and 8'-7" (2.62 m) for 601V to 1,000V. In all cases, adequate space must be provided around motors with exposed live parts, even when properly grounded, to allow for maintenance, troubleshooting, and repairs. The chart in *Table 7* summarizes *NEC®* installation rules.

A summary of *NEC Article 430* is shown in *Figure 67*. Detailed information may be found in the *NEC®* under the articles or sections indicated.

> **WARNING!**
> When a motor is received at the job site, always refer to the manufacturer's instructions and follow them to the letter. Failure to do so could result in serious injury or death. Install and ground according to *NEC®* requirements and good practices. Consult qualified personnel with any questions or problems.

Keep the following in mind when installing new motors:

- *Uncrating* – When the motor has been carefully uncrated, check to see if any damage has occurred during handling. Be sure that the motor shaft and armature turn freely. This is also a good time to check whether the motor has been exposed to dirt, grease, grit, or excessive moisture during shipment or storage. Motors in storage should have shafts turned over once each month to redistribute grease in the bearings. The measure of insulation resistance is a good dampness test. Clean the motor of any dirt or grit.

Motor Connections

On dual-voltage and/or multispeed motors, always check the wiring connection diagrams given on the motor nameplate to wire the motor for the correct voltage and/or speed.

Table 7 Summary of *NEC*® Requirements for Motor Installations

Application	Requirement	*NEC*® Reference
Location	Motors must be installed in areas with adequate ventilation. They must also be arranged so that sufficient work space is provided for replacement and maintenance.	*NEC Section 430.14(A)*
	Open motors must be located or protected so that sparks cannot reach combustible materials.	*NEC Section 430.14(B)*
	In locations where dust or flying material will collect on or in motors in such quantities as to seriously interfere with the ventilation or cooling of motors and thereby cause dangerous temperatures, suitable types of enclosed motors that will not overheat under the prevailing conditions must be used.	*NEC Section 430.16*
Disconnecting means	A motor disconnecting means must be within sight from the controller location (with exceptions) and disconnect both the motor and controller.	*NEC Section 430.102(A)*
	The disconnect must be readily accessible and clearly indicate the OFF/ON positions (open/closed).	*NEC Section 430.104*
	Motor control circuits require a disconnecting means to disconnect them from all supply sources.	*NEC Section 430.75*
	The disconnecting means must be as specified in the code.	*NEC Section 430.109*
Wiring methods	Flexible connections such as Type AC cable, flexible cord, flexible metal conduit, etc., are standard for motor connections.	*NEC Articles 300 and 430*
Motor control circuits	All conductors of a remote motor control circuit outside of the control device must be installed in a raceway or otherwise protected. The circuit must be wired so that an accidental ground in the control device will not start the motor.	*NEC Section 430.73*
Guards	Exposed live parts of motors and controllers operating at 50 volts or more must be guarded by installation in a room, enclosure, or other location so as to allow access by only qualified persons, or elevated 8 feet (2.5 m) or more above the floor.	*NEC Section 430.232*
Adjustable speed drive systems	Requirements for adjustable speed drives and their motors.	*NEC Article 430, Part X*
Motors operating over 1,000 volts	Special installation rules apply to motors operating at over 1,000 volts.	*NEC Article 430, Part XI*
Controller grounding	Motor controller enclosures must be grounded.	*NEC Section 430.244*

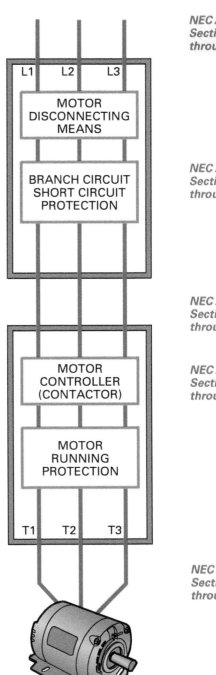

NEC Article 430, Part IX Sections 430.101 through 430.113	**Disconnects motor and controllers from circuit.** 1. Continuous rating of 115% or more of motor FLC. Also see *NEC Article 430, Part II.* 2. All disconnecting means in the motor circuit must be per *NEC Sections 430.108, 430.109, and 430.110.* 3. Must be located in sight of motor location and driven machinery. The controller disconnecting means can serve as the disconnecting means if the controller disconnect is located in sight of the motor location and driven machinery.
NEC Article 430, Part IV Sections 430.51 through 430.58	**Protects branch circuit from short circuits or grounds.** 1. Must carry starting current of motor. 2. Rating must not exceed values in *NEC Table 430.52* unless not sufficient to carry starting current of motor. 3. Values of branch circuit protective devices shall in no case exceed exceptions listed in *NEC Section 430.52.*
NEC Article 430, Part VII Sections 430.81 through 430.90	**Used to start and stop motors.** 1. Must be able to interrupt LRC. 2. Must be rated as specified in *NEC Section 430.83.*

NEC Article 430, Part III Sections 430.31 through 430.44	**Protects motor and controller against excessive heat due to motor overload.**
	1. Must trip at following percent or less of motor FLC for continuous motors rated more than one horsepower.
	  a) 125% FLC for motors with a marked service factor of not less than 1.15 or a marked temperature rise of not over 40°C.
	  b) 115% FLC for all others. (See the *NEC®* for other types of protection.)
	2. Three thermal units required for any three-phase AC motor.
	3. Must allow motor to start.
	4. Select size from FLC on motor nameplate.

NEC Article 430, Part II Sections 430.21 through 430.29	**Specifies the sizes of conductors capable of carrying the motor current without overheating.** 1. To determine the ampacity of conductors, switches, branch circuit overcurrent devices, etc., the full-load current values given in *NEC Tables 430.247 through 430.250* shall be used instead of the actual current rating marked on the motor nameplate. *(See NEC Section 430.6.)* 2. According to *NEC Section 430.22,* branch circuit conductors supplying a single motor used in a continuous duty application shall have an ampacity of not less than 125% of motor FLC, as determined by *NEC Section 430.6(A)(1).*

Figure 67 Summary of requirements for motors, motor circuits, and controllers.

> **WARNING!**
> Never start a motor that has been wet until it has been completely dried and thoroughly tested.

- *Lifting* – Eyebolts or lifting lugs on motors are intended only for lifting the motor and factory motor-mounted standard accessories. These lifting provisions should never be used when lifting or handling the motor when the motor is attached to other equipment as a single unit. The eyebolt lifting capacity rating is based on its alignment with the eyebolt center line. The capacity decreases if it is not aligned properly.
- *Guards* – Rotating parts such as pulleys, couplings, external fans, and shaft extensions must be permanently guarded against accidental contact with clothing or body extremities.
- *Requirements* – All motors must be installed, protected, and fused in accordance with *NEC Article 430*. For general information on grounding, refer to *NEC Article 250* and *NEC Article 430, Part XIII*.
- *Thermal protection information* – The motor nameplate may or may not be stamped to indicate thermal protection.

6.2.0 Motor Protection

Fuses are normally used for motor overload protection. When used, fuses must be provided in each ungrounded conductor and also the grounded conductor of a three-wire, three-phase AC system with one conductor grounded. When non-fuse overload protective devices are used, refer to *NEC Section 430.37* and *NEC Table 430.37*. Note that each motor winding must be individually protected against short circuits and ground faults.

In general, when providing overcurrent protection for motor circuits against overcurrents due to grounds and short circuits (*NEC Section 430.52*), follow the guidance listed in *Table 8*. *Table 8* provides time delay and instantaneous trip values for various types of motors as a percentage of full-load current.

When sizing the overload protection for continuous duty motors larger than 1hp, refer to *NEC Section 430.32(A)*. Motors that have a service factor of 1.15 and/or a maximum temperature rise of 40°C shall be provided with overload protection limited to 125% of the full-load current of the motor (this also applies to the secondary circuit of a wound-rotor motor). All other motors shall be limited to 115% of the full-load current. Motors that are rated at less than 1hp have separate rules found in *NEC Section 430.32(B)*.

Installing Motors

The shaft of this air conditioner motor must be aligned precisely with the shaft of the driven device. Note the micrometer attached to the motor and the load to achieve exact alignment.

Table 8 Motor Protection Devices (Data from *NEC Table 430.52*)

Type of Motor	Percent of Full-Load Current			
	Nontime Delay Fuse	Dual Element (Time Delay) Fuse**	Instantaneous Trip Breaker	Inverse Time Breaker*
Single-phase motors	300	175	800	250
AC polyphase motors other than wound rotor:				
Squirrel cage:				
Other than Design B, energy efficient	300	175	800	250
Design B, energy efficient	300	175	1,100	250
Synchronous†	300	175	800	250
Wound rotor	150	150	800	150
Direct current (constant voltage)	150	150	250	150

For certain exceptions to the values specified, see *NEC Section 430.54*.

*The values given in the last column also cover the ratings of nonadjustable inverse time types of circuit breakers that may be modified per *NEC Section 430.52*.

**The values in the Nontime Delay Fuse column apply to time-delay Class CC fuses.

†Synchronous motors of the low-torque, low-speed type (usually 450 rpm or lower), such as are used to drive reciprocating compressors, pumps, etc., that start unloaded, do not require a fuse rating or circuit breaker setting in excess of 200% of the full-load current.

Reprinted with permission from NFPA 70®-2020, *National Electrical Code*®, Copyright © 2019, National Fire Protection Association, Quincy, MA. This reprinted material is not the complete and official position of the NFPA on the referenced subject, which is represented only by the standard in its entirety which may be obtained through the NFPA website at **www.nfpa.org**.

If the desired overload ratings are not available when sizing overload protection for the motor, then use the next highest available overload rating. This is allowed provided that 140% of full-load current is not exceeded by motors that have a service factor of 1.15 and/or a maximum temperature rise of 40°C. For all other motors, this maximum value would be 130% rather than 115% as stated earlier.

Additionally, in certain situations, motors with installed overloads rated as discussed previously may not start or be able to carry system load. In these instances, it is permissible to increase the overload settings to the respective 140%/130% values. When motor starting is still a problem, the overload protective device may be shorted out during the equipment startup sequence provided that the circuit breaker or fuse protecting the branch is not set at greater than 400% of the full-load current value and the motor does not have an automatic starter.

Overload protection for adjustable speed drives is based on the rated input to the power conversion equipment. If overload protection is supplied with the equipment, then no further overload protection is required. The rating of this disconnecting means shall be no less than 115% of the power conversion equipment rated input current, and it shall be physically located in the incoming line. Overload protection, if not shunted, should allow a sufficient time delay for the motor to start and accelerate.

6.2.1 Thermal Protectors

A **thermal protector** that is integral with the motor is often used to protect the motor from overloads and starting failures. All motors with a voltage rating greater than 1,000V must have a thermal protector and its overload must not have an automatic reset feature. They shall trip no higher than the following percentage of full-load current:

- Motor full-load current not exceeding 9A–170%
- Motor full-load current between 9.1A and 20A–156%
- Motor full-load current greater than 20.1A–140%

This requirement is based on the maximum full-load motor current as listed in the tables provided in *NEC Article 430*.

Motors over 1,500hp include a device that is set to de-energize the motor when the actual temperature rise of the motor equals the rated temperature rise of the motor insulation. Thermal protectors are usually sized and installed by the motor manufacturer.

6.2.2 Branch Considerations

According to *NEC Section 430.22*, when a single motor used in a continuous duty application is supplied from a branch circuit, the ampacity of the branch circuit must be not less than 125% of the motor full-load current as determined by *NEC Section 430.6(A)(1)*, or not less than required by *NEC Section 430.22*. If a multiple-speed motor is used, then the ampacity shall be based on the highest of the full-load current ratings on the motor nameplate. Where motors have unusual duty cycle requirements, use the requirements listed in *Table 9*, as referenced in *NEC Section 430.22(E)*.

Per *NEC Section 430.24*, when sizing conductors supplying several motors, the capacity shall not be less than 125% of the largest motor plus the sum of the full-load current ratings of all other motors in the group, as determined by *NEC Section 430.6(A)*. Several motors or loads are permitted to be provided for on one branch circuit if:

- The system voltage is less than 1,000V.
- The branch protective device protects the smallest installed motor.
- All motors on the circuit must not be over 1hp and less than 20A (15A) on 120V (1,000V) circuits where each motor draws less than 6A, overloads must be installed on the motors, and short circuit current and ground fault current must not exceed the branch circuit rating.
- It is part of a factory-listed assembly.

In instances where taps are used, short circuit current and ground fault current protection may not be required for the taps used. This is true provided that the tap used has the same ampacity as the branch circuit to which it is connected. Additionally, the tap cannot be longer than 25' (7.5 m) and it must also be physically protected from damage.

Table 9 Duty Cycle Service [Data from *NEC Table 430.22(E)*]

Classification of Service	Percentages of Nameplate Current Rating			
	5-Minute Rated Motor	15-Minute Rated Motor	30- and 60- Minute Rated Motor	Continuous Rated Motor
Short-Time Duty				
Operating valves, raising or lowering rolls, etc.	110	120	150	—
Intermittent Duty				
Freight and passenger elevators, tool heads, pumps, drawbridges, turntables, etc.				
For arc welders, see NEC Section 630.11	85	85	90	140
Periodic Duty				
Rolls, ore- and coal-handling machines, etc.	85	90	95	140
Varying Duty	110	120	150	200

Any motor application shall be considered as continuous duty unless the nature of the apparatus it drives is such that the motor will not operate continuously with load under any condition of use.

Reprinted with permission from NFPA 70®-2020, *National Electrical Code®*, Copyright © 2019, National Fire Protection Association, Quincy, MA. This reprinted material is not the complete and official position of the NFPA on the referenced subject, which is represented only by the standard in its entirety which may be obtained through the NFPA website at **www.nfpa.org**.

Think About It

Putting It All Together

Count the motors in your home. The typical home may easily have more than 30 motors (including electronic equipment and tools). A century ago, a typical home might have had none. Determine which types of motors you have. Are they AC or DC? What identifying information can you determine by examining each motor nameplate?

Trade Terms Introduced in This Module

Armature: The rotating windings of a DC motor.

Branch circuits: The circuit conductors between the final overcurrent device protecting the circuit and the outlet(s).

Brush: A conductor between the stationary and rotating parts of a machine. It is usually made of carbon.

Circuit breaker: A device designed to open and close a circuit by nonautomatic means and to open the circuit automatically on a predetermined overcurrent without injury to itself when properly applied within its rating.

Commutator: A device used on electric motors or generators to maintain a unidirectional current.

Continuous duty: Operation at a substantially constant load for an indefinitely long time.

Controller: A device that serves to govern, in some predetermined manner, the electric power delivered to the apparatus to which it is connected.

Delta-connected motor: A type of motor terminal arrangement in which there are three sets of two coils connected together.

Duty: Describes the length of operation. There are four designations for circuit duty: continuous, periodic, intermittent, and varying.

Equipment: A general term including material, fittings, devices, appliances, fixtures, apparatus, and the like used as a part of, or in connection with, an electrical installation.

Field poles: The stationary portion of a DC motor that produces the magnetic field.

Horsepower: The rated output capacity of the motor. It is based on breakdown torque, which is the maximum torque a motor will develop without an abrupt drop in speed.

Hours: The duty cycle of a motor. Most fractional horsepower motors are marked continuous for around-the-clock operation at the nameplate rating in the rated ambient conditions. Motors marked one-half are for ½-hour ratings, and those marked one are for 1-hour ratings.

Intermittent duty: Operation for alternate intervals of (1) load and no load, (2) load and rest, or (3) load, no load, and rest.

Overcurrent: Any current in excess of the rated current of equipment or the ampacity of a conductor. It may result from an overload, short circuit, or ground fault.

Overload: Operation of equipment in excess of the normal, full-load rating, or of a conductor in excess of rated ampacity, which, after a sufficient length of time, will cause damage or dangerous overheating. (A fault, such as a short circuit or ground fault, is not an overload.)

Periodic duty: Intermittent operation at a substantially constant load for a short and definitely specified time.

Revolutions per minute (rpm): The approximate full-load speed at the rated power line frequency. The speed of a motor is determined by the number of poles in the winding.

Rotation: For single-phase motors, the standard rotation, unless otherwise noted, is counterclockwise facing the lead or opposite shaft end. All motors can be reconnected at the terminal board for opposite rotation unless otherwise indicated.

Synchronous speed: The speed of the revolving field of the stator, which is dependent on the supply frequency and number of poles in a stator winding.

Thermal protector: A protective device for assembly as an integral part of a motor or motor compressor that, when properly applied, protects the motor against dangerous overheating due to overload or failure to start.

Varying duty: Operation at varying loads and/or intervals of time.

Wye-connected motor: A type of motor terminal arrangement in which three coils are connected together and three coils are isolated.

Additional Resources

This module presents thorough resources for task training. The following reference material is recommended for further study.

Electric Motors and Drives: Fundamentals, Types, and Applications, Austin Hughes and Bill Drury. 5th Edition. Waltham, MA: Newnes.

National Electrical Code® Handbook, Latest Edition. Quincy, MA: National Fire Protection Association.

Figure Credits

National Electrical Manufacturers Association, Table 2

Data from *NEC® Tables 430.7(B), 430.52, and 430.22(E)*, Tables 5, 6, 8, and 9. Reprinted with permission from NFPA 70-2020, *National Electrical Code®*, Copyright © 2019, National Fire Protection Association, Quincy, MA. This reprinted material is not the complete and official position of the NFPA on the referenced subject, which is represented only by the standard in its entirety which may be obtained through the NFPA website at **www.nfpa.org**.

Section Review Answer Key

Section 1.0.0

Answer	Section Reference	Objective
1. b	1.1.0	1a
2. c	1.2.0	1b

Section 2.0.0

Answer	Section Reference	Objective
1. d*	2.1.2	2a
2. c	2.2.1	2b
3. d	2.3.0	2c
4. a	2.4.0	2d

Section 3.0.0

Answer	Section Reference	Objective
1. a	3.1.0	3a
2. d*	3.2.2	3b
3. b	3.3.2	3c

Section 4.0.0

Answer	Section Reference	Objective
1. c	4.1.2	4a
2. b	4.2.0	4b
3. d	4.3.9; Table 6	4c

Section 5.0.0

Answer	Section Reference	Objective
1. b	5.1.0	5a
2. c	5.2.0	5b

Section 6.0.0

Answer	Section Reference	Objective
1. a	6.1.0	6a
2. d	6.2.0	6b

*Calculations for these answers are provided on the following page(s).

Section Review Calculations

2.0.0 Section Review

Question 1

Multiply the frequency by 120, and then divide the result by the number of magnetic poles:

$$N = \frac{120f}{P}$$

$$N = \frac{120 \times 60Hz}{4}$$

$$N = 1,800 \text{ rpm}$$

The synchronous speed is **1,800 rpm**.

3.0.0 Section Review

Question 2

Divide the voltage by the frequency to determine the voltage-to-frequency ratio (volts-per-hertz ratio):

480V ÷ 60Hz = 8

The voltage-to-frequency ratio is **8**.

NCCER CURRICULA — USER UPDATE

NCCER makes every effort to keep its textbooks up-to-date and free of technical errors. We appreciate your help in this process. If you find an error, a typographical mistake, or an inaccuracy in NCCER's curricula, please fill out this form (or a photocopy), or complete the online form at **www.nccer.org/olf**. Be sure to include the exact module ID number, page number, a detailed description, and your recommended correction. Your input will be brought to the attention of the Authoring Team. Thank you for your assistance.

Instructors – If you have an idea for improving this textbook, or have found that additional materials were necessary to teach this module effectively, please let us know so that we may present your suggestions to the Authoring Team.

NCCER Product Development and Revision
13614 Progress Blvd., Alachua, FL 32615

Email: curriculum@nccer.org
Online: www.nccer.org/olf

❏ Trainee Guide ❏ Lesson Plans ❏ Exam ❏ PowerPoints Other _____

Craft / Level: _____ Copyright Date: _____

Module ID Number / Title: _____

Section Number(s): _____

Description: _____

_____

_____

_____

Recommended Correction: _____

_____

_____

_____

Your Name: _____

Address: _____

_____

Email: _____ Phone: _____

This page is intentionally left blank.

Electric Lighting

Overview

Electric lighting is used extensively throughout residential structures, commercial businesses, industrial plants, and outdoor sites. It provides illumination for the performance of visual tasks with a maximum of comfort and a minimum of eyestrain and fatigue, allowing individuals to perform their daily living and work-related tasks more easily. This module introduces the principles of human vision and the characteristics of light. It also covers different types of light sources and describes the operating characteristics and installation requirements of various lighting fixtures (luminaires).

Module 26203-20

Trainees with successful module completions may be eligible for credentialing through the NCCER Registry. To learn more, go to **www.nccer.org** or contact us at 1.888.622.3720. Our website, **www.nccer.org**, has information on the latest product releases and training.

Your feedback is welcome. You may email your comments to **curriculum@nccer.org**, send general comments and inquiries to **info@nccer.org**, or fill in the User Update form at the back of this module.

This information is general in nature and intended for training purposes only. Actual performance of activities described in this manual requires compliance with all applicable operating, service, maintenance, and safety procedures under the direction of qualified personnel. References in this manual to patented or proprietary devices do not constitute a recommendation of their use.

Copyright © 2020 by NCCER, Alachua, FL 32615, and published by Pearson Education, Inc. or its affiliates, 221 River Street, Hoboken, NJ 07030. All rights reserved. Printed in the United States of America. This publication is protected by Copyright, and permission should be obtained from NCCER prior to any prohibited reproduction, storage in a retrieval system, or transmission in any form or by any means, electronic, mechanical, photocopying, recording, or likewise. To obtain permission(s) to use material from this work, please submit a written request to NCCER Product Development, 13614 Progress Blvd., Alachua, FL 32615.

26203-20 V10.0

From *Electrical, Trainee Guide*. NCCER.
Copyright © 2020 by NCCER. Published by Pearson. All rights reserved.

26203-20
ELECTRIC LIGHTING

Objectives

When you have completed this module, you will be able to do the following:

1. Explain the relationship between human vision and light.
   a. Identify how the human eye operates.
   b. Identify the characteristics of light.
2. Evaluate light sources and luminaires to solve common lighting needs.
   a. Compare light sources and trends in their use.
   b. Choose auxiliary equipment needed for different light sources.
3. Select and install luminaires for various applications.
   a. Identify luminaires and their applications.
   b. Store and handle lamps and luminaires.
   c. Install luminaires.

Performance Task

Under the supervision of the instructor, you should be able to do the following:

1. Install one or more of the following luminaires and their associated lamps:
   - Surface-mounted
   - Recessed
   - Suspended
   - Track-mounted

Trade Terms

Ballast
Color rendering index (CRI)
Dip tolerance
Driver
Efficacy
Incandescence
Incident light
Lumen (lm)
Lumen maintenance
Lumens per watt (LPW)
Luminaire
Luminance
Reflection
Reflected light
Refraction
Troffers

Industry Recognized Credentials

If you are training through an NCCER-accredited sponsor, you may be eligible for credentials from NCCER's Registry. The ID number for this module is 26203-20. Note that this module may have been used in other NCCER curricula and may apply to other level completions. Contact NCCER's Registry at 888.622.3720 or go to **www.nccer.org** for more information.

Contents

- 1.0.0 Perception of Light 1
- 1.1.0 Vision 1
  - 1.2.0 Characteristics of Light 2
    - 1.2.1 Incident and Reflected Light 2
    - 1.2.2 Absorption, Reflection, and Refraction of Light 3
    - 1.2.3 Light Colors 4
- 2.0.0 Sources of Light 6
  - 2.1.0 Comparison of Sources 6
    - 2.1.1 Standard Incandescent Lamps 7
    - 2.1.2 Tungsten-Halogen Incandescent Lamps 9
    - 2.1.3 Fluorescent Lamps 10
    - 2.1.4 Light-Emitting Diode (LED) Lamps 12
    - 2.1.5 High-Intensity Discharge (HID) Lamps 14
    - 2.1.6 Magnetic Induction Lamps 16
    - 2.1.7 Xenon Arc Lamps 16
  - 2.2.0 Components of Luminaires 17
    - 2.2.1 Lamp Shapes and Dimensions 17
    - 2.2.2 Lampholders 18
    - 2.2.3 Installing Lamps 18
    - 2.2.4 Fluorescent Luminaire Ballasts 19
    - 2.2.5 HID Luminaire Ballasts 20
    - 2.2.6 LED Drivers 22
- 3.0.0 Luminaires 23
  - 3.1.0 Luminaires and their Applications 23
    - 3.1.1 Surface-Mounted Luminaires 23
    - 3.1.2 Recessed Luminaires 23
    - 3.1.3 Suspended Luminaires 26
    - 3.1.4 Track-Mounted Luminaires 26
    - 3.1.5 Pole-Mounted Luminaires 27
  - 3.2.0 Storing and Handling Lamps and Luminaires 28
  - 3.3.0 Installing Luminaires 28
    - 3.3.1 Surface-Mounted Luminaires 30
    - 3.3.2 Recessed Luminaires 35
    - 3.3.3 Suspended Luminaires 39
    - 3.3.4 Track Luminaires 40
    - 3.3.5 Making Electrical Connections to Luminaires 45
- Appendix A   Summary of Luminaire Symbols 52–53
- Appendix B   Testing Procedure and Checklist 54–57

Figures and Tables

Figure 1 Color sensitivity of rod and cone cells 2
Figure 2 The electromagnetic spectrum 3
Figure 3 Absorbed, reflected, and transmitted light 3
Figure 4 Visible light spectrum 4
Figure 5 Light being refracted into its component colors by a prism 4
Figure 6 Components of an incandescent lamp 7
Figure 7 Relationship of rated lamp voltage to watts, lumens, and lamp life 7
Figure 8 Incandescent lamp spectrum 9
Figure 9 Fluorescent lamp components 10
Figure 10 Compact fluorescent lamps 11
Figure 11 Typical LED applications 13
Figure 12 LED roadway luminaire 14
Figure 13 Two high-intensity discharge lamps 15
Figure 14 Metal halide lamp 15
Figure 15 Xenon arc lamp 16
Figure 16 Incandescent lamp shapes. (Courtesy of GE Consumer Products.) 17
Figure 17 Typical fluorescent lampholders 18
Figure 18 Simplified HID lamp ignitor circuit 21
Figure 19 Surface-mounted ceiling luminaires 23
Figure 20 Surface-mounted indoor wall luminaires 24
Figure 21 Surface-mounted outdoor wall luminaire 24
Figure 22 Typical recessed luminaires 24
Figure 23 Typical recessed fluorescent troffers 25
Figure 24 Fluorescent troffer mounted in a suspended ceiling 25
Figure 25 2x2-foot edge-lit LED luminaire 26
Figure 26 Suspended luminaires 27
Figure 27 Track lighting 27
Figure 28 Pole-mounted LED luminaire 27
Figure 29 Typical lighting floor plan and luminaire schedule 29
Figure 30 Example of luminaire installation instructions 30
Figure 31 Typical outlet boxes used with surface-mounted luminaires 31
Figure 32 Common methods of installing surface-mounted luminaires 33
Figure 33 Typical mounting of a small luminaire to an outlet box ... 34
Figure 34 Methods for attaching chandeliers to an outlet box 34
Figure 35 Typical ceiling fan/light mounting method 35
Figure 36 Installing washers and gaskets in outdoor luminaires 36
Figure 37 Summary of *NEC*® requirements for recessed luminaires 37
Figure 38 Typical mounting of a recessed luminaire 38
Figure 39 Typical clips used to fasten lay-in troffers to suspended ceiling grid systems 39
Figure 40 Examples of hangers used to attach luminaires to building structural members 40
Figure 41 Some support methods for suspended fluorescent and LED luminaires 41

Figures and Tables (continued)

Figure 42 Some luminaire supports for commercial/industrial applications 42
Figure 43 Track lighting accessories and components 44
Figure 44 Example of basic luminaire connections 45
Figure 45 Basic modular wiring system... 46
Figure 46 Strut for supporting and powering luminaires 47
Figure 47 Lighting trolley busway. 47

Table 1 Percentages of Light Reflected by Common Surface Materials 3

SECTION ONE

1.0.0 PERCEPTION OF LIGHT

Objective

Explain the relationship between human vision and light.
a. Identify how the human eye operates.
b. Identify the characteristics of light.

Trade Terms

Incident light: The light emitted from a self-luminous object such as the sun, a flame, or an electric source.

Luminaire: A complete lighting unit consisting of a light source such as a lamp or lamps, together with the parts designed to position and protect the light source and to distribute the light. It may also include components to match the incoming electricity with the needs of the light source.

Luminance: The intensity of light traveling in a given direction from any surface. It is measured in candela/meter^2. The term luminance is commonly used to express brightness.

Reflection: The bouncing of light waves or rays away from a surface.

Reflected light: Any light that comes from a surface that is not self-illuminating.

Refraction: The bending of a light wave or ray as it passes obliquely (at an angle) from one medium to another of different density, or through layers of different density in the same medium.

Electric lighting is used to supplement or replace sunlight in places where we live, work, eat, and play. The color, brightness or luminance, and contrast between lit surfaces can:

- Improve ambiance
- Provide illumination for various tasks
- Improve the efficiency of workers
- Aid in selling merchandise
- Increase productivity
- Reduce mistakes
- Increase safety
- Focus attention

Regardless of the application, well-designed electric lighting serves to provide illumination for the performance of visual tasks with a maximum of comfort and a minimum of eyestrain and fatigue, allowing individuals to perform their daily living and work-related tasks more easily.

1.1.0 Vision

Human vision is a process that occurs partly in the eye and partly in the brain. Light enters the eye through a thin, transparent layer known as the *cornea*, then passes through the pupil (the opening in the iris that controls the amount of entering light), then to the lens (to focus it), and finally to the retina. The surface of the retina is made up of two types of light-sensing cells known as *rods* and *cones*. In humans, the nearly 90 million rod cells are sensitive to low levels of light and are spread throughout the entire retina. The nearly 5 million cone cells occur in three types and are concentrated near the center of vision, providing color and detail in higher levels of light. See *Figure 1*.

The human eye is able to distinguish well over one million different colors and tints or shades of colors when they are mixed together or with the addition of white or black. At any instant, the eye can see objects whose brightness varies by nearly 10,000:1. With changes in the opening of the pupil and chemical changes in the retinal cells (dark adaptation), the total range of brightness the eye can see is in excess of 1,000,000:1. These abilities make the human eye more capable than any digital or film camera.

The rod and cone cells send their information through the optic nerve to the brain. In the brain, the image is remembered and compared with other images and memories to influence emotion, action, and thought.

Therefore, lighting plays an important role in how people feel about their surroundings and the buildings that they occupy. Different lighting techniques can be used to highlight the special architectural features of a building while providing light for functional purposes. Lighting can be used to enhance the décor of a building and its furnishings, to help people perform tasks more easily, and to make people feel safer and more comfortable both inside and outside.

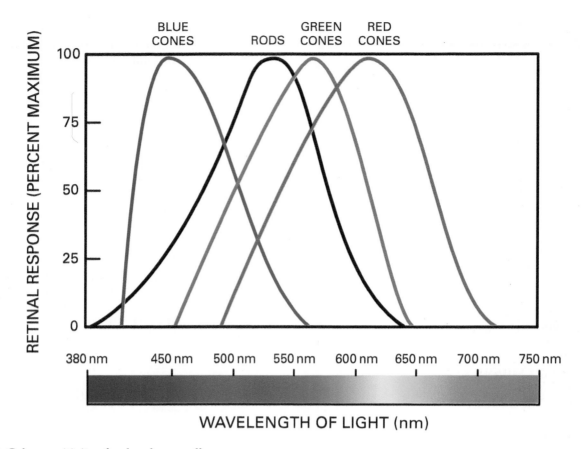

Figure 1 Color sensitivity of rod and cone cells.

Color Blindness

About 8% of males and a much smaller number of females are lacking one or more types of cone cells or have variations in a cone cell that shift its peak color sensitivity. This condition may make it difficult to distinguish between certain colors, and in some cases, to see color at all. These variations are lumped together and referred to as *color blindness*. Color blindness is usually genetic but is sometimes caused by disease or other conditions. Depending on the exact type of color blindness, a person may be unable to hold a job where distinctions between colors are essential, such as working with color codes or piloting an aircraft. Special contact lenses and glasses are available that may help color blind individuals to see the difference between colors. Smart phone apps are also available that can be used to identify colors.

1.2.0 Characteristics of Light

Visible light is a form of radiant energy and is a very small portion of the electromagnetic spectrum. Radio waves and X-rays are also parts of this same spectrum (see *Figure 2*). Because of how important light is to life, this section will discuss some of the different characteristics of light.

1.2.1 *Incident and Reflected Light*

Incident light is light that hits a surface or object, regardless of the source of this light. If an object is not self-illuminating, any light that comes from it is called reflected light. Thus, incident light hitting an object or surface includes all light from light sources such as the sun plus any light reflected from other nearby objects or surfaces. At the same time, when no incident light strikes an object or surface, there will be no reflected light.

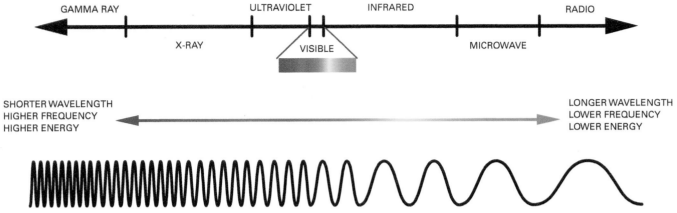

Figure 2 The electromagnetic spectrum.

Some surfaces and materials reflect light better than others. Light-colored or highly polished surfaces reflect more light than darker-colored or dull surfaces. This is because white and light-colored materials absorb less of the light energy than do dark or dull surfaces, thus leaving more energy available to be reflected. The color of the walls, ceilings, and floors and their reflecting ability are major considerations in interior lighting design. Typically, areas with darker surfaces will require the use of more artificial light. *Table 1* shows some typical examples of the percentages of light reflected by common surface materials.

1.2.2 Absorption, Reflection, and Refraction of Light

All the light striking an object's surface will be absorbed, reflected, or transmitted, as shown in *Figure 3*. The relative portions of absorption, reflection, and transmission are determined by the object's composition, surface texture, overall shape, and internal complexities, as well as the wavelengths of the incident light.

When light energy is absorbed by an object, it is stored in the form of heat. This explains why walls, floors, metals, etc. feel warm or hot when exposed to direct sunlight or light from a strong artificial light source.

When light energy is reflected from an object's surface, different wavelengths of light will have different amounts of reflection, depending on the chemicals at the surface. This is how an object's color is created.

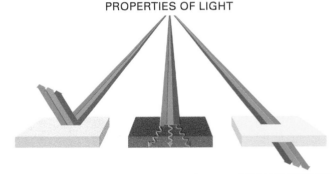

Figure 3 Absorbed, reflected, and transmitted light.

Table 1 Percentages of Light Reflected by Common Surface Materials

Surface Material	Percentage of Light Reflected
White plaster	90% to 92%
Mirrored glass	80% to 90%
White paint	75% to 85%
Polished aluminum	75% to 87%
Copper, highly polished	70% to 75%
Stainless steel	55% to 65%
Limestone	35% to 65%
Marble (white)	30% to 70%
Rough concrete	20% to 30%
Dark red bricks	10% to 15%

When light energy enters a transparent surface at an angle, it is bent at different angles according to its wavelength (red bends the least and violet bends the most). This is called refraction and has many uses in science and other professions.

1.2.3 Light Colors

Light is made up of that portion of the electromagnetic spectrum encompassing wavelengths between 380 nanometers (violet color) and 780 nanometers (red color). See *Figure 4*. A nanometer is one billionth of a meter.

When all the wavelengths of the light spectrum are presented to the eye in nearly equal proportions, white light is seen. When a narrow beam of sunlight is passed through a prism, as shown in *Figure 5*, the light spectrum is refracted at different angles according to its wavelength. This figure shows the six common names for colors. Since the spectrum is continuous, the actual number of colors is infinite, but tests show that about 125 individual colors can be identified within the visible spectrum.

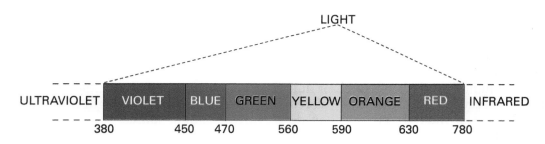

Figure 4 Visible light spectrum.

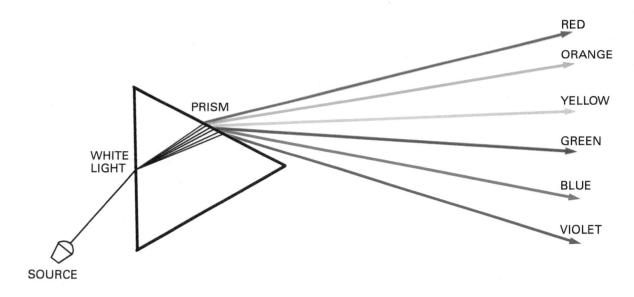

Figure 5 Light being refracted into its component colors by a prism.

As previously described, the amount of light energy absorbed by an object depends on the wavelengths of the incident light striking the object. All objects absorb light of different wavelengths in different proportions. This is called *selective absorption*. Practically all colored objects owe their color to selective absorption in some part of the visible light spectrum, with resulting reflection and transmission in other selected parts of the spectrum. This characteristic gives us our color distinction when viewing different objects, surfaces, etc. An object's appearance results from the way it reflects the light that is falling on it. For example, under white light an apple appears red because it tends to reflect light in the red portion of the spectrum and absorb light of other wavelengths (colors).

> **Think About It**
> ## Visible Light Spectrum
> Can you think of one example that sometimes occurs in nature that confirms the fact that white light is made up of different colors?

Understanding these basic concepts of light and how it is perceived will help you select the appropriate electric lamp or luminaire to provide light in any residential, commercial, and industrial application.

1.0.0 Section Review

1. Cone cells are _____.
   a. sensitive to dim light
   b. spread throughout the retina
   c. sensitive to different colors
   d. more abundant than rod cells

2. Colored objects owe their color to selective _____.
   a. absorption
   b. luminance
   c. radiance
   d. frequency

Section Two

2.0.0 Sources of Light

Objective

Evaluate light sources and luminaires to solve common lighting needs.
a. Compare light sources and trends in their use.
b. Choose auxiliary equipment needed for different light sources.

Trade Terms

Ballast: A circuit component in fluorescent and high-intensity discharge (HID) luminaires that provides the required voltage surge at startup and then controls the subsequent flow of current through the lamp during operation.

Color rendering index (CRI): A measurement of the way a light source reproduces color, with a range from 0 to 100. The higher the index number, the closer colors are to how an object appears in full sunlight or incandescent light.

Dip tolerance: The ability of an HID lamp or luminaire circuit to ride through voltage variations without the lamp extinguishing and cooling down.

Driver: The circuit components in LED lighting that take incoming power and convert it to the required stable voltage and current for the LED.

Efficacy: The visible light output of a light source divided by the total power input to that source. It is expressed in lumens per watt (LPW). The theoretical maximum efficacy is 683 LPW at 540 nanometers (a green-colored light).

Incandescence: The self-emission of visible radiant energy from an object heated above 750°K, such as occurs when an electric current is passed through the filament in an incandescent lamp.

Lumen (lm): The basic measurement of light. One lumen is defined as the amount of light cast upon one square foot of the inner surface of a hollow sphere with a one-foot radius with a light source of one candela at its center.

Lumen maintenance: A measure of how a lamp maintains its light output over time. It may be expressed either numerically or as a graph of light output versus time. A commonly used value is L_{70}, the number of hours before the lamp output drops to 70% of its initial value.

Lumens per watt (LPW): A measure of the efficiency, or, more properly, the efficacy of a light source. The efficacy is calculated by taking the lumen output of a lamp and dividing by the input wattage. For example, a 100W lamp producing 1,750 lumens has an efficacy of 17.5 lumens per watt.

From pre-history to the present, various sources have been used to create light. Each has required energy—first from fuel such as wood or gas, and now from electricity. Safety, ease of use, and cost have been the principles for selection of the light sources we use today. Because nearly every source of light uses electricity, an electrician must understand the differences between light sources, the advantages and disadvantages of each type, the requirements for their installation and use, and their *NEC®* requirements.

2.1.0 Comparison of Sources

For all electrically powered light sources, the following points are generally true:

- Lumen maintenance (light output) decreases over time due to physical and chemical changes in the source.
- When comparing two lamps of the same type, the lamp with a greater light output will have a higher efficacy.
- For non-incandescent sources, lamp life is the time until light output decreases to 70% of its initial value.
- For non-incandescent sources, auxiliary components are required, such as a starter, an ignitor, a ballast, or a driver.

Colors appear differently under various light sources. Sunlight is the source against which all other light sources are compared. It has a continuous spectrum to which the human eye is very well adapted. The color rendering index (CRI) is defined as how well a light source matches the solar spectrum, and ranges from 0 (the monochromatic yellow of a low-pressure sodium lamp) to 100 (sunlight). The CRI is used by lamp manufacturers to indicate how normal and natural a specific lamp makes objects appear. Generally, the higher the CRI, the better it makes people and objects appear. CRI differences among lamps are not usually visible unless the difference is greater than three to five points. Note that the CRI of different lamps can be compared only if the sources have approximately the same color temperature. The color temperature, expressed in degrees kelvin (°K), is commonly how lamp manufacturers describe the color tone (warmth or coolness) produced by a lamp. These differences in CT and the CRI are important in selecting and using different light sources.

Lamps can create atmospheres that are warm or cool in appearance. For example:

- Color temperatures of 3,000°K and lower are described as warm in tone and slightly enhance reds and yellows.
- A color temperature of 3,500°K is considered moderate in tone, producing a balance between warmth and coolness.
- Color temperatures of 4,100°K and higher are considered cool in tone, slightly biased toward blues and greens.

There are many ways in which electricity produces light, including incandescent heating, ionized plasma, induction, fluorescence, gaseous discharge, and electron energy transitions. These are used individually or in combination. Lamps that use these methods are made in a wide variety of shapes, sizes, finishes, and mounting bases.

2.1.1 Standard Incandescent Lamps

Incandescent lamps were first investigated over 200 years ago and became practical for use by 1880. Since then, they have remained much the same, with the only significant refinement being a change from a carbon filament to a tungsten filament around 1920. Their basic construction has not changed, although the number of sizes, uses, and other details have increased tremendously. Incandescent lamps typically consist of a thin coiled or shaped tungsten-wire filament supported inside an evacuated glass envelope (bulb) filled with an inert gas, typically a mix of argon and nitrogen (*Figure 6*). The inert gas helps to improve lamp life. The envelopes of most lamps are made of regular lead or soda lime (soft) glass. Envelopes of lamps that must withstand higher temperatures or be used outdoors are typically made of borosilicate heat-resistant (hard) glass. The lamp's base supports the lamp envelope and filament and provides the electrical connection between the lamp and its power source.

In an incandescent lamp, light is generated by passing an electric current through a filament, the resistance of which causes it to heat to the point of incandescence, releasing visible radiant

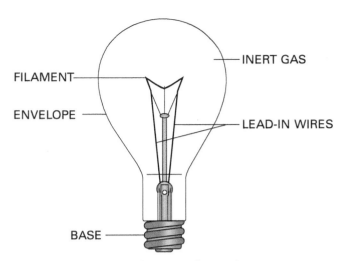

Figure 6 Components of an incandescent lamp.

energy. There is a direct relationship between voltage and light output. If the voltage to the lamp is increased by 5%, the filament gets about 2% hotter, the light output increases by about 20%, but the lifespan is cut in half. These relationships can be seen in *Figure 7*. These relationships make the design and selection of each type of lamp a tradeoff between efficiency and lamp life and explain why lamps of equal wattage may have different lumen (lm) and life ratings. The color temperature range for incandescent lamps is approximately 2,400°K–3,000°K, depending on the wattage (higher wattages run hotter).

The advantages of incandescent lamps include the following:

- They are available in many sizes, shapes, and wattages.
- No ballast is required.
- They are dimmable.

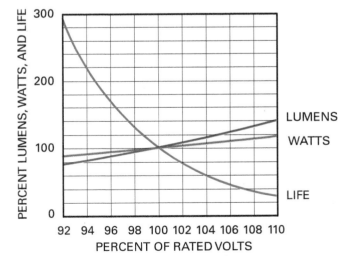

Figure 7 Relationship of rated lamp voltage to watts, lumens, and lamp life.

- They offer a very high CRI of close to 100.

The disadvantages of incandescent lamps include the following:

- They are the least efficient light source, with an efficacy of 5 to 22 **lumens per watt (LPW)**.
- Over 90% of the energy they consume is given off as heat, as seen in *Figure 8*. The heat can damage the luminaire and its wiring, as well as increase air conditioning costs.
- They have the shortest life expectancy of all lamp types, typically between 750 and 2,000 average hours, depending on type.
- The cold resistance of the filament is very low compared to its resistance when lit, so lamp failure (with a bright flash) typically comes from the high inrush current of a cold filament.
- Over time, the tungsten evaporates from the hot filament and deposits on the bulb, darkening it and reducing light output.

In 2007, the United States Congress passed the Energy Independence and Security Act (EISA), which mandated a phased elimination of general-use incandescent lamps in 40W–100W sizes. However, these lamps will continue to be made and sold for the following applications:

- Ovens and similar appliances where the temperatures exclude other types
- Studio and stage lighting where their nearly perfect CRI is critical to the recording media
- Mines, aircraft, shipping, and specialized instruments

Currently, some manufacturers are working toward producing an incandescent lamp with an infrared reflective coating that reflects much of the normally lost energy back onto the filament. This will reduce the electricity used to heat the filament, resulting in a much higher lamp efficacy. If this work has commercial success, then the phased elimination of incandescent lamps may be reconsidered.

> **CAUTION**
> It is important to be aware that replacement lamp types may not work with incandescent dimmer systems. Replacement lamps that work with dimmers are clearly marked on the product packaging and must be matched with the type of dimmer in use.

Think About It

Energy Consumed by Incandescent Lamps

How much of the energy consumed by incandescent lamps is dissipated as heat?

a. 25%
b. 50%
c. 75%
d. 90%

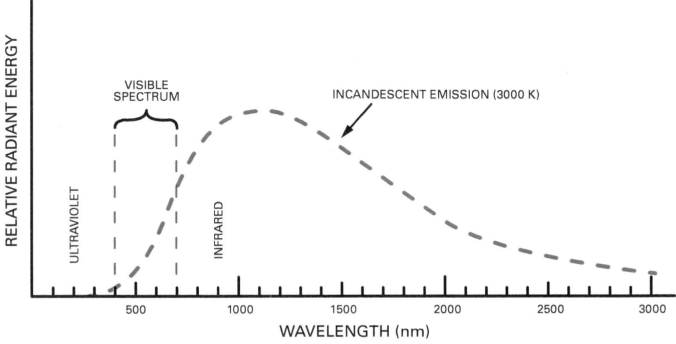

Figure 8 Incandescent lamp spectrum.

2.1.2 Tungsten-Halogen Incandescent Lamps

Tungsten-halogen incandescent lamps (halogen lamps) are a refinement over the standard incandescent lamp. Two changes allow halogen lamps to have a hotter filament and a longer life—the addition of a small amount of a halogen gas (typically iodine or bromine) and closely encasing the filament and surrounding gases in a fused silica or similar heat-resistant glass. This allows the tungsten that evaporates from the filament to be picked up from the hotter wall and recycled back onto the filament.

The advantages of halogen lamps include the following:

- Like standard incandescent lamps, halogen lamps are made in many sizes, shapes, and wattages.
- When compared to a standard incandescent lamp, they provide greater efficacy (12 to 36 LPW), longer service life (2,000 to 5,000 average hours), and improved light quality.
- The filament of a tungsten-halogen lamp runs at about 3,300°K, significantly hotter than a standard incandescent lamp, making the light appear whiter and brighter.
- Like standard incandescent lamps, halogen lamps have a CRI of nearly 100, which is equivalent to sunlight.
- Lumen maintenance is very good, and the lamp is usually much smaller than a standard incandescent lamp of similar wattage.

The disadvantages of halogen lamps include the following:

- Because of the higher filament operating temperatures, halogen lamps generate a significant amount of ultraviolet (UV) radiation. This can damage fabrics, artwork, human skin, and similar surfaces if UV filters are not used.
- Special handling and/or cleaning is required because dirt, fingerprints, and similar markings on the surface of the lamp will lead to early failure of the tube.
- Their efficacy is lower than many other lamp types, so general-use halogen lamps are being phased out or prohibited by law in many countries.
- The high temperature of halogen lamps presents a fire hazard and requires the use of suitable guards or protective luminaire designs. When these are specified by the manufacturer, they are required by *NEC Section 110.3(B)*.

Halogen lamps that look exactly like older incandescent lamps are frequently a good option, but the total lifetime cost will be more than a fluorescent or LED lamp.

> **CAUTION**
> Operating halogen lamps at voltages above and below the manufacturer's recommendations can have adverse effects on the internal chemical process because the temperature will differ from the design value. Also, it is important to follow the manufacturer's instructions as to burning position, lamp handling, and luminaire temperatures.

2.1.3 Fluorescent Lamps

Fluorescent lamps consist of a phosphor-coated glass tube containing a low-pressure mixture of mercury and an inert gas such as argon (*Figure 9*). Light is produced by passing an electric arc between two tungsten cathodes at opposite ends of the tube. This arc excites the atoms of mercury, causing them to emit UV radiation, which in turn causes the phosphor coating on the inside of the tube to fluoresce and produce visible light. Fluorescent lamps are fairly energy efficient (75 to 100 LPW) and have a long service life (12,000 to more than 24,000 average hours). They require a means to start the arc and a ballast to limit the current through the low-resistance arc.

The phosphor coating in fluorescent lamps produces a non-continuous spectrum of light that looks "white" but has a poorer CRI compared to incandescent or halogen lamps. By using different phosphor coatings, the spectral light output of a fluorescent lamp can be made to produce different colors as well as at least 20 different white tones, all with different CRIs. Higher wattages equate with longer tubes. For example, a 40W straight lamp is about twice the length of a 20W lamp.

Fluorescent lamps have two electrical requirements. To start the lamp, a high-voltage surge is needed to establish an arc in the mercury vapor. Once the lamp is started, the gas offers a decreasing amount of resistance, which means that current must be regulated to match this drop. Otherwise, the lamp would draw more and more power and rapidly burn itself out. This is why fluorescent lamps are operated in luminaire circuits containing a ballast that provides the required voltage surge at startup and then controls the subsequent flow of current to the lamp. Fluorescent lamps and luminaires are categorized into the following classes:

- *Preheat* – Preheat lamps and luminaires have a delay between when the lamp is turned on and when light is produced. They are nearly obsolete and are not used in new construction.
- *Rapid start* – A rapid-start luminaire is designed to start the lamp by continuously heating or preheating the lamp electrodes by means of heater windings built into the ballast. To start properly, the luminaire must be properly grounded. During operation, standard rapid-start lamps draw about 430mA of current. Rapid-start lamps have a bi-pin (2-pin) base at each end.

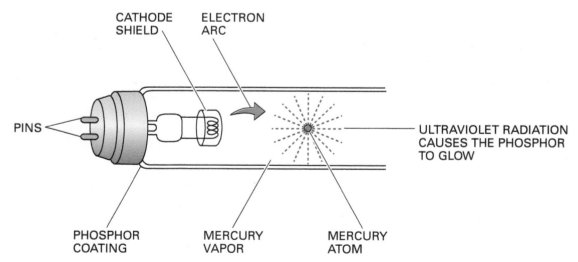

Figure 9 Fluorescent lamp components.

Fluorescent Lamp Colors

Because of the different CRI values and color temperatures of fluorescent lamps, when relamping a luminaire, make sure the replacement lamp matches the others in use. This avoids the visual distraction of different lamp colors in the same or nearby luminaires. If a change in color appearance is desired, a group relamping will be the best option.

- *Instant start* – An instant-start luminaire is used to start specially designed lamps without the aid of a starter. To strike the arc instantly, the circuit uses a high open circuit voltage of about three times the normal lamp operating voltage, much higher than is required for a rapid-start lamp of the same length. Instant-start lamps have a single pin at each end of the lamp.

Normally, lamps identified as preheat, rapid-start, or instant-start types should be used only with the corresponding type of ballast. Other terms commonly used to designate types of fluorescent lamps include the following:

- *Slimline lamps* – A group of instant-start lamps with single-pin bases.
- *High-output (HO) lamps* – A group of rapid-start lamps that produce higher light levels by operating at higher operating currents (800mA to 1,000mA). These lamps have a recessed double-contact base and require a matching lampholder. HO lamps are typically used in industrial areas and retail stores with high ceilings.
- *Very high-output (VHO) lamps* – A group of rapid-start lamps that produce even higher light levels by operating at 1,500mA. Because of the higher operating currents involved, the lamps also have a recessed double-contact base. VHO lamps are typically used in factories, warehouses, gymnasiums, and open areas.
- *Compact fluorescent lamps (CFLs)* – In the U.S., these lamps are made of single or multiple U-shaped or spiraling tubes that terminate in a plastic base and may include an outer globe that appears similar to an incandescent lamp in size and shape (*Figure 10*). The base contains the cathodes and, in some versions, a magnetic or electronic ballast. They are used both to replace incandescent lamps and in luminaires designed for their use. They can provide up to 75% energy cost savings when compared to incandescent lamps of comparable light output. Some are dimmable.
- *T-5 lamps* – T-5s are newer and more energy-efficient lamps than T-8 or T-12 lamps. They can only be used with electronic ballasts and produce nearly twice the light output of traditional T-8 or T-12 systems.

> **NOTE**
> Although the bases on HO and VHO lamps are the same, each must be matched to the correct ballast.

The life of a fluorescent lamp is affected by the number of times the lamp is started. Frequent starting of fluorescent lamps results in shorter lamp life, while continuous operation provides the longest lamp life. The typical fluorescent lamp ratings given by one major manufacturer are based on three hours per start. For example, if a lamp is rated as having an average life of 12,000 hours, this is based on the lamp being turned on 4,000 times and left on for an average of at least three hours per start.

As fluorescent lamps age, the light output drops. The tungsten evaporates from the cathodes at each end of the tube, causing darkening at the ends. This is helpful when identifying lamps that need replacement.

The advantages of fluorescent lamps include the following:

- Lower surface brightness than many other types, leading to less eyestrain or difficulty in seeing nearby objects
- More efficient than incandescent or halogen lamps

(A) CLASSIC BULB SHAPE FLUORESCENT   (B) SPIRAL COMPACT FLUORESCENT

Figure 10 Compact fluorescent lamps.

The disadvantages of fluorescent lamps include the following:

- Mercury content requires special disposal and cleanup if the tube is broken
- Lamp and ballast must be compatible
- Shortened lifespan with frequent starts
- Phosphor coating is inefficient when cold or hot, with little light output in these environments

2.1.4 Light-Emitting Diode (LED) Lamps

All diodes emit light—this has been known since 1909. Decades would pass before diodes were designed so this light could be used, and they were called light-emitting diodes (LEDs). Initially, light output was only infrared and red. New types of semiconductor materials have allowed LEDs to produce yellow, green, blue, violet, and ultraviolet (UV) colors. Like fluorescent tubes, white LEDs use a phosphor coating on a blue or violet LED to produce white light—two phosphors give a CRI of only 60–70, but more recent four-phosphor blends have produced LEDs with a CRI as high as 99.

Unlike incandescent lamps with a filament and typical energy losses greater than 90%, LEDs lose very little energy because they operate very differently, with electrons in the diodes producing light in a process known as *electroluminescence*. Because of this, LED lamps are inherently much more energy efficient than incandescent lamps. Due to their small size and a lifespan that can exceed 100,000 hours (15–30 years of normal use), LEDs are not limited to being sold as replacement lamps for existing luminaire designs—they are becoming integral parts of complete luminaires in a wide variety of shapes and applications.

The characteristics of LEDs require a very stable DC power supply of just a few volts. This low-voltage power supply is called a *driver*. Drivers can be designed for a single LED or for an array of LEDs connected in series. LED drivers only regulate the current or voltage, unlike fluorescent or HID lamp ballasts that need to both provide a high starting voltage and regulate the current through the arc. LED drivers often are designed to provide dimming from 100% to 0% light output.

LEDs are used for traffic lights, exit and emergency lighting, step lighting, and very large direct-emission RGB video displays (*Figure 11*). They are also used for accent lighting, sales display case lighting, sign lighting, light bars on emergency vehicles, flashlights, flood lights, vehicle taillights, Christmas tree lights, rope lights, backlighting for LCD television sets, infrared remote controls, optical-fiber communications, and a host of other applications.

High-output LEDs are now being used for luminaires on streets, roadways, high-bay warehouses, and sports stadiums. Filament-type LED lamps are widely marketed because of their vintage appearance. Color-tunable LED luminaires are widely used in greenhouses because their light output can be matched very closely to the brightness and spectrum best suited for different plants throughout the growth cycle. Organic LEDs (OLEDs) allow lighting within a thin film on semi-flexible and large flat surfaces. LEDs are also used in sterilization, insect control, and for specialized hospital applications.

The advantages of LED lamps include the following:

- Highly efficient, with efficacies that range from 80 to over 200 LPW
- Dimmable
- Smaller lamp size
- Low operating temperature
- Repeated starts have no effect on lifespan
- No visible warmup time and no requirement to cool down before starting again
- Shock and vibration resistant
- Changeable colors using color-changing red, green, and blue (RGB) LEDs
- Provide a variety of lighting effects through the use of special control circuits or with applications on smart phones and other digital devices
- Can be designed for use in a wide temperature range from −40°C to 100°C

Retrofitting

Going Green

For every dollar spent on the cost of lighting, 80% to 88% is spent on the electricity to run the lamp, while the remainder is spent on the purchase of lamps, lamp maintenance, etc. When doing retrofit work, there are many opportunities to incorporate more efficient lamps. This will not only result in reduced energy costs but also usually provides lighting of equal or better quality. Another possibility to consider is switching to more energy-efficient luminaires. The initial expenditure will be recovered by the resultant energy savings over the long run.

The *National Energy Policy Act (EPACT)* passed in 1992 requires that high-efficiency lamps and luminaires be used in all new construction and most retrofit construction in the United States. LEDs have the greatest efficiency.

(A) LARGE RGB VIDEO DISPLAY

(B) TRAFFIC SIGNAL

(C) EXIT AND EMERGENCY LIGHTING

(D) PATH AND STAIR STEP LIGHTING

Figure 11 Typical LED applications.

LED Applications

LED luminaires are now used to provide energy-efficient, low-maintenance lighting in a variety of applications, such as the landscape lighting shown here.

Figure Credit: Halco Lighting Technologies

The disadvantages of LED lamps include the following:

- Current and lifespan are dependent on changes in voltage. An uncontrolled spike or surge in the incoming power can burn out the driver or LEDs.
- The greater amount of blue in many white LED roadway and street luminaires increases night sky light pollution.
- Poorly designed drivers impose a very large amount of harmonic noise on the power lines, resulting in overheating of transformer high neutral currents and many other problems.
- For studio, broadcast, and film applications, including sporting games, typical LEDs have gaps in their color spectrum (with a low CRI of 60–80). LEDs with a very high CRI (95–99) are available for these uses but at a higher cost.
- When turned on, LED drivers have an inrush current as high as 100 times the steady-state operating current. It lasts for less than a few milliseconds, but when switching many LEDs at once, this inrush can trip the circuit breaker, weld relay contacts, or damage a light switch. Mitigation methods include reducing the number of luminaires on the circuit, sequenced switching, control with a zero-voltage crossing solid-state relay, installing an inrush current limiter, or changing LED driver design.

Various fixed-color LED lamps are available for use in existing luminaires. LED spotlights are also available with different beam spreads and shapes. High output LED luminaires are commonly used for high-bay, roadway (*Figure 12*), and sports stadium lighting, with efficacies as high as 100–150, CRI values as high as 95 or better, and rated lifespans of 65,000 to over 80,000 hours.

Because of their small size, very long life, high efficacy, and decreasing cost, LEDs are rapidly becoming the preferred light source. Unlike the mature technology used in most other types of lighting, LED technology is being developed at a very rapid rate by many different companies.

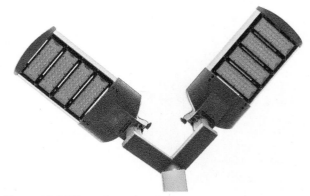

Figure 12 LED roadway luminaire.

2.1.5 High-Intensity Discharge (HID) Lamps

High-intensity discharge (HID) lamps have lifespans and efficiencies similar to those of fluorescent lamps. Unlike fluorescent lamps, the gases in the HID lamp's much shorter arc tube operate at much higher pressures, with more of the light output being in the visible portion of the spectrum. Like fluorescent lamps, they require a ballast to control the current and a source of high voltage to strike the arc (either by ballast design or using a separate ignitor), and the HID ballast and lamp must be matched exactly. In some HID lamps, a phosphor coating is used to diffuse the light and shift light colors to provide a higher CRI.

Common types of HID lamps include the following:

- *Mercury vapor lamps* – Mercury vapor lamps are the oldest of the HID lamp types. Their blue-green light output has a fairly low CRI and a modest efficacy of 50–60 LPW. They are rapidly disappearing from use because other types of lamps and luminaires have a much better CRI, improved efficacy, and a longer life. In the European Union they are now banned, with bans proposed for other countries. Replacement CFL and LED lamps for existing mercury vapor luminaires are readily available and produce more light at a lower cost. Because the mercury vapor arc

Efficiency

Different lamp types have varying abilities to convert electrical power into visible light. The quantity of light emitted (measured in lumens) is divided by the input power (in watts) to determine a lamp's efficiency (efficacy). This number is expressed in lumens per watt (LPW) and is a measure of energy efficiency. For example, a 60W lamp that produces 800 lumens of light has an efficacy of 13 LPW. The same light output from a CFL uses 13W (an efficacy of 62 LPW). In an LED, this light output uses 7W (an efficacy of 114 LPW).

produces a substantial amount of UV light, breakage of the glass outer globe results in unsafe exposure to this UV light. The typical failure mode of mercury vapor lamps is very low light output.

- *Low-pressure sodium (LPS) lamps* – Because they operate at a lower pressure than other HID lamps, LPS lamps are very closely related to fluorescent lamps, with a length that varies according to their wattage. They produce a monochromatic yellow light, with a CRI of 0 (all objects appear yellow or black), but with a very high efficacy (200 or more). Due to decreasing use, major manufacturers of LPS lamps have announced an end to production.
- *High-pressure sodium (HPS) lamps* – By using a much higher pressure for the arc tube, HPS lamps produce a golden-white light compared to the pure yellow of LPS lamps. HPS lamps have a better CRI and a higher lumen output than mercury vapor lamps. They also have the longest lifespan of any HID lamp type. The typical failure mode is blinking off and restriking at intervals of a few seconds to a minute or more because of the loss of sodium in the arc tube.
- *Metal halide lamps* – Of all the HID lamp types, metal halide lamps (also called multi-vapor lamps) have the highest CRI. Metal halide and HPS lamps are the two most common types of HID lamps (see *Figure 13*). Because of danger from arc tube rupture, *NEC Section 410.130(F)(5)* requires all metal halide lamps except thick-glass PARs to either have a containment barrier (lens) or be Type O with a protected socket. Type O lamps have an internal shroud that protects the outer bulb from damage if the arc tube is ruptured. An exclusionary base prevents the use of any lamp except a Type O lamp in a protected socket. *Figure 14* shows a metal halide lamp with a Type O base. The normal failure mode of metal halide lamps is a failure to start.

The advantages of HID lamps include the following:

- Compact fixture size
- Greater illumination than fluorescent or incandescent lamps

The disadvantages of HID lamps include the following:

- Special disposal requirements due to mercury content

(A) METAL HALIDE   (B) HIGH-PRESSURE SODIUM

Figure 13 Two high-intensity discharge lamps.

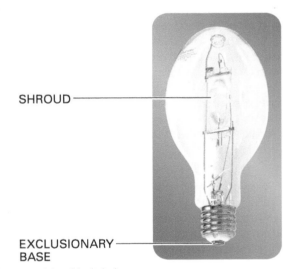

Figure 14 Metal halide lamp.

- Danger of lamp explosion due to very high pressure
- UV light damage to paints, plastics, and other items if UV filtering is not used
- Delay of 3–7 minutes from switching on until light output reaches full brightness
- Cooldown period required before restrike when turned off for any reason (e.g., a brief power failure can be followed by 10–15 minutes of near darkness)

2.1.6 Magnetic Induction Lamps

Induction lamps are essentially a type of fluorescent lamp without any electrodes. They function like a transformer with an external primary winding in a toroidal shape and a circular discharge tube filled with low-pressure gas that acts as the secondary winding. The absence of electrodes results in lifespans up to 100,000 hours. With electronic controls they can operate at frequencies above the range that causes flicker to the eye. Some types use radio-frequency induction to power the lamp, but care must be used to select frequencies that do not interfere with communications.

Induction lamps are available as screw-in replacements, retrofit kits for HID luminaires, and as complete luminaires. Their long life, wide temperature range, good efficacy, and lack of any warmup time make them a good choice for freezer rooms and hard-to-reach locations. However, LED lamps offer the same advantages as well as better efficacy.

A variation on induction lamps uses higher frequencies to energize the tube contents into a plasma. At their operating temperature, the lamp has a very high CRI. These types of lamps use a magnetron and waveguide to put the energy into the lamp, reducing their lifespan.

The advantages of magnetic induction lamps include the following:

- Long life
- Instant start and restrike operation
- No flickering, strobing, or noise
- Low-temperature operation
- Some are dimmable
- High power factor

The disadvantages of magnetic induction lamps include the following:

- Special disposal requirements due to mercury content
- Limited use

2.1.7 Xenon Arc Lamps

Early arc lamp types included carbon arc and lime lights. Modern arc lamps use a very high-pressure xenon arc to produce a pure white light (CRI 95+) with a color temperature of 6,000°K. The arc is in a very small area surrounding the lamp's cathode (see *Figure 15*). This allows the light output to be focused into a very controlled beam with uses ranging from flashlights to scientific instruments, lighthouses, searchlights, and theater projectors.

The advantages of xenon arc lamps include the following:

- Use of the significant UV light output for ozone generation, disinfection, and medical research
- No loss of light output over the lifespan of the lamp
- Lifespans rivaling those of LED lamps
- Dimmable

The disadvantages of xenon arc lamps include the following:

- Danger of lamp explosion because of the very high pressure, so special handling is required
- UV light damage to paints, plastics, and other items if UV filtering is not used

2.2.0 Components of Luminaires

Luminaire components include various types of lamp shapes, lampholders, and related wiring. Both fluorescent and HID lamps require the use of ballasts, while LED lamps incorporate drivers.

2.2.1 Lamp Shapes and Dimensions

Lamps are made in numerous sizes and shapes, with different mounting bases. *Figure 16* shows some examples of typical incandescent lamp shapes. Lamps are identified by a letter referring to their shape and a number that indicates the maximum diameter stated in eighths of an inch. For example, in the United States, an A-40 identifies a lamp with

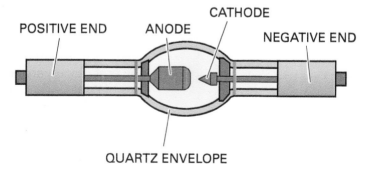

Figure 15 Xenon arc lamp.

an A-shape that is $^{40}/_8$ (5") in diameter and a T-12 is $^{12}/_8$ or $1^{1}/_{2}$" in diameter. (Metric sizes vary widely; please consult the lamp manufacturer's catalog for specific information.) Lamps with standard, tubular, and similar envelope shapes provide lighting in all directions (omni-directional). Those with shapes designated as R, ER, and PAR are all reflector-type lamps. They direct their light out in front by reflecting it from their inside walls.

Fluorescent lamps are designated by the letter F, followed by the wattage, the letter T for their tubular shape, the diameter (in eighths of an inch), the color, and any other characteristics. If the fluorescent lamp is bent in a U or a circle, the letter F is followed by a U or C. Compact fluorescent lamps do not have a standard labeling convention. HID lamps come in some of the same shapes as incandescent lamps plus BT (blown-tubular) and ED (elliptical-dimpled) shapes.

The luminaire and lamp size must work together to place the light output in the desired location. The most common mistake is to mix up when a reflector type or a non-reflector type lamp is required. Another mistake is to choose a

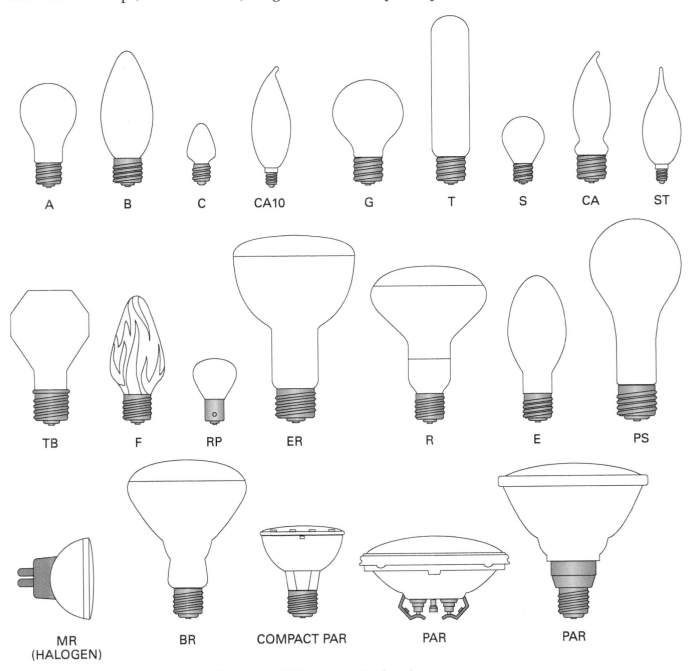

Figure 16 Incandescent lamp shapes. (Courtesy of GE Consumer Products.)

lamp that cannot fit within the luminaire because of its physical size. This is particularly true when replacing an incandescent lamp with a CFL because the plastic base containing the driver may not screw fully into the socket.

2.2.2 Lampholders

Lampholders allow the lamp to be installed into a luminaire in the correct position and orientation. In addition, the lampholder must be suitable for the lamp type and size (wattage, voltage, and pulse voltage for HID types), as well as the luminaire temperature, environmental exposure, luminaire material, wiring method, number of lamps to be held, switch control (if desired), and other considerations. Lampholders generally fail because of heat, breakage, or corrosion. The cause for lampholder failure must be identified before any replacement is made. The replacement must match as closely as possible the original unless the original lampholder was not suitable for its use.

Lampholders are generally identified by a type letter and a diameter or spacing size (in millimeters). Thus, a screw shell (Edison base) lampholder may range from an E10 (for a flashlight) to an E27 (the standard medium base), and up to an E40 (mogul). Other lampholder designations are BA for bayonet types; G, GU, GY, and GX for bipost or pin types; and W for wedge types. Some E-type lampholders have an extra contact for use with three-way lamps. Certain G types have one or two contacts for use with tubular fluorescent and replacement LED lamps and others have two or even four contacts for use with compact fluorescent and LED lamps. Three G-type fluorescent lampholders are shown in *Figure 17*.

2.2.3 Installing Lamps

Follow these general guidelines when installing lamps in their related luminaires:

- Make sure that the lamp and luminaire are compatible.
- With HID lamps, make sure that its orientation in the luminaire (such as base up) matches the burning position marked on the lamp. Also make sure the ANSI code on the lamp matches the ANSI code on the ballast.
- Make sure that the lamp or lamps being installed in a luminaire have the correct voltage rating for use with the luminaire. This is especially important when installing incandescent lamps because a small difference between the rating and the actual voltage has a great effect on a lamp's life and light (lumen) output.
- Make sure that the maximum luminaire wattage rating is not exceeded by the replacement lamp or combination of lamps; otherwise, the luminaire can overheat and may cause a fire. In recessed luminaires with built-in thermal protectors, it will cause the lamps to cycle on and off as the luminaire overheats, then cools down.
- When installing fluorescent and HID lamps, make sure that the lamp type and wattage are matched to the luminaire and ballast being used. Always install the lamps specified by the luminaire manufacturer.
- When installing bi-pin fluorescent lamps, make sure to rotate them ¼ turn to seat and connect them. The metal end caps on the tubes have marks to help with this alignment.
- Do not install lamps that have scratched or otherwise damaged envelopes.
- Use a soft cloth or gloves to handle an unencapsulated halogen tubular lamp. This prevents oil from your hands from contacting the bulb. The quartz tube walls can withstand

(A) BI-PIN SLIDE-ON AND SCREW MOUNT LAMP HOLDER

(B) HIGH-OUTPUT (HO) LAMP HOLDER

(C) SINGLE-PIN (SLIMLINE) LAMP HOLDER

Figure 17 Typical fluorescent lampholders.

high operating temperatures but will overheat and crack if contaminated with fingerprints or other materials. Also make sure that no combustible materials can come in contact with the lamp when illuminated.

- Unlike standard lamps, halogen and HID lamps may continue to light even if the exterior lamp glass is broken. If this occurs, replace the lamp immediately.
- In some applications, such as food processing plants, bakeries, dairies, schools, warehouses, etc., safety sleeves made to fit over fluorescent lamps should be used in order to retain broken glass and phosphors if the lamp ever breaks, falls, or bursts.
- In all this, do the work safely—ensure the power is off, protect your hands and eyes, use proper ladder safety, and beware of hot lamps or luminaires and any collected dust.

In general, energy legislation promotes replacing standard incandescent lamps with LED and compact fluorescent lamps, replacing incandescent reflector lamps with halogen PARs, and using higher-efficiency fluorescent lamps, such as T-5s, instead of full-wattage types. If it is necessary to relamp existing luminaires designed for use with a discontinued or inefficient type of lamp, replace the lamp with an approved energy-efficient substitute. Make sure that any dimmer used is designed for the replacement lamp. Lists of approved substitutes are available at lighting and electrical distributors.

When fluorescent lamps are used in circuits providing an open circuit voltage in excess of 300V, or in circuits that may permit a lamp to ionize and conduct current with only one end inserted in the lampholder, *NEC Section 410.130(G)* requires some automatic means for de-energizing the circuit when the lamp is removed. This is usually accomplished by the lampholder so that upon removal, the ballast primary circuit is opened. Note that the use of recessed contact bases for HO and VHO lamps has eliminated the need for this disconnect feature in the lampholders for these lamps.

2.2.4 *Fluorescent Luminaire Ballasts*

In fluorescent luminaires, the ballasts perform the following functions:

- Provide the proper voltage to establish an arc between two electrodes
- Regulate the electric current flowing through the lamp to stabilize the light output
- Supply the correct voltage required for proper lamp operation and compensate for voltage variations in the incoming power
- Provide continuous voltage to maintain heat in the lamp electrodes while the lamp operates (rapid-start circuits)

Each fluorescent lamp must be operated by a ballast that is specifically designed to provide the proper starting and operating voltage required by the lamp. Lamp and luminaire manufacturers make a wide variety of ballasts designed for use in luminaires that operate in all of the lamp starting modes. Ballasts are made for single-lamp, two-lamp, three-lamp, and four-lamp operation.

Magnetic ballast components lose 5–20% of the luminaire's input energy, so they are installed and potted into a metal case to dissipate this heat. The magnetic components vibrate at line frequency, producing an audible hum that can be disrupting in quieter environments. They are generally available to operate only one or two rapid-start or instant-start lamps. Since 1988, they have been required to be made with high-efficiency components. Because of the increasing availability, wider applicability, and energy savings of electronic ballasts, magnetic ballasts are being phased out.

A hybrid ballast type uses magnetic components to operate the lamp but electronic components to turn off the heat to the cathodes after the lamp is running, resulting in lower energy losses. However, electronic ballasts provide the greatest efficiency.

Electronic ballasts eliminate the heat loss of the magnetic components by using electronic power supplies to obtain the higher voltages needed to start the arc in the fluorescent tube while also limiting the current. They are available to operate from one to four lamps. With electronic ballasts, the power input is 50Hz to 60Hz, but the ballast operates the lamps at 20kHz to 50kHz, with resulting improvements in ballast and lamp efficacy and elimination of audible hum and lamp flicker. Some ballasts with integrated circuits are compatible for use with various dimming systems, occupancy sensors, daylight sensors, and wireless control technologies. Many electronic ballasts do not come equipped with wire leads. Instead they include plug-in harnesses or individual wires and push-in terminals.

The names used to identify the different types of ballasts can vary by manufacturer but are generally of the following types:

- *Rapid-start ballast* – These ballasts apply operating voltage to the lamp while simultaneously heating the cathodes. The lamp ignites quickly but if the cathode is not fully warmed some material sputters off, leading to slow deterioration of the cathodes and eventual failure. These ballasts are designed only for rapid-start lamps (two pins at each end). Lamp life is average.

- *Instant-start ballast* – These ballasts operate using a higher starting voltage to provide immediate illumination. They offer the lowest system wattage and highest system efficiency but at the cost of shorter lamp life.
- *Electronic programmed-start ballast* – These ballasts offer the longest lamp life due to minimal cathode damage when starting, and nearly as great energy savings as instant-start ballasts due to reduced cathode heating. Electronic ballasts are often the best choice, particularly for luminaires that turn off and on frequently, such as with occupancy sensors. They are not available in magnetic versions.
- *Emergency lighting ballasts* – Special emergency ballasts with self-contained battery-operated power packs are made for use in some fluorescent luminaires. Upon the loss of input power to the luminaire, these ballasts typically function to operate one 8' (2.5 m) lamp at emergency lighting levels for about 90 minutes, or one 4' (1.2 m) lamp for about 120 minutes.

NEC Section 410.130(E)(1) requires that all fluorescent luminaire ballasts used indoors, including any replacement ballasts, have a thermal protection device integral within the ballast (Class P). Exceptions in this section are for simple reactance ballasts used with fluorescent luminaires using straight tubular lamps, ballasts used with exit luminaires and so identified, and egress lighting energized only during an emergency. Class P ballasts with thermal protection reduce fire danger by opening the circuit at a predetermined temperature due to abnormal heat buildup caused by any fault in the luminaire.

When replacing ballasts, be sure to use an exact equivalent. In most circumstances, if the luminaire does not have a disconnect device, one must be installed, as required by *NEC Section 410.130(G)(1)*.

> **CAUTION**
>
> When installing a replacement ballast, make sure to dispose of the old ballast in a proper manner. Unless you see a label stamped *No PCBs* on the failed ballast you are disposing of, you must assume that it contains toxic PCBs and must be disposed of in accordance with the prevailing EPA and local requirements. Failure to do so can expose you and your employer to potential liability for cleanup in the event of PCB leakage.

2.2.5 HID Luminaire Ballasts

HID lamps require the use of a ballast to provide enough voltage to strike an arc in the lamp. This function may be accomplished by the ballast itself or in conjunction with a separate electronic ignitor circuit. An ignitor (*Figure 18*) is an electronic device used in the circuitry for high-pressure sodium and some metal halide HID lamps. It provides a pulse of at least 2,500V peak root-mean-square (rms) to initiate the lamp arc. When the luminaire is energized, the ignitor provides the required pulse until the lamp is completely lit and then automatically stops pulsing.

The HID lamp ballast also acts to control the arc wattage during warmup and normal operation. In addition, some ballasts may also provide a line-voltage-matching transformer function, enhance lamp wattage regulation with respect to changes in line voltage and/or lamp voltage, and sometimes provide dimming or other control interface functions.

Electronic Luminaire Ballasts

Modern electronic luminaire ballasts offer superior application flexibility. These ballasts can be rated for two, three, or four lamps. In addition, electronic ballasts may also be used to operate less than the maximum number of lamps listed. For example, a ballast rated for four lamps may be used to operate a two- or three-lamp luminaire. These ballasts also provide for the ability to use lamps of different lengths.

GOING GREEN

Power Factor

Power factor is the conversion of the phase angle between voltage and current into a percentage. A power factor of 1.00 or 100% means that the voltage and current are in phase, which is the ideal situation. Ballasts with a power factor of 0.90 (90%) or above are considered very efficient. Installing a ballast with a low power factor can reduce the efficiency of a luminaire.

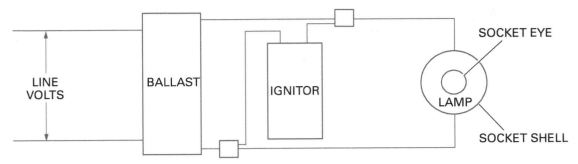

Figure 18 Simplified HID lamp ignitor circuit.

There are numerous types and sizes of ballasts used with HID lamps. To permit interchangeability regardless of manufacturer, each ballast is given an ANSI code. This ANSI code must be matched to the ANSI code for the lamp. HID ballasts can be grouped into three basic categories, according to how they perform:

- *Linear, nonregulating circuit ballasts* – These ballasts are the simplest and most basic type, with no compensation for line voltage dips or brownouts. An auto-lag variation provides some input voltage regulation.
- *Constant-wattage autotransformer (CWA) ballasts* – The use of magnetic saturation in these ballasts provides good dip tolerance and maintains better lamp wattage regulation. This avoids light loss due to a low-voltage arc extinguishment and the 10–15 minutes required for the lamp to cool down, restrike, and warm back up. CWA ballasts come in different versions designed for metal halide or mercury vapor lamps.
- *Three-coil ballasts* – These ballasts provide the best light output stability and dip tolerance during periods of voltage fluctuation. They also provide better lamp life. Two related versions are used specifically with high-pressure sodium and metal halide lamps. They cost more than the other types.

Ballast Wiring

Manufacturers color-code ballast wiring for ease of installation. Always follow the manufacturer's color code when wiring ballasts.

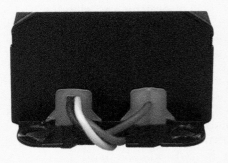

Unused Wire Taps

Some ballasts are made with primary leads that allow the ballast to be connected to different supply voltages, such as 120V, 208V, 240V, or 277V. Such ballasts are called *multi-tap ballasts*. It is extremely important that only the proper voltage lead be connected to the supply voltage, with the others being electrically insulated.

Case History

The Importance of Connecting Grounds

A painter in an industrial facility was standing on a wooden ladder and painting steel I-beams. He received a fatal shock while leaning across one of many suspended fluorescent luminaires, with his other arm touching a metal pipe. Later investigation revealed that the ground wire in the luminaire was disconnected. It is presumed that the ground wire had not been reconnected when the ballast was last replaced. Numerous burn marks were noted within the luminaire at the points where the conductors were connected to the ballast.

The Bottom Line: Don't forget to connect the green (ground) wire when replacing any electrical device.

2.2.6 LED Drivers

LED drivers provide a stable source of power with a controlled DC voltage and/or DC current. LED lamps that replace incandescent lamps usually have the driver in the base of the lamp, which is replaced as a unit. In some cases, the lamp has a special base that locks it into the luminaire with the driver mounted within the luminaire.

For fluorescent replacement lamps, the LED driver and lamp save energy and are available in the following variations:

- *Integral driver designed for 120V operation* – Requires ballast removal and luminaire rewiring.
- *Integral driver designed to operate from the existing ballast and wiring* – Simplest but uses more energy because the ballast is still being used.
- *Hybrid-type integral driver* – Most versatile and designed for ballast operation both while the ballast is operating and also for 120V operation without a ballast.
- *Separate driver mounted in the luminaire in place of the ballast* – Requires luminaire rewiring.

For new and retrofit work, a LED luminaire may have the driver within the luminaire, but frequently the driver is a separate component mounted on the nearby junction box with a special wiring whip to the luminaire. Dimming by different methods, as well as programmed or programmable color changing of RGB LEDs, are all possible when the appropriate driver (and compatible LEDs) are selected.

LED drivers use solid-state electronic components, which can be damaged by excessive heat. They are usually identified with a typical operating life (such as 50,000 hours) at a maximum operating temperature (at a spot on the driver case marked with a T_C). If the driver's operating temperature is higher than its rated maximum, each increase of 10°C (18°F) will cut the driver's lifespan in half. If used outdoors, direct exposure to sunlight can far exceed the driver's temperature limit and failure is certain.

LED drivers can and often do impose harmonic distortion on the incoming power line with resultant high temperature losses in supply transformers and conductors. LED drivers can be designed with a higher power factor and harmonic filtering in order to eliminate these effects. LED drivers can also be damaged by spikes, voltage surges, and other types of power line noise. Again, they can be designed with surge suppression or installed on feeders with surge suppression.

When an LED luminaire fails prior to the expected lifespan and has an accessible driver, the first step in troubleshooting should be to verify if the driver is working. Replacement drivers must be matched to the voltage and current needs of the luminaire, with at least a 10% safety margin on its maximum current output. When possible, replacement drivers should be obtained from the luminaire manufacturer. Some vendors of LED drivers have versions that can be programmed for a maximum current; however, if too much current is provided to the LEDs, the condition is called *overdriving* and a shortened LED life will result.

2.0.0 Section Review

1. The lamp with the shortest life expectancy is the _____.
   a. fluorescent lamp
   b. HID lamp
   c. incandescent lamp
   d. halogen lamp

2. A ballast is used to _____.
   a. prolong the life of an incandescent lamp
   b. dim an incandescent lamp
   c. operate a screw-in replacement LED lamp
   d. regulate the current in a fluorescent lamp

Section Three

3.0.0 Luminaires

Objective

Select and install luminaires for various applications.
a. Identify luminaires and their applications.
b. Store and handle lamps and luminaires.
c. Install luminaires.

Performance Task

1. Install one or more of the following luminaires and their associated lamps:
   - Surface-mounted
   - Recessed
   - Suspended
   - Track-mounted

Trade Term

Troffers: Recessed luminaires installed with the opening flush with the ceiling.

In the electrical trade, the terms *luminaire* and *lighting fixture* are used interchangeably (luminaire is an international term). A luminaire is defined as a complete lighting unit consisting of the lamp or lamps, together with the parts designed to distribute the light, position and protect the lamps, and connect the lamps to a power supply.

3.1.0 Luminaires and their Applications

Luminaires are classified in many ways, such as: light source, mounting method, aesthetics, and use. From an installer's point of view, luminaires can best be classified by the way in which they are mounted. These include the following types of fixtures:

- Surface-mounted
- Recessed
- Suspended
- Track-mounted
- Pole-mounted

3.1.1 Surface-Mounted Luminaires

Surface-mounted luminaires are directly mounted on a ceiling or wall. They can be incandescent, fluorescent, LED, or HID types. Some typical examples of surface-mounted indoor ceiling and wall luminaires are shown in *Figure 19* and *Figure 20*, respectively. *Figure 21* shows an example of a wall-mounted outdoor luminaire.

3.1.2 Recessed Luminaires

As the name implies, recessed luminaires are mounted into a ceiling so that the main body of the luminaire is not visible. *Figure 22* shows examples

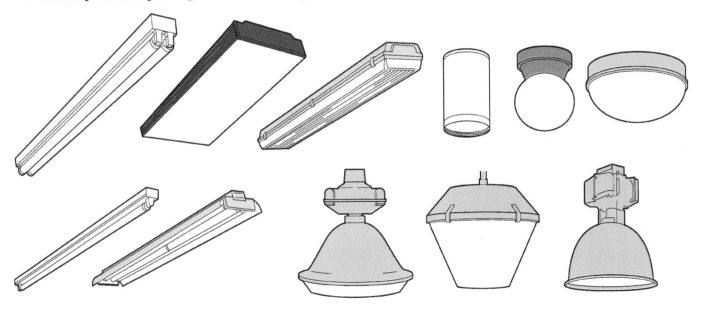

Figure 19 Surface-mounted ceiling luminaires.

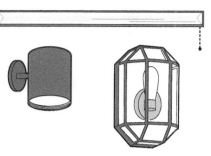

Figure 20 Surface-mounted indoor wall luminaires.

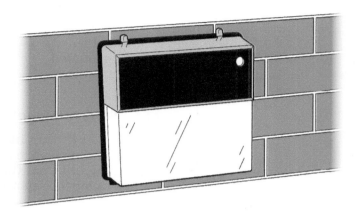

Figure 21 Surface-mounted outdoor wall luminaire.

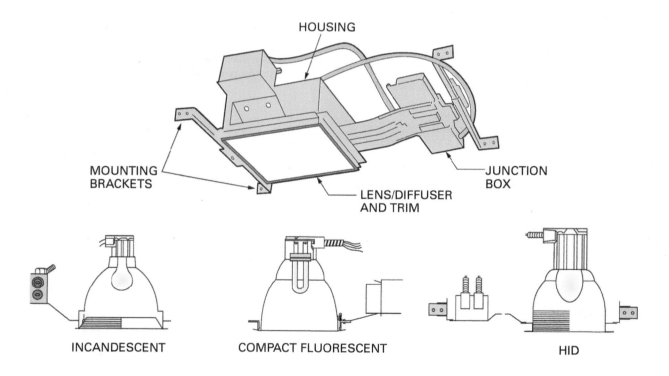

Figure 22 Typical recessed luminaires.

of recessed luminaires typical of those used with various lamps. As shown, a typical luminaire for new work consists of a housing, a junction box, and a mounting frame used to fasten it to the ceiling structure. These components of the luminaire are all located above the ceiling line. Only the lens/diffuser and trim are viewable on the ceiling surface.

Other recessed luminaires, called **troffers**, are square or rectangular units installed above the ceiling (*Figure 23*). These are typically 2' × 4' or 2' × 2' units in the United States, containing two or three T-8, T-10, or T-12 fluorescent or LED lamps. (Metric sizes vary widely but are available in similar dimensions.) Some troffers may incorporate a mounting flange or are used with a mounting flange kit, allowing them to be installed in ceiling recesses in plaster, drywall, or concealed-frame ceiling systems. Other troffers, called grid-type troffers, are designed for installation in grid T-bar suspended ceilings (*Figure 24*). LED luminaires greatly expand the freedom of mounting recessed troffers because their reduced thickness (as little as ⅝" or 16 mm) allows surface mounting or suspended mounting (with a separately mounted driver) and mounting between ceiling grid bars under ducts or piping. See *Figure 25* for one such luminaire.

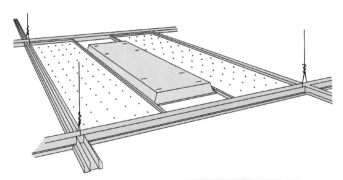

Figure 24 Fluorescent troffer mounted in a suspended ceiling.

CAUTION

Incandescent recessed luminaires generate a considerable amount of heat within their enclosure and are a definite fire hazard if not installed properly. For this reason, several precautions and *NEC®* requirements must be observed when installing them. These precautions and requirements are covered later in the installation section of this module.

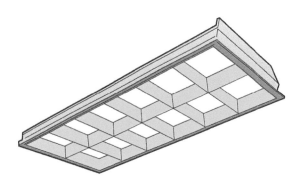

FOR USE IN INVERTED-T OR GRID CEILINGS

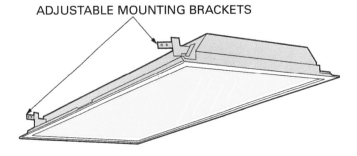

ADJUSTABLE MOUNTING BRACKETS

FOR USE IN DRYWALL OR PLASTER CEILINGS

Figure 23 Typical recessed fluorescent troffers.

Figure 25 2×2-foot LED luminaire.

LED Luminaires

LED luminaires are rapidly replacing incandescent luminaires in recessed lighting and many other applications.

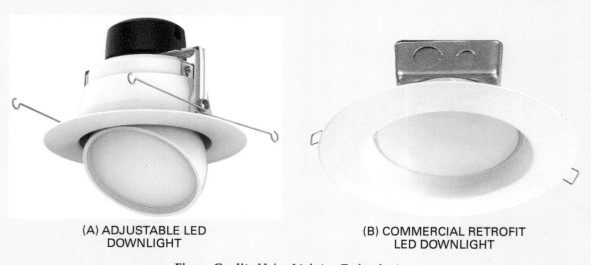

(A) ADJUSTABLE LED DOWNLIGHT

(B) COMMERCIAL RETROFIT LED DOWNLIGHT

Figure Credit: Halco Lighting Technologies

3.1.3 Suspended Luminaires

Suspended luminaires are hung from a ceiling using a chain, aircraft wire, cord, or metal stem. Included in this group are chandeliers and pendant units (*Figure 26*).

3.1.4 Track-Mounted Luminaires

Track-mounted lighting systems (*Figure 27*) include track, an electrical feed box, and one or more luminaires of various types that can be positioned along the track. Track lighting wiring is in the metal track channels. It can be single-circuit or multiple-circuit wiring to allow two or more sets of luminaires to be independently controlled. With all track lighting, the luminaires, track, and fittings must be listed for use together. Although some brands appear interchangeable, poor connections and poor mounting can cause a fire or cause the luminaire to fall.

its outlet box using the mounting hardware supplied with the luminaire and installed according to the manufacturer's instructions. If you are not thoroughly familiar with the luminaire to be installed, first read and be sure that you understand the manufacturer's product literature and installation instructions supplied either with the luminaire or printed on the shipping carton.

In buildings of all types, branch circuit wiring for surface-mounted ceiling and wall luminaire outlets normally terminates in octagonal or round boxes (*Figure 31*). Some may terminate in square or rectangular boxes. Both metal and nonmetallic types of boxes may be used. These boxes are designed to be covered by the mounting base of a luminaire with a similar shape. For concealed wiring in wooden frame construction, the outlet boxes are fastened directly to studs, joists, or other framing members using nails or screws. Some ceiling outlet boxes are mounted on an adjustable bar hanger so that the box can be installed in the space between two ceiling joists. The box is attached to the bar hanger by a luminaire stud or two bolts. When properly installed, the front face of all concealed ceiling and wall boxes should be flush with the finished ceiling or wall. For exposed conduit installed on concrete or other masonry surfaces, the outlet boxes should be fastened to the ceiling or wall with screws and expansion shields, anchors, powder-actuated fasteners, or similar appropriate fasteners. On steel work, they should be bolted or clamped to the steel. Outlet boxes embedded in concrete are held securely in position by the hardened concrete.

The size and weight of the luminaire determine what type of outlet box should be used with a luminaire. When mounting luminaires that weigh more than 50 lbs (23 kg) on a ceiling, *NEC Section 314.27(A)(2)* requires that the luminaire be supported independently of the box, unless the outlet box is listed for not less than the weight to be supported. Boxes used at luminaire outlets in a wall must be designed for the purpose and marked to indicate the maximum weight of the luminaire to be supported, if other than 50 lbs (23 kg), per *NEC Section 314.27(A)(1)*. In addition to complying strictly with the code, good judgment must also be exercised. Remember, codes only give the minimum requirements for electrical installations, not necessarily those that make for the highest-quality installation.

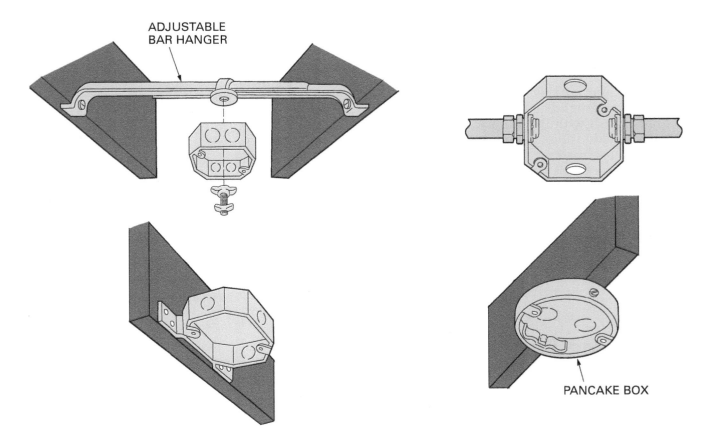

Figure 31 Typical outlet boxes used with surface-mounted luminaires.

> **NOTE**
>
> When a manufacturer supplies insulation with a surface-mounted luminaire, do not throw it away. It must be placed between the luminaire and the mounting surface to insulate the mounting surface and wiring in the fixture box from the heat produced by the luminaire. Make sure to install any such insulation as directed by the manufacturer.

Some of the installation considerations for various types of surface-mounted luminaires include the following:

- *Drum, globe, sconce, and similar luminaires* – *Figure 32* shows some common methods used to mount drum, globe, sconce, and similar types of indoor luminaires to an outlet box. Usually, these types of luminaires can be directly mounted to the outlet box with no additional support. As shown in *Figure 32 (A)*, the common porcelain-type lampholder luminaire screws directly to the ears of the outlet box. Other ceiling luminaires and wall sconces are fastened to the outlet box using a mounting bar that is attached to a threaded stud in the box, as shown in *Figure 32 (B)*. The mounting bar is positioned on the stud and fastened in place with a locknut. After the luminaire wiring has been completed, the mounting holes in the base of the luminaire are slipped over the machine screws installed in the mounting bar. Cap nuts are then installed on the machine screws and tightened until the luminaire is drawn flush with the surface. Luminaires can be mounted to an outlet box without a box stud in a similar manner by screwing a mounting bar to the ears of the box, as shown in *Figure 32 (C)*. Another common way some luminaires are installed is by screwing a reducing nut and nipple onto a box stud, as shown in *Figure 32 (D)*. Following this, the hole in the center of the luminaire base is placed over the nipple, then a cap nut is installed on the nipple and tightened until the luminaire is drawn flush with the surface. A similar method is to screw a nipple into the threaded center hole of a mounting bar attached to the ears of the outlet box.
- *Fluorescent and LED luminaires* – Some lightweight, surface-mounted luminaires can be supported directly from an outlet box. Such types typically use a stud, reducing nut, and nipple in the box, with a mounting bar inside the housing, as shown in *Figure 33*. The assembly is held in place with a locknut. Heavier and larger luminaires are supported directly from the ceiling using toggle bolts, wood screws in joists, anchors in masonry, or other appropriate fasteners. These luminaires have an access opening that must be positioned directly under the outlet box when the luminaire is in place. Many surface-mounted luminaires have bumps or ridges on their mounting surfaces to provide a slight ceiling clearance in order to dissipate heat.
- *Chandeliers and pendants* – Typical methods for attaching chandeliers to an outlet box are shown in *Figure 34*. Chandeliers are normally heavy, requiring that they be mounted on a box with a stud. A hickey (or a hickey together with a threaded adapter) is installed over the luminaire chain and wires and then screwed onto the stud. The luminaire base is then fastened to the hickey or adapter with a locknut or cap nut. Some are secured to a mounting bar by means of a nipple. Pendant luminaires are typically supported by chains, metal stems, thin cables, or retractable cords. Most pendants are designed to be wired directly into a ceiling outlet box in a similar manner as that described for chandeliers.

> **WARNING!**
>
> *NEC Section 410.10(D)(1)* prohibits pendants and other hanging-type luminaires from being installed within a zone measured 3' (900 mm) horizontally and 8' (2.5 m) vertically from the top of a bathtub rim or shower stall threshold. Per *NEC Section 410.10(D)(2)*, any luminaire within these areas must be suitable for damp locations, and if subject to shower spray, must be suitable for wet locations.

- *Ceiling fans/luminaires* – *NEC Section 422.18* covers the support of ceiling-hung fans (paddle fans). It refers to *NEC Sections 314.27(C) and (E)*. *NEC Section 314.27(C)* states that for fans weighing 70 lbs (32 kg) or less, the outlet box may be used as the only support of the fan if the box is listed and marked for such use. The code also requires that

> **Think About It**
>
> **Installing Surface-Mounted Luminaires in Closets**
>
> What precautions should you take when installing surface-mounted ceiling or wall luminaires in a clothes closet?

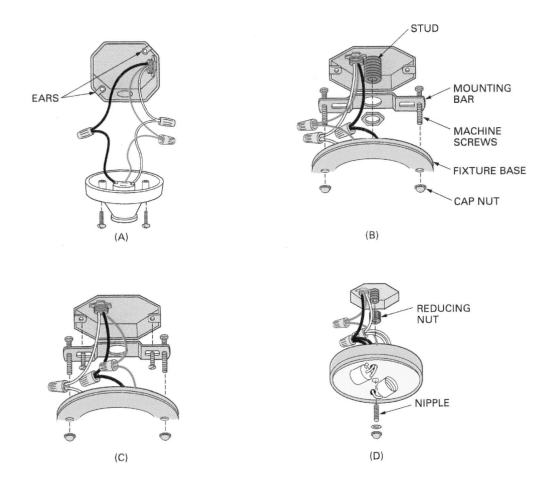

Figure 32 Common methods of installing surface-mounted luminaires.

Installing Ceiling-Mounted Fluorescent Luminaires

While standing on a ladder to mount a fluorescent luminaire on a ceiling, it is much easier to mark the mounting hole locations for the luminaire on the ceiling using a cardboard template than it is to hold the actual luminaire in place while marking the hole locations.

Use a portion of the cardboard carton that the luminaire was shipped in to make a template of the luminaire. Transfer the locations of all luminaire mounting holes, and any other required holes, to the template. Cut out these holes, then use the template to locate and mark the mounting hole locations on the ceiling.

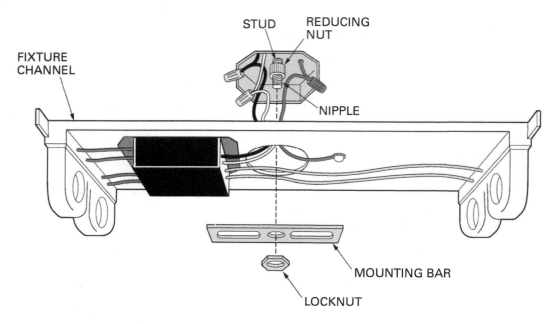

Figure 33 Typical mounting of a small luminaire to an outlet box.

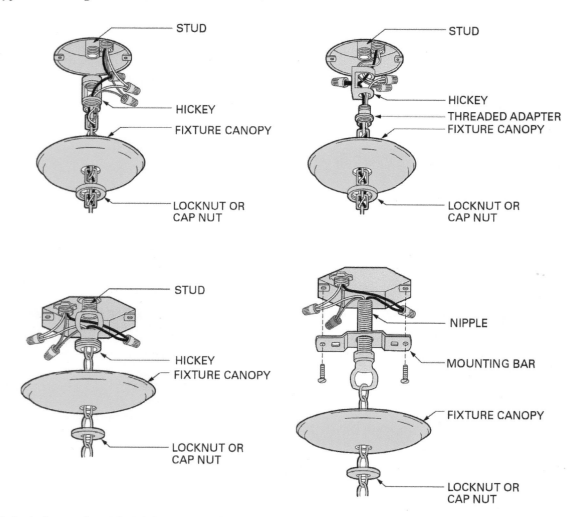

Figure 34 Methods for attaching chandeliers to an outlet box.

boxes designed to support fans weighing over 35 lbs (16 kg) must be listed for such use and marked with the maximum weight to be supported. Extra support can be provided by mounting the box on a fan joist hanger (*Figure 35*). *NEC Section 314.27(E)* permits the use of a listed locking support and mounting receptacle, in combination with a compatible factory-installed attachment fitting designed for the support of the luminaire or fan within the weight and mounting orientation limits of the listing. The specific method for mounting ceiling fans varies among ceiling fan manufacturers; therefore, the fan should be mounted to the outlet box per the manufacturer's directions. *Figure 35* shows an example of one mounting method. *NEC Section 410.10(D)(1)* also prohibits ceiling-suspended fans from being installed within a zone measured 3' (900 mm) horizontally and 8' (2.5 m) vertically from the top of a bathtub rim or shower stall threshold.

Installing Chandeliers and Pendants

It is important to make sure that a chandelier or pendant luminaire is installed at the proper height. For example, you should allow for door clearances when installing the luminaire in a foyer or entrance hallway. Such a luminaire should also be hung at a sufficient height to allow safe passage for tall people.

If the luminaire is being hung over a table, it should be narrower than the width of the table so that people do not hit their heads on the luminaire when sitting down or getting up.

- *Outdoor luminaires* – The methods for attaching outdoor surface-mounted luminaires to outlet boxes are basically the same as those described for indoor surface-mounted luminaires. However, the installation must be done so that the final assembly is watertight. All luminaires installed in wet locations must be marked as suitable for wet locations. All luminaires installed in damp locations must be marked as either suitable for wet locations or suitable for damp locations. When mounting a luminaire to an outlet box that is fully exposed to the weather, the outlet box must be a metal weatherproof-type box. All washers, gaskets, etc., required to make a watertight assembly must be properly installed as directed by the manufacturer. *Figure 36* shows an example of a common outdoor spotlight luminaire. With this luminaire, waterproofing fiber washers must be inserted into the luminaire sockets and rubber gaskets installed over the end of each socket. In addition, a gasket must be installed between the luminaire base and the face of the outlet box.

3.3.2 Recessed Luminaires

Recessed luminaires are mounted into ceiling recesses so that the main body of the luminaire is not visible. *Figure 37* lists some of the code requirements for recessed luminaires.

> **WARNING!**
> Recessed luminaires generate a considerable amount of heat within their enclosures and are a definite fire hazard if not installed properly. For this reason, the code requirements given in *NEC Article 410* must be strictly observed when installing these luminaires.

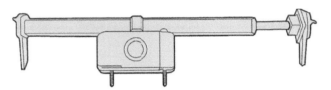

TYPICAL FAN JOIST HANGER

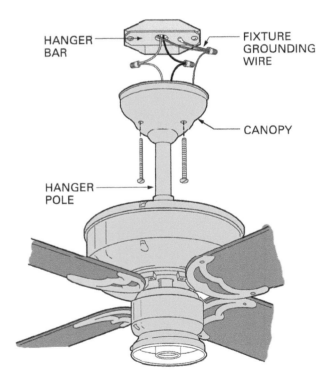

Figure 35 Typical ceiling fan/light mounting method.

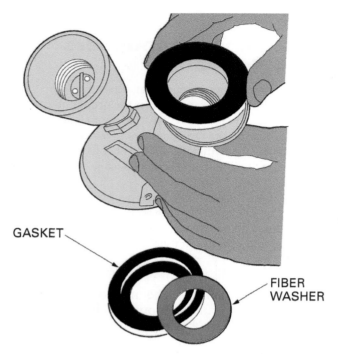

Figure 36 Installing washers and gaskets in outdoor luminaires.

Wiring for luminaires in a suspended ceiling is installed above the ceiling using electrical metallic tubing (EMT), armored cable (AC), nonmetallic-sheathed cable (NM), or other acceptable methods as listed in the *NEC®*. Properly supported outlet boxes are placed above the ceiling and separated by a minimum distance of at least 1' (300 mm) from the luminaire. A flexible connection, commonly called a *luminaire whip*, is made between the outlet box and the luminaire (shown in *Figure 39*). As appropriate, the length of the luminaire whip is 18" (450 mm) minimum to 6' maximum (1.8 m) and can be made with any of the following:

- Trade size ⅜" or Metric Designator (MD) 12 flexible metal conduit using conductors suitable for the temperature requirement stated on the luminaire label
- Armored cable containing conductors suitable for the temperature as stated on the luminaire label
- Trade size ⅜" (MD 12) flexible nonmetallic conduit with conductors and a grounding conductor suitable for at least 90°C (194°F)

> **NOTE**
> Always refer to the manufacturer's instructions when installing any luminaire.

What's wrong with this picture?

A general installation procedure involves the following steps:

Step 1 Extend the luminaire mounting bars to reach the framing members.

Step 2 Position the luminaire on the mounting bars to locate it properly.

Step 3 Align the bottom edges of the mounting bars with the bottom face of the framing members, then nail, screw, or otherwise attach the bars to the framing members.

Step 4 Remove the junction box cover and open one knockout for each cable entering the box.

Step 5 If required, install cable clamps for each open knockout and tighten the locknut.

Step 6 Run the branch circuit wiring through the cable clamps and into the luminaire connection box.

Step 7 Make up the electrical connections.

Step 8 After the finished ceiling is in place, install the lens/diffuser and/or trim on the luminaire. The luminaire trim combinations are marked on the label inside the luminaire.

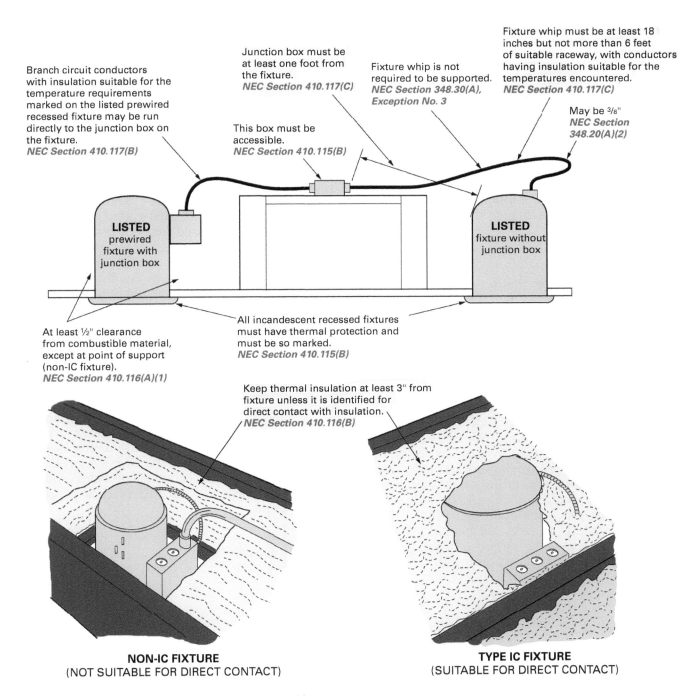

Figure 37 Summary of *NEC®* requirements for recessed luminaires.

> **CAUTION**
>
> The use of a trim type that is not listed on the luminaire can result in overheating and possible on-off cycling of the luminaire's thermal protective device. Mismatching these components violates *NEC Section 110.3(B)*, which states that listed and/or labeled equipment shall be installed and used in accordance with any instructions included in the listing or labeling.

Installation considerations for various types of recessed luminaires include the following:

- *Incandescent, compact fluorescent, and HID recessed luminaires* – These luminaires are all mounted in a ceiling in basically the same way. The rough-in for these types of luminaires must be completed before the ceiling material is installed. As shown in *Figure 38*, they typically include captive adjustable bar hangers for attachment to wooden joists or T-bar grid ceiling members. Recessed luminaires that are identified as acceptable for through wiring are equipped with feed-through junction boxes, which allow a series of fixtures to be wired from one to another or branch circuits to be fed through the housing junction box. Recessed luminaires are typically wired with conduit, nonmetallic-sheathed cable, or armored cable, depending on the application.
- *Fluorescent and LED troffers* – The code requirements for the installation of recessed fluorescent and LED troffers are basically the same as those for other types of recessed luminaires. Recessed fluorescent and LED troffers are inserted into the rough-in opening in the drywall, plaster, or other ceiling types and fastened in place using various devices. Some have mounting brackets used to fasten the luminaire to the ceiling structure that are adjustable from within the installed luminaire. If a suspended ceiling supports recessed fluorescent troffers or other luminaires, *NEC Section 410.36(B)* requires that all the ceiling framing members be securely fastened together and to the building structure. This job is normally done by the installer of the ceiling grid. The proper support of suspended luminaires is a critical safety issue for people who occupy the building and for any emergency personnel who may be called into the building in the event of an earthquake, fire, etc. For this reason, all luminaires, including any lay-in luminaires such as fluorescent or LED troffers, must also be securely fastened to the main ceiling framing members by mechanical means such as bolts, screws, rivets, or listed clips identified for use with the type of ceiling framing members and luminaires involved. These are installed by the electrician. Most manufacturers now provide locking clips in the end plates of their troffers. These clips are simply bent perpendicular to the end plate so that as the troffer is lowered into the ceiling grid, the clip locks over the top of the T. *Figure 39* shows listed clips commonly used to fasten lay-in troffers to suspended grid systems.

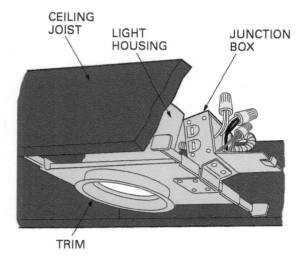
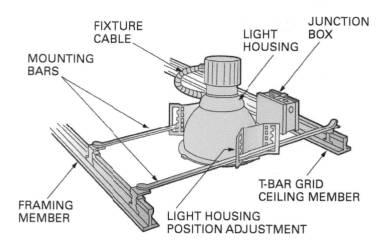

Figure 38 Typical mounting of a recessed luminaire.

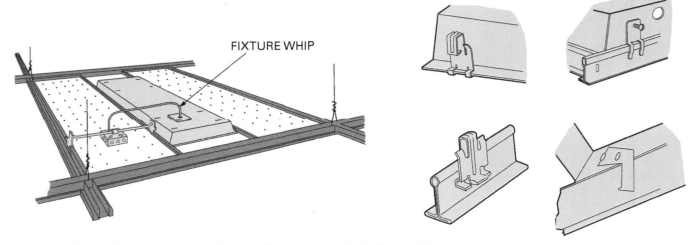

Figure 39 Typical clips used to fasten lay-in troffers to suspended ceiling grid systems.

> **Think About It**
>
> **Installing Recessed Luminaires**
>
> When installing recessed luminaires, is it permissible to wire a series of luminaires from one to another or to run branch circuit wiring through a luminaire junction box? If so, when?

3.3.3 Suspended Luminaires

Suspended luminaires used in commercial and industrial buildings can be supported in numerous ways from the building's structural members, depending on the construction of the building and the lighting application. Because the methods of support used can vary so widely, it would be impossible to describe them all here. *Figure 40* shows some types of hangers commonly used to attach supporting rods, cables, chains, etc.

Generally, on large construction jobs, the specific method for hanging suspended luminaires is shown in the construction drawings. However, on many smaller jobs, the method used for suspending luminaires is frequently left up to the electrician. This can be a problem for less experienced electricians. Fortunately, technical help is readily available. To aid customers, manufacturers of fasteners and metal framing devices produce numerous catalogs and technical product support bulletins that show example solutions to common hanging/support problems. Other good sources of technical advice are your supervisor, other qualified electricians, and fastener/framing company sales engineers or representatives.

Figure 41 shows some examples of the different kinds of support methods used with suspended fluorescent and LED luminaires.

Figure 42 shows luminaire support methods representative of those found in commercial and industrial sites.

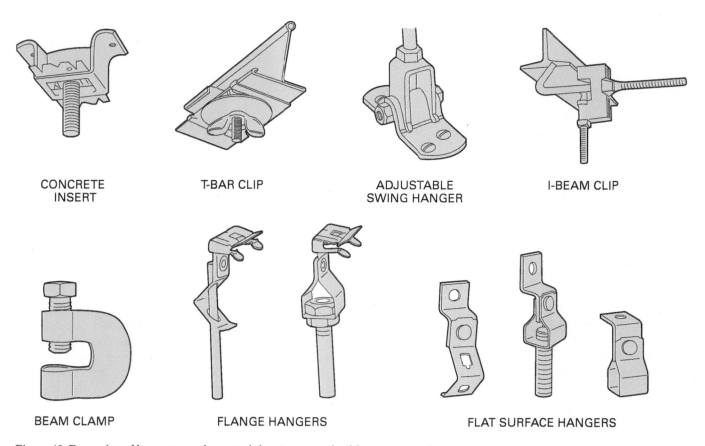

Figure 40 Examples of hangers used to attach luminaires to building structural members.

3.3.4 Track Luminaires

Track-mounted lighting systems include the track (rail), an electrical feed box, two or more luminaires that can be positioned along the track, and various types of accessories and other components that allow assembly in any pattern (*Figure 43*). The track lighting wiring is in the metal track rails. It can be either single-circuit or multiple-circuit wiring to allow two or more sets of lights to be independently controlled. The track is an extruded aluminum channel that holds two or more electrical conductors in separate insulated grooves within the channel. The inside surfaces of the conductors are bare. When a track luminaire is snapped into the track, two terminals on the luminaire contact the track conductors, supplying power to the luminaire.

Track lighting systems are made so that the proper electrical polarity is always maintained. The polarity lines inside the track must be aligned when the sections of track are installed. The luminaires will snap into the track in only one way. The *NEC*® requires that luminaires, accessories, etc., used in a track system be specifically designed for the track on which they are to be installed.

Track sections are made in various lengths and fastened together using straight, flexible, X, L, or T connectors. End caps provide a finished look. Track lighting can be connected to a branch circuit outlet box in many ways, including the use of a floating power feed, end power feed, or X, L, or T power feed connector. Some are connected to a receptacle outlet using a cord and plug connector.

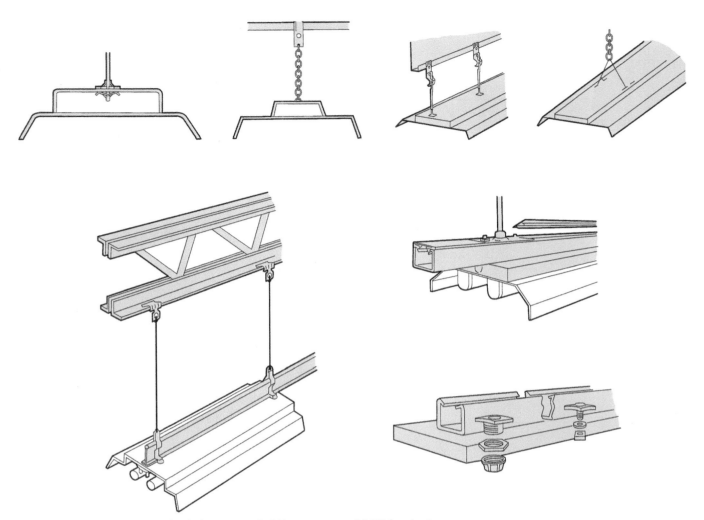

Figure 41 Some support methods for suspended fluorescent and LED luminaires.

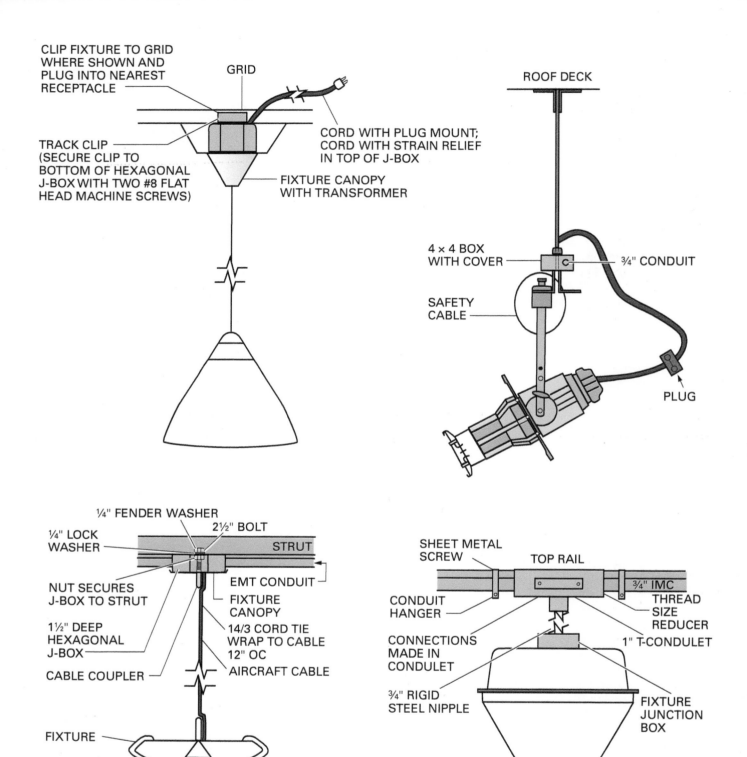

Figure 42 Some luminaire supports for commercial/industrial applications.

Case History

Installing Luminaires

A group of electricians and apprentices were wiring fluorescent luminaires in a suspended ceiling in the new wing of a hospital. Emergency luminaires on a separate circuit were to be wired first, then the existing temporary luminaires were to be de-energized. The crew decided that this process was too slow and, contrary to their supervisor's instructions, disconnected the temporary luminaires and began to wire the remaining luminaires while some circuits were still energized. Co-workers warned the crew to test the circuits to see if they were energized.

An apprentice was standing on a wooden ladder with his body extended above the ceiling grid while wiring a fluorescent luminaire. He was wet with sweat and was leaning against a ceiling grid as he worked. Co-workers heard a noise and upon investigation, found the apprentice dangling unconscious from the ceiling grid. The victim was pulled free and given CPR, but he had received a fatal shock.

The Bottom Line: This accident could have been prevented if standard safety measures had been taken:

- The crew had followed the supervisor's instructions.
- The branch circuit feeding the luminaire had been de-energized and the proper lockout/tagout procedures followed.
- The apprentice had been working under the supervision of a qualified electrician who had first checked the branch circuit wires feeding the luminaire with a voltage tester to make sure that the wires were not energized.

NEC Sections 410.151 through 410.155 cover the installation requirements for track lighting. The track is mounted directly to the ceiling or wall using screws, toggle bolts, or other appropriate fasteners installed through holes in the track. When installed on a wall, the track must be at least 5' (1.5 m) above the finished floor, except where protected from physical damage or when using low-voltage (less than 30V rms) track.

The *NEC®* requires that the track be fastened so that each fastener is suitable for supporting the maximum weight of the luminaires that can be installed. Unless identified for use with supports at greater intervals, a single section of track that is 4' (1.2 m) or less in length must be supported in at least two places. If more track is being used, at least one additional support for each individual extension that is 4' (1.2 m) long or less must be used.

A typical installation begins by first attaching the track electrical feed assembly to the designated outlet box. If using an end power feed for this purpose, the track adapter plate covers the outlet box and holds the track connector and electrical housing.

When using a wire-in connector, track clips that are spaced at the required distance apart are used to hold the track ¼" to ½" (6 mm to 13 mm) away from the ceiling or wall mounting surface. These clips are fastened in place using screws, toggle bolts, or other appropriate fasteners. It is a good idea to snap a chalkline from the center of the outlet box or center slot on the track connector to the proposed end of the track run for use as a track installation guide. Once the clips are installed, the first section of track is fitted solidly into the track connector, and then the track is snapped into the clips. Following this, setscrews in the sides of the clips are tightened to hold the track firmly in place. The remaining sections of track are installed in the same manner using the appropriate connectors and accessories. When it is necessary to cut a section of track to length, it can be cut using a metal-cutting saw. Make sure to follow the manufacturer's instructions as to where and how to make the cut and how to close off the cut with a proper end cap.

Another type of track lighting is covered in *NEC Article 393*, *Low-Voltage Suspended Ceiling Power Distribution Systems*. These operate as Class 2 circuits limited to 30VAC and 60VAC to supply power to luminaires, control, and signaling systems that operate within these limits.

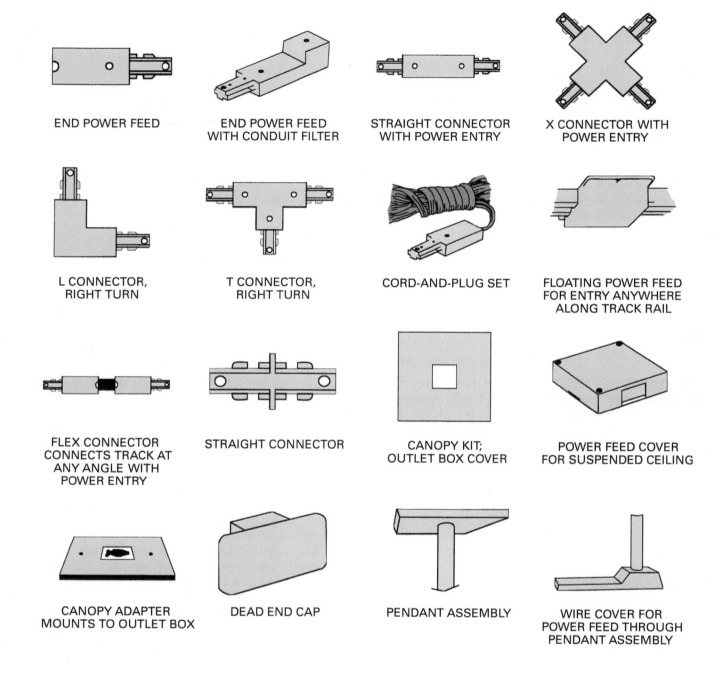

Figure 43 Track lighting accessories and components.

3.3.5 Making Electrical Connections to Luminaires

Making the electrical connections between the luminaire and related outlet box wires is done at the appropriate point during the physical installation of the luminaire. The specific way in which the supply wires in the outlet box are connected to the luminaire wires varies depending on the design of the branch circuit and the way in which the branch circuit cable or raceway wiring is routed. Therefore, you must refer to the lighting plan to determine the exact connections.

The most common type of electrical connection is a conventional hard-wired installation. The connection of the luminaire wiring to the outlet box supply wires should be done in accordance with the manufacturer's instructions. The general procedure is basically the same, no matter what type of luminaire is being connected.

> **WARNING!**
> Before attempting to connect luminaires to a branch circuit outlet box, make sure that the power is turned off to the branch circuit involved at the circuit breaker panel. Then, make sure to follow the prevailing lockout and tagout procedures.

When cutting outlet box wires in preparation for connecting the luminaire, do not cut the branch circuit wires too short or too long. *NEC Section 300.14* requires that at least 6" (150 mm) of free conductor length be left at each outlet or junction box where it emerges from its cable sheath or raceway. Furthermore, for splices or the connection of luminaires or devices, each conductor's length shall extend at least 3" (75 mm) outside any box opening that is less than 8" (200 mm) in any dimension. The 3" (75 mm) length is generally measured from the outer edge of the box. Note that leaving the wires too long can make it difficult to position the wires in the box and overcrowd the box, possibly resulting in overheating.

The luminaire's black wire(s) are connected to the outlet box hot wires (black, red, or marked hot) and the white wire(s) to white wires (*Figure 44*). If dealing with nonmetallic cable and metal boxes, the bare grounding wires should be spliced with one end of a grounding jumper. The other end of the grounding jumper should be attached to the box using either a grounding clip or the box grounding screw. If dealing with nonmetallic cable and a nonmetallic outlet box, the bare cable wires should be spliced together. In some cases, a green wire from the luminaire and the bare wires will be connected to a grounding screw provided on the mounting bar supplied with the luminaire.

All wire splices should be made using the correct wire nuts sized for the number and gauges of wire in use. If in doubt, most wire nut manufacturers mark their packages to show the maximum and minimum number of wires of different sizes that can be connected using a particular wire nut. When properly installed and tightened, wire nuts should cover all bare current-carrying conductors. Some wire nut manufacturers recommend twisting the wires together before installing the wire nut; others do not. Follow the manufacturer's instructions.

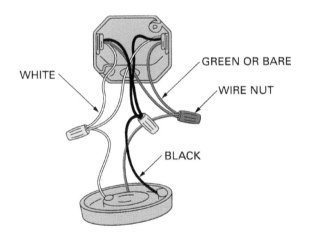

Figure 44 Example of basic luminaire connections.

The *NEC®* requires that all luminaires with exposed metal parts be grounded. If the luminaire has a separate green grounding wire, it should be spliced together with the other grounding wires. Once the luminaire box and/or metal mounting strap is grounded, the nipple or screws holding the luminaire will ground the luminaire.

In addition to conventional hard-wired systems, other types of wiring systems are also in common use. Guidelines for making other types of wiring connections include the following:

- *Modular system wiring* – In some installations, luminaires are connected using a modular wiring system (*Figure 45*). Modular wiring provides an alternative method to conventional hard wiring for power to open office partitions and luminaires in grid-type suspended ceilings. Modular system components are interchangeable and can be reused. Component labels are color-coded by voltage and the units are keyed to prevent mismatching. Lighting branch circuit wiring is run from the building panel to an outlet or junction box that serves as the starting point for the modular wiring. There, a modular system feeder adapter is wired and mounted. From this point on, all the wiring to the luminaires is modular, using the appropriate components that can be safely connected or disconnected without having to turn off the power.

- *Strut channel wiring* – Non-perforated strut and similar structural members are frequently used to both support and provide a wiring raceway for powering luminaires in industrial/commercial lighting systems. Various support and wiring methods can be used, depending on the application. Different types of hangers and other luminaire-supporting hardware used to support the luminaires from the strut sections are specially designed by the strut manufacturer for that purpose. *Figure 46* shows two typical strut-type methods for supporting and wiring luminaires. Snap-in cover strips are used to close the open side of the strut after wires and splices have been installed.

- *Lighting trolley busways* – Lighting trolley busways (*Figure 47*) use a special channel similar to strut, with insulated conductors installed similar to track lighting. Stationary plug-in devices, commonly called *twistouts*, are made to attach luminaires at fixed points. Wheeled trolleys allow attached luminaires to be moved at will. Lighting trolley busways are used in industrial and some commercial lighting applications because they allow the lighting system to be reconfigured as needed. Lighting trolley busways are made with different amperage ratings and different numbers of conductors for use at a maximum voltage of 300V (AC or DC) to ground. *NEC Article 368.56(C)* touches on their use.

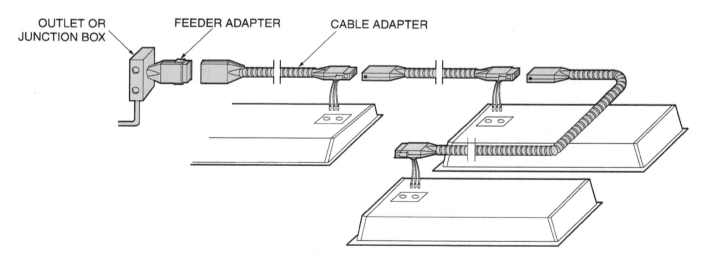

Figure 45 Basic modular wiring system.

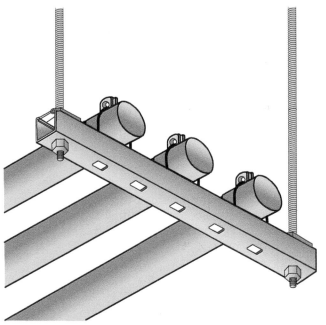

Figure 46 Strut for supporting and powering luminaires.

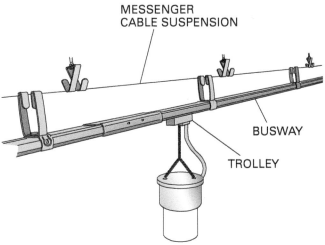

Figure 47 Lighting trolley busway.

3.0.0 Section Review

1. Fluorescent or LED troffers are typically a type of _____.
   a. surface-mounted luminaire
   b. recessed luminaire
   c. pendant luminaire
   d. track-mounted luminaire

2. Which of the following is the best tool for opening a luminaire carton?
   a. Screwdriver
   b. Long knife
   c. Utility knife
   d. Keys

3. Unless otherwise marked, boxes for indoor ceiling-mounted luminaires can support a luminaire weighing no more than _____.
   a. 25 pounds
   b. 35 pounds
   c. 50 pounds
   d. 70 pounds

Review Questions

1. Light energy that is NOT absorbed by an object can be ___.
   a. reflected from the object
   b. transmitted through the object
   c. reflected from and/or transmitted through the object
   d. dissipated as heat in the object

2. A light that is considered warm in tone is produced by lamps with color temperatures of ___.
   a. 3,000°K
   b. 3,500°K
   c. 4,100°K
   d. 5,000°K

3. The resistance of a tungsten filament is ___.
   a. much higher after the lamp is turned on
   b. lower after the lamp is turned on
   c. somewhat higher after the lamp is turned on
   d. the same whether the lamp is turned on or off

4. The filament of a tungsten-halogen lamp is encased in a capsule containing iodine or ___.
   a. bromine gas
   b. argon gas
   c. nitrogen gas
   d. xenon gas

5. During operation, a high-output (HO) fluorescent lamp typically has a current draw of ___.
   a. 400mA to 600mA
   b. 800mA to 1,000mA
   c. 1,100mA to 1,300mA
   d. 1,500mA to 1,700mA

6. The type of high-intensity discharge (HID) lamp that uses the oldest HID technology is the ___.
   a. low-pressure sodium lamp
   b. high-pressure sodium lamp
   c. mercury vapor lamp
   d. metal halide lamp

7. The diameter of a 4', T-12 fluorescent lamp is ___.
   a. 1"
   b. 1⅛"
   c. 1¼"
   d. 1½"

8. To improve both ballast efficiency and lamp efficacy, an electronic ballast typically operates the connected fluorescent lamps at a frequency of ___.
   a. 50Hz to 60Hz
   b. 10kHz to 20kHz
   c. 20kHz to 50kHz
   d. 60kHz to 100kHz

9. A driver supplies stable power to a(n) ___ lamp
   a. incandescent
   b. LED
   c. fluorescent
   d. HID

10. Most code requirements governing the installation of luminaires, lampholders, and lamps are covered in ___.
    a. *NEC Article 370*
    b. *NEC Article 410*
    c. *NEC Article 411*
    d. *NEC Article 422*

11. Luminaire whips used to connect from a junction box mounted above the ceiling to a luminaire installed in a suspended ceiling must have a minimum length of ___.
    a. 18" (450 mm)
    b. 2' (600 mm)
    c. 4' (1.2 m)
    d. 6' (1.8 m)

12. The maximum luminaire whip length when wiring suspended ceiling luminaires is ___.
    a. 4'-0" (1.2 m)
    b. 6'-0" (1.8 m)
    c. 8'-0" (2.5 m)
    d. 10'-0" (3 m)

13. Suspended luminaires used in commercial and industrial buildings are supported by the _____.
    a. conductors
    b. luminaire canopy
    c. ceiling drywall or drop ceiling
    d. building's structural members

14. When installed on a wall, the track for track lighting must be a minimum of _____.
    a. 4'-0" (1.2 m) above the finished floor
    b. 5'-0" (1.5 m) above the finished floor
    c. 6'-0" (1.8 m) above the finished floor
    d. 7'-0" (2.1 m) above the finished floor

15. When cutting the branch circuit wires in an outlet box in preparation for connecting a luminaire, the amount of free conductor length that should be left after it emerges from the cable sheath or raceway is _____.
    a. 3" (75 mm)
    b. 4" (100 mm)
    c. 6" (150 mm)
    d. 8" (200 mm)

Supplemental Exercises

1. When light strikes an object, it can be _____, transmitted through the object and _____, and/or it can be _____ from the object.
2. When light passes through a prism, the spectrum ranges from _____ on the longer wavelength to _____ on the shorter wavelength.
3. True or False? Lamps of equal wattage have the same lumen and life ratings.
4. The type of HID lamp that produces light with the best color rendition is the _____ lamp.
    a. metal halide
    b. mercury vapor
    c. high-pressure sodium
    d. low-pressure sodium
5. True or False? When installing a bi-pin fluorescent lamp, it should be rotated ¼ turn in its lampholder to seat and connect it.
6. Which of the following types of fluorescent lamp ballasts provide the best lamp life?
    a. Instant-start ballasts
    b. Rapid-start ballasts
    c. Electronic programmed-start ballasts
    d. Hybrid-start ballasts
7. A ballast supplies the power to an HID lamp; similarly, a _____ supplies power to an LED.
8. When opening a shipping carton containing a luminaire, wear _____.
9. True or False? When installing a ceiling-mounted luminaire that weighs more than 50 lbs (23 kg), it must always be supported independently of the outlet box unless the box is listed and labeled for such use.
10. A luminaire whip connected between a junction box mounted above the suspended ceiling and a luminaire installed in a suspended ceiling must have a minimum length of _____ and be no longer than _____.
11. Recessed luminaires not labeled Type IC must be installed so that they are at a minimum distance of _____ from any adjacent insulation.
12. When fluorescent or LED troffers and other luminaires are installed in a suspended ceiling, the ceiling framing members must be securely fastened together and to the _____.
13. True or False? Luminaires and/or accessories for track lighting systems can be used interchangeably between systems made by different manufacturers.
14. When cutting the branch circuit wires in preparation for connecting a luminaire, you should leave at least _____ of free conductor length at the outlet box where the wire emerges from its cable sheath or raceway.

Trade Terms Introduced in This Module

Ballast: A circuit component in fluorescent and high-intensity discharge (HID) luminaires that provides the required voltage surge at startup and then controls the subsequent flow of current through the lamp during operation.

Color rendering index (CRI): A measurement of the way a light source reproduces color, with a range from 0 to 100. The higher the index number, the closer colors are to how an object appears in full sunlight or incandescent light.

Dip tolerance: The ability of an HID lamp or luminaire circuit to ride through voltage variations without the lamp extinguishing and cooling down.

Driver: The circuit components in LED lighting that take incoming power and convert it to the required stable voltage and current for the LED.

Efficacy: The light output of a light source divided by the total power input to that source. It is expressed in lumens per watt (LPW).

Incandescence: The self-emission of visible radiant energy from an object heated above 750°K, such as occurs when an electric current is passed through the filament in an incandescent lamp.

Incident light: The light emitted from a self-luminous object such as the sun, a flame, or an electric source.

Lumen (lm): The basic measurement of light. One lumen is defined as the amount of light cast upon one square foot of the inner surface of a hollow sphere with a one-foot radius with a light source of one candela at its center.

Lumen maintenance: A measure of how a lamp maintains its light output over time. It may be expressed either numerically or as a graph of light output versus time. A commonly used value is L_{70}, the number of hours before the lamp output drops to 70% of its initial value.

Lumens per watt (LPW): A measure of the efficiency, or, more properly, the efficacy of a light source. The efficacy is calculated by taking the lumen output of a lamp and dividing by the lamp wattage. For example, a 100W lamp producing 1,750 lumens has an efficacy of 17.5 lumens per watt.

Luminaire: A complete lighting unit consisting of a light source such as a lamp or lamps, together with the parts designed to position and protect the light source and to distribute the light. It may also include components to match the incoming electricity with the needs of the light source.

Luminance: The intensity of light traveling in a given direction from any surface. It is measured in candela/meter^2. The term luminance is commonly used to express brightness.

Reflection: The bouncing back of light waves or rays by a surface.

Reflected light: Any light that comes from a surface that is not self-illuminating.

Refraction: The bending of a light wave or ray as it passes obliquely (at an angle) from one medium to another of different density, or through layers of different density in the same medium.

Troffers: Recessed luminaires installed with the opening flush with the ceiling.

Appendix A

Summary of Luminaire Symbols

□ OR ○	LUMINAIRE; POINT SOURCE, SURFACE-MOUNTED, (DRAWN TO SCALE OR LARGE ENOUGH FOR CLARITY)
▭	LUMINAIRE; EXTENDED SOURCE, SURFACE-MOUNTED (DRAWN TO SCALE)
⊢——⊣	LUMINAIRE; STRIP-TYPE (LENGTH DRAWN TO SCALE)
◁	LUMINAIRE; FLOOD-TYPE
— ‥ —	LINEAR SOURCE; i.e., LOW-VOLTAGE STRIP, NEON, FIBER OPTIC, ETC. (LENGTH DRAWN TO SCALE)
⊗ with arrow	EXIT SIGN; MOUNTING, NO. OF FACES AND ARROWS AS SHOWN
▱ OR ⊘	RECESSED
▭ OR ♀	WALL-MOUNTED
▭(• •) OR ⊙	PENDANT, CHAIN- OR STEM-MOUNTED
⌐□⌐ OR ♀	POLE-MOUNTED WITH ARM
⊡ OR ⊙	POLE-MOUNTED ON TOP
▭↓ OR ♀↓	ACCENT/DIRECTIONAL ARROW (DRAWN FROM PHOTOMETRIC CENTER IN DIRECTION OF OPTICS OR PHOTOMETRIC ORIENTATION)
○——	DIRECTIONAL AIMING LINE (DRAWN FROM PHOTOMETRIC CENTER TO ACTUAL AIMING POINT)
○—○↘	TRACK-MOUNTED; LENGTH, LUMINAIRE TYPES AND QUANTITIES AS SHOWN (TRACK LENGTH DRAWN TO SCALE)
◨ OR ◉	LUMINAIRE PROVIDING EMERGENCY ILLUMINATION (FILLED IN)
□ OR ○	DOUBLE LINE FOR INDIRECT LUMINAIRES
▦	LOUVERS

NCCER CURRICULA — USER UPDATE

NCCER makes every effort to keep its textbooks up-to-date and free of technical errors. We appreciate your help in this process. If you find an error, a typographical mistake, or an inaccuracy in NCCER's curricula, please fill out this form (or a photocopy), or complete the online form at **www.nccer.org/olf**. Be sure to include the exact module ID number, page number, a detailed description, and your recommended correction. Your input will be brought to the attention of the Authoring Team. Thank you for your assistance.

Instructors – If you have an idea for improving this textbook, or have found that additional materials were necessary to teach this module effectively, please let us know so that we may present your suggestions to the Authoring Team.

NCCER Product Development and Revision
13614 Progress Blvd., Alachua, FL 32615

Email: curriculum@nccer.org
Online: www.nccer.org/olf

❏ Trainee Guide   ❏ Lesson Plans   ❏ Exam   ❏ PowerPoints   Other _____

Craft / Level: _____   Copyright Date: _____

Module ID Number / Title: _____

Section Number(s): _____

Description: _____

_____

_____

_____

_____

Recommended Correction: _____

_____

_____

_____

_____

Your Name: _____

Address: _____

_____

Email: _____   Phone: _____

This page is intentionally left blank.

This page is intentionally left blank.

This page is intentionally left blank.

Conduit Bending

Overview

The normal installation of intermediate metal conduit (IMC), rigid metal conduit (RMC), and electrical metallic tubing (EMT) requires many changes of direction in the conduit runs, ranging from simple offsets at the point of termination at outlet boxes and cabinets to complicated angular offsets at columns, beams, cornices, and so forth. This module describes how to make conduit bends using mechanical, hydraulic, and electric benders.

Module 26204-20

Trainees with successful module completions may be eligible for credentialing through the NCCER Registry. To learn more, go to **www.nccer.org** or contact us at 1.888.622.3720. Our website, **www.nccer.org**, has information on the latest product releases and training.

Your feedback is welcome. You may email your comments to **curriculum@nccer.org**, send general comments and inquiries to **info@nccer.org**, or fill in the User Update form at the back of this module.

This information is general in nature and intended for training purposes only. Actual performance of activities described in this manual requires compliance with all applicable operating, service, maintenance, and safety procedures under the direction of qualified personnel. References in this manual to patented or proprietary devices do not constitute a recommendation of their use.

Copyright © 2020 by NCCER, Alachua, FL 32615, and published by Pearson, New York, NY 10013. All rights reserved. Printed in the United States of America. This publication is protected by Copyright, and permission should be obtained from NCCER prior to any prohibited reproduction, storage in a retrieval system, or transmission in any form or by any means, electronic, mechanical, photocopying, recording, or likewise. To obtain permission(s) to use material from this work, please submit a written request to NCCER Product Development, 13614 Progress Blvd., Alachua, FL 32615.

26204-20 V10.0

From *Electrical, Trainee Guide*. NCCER.
Copyright © 2020 by NCCER. Published by Pearson. All rights reserved.

Figure 45	24" offset	36
Figure 46	20" offset	37
Figure 47	Principles of saddle bending	38
Figure 48	36" saddle	38
Figure 49	Saddle	39
Figure 50	Cosine function	39
Figure 51	Conduit layout	39
Figure 52	PVC heating units	41
Figure 53	Plywood template	42
Figure 54	Some PVC bends may be formed by hand	42
Figure 55	After the bend is formed, wipe a wet rag over the bend to cool it	42
Figure 56	Typical plug set	43

Table 1	*NEC*® Minimum Requirements for One-Shot and Full-Shoe Benders (Data from *NEC Chapter 9, Table 2*)	4
Table 2	*NEC*® Minimum Requirements for Other Conduit Bends (Data from *NEC Chapter 9, Table 2*)	4
Table 3	Gain Factors	12
Table 4	Decimal Equivalents of Some Common Fractions	13
Table 5	Dimensions of Stub-Ups for Various Sizes of Conduit	26
Table 6	Saddle Table	38
Table 7	PVC Expansion Rates	43

This page is intentionally left blank.

SECTION ONE

1.0.0 NEC® REQUIREMENTS FOR CONDUIT BENDS

Objective

Identify the *NEC®* requirements for conduit bends.
 a. Identify the minimum radius requirements for various types of conduit.
 b. Calculate the number of bends per run.

Trade Terms

Back-to-back bend: Any bend formed by two 90° bends with a straight section of conduit between the bends.

Conduit: Piping designed especially for pulling electrical conductors. Types include RMC, IMC, EMT, PVC, aluminum, and other materials.

Elbow: A 90° bend.

Kick: A bend made to change the direction of the conduit. Kicks are typically less than 90°.

Offset: Two equal bends made to avoid an obstruction blocking the run of the conduit.

Radius: The relative size of the bent portion of a pipe.

Sweep bend: A 90° bend with a radius larger than that produced by a standard one-shot shoe.

The normal installation of intermediate metal conduit (IMC), rigid metal conduit (RMC), and electrical metallic tubing (EMT) requires many changes of direction in the conduit runs, ranging from a simple offset at the point of termination at outlet boxes and cabinets to complicated angular offsets at columns, beams, cornices, and other obstructions. Unless the contract specifications dictate otherwise, such changes in direction, particularly in the case of smaller sizes, are made by bending the conduit or tubing as is required. In the case of larger sizes, right angle changes of direction are sometimes accomplished with the use of a factory elbow or conduit body. In most cases, however, it is more economical to make these conduit bends in the field.

On-the-job conduit bends are also performed when multiple runs of the larger conduit sizes are installed. Truer parallel alignment of multiple runs is maintained by using on-the-job conduit bends rather than factory elbows. Such bends can all be made from the same center, using the bends of the largest conduit in the run as the pattern for all other bends. This is just one of the useful techniques covered in this module.

Exposed conduit work is one area of the trade that puts electricians' skills on display. Exposed conduit directly reflects on the ability of the installer. With these thoughts in mind, it will benefit you to learn several methods of bending conduit that will ensure accurate and precisely bent conduit—conduit that you can step back and look at with pride because it was bent right the first time.

1.1.0 Minimum Radius Requirements

NEC Chapter 3 contains the installation requirements for various types of conduit. *NEC Section 344.24* requires that all rigid metal conduit bends be made so that the conduit will not be damaged and the internal diameter of the conduit will not be effectively reduced. To accomplish this, the *NEC®* further specifies that the minimum radius (*Figure 1*) to the center line of the conduit shall not be less than that listed in *Table 1*. There is a good reason for this rule. When bends are too tight,

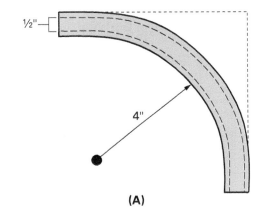

(A)

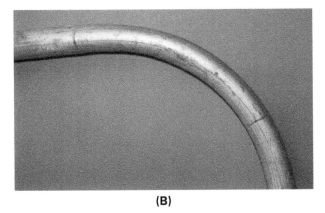

(B)

Figure 1 Inside radius requirements.

pulling becomes extremely difficult and the insulation on the conductors may be damaged. The *NEC®* requirements for other bends are shown in *Table 2*.

> **NOTE**
> The bending radii requirements and maximum number of bends per conduit run vary by country. This module applies the *NEC®* requirements.

1.2.0 Number of Bends per Run

Every change of direction in a conduit run adds to the difficulty of the pull. The *NEC®* specifically states that no more than four quarter bends (360° total) may be made in any one conduit run between outlet boxes, cabinets, panels, junction boxes, or pull points. Types of conduit bends include the following:

- *Elbow* – An elbow, or ell, is a 90° bend that is used when a conduit must turn at a 90° angle. Two elbows are often used to create a back-to-back bend. In single conduit runs when the larger sizes of conduit are being installed, factory elbows are frequently used to save labor on setting up a power bending machine, calculating and marking the conduit for bending, and finally, making the bend. However, in multiple conduit runs, a neater job will result if on-the-job multiple sets of the sweep bend or concentric bend are properly calculated and installed (*Figure 2*).
- *Offset* – An offset consists of two equal bends and is used when the conduit run must go over, under, or around an obstacle. An offset is also used at outlet boxes, cabinets, panelboards, and pull boxes (*Figure 3*).
- *Saddle* – A saddle is used to cross a small obstruction or other runs of conduit. A saddle is made by marking the conduit at a point where the saddle is required and placing a bender a few inches ahead of this point. Bends are made as shown in *Figure 4*. Both three-bend and four-bend saddles can be used, depending on the type of obstruction.
- *Kick* – A kick is a single change in direction of a conduit run of less than 90°. It is used mostly where the conduit run will be concealed in deck work. The first bend in an offset, for example, is really a kick, as shown in *Figure 5*; another kick in the opposite direction transforms the bend into an offset.

Some electricians believe that offsets, kicks, and saddles are not bends, especially in areas where the electrical inspectors are lax. These electricians count only those bends that are actually a quarter circle (90°). The misconception of

Table 1 *NEC®* Minimum Requirements for One-Shot and Full-Shoe Benders (Data from *NEC Chapter 9, Table 2*)

Trade Size (Inches)	Radius to Center of Conduit (Inches)
½	4
¾	4½
1	5¾
1¼	7¼
1½	8¼
2	9½
2½	10½
3	13
3½	15
4	16
5	24
6	30

Reprinted with permission from NFPA 70®-2020, *National Electrical Code®*, Copyright © 2019, National Fire Protection Association, Quincy, MA. This reprinted material is not the complete and official position of the NFPA on the referenced subject, which is represented only by the standard in its entirety which may be obtained through the NFPA website at **www.nfpa.org**.

Table 2 *NEC®* Minimum Requirements for Other Conduit Bends (Data from *NEC Chapter 9, Table 2*)

Trade Size (Inches)	Other Bends (Inches)
½	4
¾	5
1	6
1¼	8
1½	10
2	12
2½	15
3	18
3½	21
4	24
5	30
6	36

Reprinted with permission from NFPA 70®-2020, *National Electrical Code®*, Copyright © 2019, National Fire Protection Association, Quincy, MA. This reprinted material is not the complete and official position of the NFPA on the referenced subject, which is represented only by the standard in its entirety which may be obtained through the NFPA website at **www.nfpa.org**.

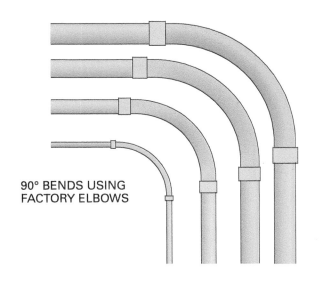

Figure 2 Typical 90° bends.

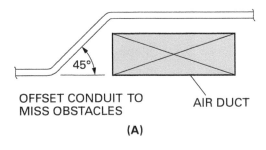

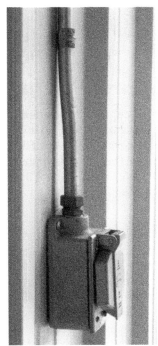

Figure 3 Applications of conduit offsets.

Take Pride in Your Work

Even though the conduit in this installation will be hidden by the flooring and the final wall finish, notice how the installing electrician made a special effort to keep the bends perfectly spaced and aligned. Doing a good job even when it might never be seen or admired is the mark of a true professional.

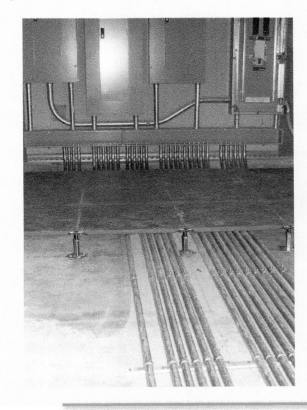

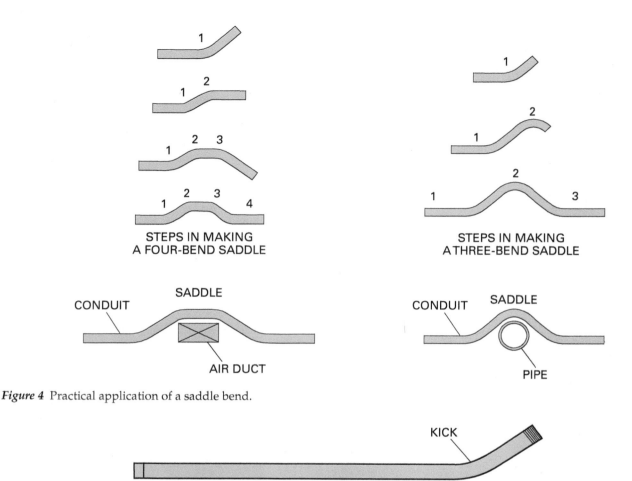

Figure 4 Practical application of a saddle bend.

Figure 5 Kick.

this is quickly apparent when wires are pulled. Offsets and saddles add just as much resistance to pulling conductors as any 90° elbow. A 45° offset, for example, takes two 45° bends, which equal one 90° bend. A saddle may be as low as 60° or as high as 180°, depending on the types of bends used.

> **NOTE**
> The acronym MD is used as the metric trade size designator for conduit. It is important to understand that these numbers are size identifiers only. They are not actual dimensions of the conduit.

A 15° kick in a conduit run may seem insignificant, but after several of these are incorporated into the run, the difficulty of pulling wire becomes apparent. The number of degrees in each kick should be included in the total count, and in no case should the total number (number of bends × number of degrees in each bend) exceed 360°.

International Differences

In other countries, conduit is measured using the metric designator (MD) system, so it might be helpful to familiarize yourself with these values, especially if you live near the Canadian or Mexican border. The tables in *NEC Chapter 9* list both the MD values and trade size (US) values.

This is the maximum number of degrees allowed. For example, you could have one 45° offset, a 90° saddle, and two elbows. Many electricians prefer to install pull boxes at closer intervals to reduce the number of bends, especially when the larger conductor sizes are being pulled. The additional cost of the pull boxes and the labor to install them is often offset by the labor saved in pulling the conductors.

Don't Ignore the Outside Diameter

Although conduit measurements are based on the inside diameter, don't forget that the outside diameter determines the size of the opening that the conduit must penetrate.

Check It Out

When making complex bends, it is a good idea to test your bends using a piece of wire first, then use the bent wire as a template for your bend. This will give you an idea of whether or not the bend will do the job before wasting expensive conduit.

1.0.0 Section Review

1. The minimum radius for 3½" conduit using a one-shot bender is _____.
   a. 5"
   b. 10"
   c. 15"
   d. 30"

2. The minimum radius for MD 91 conduit using a one-shot bender is _____.
   a. 127 mm
   b. 254 mm
   c. 381 mm
   d. 762 mm

3. The type of bend required where conduit enters a panelboard is most likely a(n) _____.
   a. elbow
   b. offset
   c. kick
   d. saddle

Section Two

2.0.0 Bend Distances

Objective

Use equations to find bend distances.
a. Use right-angle mathematics to find bend distances.
b. Use the circumference of a circle to determine bend distances.

Trade Terms

Developed length: The amount of straight pipe needed to bend a given radius. Also, the actual length of the conduit that will be bent.

Gain: The amount of pipe saved by bending on a radius as opposed to right angles. Because conduit bends in a radius and not at right angles, the length of conduit needed for a bend will not equal the total determined length. Gain is the difference between the right angle distances A and B and the shorter distance C—the length of conduit actually needed for the bend.

Rise: The length of the bent section of conduit measured from the bottom, center line, or top of the straight section to the end of the bent section.

Stub-up: Another name for the rise in a section of conduit.

Take-up (deduct): The amount that must be subtracted from the desired stub length to make the bend come out correctly using a point of reference on the bender or bending shoe.

Learning to bend conduit involves using a few simple equations to find various distances. The 90° or right angle bend is probably the most basic of all and is used much of the time, regardless of the type of conduit being installed.

2.1.0 Finding Bend Distances with Right-Angle Math

A right triangle, as shown in *Figure 6 (A)*, is defined as any triangle with one 90° angle. The side directly opposite the 90° angle is called the hypotenuse and the side on which the triangle sits is the base. The vertical side is called the height or altitude. For offset bends, right triangle mathematics can also be applied because the offset forms the hypotenuse of a right triangle, as shown in *Figure 6 (B)*. There are reference tables available for sizing offset bends based on these characteristics.

Right triangles are used to develop trigonometric equations (*Figure 7*). These equations can be used to determine the rise or stub-up distance between bend points, the distance between bend joints, the distance a kick needs to be above the surface, or the amount of additional conduit required. Six basic trigonometric functions will be required:

- Sine
- Cosine
- Tangent
- Cotangent
- Secant
- Cosecant

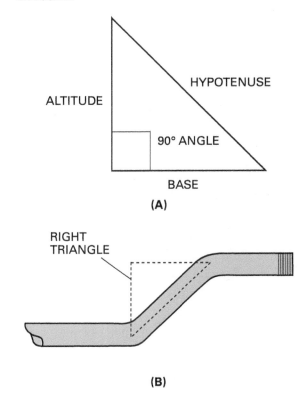

Figure 6 A right triangle and its relationship to a conduit offset.

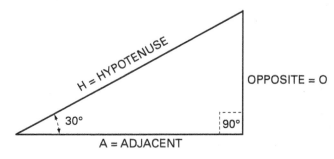

Figure 7 Trigonometry fundamentals of a right triangle.

The sine function can be computed by dividing the hypotenuse by the side opposite the angle being considered. This function can be represented by the following equation:

$$\text{Sine} = \frac{\text{opposite}}{\text{hypotenuse}} = \frac{O}{H}$$

The equations for the other functions are as follows:

$$\text{Cosine} = \frac{\text{adjacent}}{\text{hypotenuse}} = \frac{A}{H}$$

$$\text{Tangent} = \frac{\text{opposite}}{\text{adjacent}} = \frac{O}{A}$$

$$\text{Cotangent} = \frac{\text{adjacent}}{\text{opposite}} = \frac{A}{O}$$

$$\text{Secant} = \frac{\text{hypotenuse}}{\text{adjacent}} = \frac{H}{A}$$

$$\text{Cosecant} = \frac{\text{hypotenuse}}{\text{opposite}} = \frac{H}{O}$$

Where:
- H = hypotenuse, side facing the right (90°) angle
- O = side opposite the angle you are working with
- A = side adjacent to the angle you are working with, but not the hypotenuse

Therefore, in *Figure 6 (B)*, the hypotenuse or distance between bends could be computed by using the following cosecant trigonometric function:

$$\text{Cosecant} = \frac{\text{hypotenuse}}{\text{opposite}} = \frac{H}{O}$$

Let's say that the angle of the first bend is 30° and the rise or side opposite is 12". Therefore:

$$\text{Cosecant } 30° = \frac{H}{12"}$$

The cosecant of 30° is 2 (check using your calculator).

$$2 = \frac{H}{12"}$$

Cross multiply:

$$H = 2 \times 12"$$
$$H = 24"$$

Therefore, the hypotenuse or distance between bends would be 24".

Here is another example. In this case, a kick is to be made on the end of a piece of conduit (*Figure 8*). The angle of the kick is to be 30°. The hypotenuse of the kick is 20". Determine how far off the surface the end of the kick needs to be.

When calculating the hypotenuse of a bend, the cosecant of the angle is multiplied by the side opposite. If the angle and the hypotenuse are known, then the inverse can be used to determine the side opposite. Therefore, divide 20" by the cosecant of 30°, or:

$$20" \div 2 = 10"$$

The end of the conduit needs to be brought 10" off the surface to acquire a 30° bend. This process eliminates the need for a protractor level.

These equations also apply to metric examples. For instance, if the hypotenuse were 400 mm for a 30° kick, divide 400 mm by the cosecant of 30°, or:

$$400 \text{ mm} \div 2 = 200 \text{ mm}$$

The end of the conduit needs to be brought 200 mm off the surface for a 30° bend.

2.2.0 Finding Bend Distances Using Circumference

A circle is defined as a closed curved line whose points are all the same distance from its center, as shown in *Figure 9 (A)*. The distance from the center point to the edge of the circle is called the radius, and the length of a straight line from one edge through the center to the other edge is called the diameter. The distance around the circle is called the circumference. A circle can be divided into four equal quadrants, as shown in *Figure 9 (B)*. Each quadrant accounts for 90°, making a total of 360°. A 90° bend is based on $\frac{1}{4}$ of a circle, or one quadrant.

Concentric circles, shown in *Figure 9 (C)*, are several circles that have a common center but different radii. The concept of concentric circles can be applied to concentric 90° bends in conduit. Such bends have the same center point, but the radius of each is different. *Figure 10* shows how parts of a circle relate to a 90° conduit bend.

For bending conduit, it is necessary to understand the dynamics of the unit circle (*Figure 11*). The unit circle is a circle with a given radius of 1.

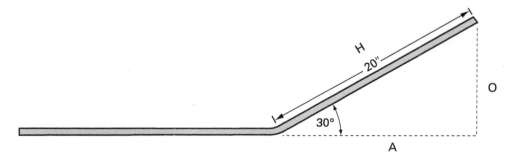

Figure 8 Kick example.

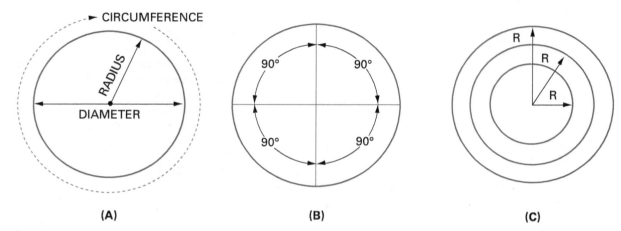

Figure 9 Characteristics of a circle.

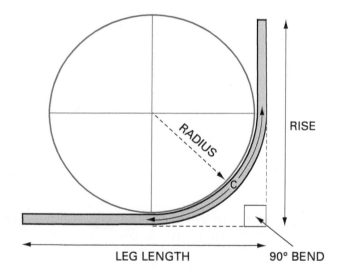

Figure 10 Parts of a circle related to conduit bending.

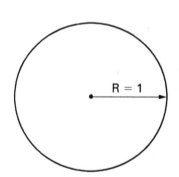

Figure 11 Unit circle.

2.2.1 Developed Length

When calculating the circumference of a circle, the following formula is used:

$$C = 2\pi R$$

Where:

C = circumference
π = 3.14 (pi)
R = radius

By taking the radius of the unit circle and substituting it into the formula, the result is:

$$C = 2\pi \times 1$$

Any number multiplied by the number 1 is equal to that number, or:

$$1 \times 2 = 2$$

Therefore, $2\pi \times 1 = 2\pi$. This means that in terms of pi, the circumference of the circle is 2π; 360° is 2π and 180° = π (*Figure 12*).

If you look at 90° in terms of π, 90° is half of 180° or half of π, so 90° is equal to $\frac{1}{2}\pi$ or $\pi \div 2$. Again, π is a symbol for the numerical value 3.14 (rounded). $\pi \div 2$ is $3.14 \div 2$ or 1.57. Therefore, 90° is represented by the numerical value 1.57. Looking back at the circumference formula shows that multiplying $2 \times \pi$ or 6.28 times the given radius of the circle gives the linear distance around the circle. With that in mind, if you multiply 1.57 times the given radius, it will provide the linear distance from 0° to 90° on the circle. This linear distance is known as the developed length in regard to the amount of conduit that is required to make a 90° sweep.

For example, if the radius in a circle measures two feet, the developed length may be found as follows:

Length of arc = 1.57R
Length of arc = 1.57 × 2' = 3.14'

Using a metric example with a radius of 200 mm, the developed length may be found as follows:

Length of arc = 1.57 × 200 mm = 314 mm

2.2.2 Gain

Conduit bends with the circumference of the circle (*Figure 13*) and not at right angles. Therefore, the length of the conduit needed for a bend will not equal the right angle distances A and B. Gain is the difference between the right angle distances A and B and the shorter distance C, the length of conduit actually needed for the bend.

The gain for a bend is found by multiplying the radius of the bend times the gain factor listed in *Table 3*. For example, the gain for a 90° bend is found by multiplying the radius of the bend by 0.43 (rounded up from the value of 0.4292 in *Table 3*). Therefore, if the radius of the bend in *Figure 13* is 2', the gain may be found as follows:

Gain = radius × gain factor
Gain = 2' × 0.43 = 0.86' = 10.32"

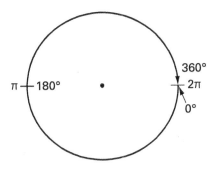

Figure 12 π and 2π.

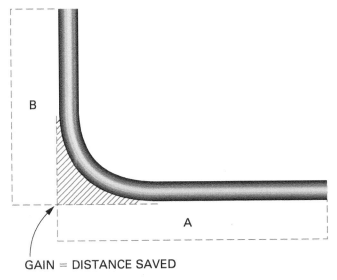

GAIN = DISTANCE SAVED

Figure 13 Gain.

2.2.3 Applying Math to a 90° Bend

The 90° stub bend is probably the most basic bend of all. The stub bend is used much of the time, regardless of the type of conduit being installed. Before beginning to make the bend, you need to know two measurements:

- The desired rise or stub-up
- The take-up (deduct) distance of the bender

The desired rise is the height of the stub-up. The take-up is the amount of conduit the bender will use to form the bend. Take-up distances are usually listed in the manufacturer's instruction manual. When the take-up has been determined, subtract it from the stub-up height. Mark that distance on the conduit (all the way around) at that distance from the end. The mark will indicate the point at which you will begin to bend the conduit. Line up the starting point on the conduit with the starting point on the bender. Most benders have an arrow or other mark to indicate the starting point.

> **NOTE:** When bending conduit, the conduit is placed in the bender and the bend is made facing the end of the conduit from which the measurements were taken.

As discussed above, if the radius of the bend in *Figure 13* is 2', the gain is 10.32".

Table 4 shows the equivalent fractions for many decimals. For 10.32", search the decimal column for 0.32. The closest decimal in the table is 0.3125, equivalent to $5/16$. Therefore, the number in question becomes $10\,5/16"$.

A decimal may also be mathematically converted to a fraction. A decimal whose denominator is contained in the numerator without a remainder can easily be converted to a fraction by removing the decimal point from the numeral, which then becomes the numerator (the top numeral of the fraction). The denominator is always one plus as many zeros as there are decimal places in the decimal. For example:

> **NOTE:** Decimals may also be converted to fractions by using a calculator, provided the calculator has a fraction key. The exact procedure will vary with the different models of calculators.

$$0.75 = 75/100 = 3/4$$

or:

$$0.375 = 375/1000 = 3/8$$

Gain factors for 0° to 90° bends are shown in *Table 3*. To demonstrate the use of this table, assume that you want to find the gain on a 45° conduit bend with a 15" center line radius. Referring to *Table 3*, look in the left-hand column; glance down the column until the number 40° is found. Because the bend is 45°, read to the right in this row until the column titled 5° is found; note the figure, 0.043. Therefore, the gain factor for a 45° bend is 0.043.

Multiply the gain factor by the center line radius (15") to obtain the full gain of a 45° bend.

$$0.043 \times 15" = 0.645"$$

To convert this figure to a readable figure on the foot rule, convert the decimal to a common fraction: $645/1000$ = approximately $5/8$. Thus, the full gain of the 45° bend is $5/8$".

For example, the gain for a given bend is $2\,1/2"$. In *Figure 14*, there are two back-to-back 90° bends. The rise for the first 90° bend is 2', and the rise for the second 90° bend is 3'. The back-to-back

Table 3 Gain Factors

	—	1°	2°	3°	4°	5°	6°	7°	8°	9°
0°	0	0	0	0	0	0	0.0001	0.0001	0.0003	0.0003
10°	0.0005	0.0006	0.0008	0.001	0.0013	0.0015	0.0018	0.0022	0.0026	0.0031
20°	0.0036	0.0042	0.0048	0.0055	0.0062	0.0071	0.0079	0.009	0.01	0.0111
30°	0.0126	0.0136	0.015	0.0165	0.0181	0.0197	0.0215	0.0234	0.0254	0.0276
40°	0.0298	0.0322	0.0347	0.0373	0.04	0.043	0.0461	0.0493	0.0527	0.0562
50°	0.06	0.0637	0.0679	0.0721	0.0766	0.0812	0.086	0.0911	0.0963	0.1018
60°	0.1075	0.1134	0.1196	0.126	0.1327	0.1397	0.1469	0.1544	0.1622	0.1703
70°	0.1787	0.1874	0.1964	0.2058	0.2156	0.2257	0.2361	0.247	0.2582	0.2699
80°	0.2819	0.2944	0.3074	0.3208	0.3347	0.3491	0.364	0.3795	0.3955	0.4121
90°	0.4292	—	—	—	—	—	—	—	—	—

Table 4 Decimal Equivalents of Some Common Fractions

Fraction	Decimal	MM	Fraction	Decimal	MM
1/64	0.015625	0.397	33/64	0.515625	13.097
1/32	0.03125	0.794	17/32	0.53125	13.494
3/64	0.046875	1.191	35/64	0.546875	13.891
1/16	0.0625	1.588	9/16	0.5625	14.288
5/64	0.078125	1.984	37/64	0.578125	14.684
3/32	0.09375	2.381	19/32	0.59375	15.081
7/64	0.109375	2.778	39/64	0.609375	15.478
1/8	0.125	3.175	5/8	0.625	15.875
9/64	0.140625	3.572	41/64	0.640625	16.272
5/32	0.15625	3.969	21/32	0.65625	16.669
11/64	0.171875	4.366	43/64	0.671875	17.066
3/16	0.1875	4.763	11/16	0.6875	1.7463
13/64	0.203125	5.159	45/64	0.703125	17.859
7/32	0.21875	5.556	23/32	0.71875	18.256
15/64	0.234375	5.953	47/64	0.734375	18.653
1/4	0.25	6.35	3/4	0.75	19.05
17/64	0.265625	6.747	49/64	0.765625	19.447
9/32	0.28125	7.144	25/32	0.78125	19.844
19/64	0.296875	7.54	51/64	0.796875	20.241
5/16	0.3125	7.938	13/16	0.8125	20.638
21/64	0.32812	8.334	53/64	0.828125	21.034
11/32	0.34375	8.731	27/32	0.84375	21.431
23/64	0.359375	9.128	55/64	0.859375	21.828
3/8	0.375	9.525	7/8	0.875	22.225
25/64	0.390625	9.922	57/64	0.890625	22.622
13/32	0.40625	10.319	29/32	0.90625	23.019
27/64	0.421875	10.716	59/64	0.921875	23.416
7/16	0.4375	11.113	15/16	0.9375	23.813
29/64	0.453125	11.509	61/64	0.953125	24.209
15/32	0.46875	11.906	31/32	0.96875	24.606
31/64	0.484375	12.303	63/64	0.984375	25.003
1/2	0.5	12.7	1	1	25.400

measurement for these bends is 4'. You need to determine the length of straight conduit required to accomplish this task. The conduit is to be cut to length and threaded before bending.

Remember that the gain is the amount of conduit that is saved by the radius of the bend. The amount of conduit saved is 2½" per 90° bend. In this situation, there are two 90° bends, or a savings of 5". The total straight lengths add up to 9' of conduit. Therefore:

$$9' - 5" = 8'-7"$$

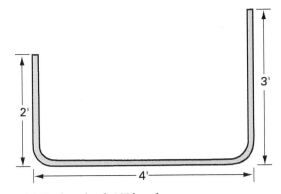

Figure 14 Back-to-back 90° bends.

Degrees, Radians, and Gradients

In addition to degree measurements from 0° to 360°, angles can also be stated in radians or gradients. Radians range from 0 to 6.28 or 0 to 2π. Gradients range from 0 to 400 gradients (grads). These measurements can be equated as follows:

90° = π ÷ 2 = 1.57 radians = 100 grads   270° = 3π ÷ 2 = 4.71 radians = 300 grads
180° = π = 3.14 radians = 200 grads   360° = 2π = 6.28 radians = 400 grads

2.0.0 Section Review

1. When applying right-angle mathematics to conduit bends, the distance between bends is represented by the triangle's _____.
   a. hypotenuse
   b. side opposite
   c. side adjacent
   d. base

2. The gain for a 90° bend with a 6" radius is _____.
   a. 1.08"
   b. 2.58"
   c. 4.72"
   d. 6.64"

   0.43 × 6

Section Three

3.0.0 Mechanical Benders

Objective

Use mechanical benders.
a. Chart a mechanical bender.
b. Make mechanical bends.

Trade Terms

Bending protractor: Made for use with benders mounted on a bending table and used to measure degrees; also has a scale for 18, 20, 21, and 22 shots when using it to make a large sweep bend.

Concentric bending: The process of making 90° bends in parallel runs of conduit. This requires increasing the radius of each conduit from the inside of the bend toward the outside.

Conduit bends are normally made in the smaller sizes of conduit and tubing by hand with the use of hickeys or EMT bending tools. However, on many projects, using mechanical bending equipment with suitable adjustable stops and guides has advantages. With this equipment, the exact bend can be duplicated in quantity with a minimum of effort. The angle of the bend and the location of the bend in relation to the end of the length of conduit are preset.

A widely used mechanical bender is shown in *Figure 15*. This device was originally called the Chicago bender because it was made by the Chicago Equipment and Manufacturing Company. Today, however, this type of bender is manufactured by several different companies and the correct name is *portable mechanical conduit bender*. In any event, you may still hear the term Chicago bender on many jobs.

This type of mechanical bender is very common among many electrical contractors. To use it, a length of conduit is placed in position and secured in place; a long bending handle is then pulled around and the bend completed. This type of bender may be used as a one-shot bender for the smaller sizes of conduit (bypassing the ratchet mechanism). The ratchet mechanism, however, is usually activated when bending the larger sizes of conduit to make the work easier. It is suitable for making bends in conduit sizes up to 2" EMT (MD 53), 1¼" IMC (MD 35), and 1½" RMC (MD 41), provided the proper bending accessories are used (e.g., bending shoes and follow bars).

Bending shoes and follow bars are designed to form a particular radius bend for a certain type and size of conduit (EMT, IMC, or RMC). These accessories should be treated as precision instruments; any damage to them will result in inaccurate bends, kinks, and other defects. The first consideration is to use only the proper shoe and follow bar for the type and size of conduit being bent. For example, never use an EMT shoe for bending RMC. In general, make certain that the bending shoes and follow bar are compatible with the type and size of conduit to be bent; to do otherwise may damage the tool and result in inaccurate bends.

The ratchet feature is normally engaged for the larger sizes of conduit, whereas a spring-loaded pawl engages the ratchet for easier bending in segments. For the smaller sizes of conduit, the ratchet may be bypassed so that the bend can be made in one shot.

The following are some general tips to keep in mind when bending with a portable mechanical conduit bender:

- An engineer's rule marked in hundredths of an inch will simplify formula bending by eliminating the need to convert to fractions.
- When minimum length stubs are being bent, the shoe tends to creep and deforms the end of the conduit and threads. Screwing a coupling onto the pipe stops the shoe from creeping forward and protects the threads.
- When bending offsets, the front of the bender can be temporarily elevated for clearance requirements.
- Most bender shoes are made of cast aluminum and are easily pitted and gouged if foreign material gets between the shoe and the pipe. For longer shoe life, keep the pipe clean and the shoes wiped down.
- When the remaining pipe length is too short to reach the roller or pipe support, a larger diameter conduit can be slid over the pipe being bent to complete the bend; or, if the pipe has threads, screw on a coupling and a short piece of scrap pipe.
- Segment and concentric bending of smaller sizes of pipe can be performed with this bender. Bend a scrap piece of pipe and measure from the center of the bend to the front of the bending shoe. Use this measurement to adjust the start mark using the segment and concentric bending procedures.

Roof top Bar

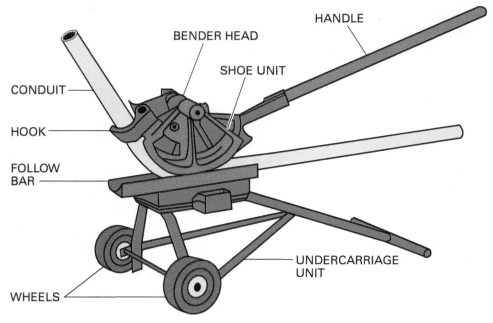

Figure 15 Typical mechanical bender.

- A bending gauge with an adjustable pointer on the bender is helpful when making multiple bends at the same angle. This pointer is set at the desired angle and then the setscrew is tightened. As the bend is being made, the handle is operated until the pointer reaches the index mark. To ensure the correct angle of bend, the first bend should be checked with a **bending protractor** (*Figure 16*) and any necessary adjustments made to the bending gauge pointer before continuing. All successive bends will be exactly the same as the first.
- To make matching bends in two different sizes of conduit when using a mechanical bender, make both bends using the larger shoe.

> **NOTE**
> The metric equivalents for various bending values vary depending on the bender in use. Refer to the charts provided by the bender manufacturer when making any calculations.

3.1.0 Charting a Mechanical Bender

Benders are typically supplied with charts or decals that indicate the gain and center line distances for various bends. This is critical data that varies depending on the type of bender in use. If this information is unavailable, the bender can be charted using a scrap piece of conduit. This chart will contain minimum size, gain, and center line distances from the arrow on the bender to the center of the bend for 15°, 30°, 45°, and 60° bends.

For this discussion, the scrap piece will be $\frac{1}{2}$" conduit and will be 3' in length (*Figure 17*).

Place a mark on the conduit at a given distance from the end of the conduit. For this discussion, the mark will be 10" in from the end of the conduit (*Figure 18*).

Take a $\frac{1}{2}$" conduit bender and place the arrow, which represents the take-up or minimum rise of the bender, on the mark 10" in from the end of the bender.

Place a protractor level on the conduit in front of the bender. This will be on the piece of conduit marked for 10" back from the end. Bend the 10" portion until the protractor level reads 15°. Remove the bender from the conduit (*Figure 19*).

Now take a ruler, torpedo level or other straightedge, and lay it on the inside of the bend so the straightedge lies across the bend and rests against the straight portion of the conduit (*Figure 20*).

With a sharp pencil, scribe a line across the bend of the conduit.

Now take the straightedge and lay it across the conduit so the straightedge is against the side adjacent to the previous position. It should extend across the bend once again (*Figure 21*).

Now scribe a line across the bend of the conduit. The two pencil lines should cross to form an X on the conduit. This represents the center of the bend (*Figure 22*).

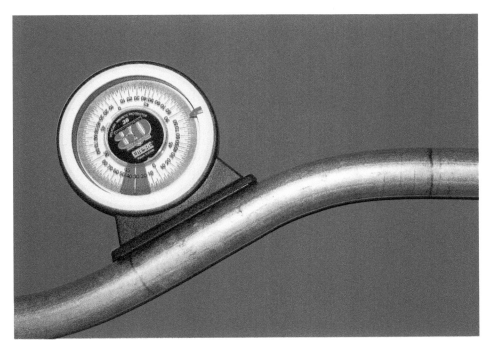

Figure 16 Bending protractor.

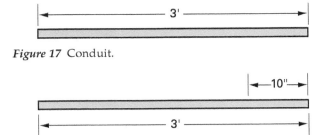

Figure 17 Conduit.

Figure 18 Conduit with 10" mark.

Figure 19 Kick of 15°.

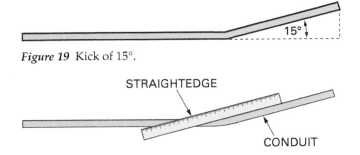

Figure 20 Conduit and straightedge.

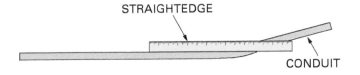

Figure 21 Conduit and horizontal straightedge.

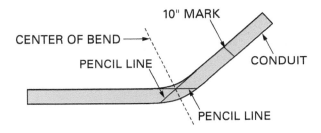

Figure 22 Center of bend.

Now take a tape measure and measure from the 10" mark to the center line of the bend. Record this distance as follows:

15° − 1" (assuming 1" is the measured distance)

Now place the bender back on the conduit so that the take-up arrow is on the 10" mark on the conduit. Place the protractor level back on the conduit. Bend the conduit until the protractor level reads 30°. Take the bender off the conduit and repeat the line crossing process that was discussed for 15°, using the same straightedge. Once again, measure from the 10" mark to the point where the pencil lines cross in the center of the bend. Record this measurement.

15° − 1"
30° − 1½" (assuming this is the measured distance)

Repeat the previous process for 45°. Record this measurement.

> 15° — 1"
> 30° — 1½"
> 45° — 2" (assuming this is the measured distance)

Repeat the previous process for 60°. Record this measurement.

> 15° — 1"
> 30° — 1½"
> 45° — 2"
> 60° — 2¼" (assuming this is the measured distance)

The center line distances for the different bends from the take-up mark of the bend have now been recorded. This portion of the chart is now completed.

The next step is to determine the minimum rise of the bender. Place the bender back on the conduit so the take-up mark of the bender is on the 10" mark on the conduit. Place the protractor level back on the conduit, and bend the conduit until the protractor level reads 90°. Take the bender off the conduit. Measure from the back of the conduit to the 10" mark. The reading on the tape is 5". This is the minimum 90° stub length (*Figure 23*). Record this measurement.

> 15° — 1"
> 30° — 1½"
> 45° — 2"
> 60° — 2½"
> Minimum rise — 5" (assuming this is the measured distance)

The next step is to determine the gain of the bender for 90°. Measure the length of both sides of the scrap piece of conduit (*Figure 24*).

Add the two measured stub lengths together.

> 15" + 1'-11⅝" = 3'-2⅝"

Subtract the original length of the conduit, which was 3', from 3'-2⅝".

> 3'-2⅝" − 3' = 2⅝"

This is the gain for this particular ½" conduit bender. Record this information.

> 15° — 1"
> 30° — 1½"
> 45° — 2"
> 60° — 2½"
> Minimum rise — 5"
> Gain — 2⅝" (assuming this is the measured distance)

The ½" bender has now been charted. This same process could be repeated for any other type of wrap-around bender.

This charting process should help you to understand how the manufacturer of a bender comes up with the marks that determine the ability to bend exact 90° bends of any length, which is the minimum rise mark. The star mark seen on many benders is used for back-to-back bends. The star mark is 2⅝" back from the minimum rise mark (take-up). In other words, it is the measured gain distance back from the minimum rise mark. The center line marks are used in lining up the centers of bends for various sizes of conduit. The center line marks can also be used in lining up saddles on the center lines of I-beams and process piping.

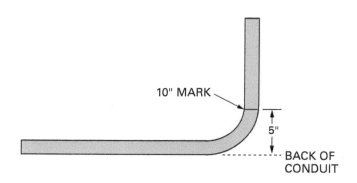

Figure 23 90° stub-up.

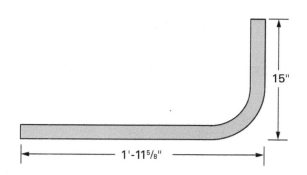

Figure 24 90° elbow.

Making Accurate Offsets

Be sure to use the same angle for both bends of an offset. To avoid crooked bends, known as dog legs, make sure your first bend lines up evenly with the rest of the conduit, the handle, and the bender. Take your time, and be sure of proper alignment before making the second bend. A No-Dog® is a simple, pocket-sized device that may be used to prevent crooked bends. It is screwed onto the end of the conduit and has a built-in level to ensure straight bends.

For more information about maintaining bend accuracy, see the section titled "Eliminating Dog Legs."

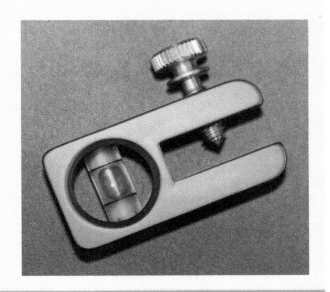

3.2.0 Making Mechanical Bends

Two of the most common bends are stub-ups and offsets. Both types of bends are quickly and easily made using mechanical benders.

3.2.1 Bending Stub-Ups

Making a stub-up requires knowing the bender take-up or deduct for the conduit type and size. A deduct decal is provided on many benders, but sometimes these decals become damaged, making them difficult to read. Therefore, backup charts should be provided on all jobs. The following is an example of making a 90° stub-up to a given height.

Assume that you are working on a deck job and need a number of 1" rigid stub-ups with a rise of 15" each. When you check the deduct chart on the bender for a 15" stub-up using 1" rigid conduit, you note that 11" should be deducted from the total rise of 15". Because 15" – 11" = 4", measure back from the end of the conduit by 4" and make a mark. Encircle the entire conduit at this point so you will not lose the mark when the conduit is placed in the bender. Many electricians like to use a black felt-tip marker for marking conduit.

Load the conduit into the bender with the mark lined up with the front of the bender hook. Engage the ratchet and start pumping the bender handle until the bender pointer reaches the preset index mark for 90°. Move the bender handle forward, then remove the conduit from the bender and check its height. It should be exactly 15". If the height of the bend is slightly off, make the necessary adjustments before continuing. When the correct height is reached, the remaining bends will also be correct.

3.2.2 Bending Offsets

The decal chart on the bender that provides deduct information for the stub-ups also contains data for making offsets that require 20°, 30°, and 45° bends. This chart is necessary to make perfect offsets every time using the mechanical bender.

Assume that you are running a raceway system with $\frac{1}{2}$" rigid conduit and an air duct must be bypassed, requiring the conduit run to be offset. After taking measurements on the job, you find that an offset of 12" is needed to clear the air duct.

Measure the distance from the end of the conduit to the start of the first bend; mark the conduit as before. Referring to the chart on the bender with offset information, you decide to make the offset with two 45° bends. The chart indicates that the distance between bends is $16\frac{5}{16}$". Therefore, measure and mark this distance back from the first mark.

Insert the conduit into the bender and line up the first mark with the front of the bender hook. The ratchet may be used, but for ½" conduit the ratchet override on the front of the bender shoe is normally employed. Make the first 45° bend. Move the bender handle forward to release the conduit. Now slide the conduit forward through the bender hook until the second mark lines up with the front of the bender hook, and then turn the bend over so the end of the conduit is pointing downward toward the deck. Also, make sure that the first bend lines up with the next bend to be made to prevent a crooked bend (commonly referred to as a dog leg) in the conduit. When everything is aligned, engage the bender handle and make another 45° bend. The height of the offset should be exactly 12" (see *Figure 25*).

The distances between marks for offset bends will vary depending upon the size and type of conduit being bent, so always check the offset information on the bender.

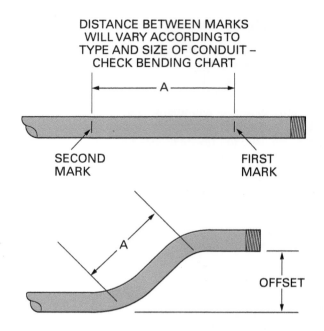

Figure 25 Bending offsets in conduit.

3.0.0 Section Review

1. If you are using a bender and the label has worn off, you should _____.
   a. throw it away as it can no longer be used
   b. chart the bender
   c. use the gain and center line distances from a similar bender
   d. estimate these values based on prior experience

2. You are making a 12" stub-up in 1" EMT, and the bender deduct is listed as 8". Where do you make your first mark?
   a. 2" from the end of the conduit
   b. 3" from the end of the conduit
   c. 4" from the end of the conduit
   d. 5" from the end of the conduit

Hydraulic Bender Safety

A hydraulic bender generates a tremendous amount of power and operates under dangerous pressures. Do not attempt to use a hydraulic bender unless you have been properly trained and are thoroughly familiar with the unit's operating and safety instructions. Even then, exercise extreme caution and always work in conjunction with a bending partner.

apart to allow the first bend to be rolled 180° and advanced enough to clear the shoe.

Segment shoes are shorter and have a radius that is far less than 90°. A 90° bend cannot be made in one operation, as the conduit walls would collapse. Bends, then, must be made in several steps (as few as four and as many as 30) to form a smooth radius. The segmented bending shoe allows pipe to be bent to larger size radii.

Segmented shoes are used for concentric bending (bending several conduits with increasing or decreasing radii). One-shot shoes can be used for a segment bend but are not as convenient as segmented bending shoes. Concentric bending is covered in detail later in this module.

Accurate bending of large conduit is possible but requires practice, patience, and ability. With few exceptions, all formulas and bending techniques discussed to this point will apply.

4.2.1 Bending Rigid Aluminum

Rigid aluminum conduit is available in all trade sizes from $\frac{1}{2}$" through 6" (MD 12 through MD 155). It is lightweight and corrosion resistant, and it has low ground impedance. Rigid aluminum conduit is, however, difficult to bend consistently and accurately. Two pipes from the same bundle will act differently when bent. Even if two pipes are bent using the same layout, they do not always come out the same. Do not be discouraged by this; it is the nature of the metal, and it cannot be helped.

Another disadvantage of aluminum is that a one-shot bending shoe will dig in and score the pipe. Also, where the pipe rides on the shoe, it is prone to wrinkling and scoring. Applying petroleum jelly or a lubricant such as WD-40® to the shoe will allow the pipe to slide without the shoe digging in. Petroleum jelly will also make it easier to remove the conduit when the bend is complete.

4.2.2 One-Shot Bending

Accurate one-shot stub-ups are easily made on hydraulic benders by applying a little basic geometry in making calculations and then knowing the operating principles of the bender. *Figure 31* shows the reference points of a common 90° bend. To make one-shot 90° bends, first determine the leg length and rise, the gain, the radius of the bend, and the half-gain.

Use the following procedure for laying out accurate stub-ups:

Step 1 Determine lengths A and B.

Step 2 Add lengths A and B. Subtract X for the length of pipe required.

Step 3 Subtract Y from length A or B to get the center of the bend.

Step 4 Calculate the developed length and the length of conduit required.

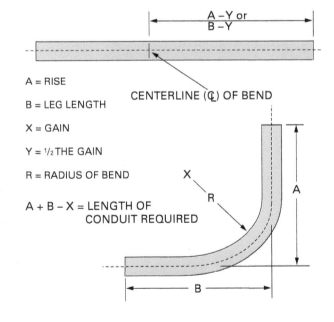

Figure 31 Laying out stub-ups.

Step 5 Determine the center of the bend. This can be done by taking the half-gain value from the distance A or B. Refer to *Table 5*.

4.2.3 90° Segment Bends

When bending conduit in segments with a hydraulic bender, the following factors must be determined:

- The size of conduit to be bent
- The radius of the bend
- The total number of degrees in the bend
- The developed length
- The gain of the bend

To determine the developed length for a 90° bend, multiply the radius by 1.57. The next step is to locate the center of the bend. Most benders have the center mark indicated on the bending shoes. When the center mark on the conduit is found, it is easy to locate the other bend marks (*Figure 32*).

You must now determine how many **bending shots** will make the bend to suit the requirements, preferably an odd number so that there is an equal number of bends on each side of the center mark. Next, calculate the width of the spaces for each segment bend and make the layout on the conduit. Make an equal number of spaces on each side of the center mark. The gain need only be determined if the bend is being fitted between two existing conduit runs or junction boxes.

To determine the bending data for a 90° bend using 3" conduit with a rise of 48", a leg length of 46", and a center line radius of 30", proceed as follows (*Figure 33*).

Step 1 Multiply the radius by 1.57 to determine the developed length.

$30" \times 1.57 = 47.10"$ ($47\frac{1}{8}"$) developed length

Step 2 Determine the gain for a 90° bend.

Gain = (2 × R) − developed length
Gain = (2 × 30") − $47\frac{1}{8}"$ = $12\frac{7}{8}"$

Table 5 Dimensions of Stub-Ups for Various Sizes of Conduit

Pipe and Conduit Size	Radius of Bend R	Minimum Developed Length 90°	Gain X	½ Gain Y
½"	4"	$6\frac{5}{16}"$	$1\frac{11}{16}"$	$\frac{27}{32}"$
¾"	$4\frac{1}{2}"$	$7\frac{7}{16}"$	$1\frac{15}{16}"$	$\frac{31}{32}"$
1"	$5\frac{3}{4}"$	9"	$2\frac{1}{2}"$	$1\frac{1}{4}"$
$1\frac{1}{4}"$	$7\frac{1}{4}"$	$11\frac{3}{8}"$	$3\frac{1}{8}"$	$1\frac{9}{16}"$
$1\frac{1}{2}"$	$8\frac{1}{4}"$	13"	$3\frac{1}{2}"$	$1\frac{3}{4}"$
2"	$9\frac{1}{2}"$	$14\frac{15}{16}"$	$4\frac{1}{16}"$	$2\frac{1}{32}"$

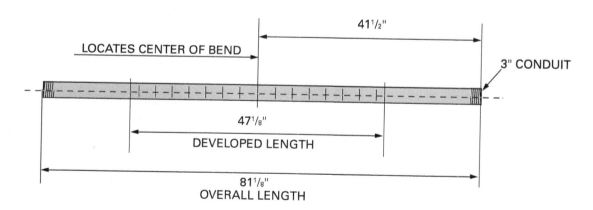

Figure 32 Laying out segment bends.

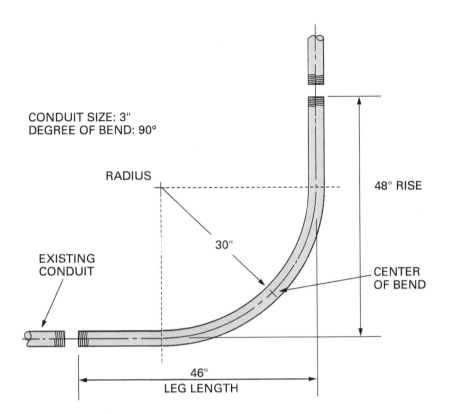

Figure 33 Specifications for sample bend.

Step 3 To calculate the overall length (OL) of the conduit, add the leg and stub-up lengths and subtract the gain. See *Figure 33*.

OL = leg length + rise − gain
OL = 46" + 48" − 12⅞"
OL = 81⅛"

Step 4 Now locate the center of the required bend. First, determine one-half of the developed length:

½ (47⅛") = 23.56" = 23½"

Use the rise or stub-up dimension of 48". Subtract the radius (30") and add one-half of the developed length:

48" − 30" + 23½" = 41½"

Step 5 As a rule, 6° or less per bend will produce a good bend for a 30" radius. In this case, 6° per bend will be used, making 15 segment bends (90 ÷ 6 = 15). An odd number of segment bends is easy to lay out after finding the center mark because there will be an equal number of spaces on each side of the center mark.

Step 6 To determine the space between the segment marks, divide the developed length by the total number of segments:

47⅛" = 47.125"
47.125" ÷ 15 = 3.14"
3.14" = 3⅛"

Step 7 Position the conduit in the pipe holders, making sure to clamp them securely.

Step 8 Place the center mark 41½" from one end of the conduit. Next, mark seven points on each side of the center point, 3⅛" apart, for a total of 15 marks. These are the centers of the segment bends.

Step 9 It is a good idea to check the distance between the first and last bend marks to be sure the layout is correct before starting the first bend. The distance from the first mark to the last is the developed length minus the length of one bend. (Actually, you are subtracting one-half of a segment bend from each end of the conduit.)

47⅛" − 3⅛" = 44"

Step 10 After positioning the conduit in the bender (*Figure 34*), attach the pipe bending degree indicator in a convenient location.

Step 11 Attach the pipe supports with the proper face toward the conduit, and insert the pipe support pins. Lock them in position by turning the small lock pin. Now, proceed to make the series of bends.

Step 12 Begin by bending 6° on the first mark. When this is done, the indicator will read 6°. Release the pressure, and check the springback; if any is found, overbend by the same amount.

Step 13 When using a bender with a rigid frame, move the pipe support one hole position in (toward the ram) on the side that you have bent the conduit.

Step 14 Continue to bend to 12° on the second mark. Check for springback. When the first bend in the conduit is moved past the one pipe support, the approximate ram travel for the remaining bends will be exactly the same.

Step 15 Follow this procedure until you get to the last mark, where you will be bending to 90°. Stop at exactly 90°, release the pressure, and check for springback, correcting if necessary. The result will be a 90° bend without any bows or twists.

For example, suppose the task is to bend a 90° sweep with a given radius of 25". This particular sweep has no definite height. This is to be done on a hydraulic bender using a segmented bending process.

The first step is to determine the linear length or developed length of the conduit to be used for the bend. This can be done by multiplying the radius of 25" by $\pi \div 2$ or $3.14 \div 2 = 1.57$. The answer is 39.25, as shown below:

$$1.57 \times 25" = 39.25"$$

Therefore, it takes 39.25" of conduit to accomplish a 90° bend with a 25" radius.

The next step is to decide how to lay out the number of segments to be bent to attain the 90° sweep. A rule of thumb is that there will be 20 segments or 21 shots. A shot is the actual bending process. There are 21 shots because the sweep is always laid out from the center of the developed length (*Figure 35*).

When the center of the developed length is established, half of the developed length is measured out in each direction from the center line.

To establish the linear distance between segments, divide the developed length by the number of shots ($39.25 \div 20$) or 1.9625" between segments (*Figure 36*).

When the 20 segments are laid out, the number of degrees per shot (bend) needs to be determined. There are 21 shots, so dividing 90° by 21 will give you the number of degrees per shot:

$$90° \div 21 = 4.29°$$

You are now ready to do the bending. Starting at one end of the developed length, place the center line of the bending shoe on the first shot

Figure 34 Conduit placed in hydraulic bender for segment bends.

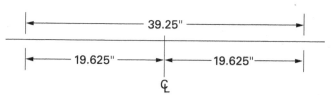

Figure 35 Conduit center.

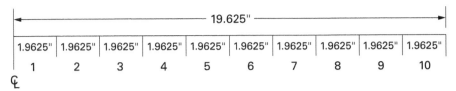

Figure 36 Conduit segments.

Think About It
Developed Length

Why is the radius multiplied by a factor of 1.57 to determine the developed length for a 90° bend?

mark and bend to 4.29°. Repeat the process for the remaining 20 bend marks. If the bend is not quite 90°, make the final adjustment in the last bend.

There is another way to lay out the segment marks for the developed length of a bend. Cut a piece of white elastic band to the length of the outside radius of the innermost bend and lay it out on a table. Make sure that it is not stretched. From one end of the elastic, measure 15 marks spaced evenly apart and mark these points with a fine-tip ink pen. Now place the end of the elastic tape on the center line of the developed bend length. Stretch out the elastic until the 15th mark is on the end of the developed length. Place a mark on the conduit beside each mark on the elastic tape. Repeat this process on the other half of the developed length. You are now ready to do the segmented bends.

Now that you understand the procedure for laying out the conduit with bending marks, your next step is bending. Because segment bending requires several small angle bends to complete a 90° stub, some method to measure the amount of bend will be required. This can be done in four ways:

- Bend degree protractor
- Magnetic angle finder
- Amount of travel method
- Number of pumps method

Bend degree protractor – This is a device that hooks onto the pipe being bent. The circular face is divided into four sections (18, 20, 21, and 30 shots) and is capable of being rotated to whichever scale is to be used. The indicating pointer is weighted and swings free. To use this device, proceed as follows:

Step 1 Level the conduit, and secure it using a pipe vise or other means.

Step 2 Rotate the face to the desired scale that corresponds to the number of shots being used.

Step 3 Adjust the scale so the pointer is on 0.

Step 4 Bend the pipe until the pointer reaches the first mark. (Bend a little past to compensate for springback, release the pressure, and then check the pointer. Only a few bends will be needed to find out how much you must bend past the mark to account for springback.)

Step 5 Move the pipe forward in the bender to the second bend mark, and bend until the pointer reaches the second mark on the protractor face. (Again, bend past for springback.)

Step 6 Move the conduit to the third mark, and bend the pipe so the pointer is at the third mark (after allowing for springback). Follow this procedure at each bend mark until the 90° stub is achieved, the pipe is level, and the bender is in a vertical position.

> **NOTE**
> It is a good idea to check the developing stub length before the last few bends are made. Make spacing corrections as required (e.g., shorten the spacing if the stub is coming up short). If the stub length is reached before the stub is plumb, do not bend at any of the remaining marks. Instead, move the pipe in the bender and bend at the start mark. This will make the stub plumb without adding to the stub length.

Magnetic angle finder – With the magnetic angle finder, the pipe must be kept level and bent vertically (the bender is in vertical position). When a magnetic angle finder is used, you must take care with each bend because a very small error may become multiplied by 15 to 30 times, becoming a large error. To use the angle finder:

Step 1 Level the conduit, and place the angle finder on the stub end. The angle finder will indicate the number of degrees in each bend as determined by the number of shots (e.g., for 20 shots, each bend is about 4.5°).

Step 2 Bend the pipe until the angle finder indicates that the bend is just past the desired degree of bend. This allows for springback. Release hydraulic pressure. If you made the right amount of overbend to allow for springback, the angle finder should read the desired angle of bend. If you bent too much or too little, make the adjustment on the next bend. In two or three bends, you will find the right amount to overbend at each mark to allow for springback.

Step 3 Move the pipe to the second bend mark. Bend at this mark until the next setting on the angle finder (allowing for springback) is indicated by the angle finder pointer. (For example, first bend $4\frac{1}{2}°$, second bend 9°, third bend $13\frac{1}{2}°$, etc.)

Step 4 Bend the pipe at each successive bend mark using the angle finder to indicate the proper degree of bend.

> **NOTE:** Again, check the developing stub length before bending the last few bends. Adjust the spacing or amount of bend as required.

Amount of travel method – The pipe may be bent in any position. To find the amount of theoretical travel for a 90° bend, proceed as follows:

Step 1 Set up the bender with the pivot shoes in the proper holes for the conduit to be bent.

Step 2 Measure the distance (D) center-to-center between the pivot shoe pins (i.e., from center of pin A to center of pin B). The plunger (also called the ram) will have to travel half this distance to bend a full 90° stub.

Step 3 The travel per shot is one-half the distance from pin to pin divided by the number of shots. For example, the distance from the center of the pins is 24", and the pipe is to be bent in 18 shots. The travel per shot equals one-half the distance from the center of the pins (12") divided by the number of shots (18), which equals 0.666 or approximately $\frac{2}{3}$" travel per shot.

Step 4 Bending with this method will follow a slightly different procedure.

Place the center of the bending shoe on the first bending mark (not the start mark), and activate the pump until $\frac{2}{3}$" of ram travel is measured. Do not allow for springback with this method.

Move the pipe to the next mark and activate the pump until another $\frac{2}{3}$" of ram travel has been measured. Continue to move the pipe and measure ram travel.

As you approach the last few marks (4 or 5), check both the developing stub length and the angle of the bend. The spacing and/or amount of travel can be adjusted on these last marks, as required.

> **NOTE:** When you have found the amount of travel for 90°, you can also find the amount of travel for any other angle for offsets, kicks, etc.

Number of pumps method – The pipe may be bent in any position. This method depends on the fact that a given hydraulic pump will produce the same amount of bend for a given number of pumps of the handle; however, it does not take into account that the number of pumps will change with a change in fluid level, the condition of the pump, and the condition of the O-rings.

For example, if it takes 40 pumps to bend a 90° stub, it should only take 20 pumps to achieve 45°, 10 pumps for $22\frac{1}{2}°$, 2 pumps for $4\frac{1}{2}°$, etc. As you can see, this method should work very well, but it will require additional time to determine pump/degree values. However, this can be offset by not having to measure ram travel at each bend as required by the amount of travel method.

Use the same procedure for bending as outlined in the amount of travel method. Check the developing stub length and degree of bend before bending the last few shots. Use these remaining shots for final adjustments in stub length and for checking the 90° bend.

> **NOTE**
> Use whichever one of the four methods is the most convenient, but bear in mind that extreme care and attention to detail are required in all methods. A small amount of error at each bend will compound itself, and accuracy in bending will be impossible to achieve.

Using a bending table – To make hydraulic bending easier, a bending table, either a commercial model or one constructed on the job, is a necessity. The table will hold the conduit and the bender, make leveling and plumbing easier, and produce more accurate bends. The table will also eliminate the need to continually wrestle with the conduit and bender.

Hydraulic bending example – For example, suppose that the task is to bend an offset containing two sweeping bends. The radius for the two sweeping bends is 30". This is to be done on a hydraulic bender using a segmented bending process. The angles of the sweeping bends are to be 30° each. The height of the offset is to be 30" (*Figure 37*).

This problem incorporates the concept of the unit circle, along with the cosecant trigonometric function.

In continuing with the concept of the unit circle and the linear distance on the circumference of the circle in terms of pi, it is necessary to understand the development of the pi relationship from 0° to 90°. All of these relationships from 0° to 90° and other interval angles in the circle are always in reference to 180°. For example: $360° = 2\pi$ or $2 \times 180°$; $180° = \pi$ or $1 \times 180°$; and $90° = \pi \div 2$ or $\frac{1}{2} \times 180°$.

The same process holds true for any angle from 0° through 90°. For example: $45° = \pi \div 4$ or $\frac{1}{4}\pi \times 180°$; $60° = \frac{1}{3}\pi$ or $\frac{1}{3} \times 180°$; $30° = \frac{1}{6}\pi$ or $\frac{1}{6} \times 180°$; and $15° = \frac{1}{12}\pi$ or $\frac{1}{12} \times 180°$ (*Figure 38*).

To figure the developed length of the conduit to be used by one swing in the 30" radius, simply multiply $\pi \div 6$, which represents pi at 30°, times the 30" radius. It is nothing more than a converted circumference formula.

$C = \frac{\pi}{6}R$ (linear distance from 0° to 30°)
$C = 0.523 \times R$
$C = 0.523 \times 30"$
$C = 15.7"$

The developed length for the first radius is 15.7". That means that the developed length for the second sweeping radius is also 15.7". The total amount of conduit used to develop the two sweeps is 31.4".

The next step is to figure out the distance between the center line of each of these bends on the hypotenuse of the imaginary triangle. To do this, simply use the cosecant of the angle trigonometric function or Cosecant = H ÷ O. Converting the formula gives H = O × Cosecant or H = 30" × 2 or 60". The layout of these measurements is shown in *Figure 39*.

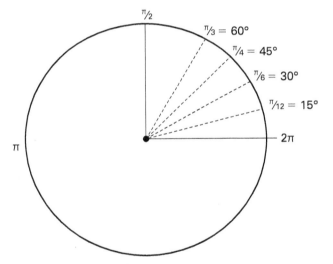

Figure 38 Radians and degrees.

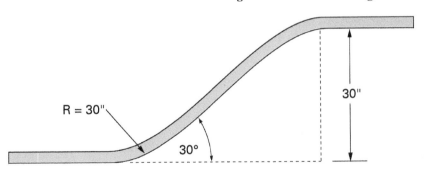

Figure 37 Two 30° sweeping bends.

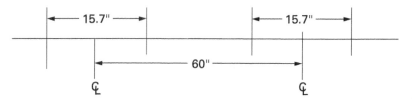

Figure 39 Bend center line distance.

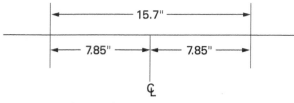

Figure 40 Bend center line.

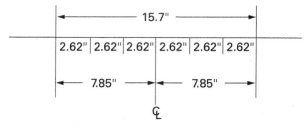

Figure 41 Bend segments.

4.2.4 Concentric Bends

If two or more parallel runs of conduit must be bent in the same direction, as shown in *Figure 42*, the best results will be obtained by using concentric bends. When laying out concentric bends, the bend for the innermost conduit is calculated first. In the example in *Figure 42*, the first bend has a radius of 20". If this dimension is multiplied by 1.57, it yields a value of $31^{13}/_{32}$". This is the developed length of the shortest radius bend.

The second radius is found by adding the **outside diameter (OD)** size of the first pipe to the

Again, laying out the segments for the bending process is done from the center line of the developed length of each sweeping bend (*Figure 40*). Therefore:

$$15.7" \div 2 = 7.85"$$

When building a 90° sweep, the rule of thumb is 21 shots or 20 segments. This gives the conduit the appearance of being a natural sweep and not a line of straight segments. (The same odd shot/even segment process will appear again.) If you choose to have seven shots or six segments within the 15.7" of developed length, each segment will be 15.7" ÷ 6 or 2.62" long. Because there are seven shots (bends), each shot will be 30 ÷ 7 or 4.29°. The layout of the bend is shown in *Figure 41*.

When you have both sweeps laid out as described previously and the center lines of these sweeps are 60" apart, you are ready to start the bending process. Again, start at one end of the offset and swing the bender segmented shoe down on the conduit. Because you are working an offset, an anti-dog device should be attached to one end of the conduit. This device maintains the center line for each bend. Begin bending the first sweep, and when this is accomplished, roll the conduit in the bender and re-level the anti-dog device. Bend the second sweep, and your offset is built. The final measurements should be as follows:

- Developed length of each sweep = 15.7"
- Distance between center line of each sweep = 60"
- Distance between shot marks in each sweep = 2.62"
- Degrees per shot in each sweep = 4.29°

Magnetic Angle Finders

A magnetic angle finder is an invaluable tool when making segmented bends.

Figure Credit: Mike Powers

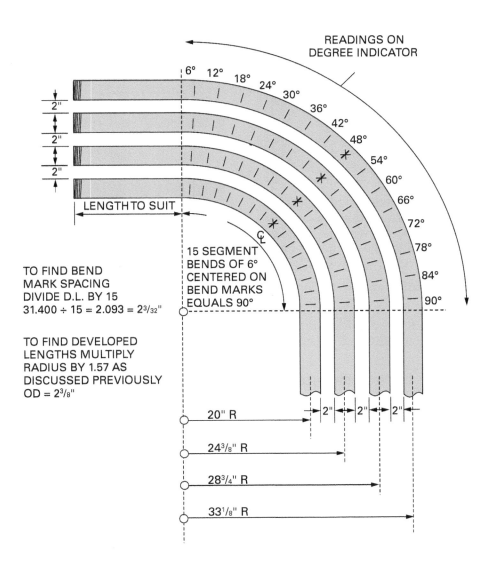

Figure 42 Principles of concentric bending.

radius of the first pipe and the desired spacing between pipes. In this example, the radius of the second bend would be equal to 24 3/8" (*Figure 42*).

Radius of first pipe = 20"

OD of first pipe = 2 3/8"

Spacing between pipes = 2"

Radius of second pipe:

20" + 2 3/8" + 2" = 24 3/8" or 24.375"

When the developed length of the second radius is found, the spacing between marks can be determined. The start mark must be the same distance from the end of the second pipe as it was for the first pipe (*Figure 42*). This will be true for each additional pipe as well. The spacing will increase between marks as the radius of the pipes increases. This is because of increasing developed length. The pipes in *Figure 42* are all laid out for 15 shots.

The equation for the developed length (DL) is 1.57 × R (radius). The developed length for the second bend is found by multiplying 1.57 × 24.375" = 38.27".

The radius and developed length of each successive bend are found in a similar manner. After determining the developed length, you must establish the number of segment bends needed to form each 90° bend. In concentric bending, every bend must receive the same number of segment bends to maintain concentricity. As illustrated, 15 segment bends of 6° each will total 90°.

> **NOTE**: The radius change from one bend to another affects the spacing of segment bends. To find the segment bend spacing, divide the developed length of each bend by 15. For the first bend illustrated, the spacing for each segment bend is $2\frac{3}{32}"$; for the second bend, it is $2\frac{9}{16}"$, and so on, as shown in *Figure 42*.

If the legs of the bends have to be a certain length, the gain must be considered just as in any segment bending procedure. When the runs of conduit are not the same size, the radius of each successive bend can be found as follows:

Step 1 Determine the radius of the innermost bend.

Step 2 Calculate one-half the outside diameter of the innermost conduit and of the next adjacent conduit.

Step 3 Note the distance between the two runs of conduit.

Step 4 Add these quantities.

4.2.5 Offset Bends

Many situations require a conduit to be bent so that it can pass by or over objects such as beams and other conduit or to enter panelboards and junction boxes. The bends used for this purpose are called *offsets*. To produce an offset, two equal bends of less than 90° are required at a specified distance apart. This distance is determined by the angle of the two bends and can be calculated by using the following procedure and the table in *Figure 43*.

First, determine the offset needed, then find the degree of bend. Next, multiply the offset measurements by the figure directly under the degree of bend (*Figure 43*). This applies to all sizes of conduit. For example, to form an 18" offset with two 45° bends, first make the following calculation to determine the distance between bends:

$$18" \times 1.414 = 25\frac{1}{2}"$$

This is the distance between bends and is labeled side L (*Figure 43*). To connect the two ends of an offset to two pieces of conduit that are already in place, it is necessary to know the total or overall length (OL) of the offset from end to end before bending.

Note the following equation:

$$OL = (A + L + B) - (2 \times gain)$$

Think About It
Concentric Bends

What does concentric mean? Why can't you just bend all your conduit sections to the same radius and lay them side by side?

Figure Credit: Tim Dean

Where:
  OL = overall length
  A = 36"
  L = $25\frac{1}{2}"$
  B = 48"
  2 × gain = $1\frac{9}{32}"$

The gain is calculated by multiplying the shoe radius by the decimal figures shown on the last line under Degrees of Bend in *Figure 43*. For example, to calculate how an offset might be made in a length of 3" conduit using 45° bends:

> **NOTE**: This offset uses a 15" radius.

$$OL = (A + L + B) - (2 \times gain)$$
$$OL = 109\frac{1}{2}" - 1\frac{9}{32}" = 108\frac{7}{32}"$$
$$15" \times 0.0430 = 0.645"$$

For two gains, you have:

$$0.645" \times 2 = 1.290" = 1\frac{9}{32}"$$

Because you already know the long dimension of the offset (25"), to find the amount of offset, refer to the table in *Figure 43*, second line down

Section Five

5.0.0 Installing PVC Conduit

Objective

Install PVC conduit.
a. Join PVC conduit.
b. Bend PVC conduit.

Rigid nonmetallic conduit and fittings (PVC electric conduit) may be used in direct earth burial; in walls, floors, and ceilings of buildings; in cinder fill; and in damp and dry locations. Its use is prohibited by the *NEC®* in certain hazardous locations, for support of fixtures or other equipment, where subject to physical damage, and under certain other conditions. Refer to *NEC Article 352*.

PVC conduit can be cut easily at the job site without special tools, although PVC cutters help in cutting square ends. Sizes from $1/2$" through $1\frac{1}{2}$" (MD 16 through MD 41) can be cut with a fine-tooth saw. For sizes 2" through 6" (MD 53 through MD 155), a miter box or similar saw guide should be used to keep the conduit steady and ensure a square cut. To ensure satisfactory joining, care should be taken not to distort the end of the conduit when cutting.

After cutting, deburr the pipe ends and wipe clean of dust, dirt, and plastic shavings. Deburring is accomplished easily using a pocket knife or file.

5.1.0 Joining PVC

An important advantage of PVC conduit, compared to other rigid conduit materials, is the ease and speed with which solvent-cemented joints can be made. The following steps are required for a proper joint:

Step 1 The conduit should be wiped clean and dry.

Step 2 Apply a full, even coat of PVC cement to the end of the conduit and the fitting. The cement should cover the area that will be inserted in the fitting.

Step 3 Push the conduit and fitting firmly together with a slight twisting action until it bottoms and then rotate the conduit in the fitting (about a half turn) to distribute the cement evenly. Avoid cement buildup inside the conduit. The cementing and joining operation should not exceed more than 20 seconds.

Step 4 When the proper amount of cement has been applied, a bead of cement will form at the joint. Wipe the joint with a brush to remove any excess cement. The joint should not be disturbed for 10 minutes.

> **WARNING!** The cement used with PVC can be hazardous. Provide adequate ventilation, avoid skin contact, and always refer to the SDS and follow the manufacturer's usage and safety instructions.

5.2.0 Bending PVC

Most manufacturers of PVC conduit offer various radius bends in a number of segments. Where special bends are required, PVC conduit is easy to bend on the job. Stub-ups, saddles, concentric bends, offsets, and kicks are all possible with PVC conduit, just as with metallic conduit.

PVC conduit is bent with the aid of a heating unit. The PVC must be heated evenly over the entire length of the curve. Heating units are available from various sources that are designed specifically for the purpose in sizes to accommodate all conduit diameters (*Figure 52*). While some heaters use gas for the heat source, most employ infrared heat energy, which is more quickly absorbed in the conduit. Small sizes are ready to bend after a few seconds in the hotbox. Larger diameters require two or three minutes, depending on the conditions. Other methods of heating PVC conduit for bending include heating blankets and hot air blowers. The use of torches or other flame-type devices is not recommended. PVC conduit exposed to excessively high temperatures may take on a brownish color. Sections showing evidence of such scorching should be discarded.

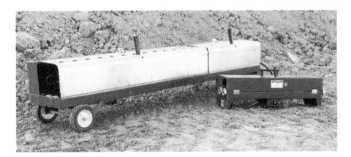

Figure 52 PVC heating units.

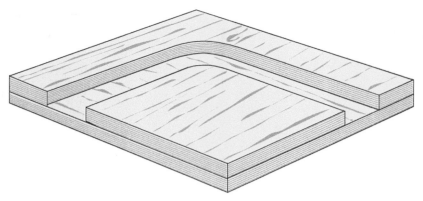

Figure 53 Plywood template.

If a number of identical bends are required, a template can be helpful (*Figure 53*). A simple template can be made by sawing a sheet of plywood to match the desired bend. Nail to a second sheet of plywood. The heated conduit section is placed in the template, sponged with water to cool, and is then ready to install. Care should be taken to fully maintain the ID of the conduit when bending.

WARNING! Always wear gloves when working with heat.

If only a few bends are needed, scribe a chalk line on the floor or workbench. Then, match the heated conduit to the chalk line and cool. The conduit must be held in the desired position until it is relatively cool because the PVC material will tend to revert to its original shape. Templates are also available for many bends.

Another method is to take the heated conduit section to the point of installation and form it to fit the actual installation (*Figure 54*). This method is especially effective for making blind bends or compound bends using smaller sizes of PVC. After bending, wipe a wet rag over the bend (*Figure 55*) to cool it.

Bends in small-diameter PVC conduit (up to 1½" or MD 41) require no filling for code-approved radii. When bending PVC of 2" (MD 53) or larger diameter, there is a risk of wrinkling or flattening the bend. To help eliminate this problem, a plug set is used. A plug is inserted in each end of the piece of PVC being bent, and then the conduit is pressurized using either a hand pump or the heat of the bender, depending on the type of plug used.

Place airtight plugs (*Figure 56*) in each end of the conduit section before heating. The retained air will expand during the heating process and hold the conduit open during bending. Do not remove the plugs until the conduit has cooled.

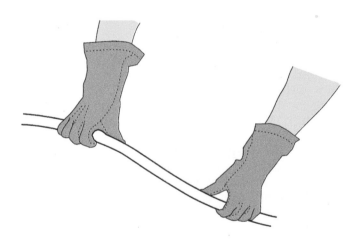

Figure 54 Some PVC bends may be formed by hand.

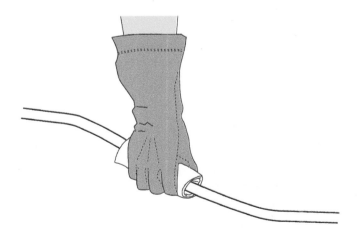

Figure 55 After the bend is formed, wipe a wet rag over the bend to cool it.

In applications where the conduit installation is subject to constantly changing temperatures and the runs are long, precautions should be taken to allow for expansion and contraction of PVC conduit. When expansion and contraction are factors, an O-ring expansion coupling should be installed near the fixed end of the run, or fixture, to take up any expansion or contraction that may occur.

Confirm the expansion and contraction lengths available in these fittings as they may vary by manufacturer. Charts are available that indicate how much expansion can be expected at various temperature levels. *Table 7* lists the expansion rates for various temperatures.

Expansion couplings are seldom required in underground or slab applications. Expansion and contraction may generally be controlled by bowing the conduit slightly or burying the conduit immediately. After the conduit is buried, expansion and contraction cease to be factors. Care should be taken, however, in constructing a buried installation. If the conduit should be left exposed for an extended period during variable temperature conditions, an allowance should be made for expansion and contraction.

In above-ground installations, care should be taken to provide proper support of PVC conduit because of its semi-rigidity. This is particularly important at high temperatures. The distance between supports should be based on temperatures encountered at the specific installation. Charts are available that clearly outline at which intervals support is required for PVC conduit at various temperature levels.

Figure 56 Typical plug set.

Table 7 PVC Expansion Rates

Temperature Change in °F	Length Change in Inches per 100 Feet of PVC Conduit	Temperature Change in °F	Length Change in Inches per 100 Feet of PVC Conduit
5	0.2	105	4.2
10	0.4	110	4.5
15	0.6	115	4.7
20	0.8	120	4.9
25	1.0	125	5.1
30	1.2	130	5.3
35	1.4	135	5.5
40	1.6	140	5.7
45	1.8	145	5.9
50	2.0	150	6.1
55	2.2	155	6.3
60	2.4	160	6.5
65	2.6	165	6.7
70	2.8	170	6.9
75	3.0	175	7.1
80	3.2	180	7.3
85	3.4	185	7.5
90	3.6	190	7.7
95	3.8	195	7.9
100	4.1	200	8.1

PVC Conduit Bending

PVC conduit has a memory. As soon as you have finished bending the PVC, cool it down with cold, wet towels or hose it down with cold water. Otherwise, it will begin to return to its original position.

Think About It
Putting It All Together

Examine the visible conduit bends in your home or workplace. Are they neat and professionally made? If not, what might you have done differently to improve the installation?

5.0.0 Section Review

1. When joining PVC, apply PVC cement to _____.
    a. the end of the conduit and the fitting
    b. the end of the conduit only
    c. the fitting only
    d. the end of the conduit and primer to the fitting

2. In preparation for bending, large-diameter PVC should be heated for _____.
    a. 2 to 3 seconds
    b. 2 to 3 minutes
    c. 20 to 30 minutes
    d. 2 to 3 hours

Review Questions

1. The minimum bending radius for 3" rigid conduit is ____.
   a. 8"
   b. 10"
   c. 11"
   d. 13"

2. The minimum bending radius for ½" rigid conduit is ____.
   a. 4"
   b. 6"
   c. 8"
   d. 10"

3. The maximum number of 90° bends allowed between pull points in a conduit system is ____.
   a. one
   b. two
   c. three
   d. four

4. A saddle bend is counted as ____.
   a. 90°
   b. 120°
   c. 180°
   d. 60°–180°, depending on the type of bends

5. When referring to the two bends required to make an offset in a length of conduit, which of the following is always *true*?
   a. The degree of bend for each must be exactly 45°.
   b. The degree of bend for each must be exactly 30°.
   c. The degree of bend for each must be equal.
   d. The degree of bend for each must be unequal.

6. The reason for making a saddle bend in a run of conduit is to ____.
   a. change direction in the height of a conduit run
   b. make a 90° bend
   c. cross a small obstacle
   d. make a conduit termination in an outlet box

7. Which of the following formulas should be used to find the side adjacent?
   a. $S = O \times H$
   b. $S = O \div H$
   c. $C = A \div H$
   d. $C = A \times H$

8. The equation used to calculate the circumference of a circle is ____.
   a. $C = 2\pi R$
   b. $C = \pi R$
   c. $C = \pi R^2$
   d. $C = \pi^2$

9. The fractional equivalent of the decimal 0.015625 is ____.
   a. $\frac{1}{64}$
   b. $\frac{1}{32}$
   c. $\frac{3}{64}$
   d. $\frac{1}{16}$

10. Two or more parallel bends that are bent in the same direction are known as ____.
    a. stubs
    b. perpendicular bends
    c. concentric bends
    d. axial bends

11. You are making a concentric bend and the radius of the first pipe is 16", the OD of the first pipe is 2", and the spacing between pipes is 3". What is the radius of the second pipe? 4.2.4.
    a. 16"
    b. 18"
    c. 19"
    d. 21"

12. You are making a 20" offset with two 30° bends. The distance between bends is 4.2.5.
    a. 10"
    b. 15"
    c. 40"
    d. 50"

13. A saddle to cross a pipe requires 4.2.6.
    a. two bends
    b. three bends
    c. four bends
    d. five bends

14. An important safety precaution to use when working with PVC conduit is to 5.1.0.
    a. provide proper ventilation to carry off fumes from the joint cement
    b. use lengths of PVC of not more than 8'
    c. use inside diameters of 4" or less
    d. never wear gloves that may adhere to the joint cement

15. To prevent flattening when bending larger diameter PVC pipe, use 5.2.0.
    a. plugs and air pressure
    b. sand fill
    c. water fill
    d. a higher temperature

Supplemental Exercises

1. In which chapter of the *NEC*® will you find the installation requirements for conduit?
   a. Two
   b. Three
   c. Four
   d. Five
2. The maximum number of degrees allowed between pull points in a conduit run is __360°__.
3. Offsets and saddles _____.
   a. do not need to be counted if less than 15°
   b. must be included when calculating total degrees between pull points
   c. add no resistance
   d. all count as 30°
4. A conduit run with _____ is acceptable to pull wire.
   a. two 90° bends, four 45° bends, and one box offset
   b. two 90° bends, three 45° bends, and one saddle
   c. two 90° bends, two 45° bends, and two 30° bends
   d. four 90° bends and two box offsets
5. Would 4" rigid metal conduit be bent using a mechanical bender?
6. The maximum size IMC that can be bent using a portable mechanical conduit bender is __1¼__.
7. The code requires a minimum radius for rigid conduit bends because when bends are too tight, _____.
8. A bend used for a change in direction of less than 90° is called a(n) _____.
   a. kick
   b. saddle
   c. stub-up
   d. offset
9. A mechanical bender may be used to make bends in _____.
   a. 2" EMT
   b. 4" rigid conduit
   c. 1½" PVC
   d. IMC conduit only
10. To make matching bends in both ½" and 1" IMC when using a mechanical bender, _____.
    a. adjust the deduct length for the ½" conduit
    b. use the ½" shoe for the ½" conduit and the 1" shoe for the 1" conduit
    c. use the 1" shoe for both conduits
    d. get matched pairs of conduit shoes from the manufacturer

11. Speed benders can bend conduit up to _____.
12. True or False? The one-shot shoes used with hydraulic benders can make 90° bends in a single operation.
13. One method of bending segments is to use _____.
    a. an elapsed time method to keep track of each bend
    b. a stress meter to monitor each bend
    c. a level to monitor each bend
    d. the amount of travel method to monitor each bend
14. Airtight plugs are installed in the ends of large sizes of PVC conduit during bending to _____.
    a. prevent the conduit from wrinkling or flattening the bend
    b. hold the heat inside the conduit to keep it flexible
    c. prevent condensation within the conduit
    d. prevent deadly PVC gases from escaping
15. A disadvantage of aluminum conduit is that _____.
    a. it cannot be bent with one-shot benders
    b. it is difficult to bend consistently and accurately
    c. special bending shoes must be used
    d. aluminum kinks when bent

Trade Terms Introduced in This Module

Approximate ram travel: The distance the ram of a hydraulic bender travels to make a bend. To simplify and speed bending operations, many benders are equipped with a scale that shows ram travel. Using a simple table (supplied with many benders), the degree of bend can easily be converted to inches of ram travel. This measurement, however, can only be approximated because of the variation in springback of the conduit being bent.

Back-to-back bend: Any bend formed by two 90° bends with a straight section of conduit between the bends.

Bending protractor: Made for use with benders mounted on a bending table and used to measure degrees; also has a scale for 18, 20, 21, and 22 shots when using it to make a large sweep bend.

Bending shots: The number of shots needed to produce a specific bend.

Concentric bending: The process of making 90° bends in parallel runs of conduit. This requires increasing the radius of each conduit from the inside of the bend toward the outside.

Conduit: Piping designed especially for pulling electrical conductors. Types include RMC, IMC, EMT, PVC, aluminum, and other materials.

Degree indicator: An instrument designed to indicate the exact degree of bend while it is being made.

Developed length: The amount of straight pipe needed to bend a given radius. Also, the actual length of the conduit that will be bent.

Elbow: A 90° bend.

Gain: The amount of pipe saved by bending on a radius as opposed to right angles. Because conduit bends in a radius and not at right angles, the length of conduit needed for a bend will not equal the total determined length. Gain is the difference between the right angle distances A and B and the shorter distance C—the length of conduit actually needed for the bend.

Inside diameter (ID): The inside diameter of conduit. All electrical conduit is measured in this manner. The outside dimensions, however, will vary with the type of conduit used.

Kick: A bend made to change the direction of the conduit. Kicks are typically less than 90°.

Leg length: The distance from the end of the straight section of conduit to the bend, measured to the center line or to the inside or outside of the bend or rise.

Offset: Two equal bends made to avoid an obstruction blocking the run of the conduit.

One-shot shoe: A large bending shoe that is designed to make 90° bends in conduit.

Outside diameter (OD): The size of any piece of conduit measured on the outside diameter.

Radius: The relative size of the bent portion of a pipe.

Rise: The length of the bent section of conduit measured from the bottom, center line, or top of the straight section to the end of the bent section.

Segment bend: Any bend formed by a series of bends of a few degrees each, rather than a single one-shot bend.

Segmented bending shoe: A smaller type of shoe designed for bending segmented bends only (always less than 15°).

Springback: The amount, measured in degrees, that a bent conduit tends to straighten after pressure is released on the bender ram. For example, a 90° bend, after pressure is released, will pull back about 2° to 88°.

Stub-up: Another name for the rise in a section of conduit.

Sweep bend: A 90° bend with a radius larger than that produced by a standard one-shot shoe.

Take-up (deduct): The amount that must be subtracted from the desired stub length to make the bend come out correctly using a point of reference on the bender or bending shoe.

Additional Resources

This module presents thorough resources for task training. The following resource material is suggested for further study.

Benfield Conduit Bending Manual, 2nd Edition. Overland Park, KS: EC&M Books.
National Electrical Code® Handbook, Latest Edition. Quincy, MA: National Fire Protection Association.
Tom Henry's Conduit Bending Package (includes video, book, and bending chart). Winter Park, FL: Code Electrical Classes, Inc.

Figure Credits

Greenlee / A Textron Company, Module Opener, Figures 15, 25, 26, 30, 31, 34, 42–45, 47

Data from *NEC® Chapter 9, Table 2*, Tables 1 and 2. Reprinted with permission from NFPA 70-2020, *National Electrical Code®*, Copyright © 2019, National Fire Protection Association, Quincy, MA. This reprinted material is not the complete and official position of the NFPA on the referenced subject, which is represented only by the standard in its entirety which may be obtained through the NFPA website at **www.nfpa.org**.

John Traister, Figures 2, 9, 10

Tim Ely, Figure 3B

Section Review Answer Key

SECTION 1.0.0

Answer	Section Reference	Objective
1. c	1.1.0; *Table 1*	1a
2. c	1.1.0; *NEC Chapter 9, Table 2*	1a
3. b	1.2.0	1b

SECTION 2.0.0

Answer	Section Reference	Objective
1. a	2.1.0	2a
2. b	2.2.2	2b

SECTION 3.0.0

Answer	Section Reference	Objective
1. b	3.1.0	3a
2. c	3.2.1	3b

SECTION 4.0.0

Answer	Section Reference	Objective
1. d	4.1.0	4a
2. c	4.2.0	4b

SECTION 5.0.0

Answer	Section Reference	Objective
1. a	5.1.0	5a
2. b	5.2.0	5b

Section Review Calculations

2.0.0 SECTION REVIEW

Question 2

Find the bend by multiplying the radius of the bend times the gain factor listed in *Table 3*:

6" × 0.43 = 2.58"

26205-20
PULL AND JUNCTION BOXES

Objectives

When you have completed this module, you will be able to do the following:

1. Identify boxes and fittings.
   a. Select pull and junction boxes.
   b. Select and install fittings.
2. Size pull and junction boxes.
   a. Size pull and junction boxes for systems under 1,000V.
   b. Size pull and junction boxes for systems over 1,000V.
3. Identify specialty enclosures.
   a. Identify conduit bodies and other cast enclosures.
   b. Select and install handholes.

Performance Tasks

Under the supervision of the instructor, you should be able to do the following:

1. Identify various NEMA boxes.
2. Properly select, install, and support pull and junction boxes over 100 cu in (1,650 cu cm) in size.
3. Identify various conduit bodies and fittings.

Trade Terms

Conduit body
Explosion-proof
Handhole enclosure
Junction box
Mogul

Pull box
Raintight
Watertight
Weatherproof

Industry Recognized Credentials

If you are training through an NCCER-accredited sponsor, you may be eligible for credentials from NCCER's Registry. The ID number for this module is 26205-20. Note that this module may have been used in other NCCER curricula and may apply to other level completions. Contact NCCER's Registry at 888.622.3720 or go to **www.nccer.org** for more information.

> **NOTE**
> NFPA 70®, *National Electrical Code*® and *NEC*® are registered trademarks of the National Fire Protection Association, Quincy, MA.

Contents

1.0.0 Boxes and Fittings 1
  1.1.0 Pull and Junction Boxes 1
    1.1.1 Boxes for Damp and Wet Locations 2
    1.1.2 NEMA and IP Enclosure Classifications 2
  1.2.0 Fittings 3
    1.2.1 EMT Fittings 3
    1.2.2 Rigid, Aluminum, and IMC Fittings 4
    1.2.3 Locknuts and Bushings 5
2.0.0 Sizing Pull and Junction Boxes 9
  2.1.0 Sizing Pull and Junction Boxes for Systems Under 1,000V 9
  2.2.0 Sizing Pull and Junction Boxes for Systems Over 1,000V 11
3.0.0 Specialty Enclosures 13
  3.1.0 Conduit Bodies and Other Cast Enclosures 13
    3.1.1 Type C Conduit Bodies 13
    3.1.2 Type L Conduit Bodies 13
    3.1.3 Type T Conduit Bodies 13
    3.1.4 Type X Conduit Bodies 14
    3.1.5 FS and FD Boxes 14
    3.1.6 Pulling Elbows 15
    3.1.7 Entrance Ells (SLBs) 15
    3.1.8 Moguls 15
  3.2.0 Handholes 15
    3.2.1 Handhole Construction 17
    3.2.2 ANSI/SCTE Requirements 17

Figures and Tables

Figure 1   Pull and junction boxes 1
Figure 2   EMT compression fittings 4
Figure 3   Setscrew fittings 4
Figure 4   Rigid metal conduit with coupling 4
Figure 5   Flexible-to-rigid coupling 4
Figure 6   Metal conduit couplings 5
Figure 7   Common types of locknuts 5
Figure 8   Typical bushings used at termination points 5
Figure 9   Insulating bushings 6
Figure 10   Myers-type (gasketed) hub 7
Figure 11   Knockout punch kit 7
Figure 12   Battery-powered knockout kit 7
Figure 13   Hydraulic knockout kit 8
Figure 14   Pull box with straight conduit runs 10
Figure 15   Pull box with conduit runs entering at right angles 11
Figure 16   Identifying conduit bodies 14
Figure 17   Type C conduit body 14
Figure 18   Type L conduit bodies 14
Figure 19   Type T conduit body 14
Figure 20   Type X conduit body 14
Figure 21   Elbows 16
Figure 22   Mogul 16
Figure 23   Handhole 16
Figure 24   Handhole containing traffic signal wiring 17

Table 1   IP Classification System 3
Table 2   NEMA and IP Enclosures 3

This page is intentionally left blank.

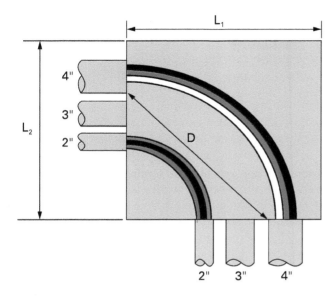

Figure 15 Pull box with conduit runs entering at right angles.

2.2.0 Sizing Pull and Junction Boxes for Systems Over 1,000V

NEC Article 314, Part IV covers requirements for pull and junction boxes, conduit bodies, and handholes used on systems over 1,000V. Because the conductors used with these systems are larger and heavier than those used on lower-voltage systems, the *NEC®* requires additional pulling and bending space. The *NEC®* requirements for these systems include the following:

- For straight pulls, the length of the box must be at least 48 times the outside diameter (OD) of the largest shielded or lead-covered conductor entering the box. See *NEC Section 314.71(A)*. When the cable is unshielded, the length may be reduced to 32 times the OD of the largest nonshielded conductor entering the box.
- For angle or U pulls, the distance between each shielded cable or conductor entry inside the box and the opposite wall of the box must be at least 36 times the OD, over sheath, of the largest cable or conductor. See *NEC Section 314.71(B)(1)*. The distance must be increased by the sum of the ODs, over sheath, of all other cables or conductors entering through the same wall of the box. When the cable is unshielded and not lead covered, the length may be reduced to 24 times the OD of the largest nonshielded conductor entering the box.
- The distance between a cable or conductor entry and its exit must be at least 36 times the OD, over sheath, of that cable or conductor. See *NEC Section 314.71(B)(2)*. When the cable is both unshielded and not lead covered, the distance may be reduced to 24 times the OD.
- In addition to a suitable cover, pull boxes containing conductors over 1,000V must be provided with at least one removable side.
- Per *NEC Section 314.72(E)*, boxes housing conductors over 1,000V must be completely enclosed and permanently marked DANGER – HIGH VOLTAGE – KEEP OUT.

Label Junction Boxes

It's a good idea to label every junction box plate with the circuit number, the panel it came from, and its destination. The next person to service the installation will be grateful for this extra help.

Using Pull Boxes

Pull boxes make it easier to install conductors. They can also be installed to avoid having more than 360° worth of bends in a single run. (Remember, if a pull box is used, it is considered the end of the run for the purposes of the *NEC®* 360° rule.)

2.0.0 Section Review

1. A 2" raceway runs alongside a 3" raceway into a junction box with an angled pull. Assuming the system operates at 600V, the required distance between these two raceway entries is _____.

    a. 14"
    b. 20"
    c. 24"
    d. 32"

    3" × 6 + 2" =

2. An MD 53 raceway runs alongside an MD 78 raceway into a junction box with an angled pull. Assuming the system operates at 600V, the required distance between these two raceway entries is _____.

    a. 131 mm
    b. 396 mm
    c. 521 mm
    d. 786 mm

    6 × 78 mm + 53 mm

3. A shielded cable with an OD of 2" enters a junction box with an angled pull. Assuming the system operates at 5,000V, the minimum length of the box _____.

    a. 12"
    b. 16"
    c. 48"
    d. 72"

    36 × 2"

4. A shielded cable with an OD of 50 mm enters a junction box with an angled pull. Assuming the system operates at 5,000V, the minimum length of the box _____.

    a. 1,200 mm
    b. 1,600 mm
    c. 1,800 mm
    d. 4,800 mm

    50 mm × 36

Section Three

3.0.0 Specialty Enclosures

Objective

Identify specialty enclosures.
  a. Identify conduit bodies and other cast enclosures.
  b. Select and install handholes.

Performance Task

3. Identify various conduit bodies and fittings.

Trade Terms

Conduit body: A separate portion of a conduit or tubing system that provides access through a removable cover (or covers) to the interior of the system at a junction of two or more sections of the system or at a terminal point of the system. Boxes such as FS and FD or larger cast or sheet metal boxes are not classified as conduit bodies.

Handhole enclosure: An enclosure used with underground systems to provide access for installation and maintenance.

Mogul: A type of conduit body with a raised cover to provide additional space for large conductors.

In addition to traditional pressed steel pull and junction boxes, a variety of other enclosures are used to provide pull and junction points in conduit runs. These specialty enclosures include conduit bodies, Type FS and FD boxes, moguls, and handholes.

3.1.0 Conduit Bodies and Other Cast Enclosures

Conduit bodies, also called condulets, are defined in *NEC Article 100* as a separate portion of a conduit or tubing system that provides access through a removable cover to the interior of the system at a junction of two or more sections of the system or at a terminal point of the system. Conduit bodies are usually used with RMC and IMC. The cost of conduit bodies, because they are cast, is significantly higher than the stamped steel boxes. Splicing in conduit bodies is typically not recommended; however, it is permitted under certain conditions as specified in *NEC Section 314.16(C)(2)*. The cover and gasket for conduit body fittings must be ordered separately; do not assume that these parts come with the conduit body when it is ordered.

As an electrical trainee, you will hear such terms as LL, LR, and other letters to distinguish between the various types of conduit bodies. To identify certain conduit bodies, an old trick of the trade is to hold the conduit body like a pistol (*Figure 16*). When doing so, if the oval-shaped opening of the conduit body is to your left, it is called an LL—the first L stands for elbow and the second L stands for left. If the opening is on your right, it is called an LR—the R stands for right. If the opening is facing upward, this type of conduit body is called an LB—the B stands for back. If there is an opening on both sides, it is called an LRL—for both left and right. The other popular shapes are named for their letter look-alikes; that is, T and X. The only exception is the C conduit body. Let us take a closer look at each of these.

3.1.1 Type C Conduit Bodies

A Type C conduit body (*Figure 17*) may be used to provide a pull point in a long conduit run or a conduit run that has bends totaling more than 360° (refer to *NEC Sections 342.26, 344.26, 352.26,* and *358.26*). In this application, the Type C conduit body is used as a pull point.

3.1.2 Type L Conduit Bodies

A Type L conduit body (*Figure 18*) is used as a pulling point for conduit that requires a 90° change in direction. (Again, the letter L is short for elbow.) To use a Type L conduit body, the cover is removed, the wire is pulled out and coiled on the ground (or floor), and then it is reinserted into the other opening and pulled. Type L conduit bodies are available with the cover on the back (Type LB), on the sides (Type LL and Type LR), or on both sides (Type LRL).

3.1.3 Type T Conduit Bodies

A Type T conduit body, also known as a tee, is used to provide a junction point for three intersecting conduits and low point drains (*Figure 19*). Tees are used extensively in rigid conduit systems. The cost of a tee conduit body is more than twice that of a standard square box with a cover. Therefore, the use of Type T conduit bodies with EMT is limited.

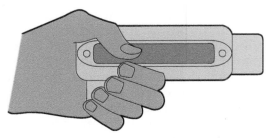
TYPE LR – OPENING ON RIGHT

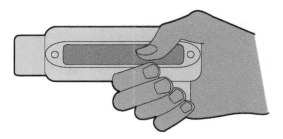
TYPE LL – OPENING ON LEFT

Figure 16 Identifying conduit bodies.

Figure 17 Type C conduit body.

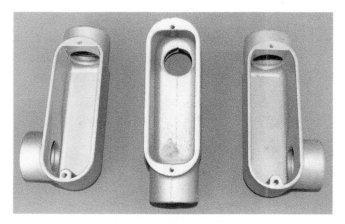

Figure 18 Type L conduit bodies.

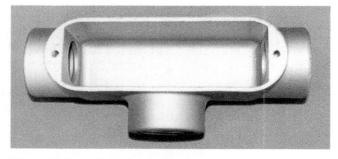

Figure 19 Type T conduit body.

3.1.4 Type X Conduit Bodies

A Type X conduit body is used to provide a junction point for four intersecting conduits. The removable cover provides access to the interior of the X so that wire pulling and splicing may be performed (*Figure 20*).

Figure 20 Type X conduit body.

3.1.5 FS and FD Boxes

FS boxes are cast boxes available in single-gang, two-gang, and three-gang configurations. They are sized to permit the installation of switches and receptacles. Covers for switches and receptacles are available for FS boxes that have formed openings much like switch and receptacle plates. FD boxes are similar to FS boxes. The letter D in the FD box indicates it is a deeper box (about $5/8$" deeper or 16 mm). Neither FS nor FD boxes are considered by the *NEC*® to be conduit bodies. FS and FD boxes may be used in environments defined by NEMA 1 (dry, clean environments); NEMA 3R (outdoor, wet environments); and NEMA 12 (dusty, oily environments).

FS and FD boxes are precision-molded boxes with a cover and gasket designed to provide a tight seal. Because they cost significantly more than standard boxes, they are typically used only in industrial environments where the boxes are subjected to moisture, dirt, dust, and corrosion.

Engineers specify and electricians install FS and FD boxes for a reason. Never alter these boxes by drilling mounting holes in them. Most are provided with cast-in mounting eyes for this purpose. Mount these boxes only as recommended by the manufacturer.

All-in-One Conduit Bodies

All-in-one conduit bodies are also available. The conduit body shown here easily converts between LB, LL, LR, T, and C types by removing the appropriate covers. It is threaded for use with rigid conduit but also includes setscrews for use with EMT.

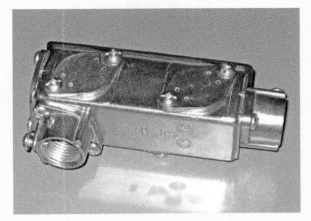

Figure Credit: Tim Ely

3.1.6 Pulling Elbows

Pulling elbows are used exclusively for pulling wire at a corner point of a conduit run. The volume of a pulling elbow is too low to permit splicing wire, as shown in *Figure 21 (A)*.

3.1.7 Entrance Ells (SLBs)

An entrance ell, or SLB, is built with an offset so that it may be attached directly to the surface that is to have a conduit penetration. A cover on the back of the SLB permits wire to be pulled out and reinserted into the conduit that penetrates the support surface, as shown in *Figure 21 (B)*.

> **CAUTION**
> Never make splices in an entrance ell (SLB). Splices can be made in a conduit body that has the volume marked, but never in an SLB. Refer to *NEC Section 314.16(C)(3)*.

3.1.8 Moguls

Mogul conduit bodies (*Figure 22*) are available in the same types as standard conduit bodies (Type L, Type T, and so on). They have raised covers to provide better access to large conductors during conductor splicing, installation, or maintenance. Moguls also allow right angle bends where splices, pulls, and taps are needed in a weatherproof chamber. Like regular boxes, they must comply with the bending space requirements of *NEC Section 314.28*. Larger moguls may contain built-in cable-pulling rollers to facilitate installation.

> **NOTE**
> When installing conduit bodies, remember that just because the body matches the conduit size does not mean it matches the conductor size. For example, if the conductor fill is near the limit for a specific conduit, it may not be acceptable for the matching size conduit body per *NEC Section 314.28*. If this is the case, a larger conduit body would be installed, along with the appropriate reducing fitting.

3.2.0 Handholes

According to *NEC Article 100*, a handhole enclosure, also referred to simply as a *handhole* (*Figure 23*) is "an enclosure for use in underground systems, provided with an open or closed bottom, and sized to allow personnel to reach into, but not enter, for the purpose of installing, operating, or maintaining equipment or wiring or both." Handholes are often used with underground PVC conduit for the installation of landscape lighting, light poles, traffic lights, and other applications. Because handholes are exposed to various levels of foot traffic and/or vehicular loading, it is critical to select the correct enclosure for the application. *NEC Section 314.30, Informational Note* refers to ANSI/SCTE 77-2013, *Specification for Underground Enclosure Integrity*, for additional information on loading for underground enclosures. Some of the *NEC®* requirements for handholes are as follows:

- Like boxes and conduit bodies, handholes must be installed in such a way that the enclosed wiring remains accessible without removing any portion of the building, such as wall coverings, or without requiring excavation, such as for enclosures serving underground wiring installations. See *NEC Section 314.29*.

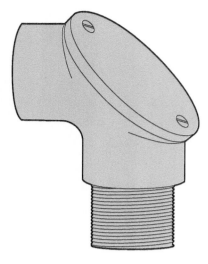
(A) PULLING ELBOW

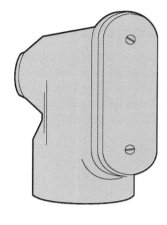
(B) ENTRANCE ELL

Figure 21 Elbows.

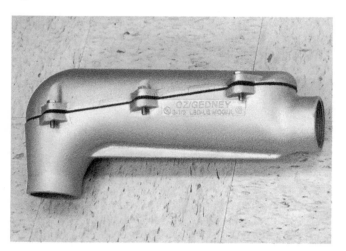

Figure 22 Mogul.

- Per *NEC Section 314.30(A)*, handhole enclosures must be sized in accordance with *NEC Section 314.28(A)* for conductors operating at 1,000 volts or below, and in accordance with *NEC Section 314.71* for conductors operating at over 1,000 volts.
- Underground raceways and cable assemblies entering a handhole enclosure must extend into the enclosure, but they are not required to be mechanically connected to the enclosure per *NEC Section 314.30(B)*.
- All enclosed conductors and any splices or terminations, if present, must be listed as suitable for wet locations per *NEC Section 314.30(C)*.
- Per *NEC Section 314.30(D)*, handhole enclosure covers shall have an identifying mark or logo that prominently identifies the function of the enclosure, such as ELECTRIC (*Figure 23*) or TRAFFIC SIGNALS (*Figure 24*). To discourage

Figure 23 Handhole.

unauthorized entry, handhole enclosure covers must either weigh over 100 lbs (45 kg) or require the use of tools to open. Metal covers and other exposed conductive surfaces must be bonded in accordance with *NEC Sections 250.92* and *250.96(A)*.

Figure 24 Handhole containing traffic signal wiring.

3.2.1 Handhole Construction

Handhole enclosures may be constructed of a variety of materials, including precast portland cement concrete, precast polymer concrete, thermoplastic, fiberglass-reinforced resin, steel, and aluminum. The type of material selected depends on the environment and the expected load.

> **CAUTION**
> Some enclosures use dissimilar materials for the body and cover of the enclosure. Dissimilar materials are likely to have different responses to temperature and pressure variations, which may result in cracking or other performance failures.

3.2.2 ANSI/SCTE Requirements

According to *ANSI/SCTE 77-2013*, handhole enclosures must be constructed in such a way that they withstand all loads likely to be imposed and remain safe and reliable for the intended application. To meet the ANSI standard, enclosures must pass various physical, environmental, and internal equipment protection tests. The ANSI standard defines loading requirements for enclosures based upon anticipated loads and separates these requirements into various tiers defined by the application. Some manufacturers mark tier designations on the enclosure cover.

While the *NEC*® does not explicitly require third party listing, some enclosure manufacturers have submitted products to UL for testing and enclosures that meet the testing requirements may now be UL Listed to the ANSI standard.

> **WARNING!**
> Without a solid footing, a handhole can sink and result in structural failure. A solid footing is essential and must be designed by a qualified individual in accordance with good engineering practices. Refer to the manufacturer's instructions for the proper installation of underground enclosures.

Lighting Poles

While the *NEC*® states that handhole enclosures are used in underground systems, the access holes on lighting poles are also typically referred to as *handholes*.

Case History

Handhole Failure

In July 2003, an eight-year-old girl was electrocuted when she stepped on the cover of an in-ground pull box (handhole) in a ballpark in Ohio. Over the years, the unsupported box sank, bringing the lid in contact with the conductors, one of which contained a damaged splice. The exposed conductor energized the lid of the handhole. When the box was installed, there were no performance standards for the construction of underground enclosures and the *NEC*® did not require bonding.

The Bottom Line: Because of this and similar incidents involving foot traffic over energized handholes, the *NEC*® now requires that handholes be effectively bonded and designed to support the expected load.

3.0.0 Section Review

1. A receptacle in a wet location would most likely be installed in a(n) _____.
    a. SLB
    b. handhole
    c. mogul
    d. FS box

2. A handhole enclosure must be _____.
    a. fully sealed on all four sides regardless of type
    b. installed so that excavation is required for access
    c. reinforced with iron rebar
    d. provided with a cover that weighs over 100 lbs (45 kg) or requires tools to open

Review Questions

1. Where in the *NEC®* will you find the most information on pull and junction boxes? _____
   a. *NEC Article 240*
   b. *NEC Article 250*
   c. *NEC Article 314*
   d. *NEC Article 320*

Figure RQ01

2. The combination coupling in *Figure RQ01* is used to connect _____.
   a. two sections of flexible conduit
   b. flexible conduit to EMT
   c. EMT to rigid conduit
   d. flexible conduit to rigid conduit

3. When calculating the pull box size for straight pulls on systems operating at 1,000V or less, the length of the box must not be less than _____.
   a. two times the trade diameter of the largest raceway
   b. four times the trade diameter of the largest raceway
   c. six times the trade diameter of the largest raceway
   d. eight times the trade diameter of the largest raceway

4. If the largest trade diameter of a raceway entering a pull box is 3", and it is a straight pull for a system operating at 1,000V or less, the minimum size box allowed is _____.
   a. 20"
   b. 24"
   c. 30"
   d. 36"

5. For a straight pull, the length of a pull box housing unshielded conductors carrying over 1,000V must be _____.
   a. 24 times the OD of the largest conductor
   b. 32 times the OD of the largest conductor
   c. 36 times the OD of the largest conductor
   d. 48 times the OD of the largest conductor

6. For a straight pull, the length of a pull box housing shielded conductors carrying over 1,000V must be _____.
   a. 24 times the OD of the largest conductor
   b. 32 times the OD of the largest conductor
   c. 36 times the OD of the largest conductor
   d. 48 times the OD of the largest conductor

7. For an angled pull, the length of a pull box housing shielded conductors carrying over 1,000V must be _____.
   a. 24 times the OD of the largest conductor
   b. 32 times the OD of the largest conductor
   c. 36 times the OD of the largest conductor
   d. 48 times the OD of the largest conductor

8. A conduit body that has openings on four different sides plus an access opening is called a _____.
   a. Type X
   b. Type C
   c. Type T
   d. Type LL

9. Which of the following best describes how Type FS boxes should be installed?
   a. Holes should be drilled in back of the box for mounting screws.
   b. Holes should be drilled on the sides of the box only for mounting screws.
   c. No holes should be drilled in the box for mounting.
   d. Holes may be drilled only in existing installations.

10. Which of the following enclosures is best suited for splicing conductors?
    a. Pulling ell
    b. SLB
    c. Conduit body with no volume mark
    d. Mogul

Supplemental Exercises

1. A Type T conduit body is used to provide a junction point for __3__ intersecting conduits.
2. Boxes for use in hazardous locations must be rated as _explo. proof_.
3. Box installations in concrete slabs are considered __wet__ locations.
4. Conduit bodies, also called _____, provide access through a removable cover or covers to the interior of the system.
5. To distinguish between Type LL and Type LR conduit bodies, hold the fitting as if it were a(n) _pistol_ and see where the cover is situated.
6. For a straight pull using 600V conductors, the minimum length of the box must be at least _____ times the trade diameter of the largest raceway.
7. The minimum length for a junction box in which the 600V conductors are pulled at an angle and that contains two 3" conduits and one 4" conduit entering each side is _____.
8. A(n) _grounding_ locknut may be required if bonding jumpers are used inside of the box or enclosure.
9. The minimum length for a junction box with one 4" conduit on each side in which the 600V conductors are pulled straight through the box is __32__. 4"x8
10. When sizing pull and/or junction boxes for systems over 1,000V, you need to know the outside diameter of the _conductor_.

Trade Terms Introduced in This Module

Conduit body: A separate portion of a conduit or tubing system that provides access through a removable cover (or covers) to the interior of the system at a junction of two or more sections of the system or at a terminal point of the system. Boxes such as FS and FD or larger cast or sheet metal boxes are not classified as conduit bodies.

Explosion-proof: Designed and constructed to withstand an internal explosion without creating an external explosion or fire.

Handhole enclosure: An enclosure used with underground systems to provide access for installation and maintenance.

Junction box: An enclosure where one or more raceways or cables enter, and in which electrical conductors can be, or are, spliced.

Mogul: A type of conduit body with a raised cover to provide additional space for large conductors.

Pull box: A sheet metal box-like enclosure used in conduit runs to facilitate the pulling of cables from point to point in long runs, or to provide for the installation of conduit support bushings needed to support the weight of long riser cables, or to provide for turns in multiple-conduit runs.

Raintight: Constructed or protected so that exposure to a beating rain will not result in the entrance of water under specified test conditions.

Watertight: Constructed so that water will not enter the enclosure under specified test conditions.

Weatherproof: Constructed or protected so that exposure to the weather will not interfere with successful operation.

Additional Resources

This module presents thorough resources for task training. The following resource material is suggested for further study.

National Electrical Code® Handbook, Latest Edition. Quincy, MA: National Fire Protection Association.

Figure Credits

Greenlee / A Textron Company, Module Opener, Figures 12 and 13
Hubbell Wiegmann, Figure 1
Eaton's Crouse-Hinds Business, Figures 2, 5, and RQ01
Tim Ely, Figures 3, 10, 22, 23
John Traister, Figures 7, 14

Section Review Answer Key

Section 1.0.0

Answer	Section Reference	Objective
1. d	1.1.2	1a
2. d	1.2.3	1b

Section 2.0.0

Answer	Section Reference	Objective
1. b	2.1.0	2a
2. c	2.1.0	2a
3. d	2.2.0	2b
4. c	2.2.0	2b

Section 3.0.0

Answer	Section Reference	Objective
1. d	3.1.5	3a
2. d	3.2.0	3b

Section Review Calculations

2.0.0 Section Review

Question 1

(3" × 6) + 2" = 20"

The required distance is **20"**.

Question 2

(MD 78 × 6) + MD 53 = 521 mm

The required distance is **521 mm**.

Question 3

For an angled pull, multiply the OD by 36 to find the minimum length of the box:
2" × 36 = 72"

The minimum length is **72"**.

Question 4

For an angled pull, multiply the OD by 36 to find the minimum length of the box:
50 mm × 36 = 1,800 mm

The minimum length is **1,800 mm**.

26206-20
CONDUCTOR INSTALLATIONS

Objectives

When you have completed this module, you will be able to do the following:

1. Install cable in conduit systems.
   a. Plan the installation.
   b. Identify a pulling location and set up the cable reels.
   c. Prepare raceways for conductors.
   d. Install a pull line.
   e. Prepare the cable ends for pulling.
   f. Select cable-pulling equipment.
2. Set up for high-force cable pulling.
   a. Set up the feeding end.
   b. Support conductors.
   c. Pull cable in cable trays.
3. Identify cable limitations when pulling.
   a. Calculate the allowable tension on pulling devices.
   b. Calculate the allowable tension on conductors.
   c. Calculate the sidewall loading.

Performance Tasks

Under the supervision of the instructor, you should be able to do the following:

1. Prepare multiple conductors for pulling in a raceway system.
2. Prepare multiple conductors for pulling using a wire-pulling basket.

Trade Terms

Basket grip
Cable grip
Capstan
Clevis
Conductor support
Conduit piston

Fish line
Fish tape
Setscrew grip
Sheave
Soap
Tensile strength

Industry Recognized Credentials

If you are training through an NCCER-accredited sponsor, you may be eligible for credentials from NCCER's Registry. The ID number for this module is 26206-20. Note that this module may have been used in other NCCER curricula and may apply to other level completions. Contact NCCER's Registry at 888.622.3720 or go to www.nccer.org for more information.

> **NOTE:** NFPA 70®, *National Electrical Code*® and *NEC*® are registered trademarks of the National Fire Protection Association, Quincy, MA.

Contents

1.0.0 Installing Cable in Conduit Systems 1
   1.1.0 Planning the Installation 2
   1.2.0 Pulling Setups and Locations 4
   1.3.0 Preparing Raceways 6
   1.4.0 Installing a Pull Line 9
   1.5.0 Preparing Cable Ends 11
   1.6.0 Selecting Cable-Pulling Equipment 12
      1.6.1 Pulling Safety 16
      1.6.2 Types of Cable Pullers 16
      1.6.3 Cable-Pulling Instruments 18
2.0.0 High-Force Cable Pulling 23
   2.1.0 Feeding End 24
   2.2.0 Supporting Conductors 26
   2.3.0 Pulling Cable in Cable Trays 27
3.0.0 Cable Limitations When Pulling 31
   3.1.0 Allowable Tension on Pulling Devices 31
   3.2.0 Allowable Tension on Conductors 31
   3.3.0 Sidewall Loading 32

Figures and Tables

Figure 1	Basket grip	2
Figure 2	Wire dispensers	2
Figure 3	Multi-cable pull	3
Figure 4	Basic steps of a wire-pulling operation	4
Figure 5	Wire-pulling equipment checklist	5
Figure 6	Two methods of transporting cable reels	6
Figure 7	Proper and improper ways of transporting cable reels	7
Figure 8	Typical reel stands	8
Figure 9	Faults that may be detected with a conduit mandrel	8
Figure 10	Devices used to inspect, clean, and lubricate raceway systems	8
Figure 11	Obtaining the greatest possible conductor sweep in a pull box	9
Figure 12	Power fishing system	10
Figure 13	Blower/vacuum fish tape system used to blow a pull line in conduit	11
Figure 14	Types of pistons in common use	12
Figure 15	Various types of pulling grips used during conductor installation	13
Figure 16	Power cable puller capstan	15
Figure 17	Cable pullers	17
Figure 18	Puller setup for a down pull	19
Figure 19	Puller setup for an up pull	20
Figure 20	Cable length meter	20
Figure 21	Circuit tester	21
Figure 22	Typical cable-pulling setup	24
Figure 23	Unreel the cable along its natural curvature	25
Figure 24	Cable feed-in setups	25
Figure 25	Cable sheaves	26
Figure 26	Never allow a polygon curvature to occur in a cable-pulling operation	26
Figure 27	Conductors supported with wedges	27
Figure 28	Typical cable tray cable-pulling arrangement	28
Figure 29	Sidewall loading	34
Figure 30	Sample conduit run	34
Table 1	Example Pulling Forces for Various Wraps	16
Table 2	Spacings for Conductor Supports in Vertical Raceways [Data from *NEC Table 300.19(A)*]	27
Table 3	Physical Limitations of Cable	32
Table 4	Angle of Bend vs. Coefficients of Friction	32
Table 5	Maximum Sidewall Pressures for Various Types of Cable	34

This page is intentionally left blank.

Section One

1.0.0 Installing Cable in Conduit Systems

Objective

Install cable in conduit systems.
  a. Plan the installation.
  b. Identify a pulling location and set up the cable reels.
  c. Prepare raceways for conductors.
  d. Install a pull line.
  e. Prepare the cable ends for pulling.
  f. Select cable-pulling equipment.

Trade Terms

Basket grip: A flexible steel mesh grip that is used on the ends of cable and conductors for attaching the pulling rope. The more force exerted on the pull, the tighter the grip wraps around the cable.

Cable grip: A device used to secure ends of cables to a pulling rope during cable pulls.

Capstan: The turning drum of a cable puller on which the rope is wrapped and pulled. An increase in the number of wraps increases the pulling force of the cable puller.

Clevis: A device used in cable pulls to facilitate connecting the pulling rope to the cable grip.

Conduit piston: A cylinder of foam rubber that fits inside the conduit and is then propelled by compressed air or vacuumed through the conduit run to pull a line, rope, or measuring tape. Also called a *mouse*.

Fish line: Light cord used in conjunction with vacuum/blower power fishing systems that attaches to the conduit piston to be pushed or pulled through the conduit. Once through, a pulling rope is attached to one end and pulled back through the conduit for use in pulling conductors.

Fish tape: A flat iron wire or fiber cord used to pull conductors or a pulling rope through conduit.

Setscrew grip: A cable grip, usually with built-in clevis, in which the cable ends are inserted in holes and secured with one or more setscrews.

Soap: Slang for wire-pulling lubricant.

After the conduit system has been installed, the next step is to pull in the conductors. The simplest method of cable installation involves running stiff but flexible fish tape through a conduit run and then attaching and pulling in the conductors.

There are three types of fish tape: steel, nylon, and fiberglass. They also come in different weights for various applications. The proper size and length of the fish tape, as well as the type, should be one of the first considerations. For example, if most of the runs between branch circuit outlets are short, a shorter fish tape can be used and it will not have the weight and bulk of a larger tape. When longer runs are encountered, the required length of the fish tape should be enclosed in one of the metal or plastic fish tape reels. This way, the fish tape can be rewound on the reel as the pull is being made to avoid having an excessive length of tape lying around on the floor or deck.

> **WARNING!**
> Never fish a steel tape through or into enclosures or raceways containing an energized conductor.

When several bends are present in the raceway system, the insertion of the fish tape may be made easier by using flexible fish tape leaders on the end of the fish tape.

After the fish tape is inserted in the raceway system, the conductors must be firmly attached by some approved means. On short runs, where only a few conductors are involved, all that is necessary is to strip the insulation from the ends of the wires, bend these ends around the hook in the fish tape, and securely tape them in place. Where several wires are to be pulled together, the wires should be staggered and the fish tape securely taped at the point of attachment so that the overall diameter is not increased any more than is absolutely necessary.

The basket grip (*Figure 1*) is available in many sizes for almost any size and combination of conductors. It is designed to hold the conductors firmly to the fish tape and can save time and trouble that would be required when taping wires.

In all but very short runs, the wires should be lubricated with wire lubricant before attempting the pull, as well as during the pull. Some of this lubricant should also be applied to the inside of

Figure 1 Basket grip.

the conduit. Always follow the wire manufacturer's recommendations when selecting lubricants. Many types of cable now have no-lube insulation that provides a very slick surface and does not require lubrication. However, using the proper lubricant decreases the force needed even when no-lube conductors are being pulled.

Wire dispensers are great aids in keeping the conductors straight and facilitating the pull. Many different types of wire dispensers are available to handle virtually any size spool of wire or cable. Some dispensers are stationary, while others have casters that make it easy to move heavy spools to the pulling location (*Figure 2*). Wheeled dispensers are sometimes called wire caddies.

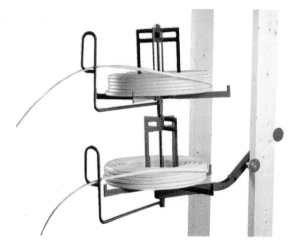

CABLE DISPENSER

WIRE CADDY

Figure 2 Wire dispensers.

Using Fish Tape

Metal fish tape can have sharp edges or burrs. To avoid injury, wear heavy-duty gloves when using fish tape. To use the tape, hold the reel in one hand and pull the tape out with the other hand. Make sure the fish tape has no kinks or bends.

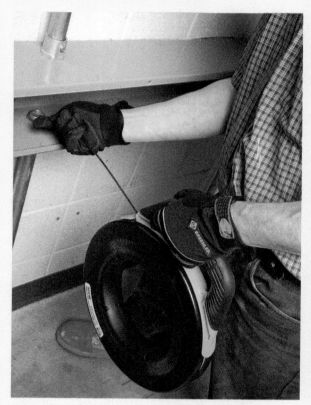

Figure Credit: Greenlee / A Textron Company

1.1.0 Planning the Installation

The importance of planning any wire-pulling installation cannot be over-stressed. Proper planning will make the work go easier and labor will be saved.

Large sizes of conductors are usually shipped on reels, involving considerable weight and bulk. Consequently, setting up these reels for the pull, measuring cable run lengths, and similar preliminary steps will often involve a relatively large amount of the total cable installation time. Therefore, consideration must be given to reel setup space, proper equipment, and moving the cable reels into place.

Whenever possible, the conductors should be pulled directly from the shipping reels without pre-handling them. This can usually be done through proper coordination of the ordering of

Nonconductive Fish Tape

When fishing conductors in a conduit or raceway that already contains existing energized (live) conductors, the safest method is to turn off and lock out/tag the power sources for all the live conductors. However, in some rare instances, fishing a conductor through a conduit or raceway containing other live conductors may be unavoidable. In this case, always request your supervisor's approval before proceeding and always use a nonconductive fish tape made of nylon or fiberglass.

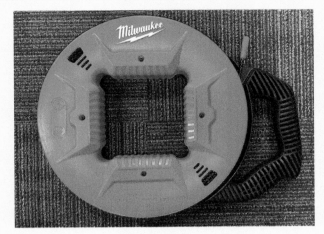

Figure Credit: Photo by Josiha Schuh

the conductors with the job requirements. While doing so requires extremely close checking of the drawings and on-the-job measurements (allowing for adequate lengths of conductors in pull boxes, elbows, troughs, connections, and splices), the extra effort is well worth the time to all involved.

When the lengths of cable have been established, the length of cable per reel can be ordered so that the total length per reel will be equal to the total of a given number of raceway lengths and the reel so identified.

In most cases, the individual cables of the proper length for a given number of runs are reeled separately onto two or more reels at the factory, depending on the number of conductors in the runs.

When individual conductors are shipped on separate reels, it is necessary to set up for the same number of reels as the number of conductors to be pulled into a given run, as shown in *Figure 3*.

As an extra precaution against error in calculating the lengths of conductors involved, it is a good idea to measure all runs with a fish tape before starting the cable pull, adding for makeup to reach the terminations, and accounting for discarding the cable underneath the pulling sleeve or pulling wrap, as it will have been excessively stressed during the pull. Check these totals against the totals indicated on the reels. If the feeder raceways have been installed relatively early during building construction, waiting to order the cables until the raceways can be measured may not delay the final completion of the installation Several fish tapes with laser foot markings are now available.

When pulling conductors directly from the reels, care must be taken that each given run be cut from the reel so that there is minimal waste. In other words, preclude the possibility of the final run of cable taken from the reel being too short for that run. The basic procedure for a wire-pulling operation is shown in *Figure 4*.

Figure 3 Multi-cable pull.

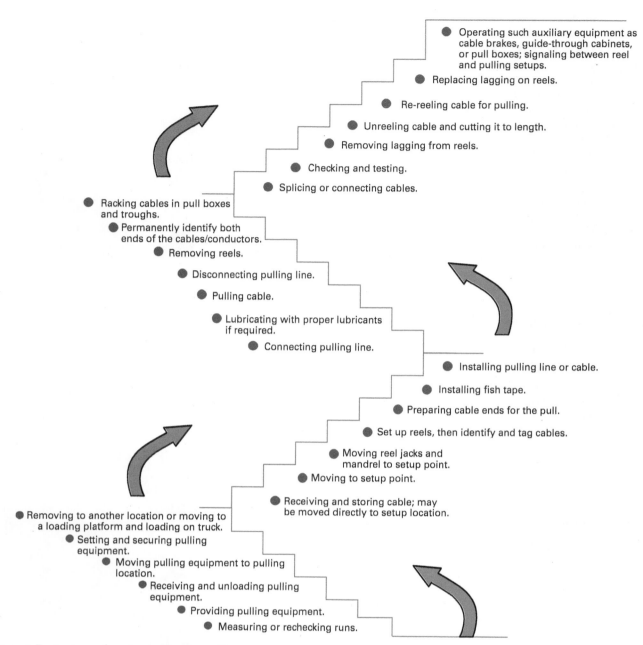

Figure 4 Basic steps of a wire-pulling operation.

The proper use of appropriate equipment is crucial to a successful cable installation. The equipment needed for most installations is shown in the checklist in *Figure 5*. Some projects may require all of these items, while others may require only some of them. Each cable-pulling project must be taken on an individual basis and analyzed accordingly. Seldom will two pulls require identical procedures.

> **NOTE**
> Think of everything that can go wrong and take every precaution.

1.2.0 Pulling Setups and Locations

Each job will have to be judged separately as to the best location for pulling setups; the number of setups should be reduced to a minimum in line with the best direction of the pull. The pulling location for a particular job is determined by the weight of the cable, height of the pull, practicality of moving the equipment to the pulling location, number of setups required, and the location of any bends. Also, a separate setup might have to be made at the top of each rise, whereas a single setup might be made at a ground floor pull box location from which several feeders are served with the same size and number of conductors.

EQUIPMENT CHECKLIST

- ☐ PORTABLE ELECTRIC GENERATOR
- ☐ EXTENSION CORDS AND GFCI
- ☐ PUMP, DIAPHRAGM
- ☐ MAKEUP BLOWER AND HOSE
- ☐ MANHOLE COVER HOOKS
- ☐ WARNING FLAGS, SIGNS
- ☐ ELECTROSTATIC kV TESTER
- ☐ ELECTRIC SAFETY BLANKETS AND CLAMPS
- ☐ RADIOS OR TELEPHONES
- ☐ GLOVES
- ☐ FLOOD LAMPS
- ☐ FISH TAPE OR STRING BLOWER/VACUUM
- ☐ HAND LINE
- ☐ DUCT-CLEANING MANDRELS
- ☐ DUCT-TESTING MANDRELS
- ☐ CAPSTAN-TYPE PULLER
- ☐ SNATCH BLOCKS
- ☐ SHORT ROPES FOR TEMPORARY TIE-OFFS
- ☐ GUIDE-IN FLEXIBLE TUBING (ELEPHANT TRUNKS)
- ☐ SEVERAL WIRE ROPE SLINGS OF VARIOUS LENGTHS
- ☐ SHACKLES/ROPE CLEVIS
- ☐ GANG ROLLERS WITH AT LEAST 4' EFFECTIVE RADIUS
- ☐ HAND WINCHES
- ☐ MANHOLE EDGE SHEAVE
- ☐ PULLING ROPE
- ☐ SWIVELS
- ☐ BASKET GRIP PULLERS
- ☐ 0-1/5/10 KIP DYNAMOMETER
- ☐ REEL ANCHOR
- ☐ REEL JACKS
- ☐ CABLE CUTTERS
- ☐ LINT-FREE RAGS
- ☐ CABLE-PULLING LUBRICANTS
- ☐ PRELUBING DEVICES
- ☐ PLYWOOD SHEETS
- ☐ DIAMETER TAPE
- ☐ 50' MEASURING TAPE
- ☐ SILICONE CAULKING (TO SEAL CABLE ENDS)

Figure 5 Wire-pulling equipment checklist.

The weight of larger conductors is frequently the most important consideration when locating equipment. When pulling wire up, the wire weight is working against you. However, if you place the wire at the top of the pull so the weight works for you, enough braking must be available to stop the pull when necessary.

The location of the pulling equipment determines the number of workers required for the job. A piece of equipment that can be moved in and set up on the first floor by four workers in an hour's time may require six workers working two hours when set up in basements or parking levels of buildings. Therefore, when planning a cable pull, make certain enough workers are on hand to adequately handle the installation.

When reels of cable arrive at the job site, it is best to move them directly to the setup location if at all possible. This prevents having to handle the reels more than necessary. However, if this is not practical, arrangements must be made for storage until the cable is needed.

The exact method of handling reels of cable depends on their size and the available tools and equipment. It is a simple operation for a few workers to roll cable reels from a loading platform to a first floor setup, whereas moving them to upper floors involves more handling and usually requires a crane or other hoists. In addition, the reel jacks have to be moved to the setup point when a downward pull is made. For reels up to 20" (508 mm) wide × 40" (1,016 mm) in diameter, a cable reel transporter can be used to transport the cable

Measuring Tape

A waterproof polyester tape with permanent markings is available on reels for use in measuring conduit/raceway runs. It can be fished through the run manually or using a power fishing system.

Figure Credit: Greenlee / A Textron Company

Think About It
Multiple Reels

Why are multiple reel pulling setups used?

reel; it also acts as a dispenser during the pulling operation. When available, a forklift is ideal for lifting and transporting cable reels (*Figure 6*).

However, for very large reels (48" or 1,219 mm in diameter), a crane or similar hoisting apparatus is usually necessary for lifting the reels onto reel jacks supported by jack stands to acquire the necessary height. *Figure 7* shows a summary of proper and improper ways to transport reels of wire or cable on the job site.

Figure 8 shows a reel stand. For a complete setup, two stands and a spindle are required for each reel. The reel stands or jacks are available in various sizes to accommodate reel diameters up to 96" (2,438 mm). Extension stands used in conjunction with reel stands can accommodate larger reels.

Reel-stand spindles are available in various lengths and diameters for carrying loads up to 15,000 lbs (6,804 kg).

1.3.0 Preparing Raceways

Another preliminary step before pulling conductors in raceway systems is to inspect the raceway itself. Few things are more frustrating than to pull four 1,000 kcmil conductors through a conduit and find out when the pull is almost done that the conduit is blocked or damaged. Such a situation usually requires pulling the conductors back out, repairing the fault, and starting all over again.

These problems are not too serious with exposed banks of conduit, as the conduit can usually be separated at the fault, the fault corrected, and another piece of conduit installed, using unions if necessary. However, in underground conduit runs—especially those encased in concrete—the situation can be both time-consuming and costly.

A test pull will detect any hidden obstructions in the conduit before the pull (*Figure 9*). Go and no-go steel and aluminum mandrels are available for pulling through runs of conduit before the cable installation. Mandrels should be approximately 80% of the conduit size (twice the 40% fill factor). If any obstructions are found, they can be corrected before wasting time on an installation that might result in conductor damage and the possibility of having to re-pull the conductors.

Figure 10 shows several devices used to inspect raceway systems, as well as to prepare raceways for easier and safer conductor pulls. The conduit swab in *Figure 10* may be used ahead of the conductors during a pull. Its main purpose is to swab out debris from the raceway and spread a uniform film of pulling compound inside the conduit for easier pulling. The flexible mandrel is also used to clean the conduit and check the run for obstructions. The flexible discs make it easier to pull around tight bends.

The conduit brush in *Figure 10* helps clean and polish the interior of conduit before pulling the cable. Such brushes remove sand and other light obstructions. Note that this brush has a pulling eye on one end and a twisted eye on the opposite end, enabling it to be pushed or pulled through the conduit.

CABLE REEL TRANSPORTER   FORKLIFT

Figure 6 Two methods of transporting cable reels.

ON THE RIMS OF THE SPOOL (MOVING EQUIPMENT DOES NOT COME INTO CONTACT WITH CABLE)

ON THE FLAT SIDE OF THE SPOOL OR ON THE CABLE (MOVING EQUIPMENT COMPRESSESS INSULATION AND MAY DAMAGE CABLE)

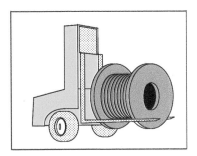
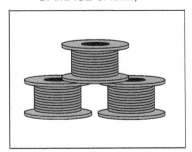
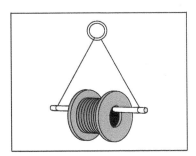

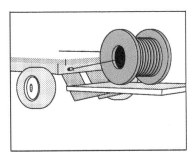
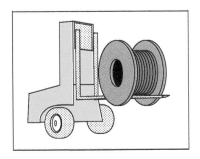
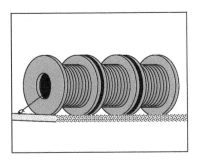
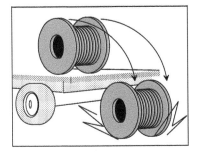

Figure 7 Proper and improper ways of transporting cable reels.

Figure 8 Typical reel stand.

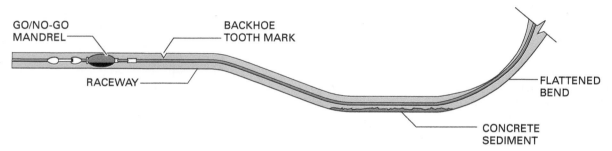

Figure 9 Faults that may be detected with a conduit mandrel.

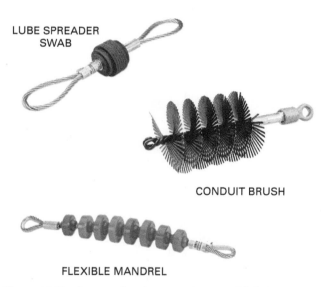

Figure 10 Devices used to inspect, clean, and lubricate raceway systems.

One final step before starting the pull is to measure the length of the raceway, including all turns in junction boxes and similar equipment. A fish tape may be pushed through the raceway system and a piece of tape used to mark the end. When it is pulled back out, a tape measure may be used to measure the exact length. An easier way, however, is to use a power fishing system to push or pull a measuring tape through the conduit run. Details of this operation are explained in the section titled "Installing a Pull Line."

When measuring the conductor length, be sure to allow sufficient room where measurements are made through a pull box. Conductors should enter and leave pull boxes in such a manner as to allow the greatest possible sweep for the conductors. Large conductors are especially difficult to bend but with proper planning, you can simplify the feeding of these conductors from one conduit to another.

For example, if a conduit run makes a right-angle bend through a pull box, the conduit for a given feeder should come into the box at the lower left-hand corner and leave diagonally opposite at the upper right-hand corner, as shown in *Figure 11*. This gives the conductors the greatest possible sweep with the box, eliminating sharp bends and consequent damage to the conductor insulation.

Runs should also be calculated to allow for splices and terminations in junction boxes, panelboards, motor-control centers, the cable discarded under the sleeve or pulling wrap, and other needs that increase the cable length.

1.4.0 Installing a Pull Line

Wire-pulling ropes have come a long way from the hemp ropes used by electricians a couple of decades ago. The most common wire-pulling ropes on the market include the following:

- Nylon
- Polypropylene
- Multiplex polyester
- Double-braided polyester composite

The type of rope selected will depend mainly on the pulling load; that is, the weight of the cable, the length of the pull, and the total resistance to the pull. For example, Greenlee's Multiplex cable-pulling rope is designed for low-force cable pullers. It has a low stretch characteristic that makes it suitable for pulls up to 2,000 lbs (8.9 kilonewtons or kN). Lengths are available from 100' to 1,200' (30 m to 365 m). Greenlee's double-braided composite rope for high-force cable pullers is designed for pulls up to 6,500 lbs (28.9 kN).

> **CAUTION**
> Any equipment associated with the pull must have a working load rating in excess of the force applied during the pull. All equipment must be used and mounted in strict accordance with the manufacturer's instructions.

Care must be used in selecting the proper rope for the pull, and then every precaution must be taken to make sure that the cable-pulling force does not exceed the rope capacity. There are several reasons for this, but the main one is safety. For example, a typical nylon rope can stretch 40' (12.2 m) before breaking, releasing 200,000 foot-pounds (271,164 newton-meters) of energy in the process. Think of the damage this amount of energy could do to a raceway system and to nearby workers if the rope broke under this amount of force.

Pull lines can be manually fished in with a steel fish tape, but much time can be saved by using a blower/vacuum fish tape system. In this type of system, a reusable conduit piston—sometimes referred to as a mouse or missile—is blown with air pressure or vacuumed through the run. The foam piston is sized to the conduit and has a loop on both ends. In most cases, fish line or measuring tape is attached to the piston as it is blown or vacuumed through the conduit run. The measuring tape serves two purposes: it provides an accurate measurement of the conduit run, and the tape is used to pull the cable-pulling rope into the conduit run. In some cases, if the run is suitable, the pulling rope is attached directly to the piston and vacuumed into the run. *Figure 12* (C) shows a blower/vacuum fish tape system being used to vacuum a pull line in a conduit while *Figure 13* shows the same apparatus blowing the piston through. Most of these units provide enough pressure to clean dirt or water from conduit during the fishing operation. *Figure 14* shows two types of pistons used with this system. The one on the right utilizes air-guide vanes to prevent the piston from tumbling inside the larger sizes of conduit.

There are certain precautions that should be taken when using a power fish tape system:

- Read and understand all instructions and warnings before using the tool.
- Never fish in runs that might contain live power.
- Be prepared for the unexpected. Make sure your footing and body position are such that you will not lose your balance in any unexpected event.
- Use blower/vacuum systems only for specified light fishing and exploring the raceway system.

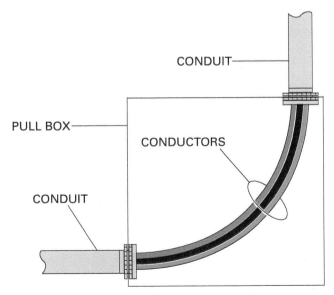

Figure 11 Obtaining the greatest possible conductor sweep in a pull box.

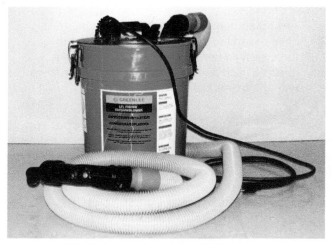
(A) BLOWER/VACUUM FISH TAPE SYSTEM

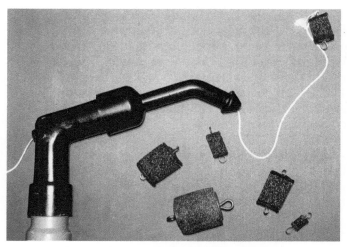
(B) BLOWER/VACUUM FISH TAPE SYSTEM AND CONDUIT PISTON

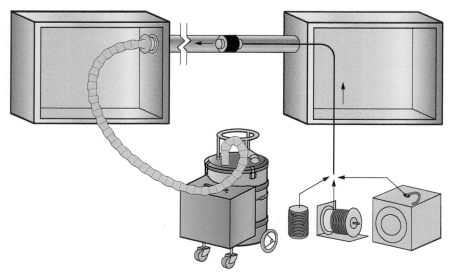
(C) BLOWER/VACUUM FISH TAPE SYSTEM IN USE

Figure 12 Power fishing system.

Pulling Eyes

Smaller conductors may require only a pulling eye attached to the fish tape or pulling line. To attach the pulling eye, the conductors are first stripped, exposing a length of bare wire. These bare wires are inserted through the eye and twisted back onto themselves. The exposed twisted wires of the conductors can then be wrapped with smaller copper wire to prevent them from untwisting. They are then completely taped with three layers of electrical tape, starting from the conductor insulation, to prevent snagging as they are drawn through the conduit run.

- Never use pliers or other devices that are not designed to pull a fish tape. They can kink or nick the tape, creating a weak spot.

1.5.0 Preparing Cable Ends

The pulling-in line or cable must be attached to the cable or conductors in such a manner that it cannot part from the cable during the pull. Two common methods include direct connection with the cable conductors themselves and connection by means of pulling grips or baskets placed over the cable or group of conductors. The use of the proper type of grip or basket will facilitate the pull, but in many cases—especially on long pulls—workers prefer to use a three- or a four-hole cable grip with setscrews that secure each conductor to the pulling block. *Figure 15* shows several types of pulling grips.

Basket grips are easy to use and quick to connect, especially for smaller wires. Note that the no-lube insulation on certain types of cable requires that the wires be stripped and the basket applied to the conductors rather than the insulation.

Most pulling blocks have a rope clevis as an integral part of the block. However, when using pulling grips or baskets in 2" or larger conduit, a rope clevis is normally used to facilitate connecting the pulling rope to the wire grip. Two types are currently used: the straight clevis and the swivel clevis. When the type of grip being used requires that the ends of the cable insulation be stripped from the conductors, conventional methods are used—the same as for terminating conductors for splices or connections to terminal lugs in panelboards, switchgear, etc.

In general, the ends of conductors should first be trimmed. Cable cutters capable of cutting conductors through 1,000 kcmil save workers time over using a hacksaw. There are also cable strippers, adjustable from 1/0 AWG through 1,000 kcmil, that handle midspan and termination stripping of THHN, THWN, XHHW, and similar insulation. These tools are excellent for stripping conductors for use in setscrew clamp-type pulling grips.

To use a stripping tool, first mark the required distance from the ends of the conductors, using the pulling grip as a gauge. Close the jaws of the stripping tool on the cable and twist. These self-feeding devices ensure positive progression down the cable to any position desired. To stop stripping, apply back pressure to the stripper until a full circle has been completed.

When the conductor ends have been stripped, insert one conductor at a time into the setscrew grip. Make sure the end of the bare conductor is firmly in place, and then tighten the setscrews with a hex wrench. Continue to the next conductors until all conductors are secured in the pulling grip.

When using a stripping tool for the first time, make sure that you read and understand all instructions and warnings before using the tool. The following should also be observed when working with cable terminators:

- Wear eye protection.
- Inspect tools before using, and replace damaged, missing, or worn parts.
- Be prepared for the unexpected. Make sure your footing and body position are such that you will not lose your balance.
- Use only the type and size material in the stated capacity.
- Do not use the tool on or near live circuits.

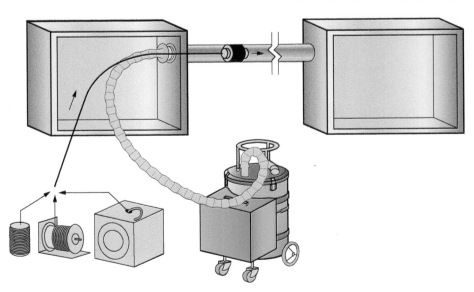

Figure 13 Blower/vacuum fish tape system used to blow a pull line in conduit.

1.6.0 Selecting Cable-Pulling Equipment

Except for short cable pulls, hand-operated or power-operated cable pullers or winches are used to furnish the pulling power. In general, cable reels are set in place at one end of the raceway system, and the cable puller is set up at the opposite end. One end of the previously installed pulling rope is attached to a clevis, basket, or other cable grip to which the conductors are attached. The other end of the rope (at the cable puller) is wrapped around the rotating drum capstan on the cable puller (*Figure 16*).

> **WARNING!** Always use a wire-pulling lubricant that is compatible with the type of cable being pulled. Failure to do so can result in unsafe pulling forces and cable damage. Check the cable manufacturer's recommendations, and always contact the lubricant manufacturer about the compatibility of lubricants with specific cables. Also check the product's safety data sheet (SDS) for any applicable safety requirements.

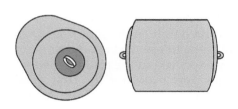
FLEXIBLE FOAM PISTON FOR AIRTIGHT SEAL

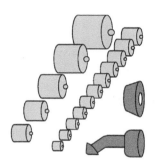
PISTONS ARE AVAILABLE IN SIZES FROM ½" TO 6"

FINS ARE SOMETIMES UTILIZED ON PISTONS FOR LARGER SIZES OF CONDUIT TO KEEP THE PISTONS FROM TUMBLING

(A) TYPES OF PISTONS

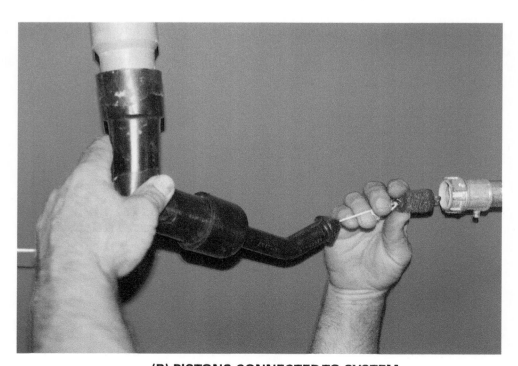

(B) PISTONS CONNECTED TO SYSTEM

Figure 14 Types of pistons in common use.

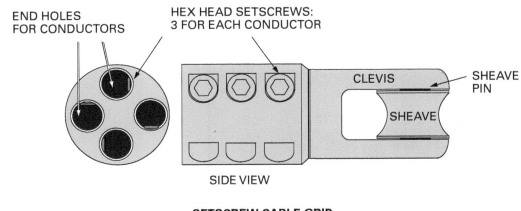

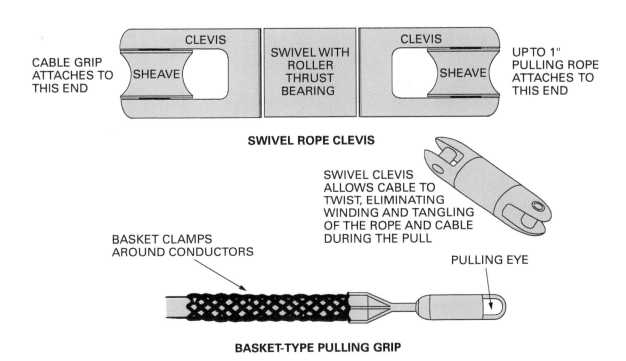

Figure 15 Various types of pulling grips used during conductor installation.

The appropriate wire-pulling lubricant—sometimes referred to as soap—is inserted into the empty conduit as well as wiped thoroughly onto the front of the cable. One or more operators are on hand to feed the cable, while one worker is usually all that is required on the pulling end.

The number of wraps on the puller drum decides the amount of force applied to the pull. For example, the operator needs to apply only 10 lbs (0.04 kN) of force to the pulling rope in all cases. With this amount of force applied by the operator, and with one wrap around the rotating drum, 21 lbs (0.09 kN) of pulling force will be applied to the pulling rope; 2 wraps, 48 lbs (0.21 kN); 3 wraps, 106 lbs (0.47 kN), etc. This principle is known as the capstan theory and is the same principle applied to block-and-tackle hoists or the lone cowboy who is able to rope and hold a large bronco by wrapping his lariat around the center post in a corral. *Table 1* gives the amount of pulling force with various numbers of wraps when the operator applies only 10 lbs (0.04 kN) of tailing force for a particular model puller.

Fish Poles

Rigid fish poles can be used in areas such as over drop-in ceilings where traditional fish tape might get caught up on ductwork or the ceiling grid. Glow-in-the-dark fish poles are also available.

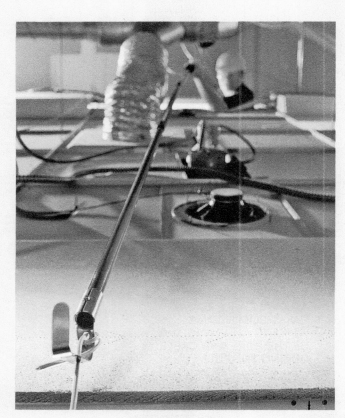

FISH POLE IN USE

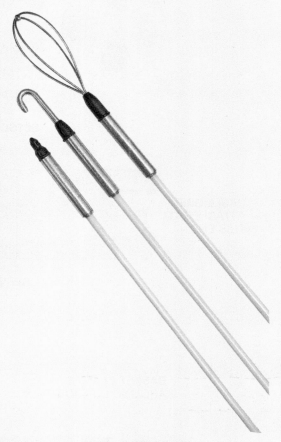

GLOW-IN-THE-DARK FISH POLES

Figure Credit: Greenlee / A Textron Company

Cable Blowing

Lightweight cables, especially fiber optic cable, can be floated through conduit using a special high-pressure blower unit.

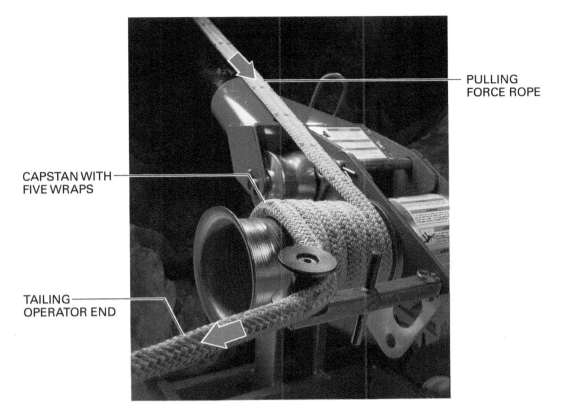

Figure 16 Power cable puller capstan.

Grips and Swivels

Conductors can be attached to a pulling line using various methods, including setscrew grips or basket grips. No matter which method is used, always insert a swivel of some sort between the pulling line or fish tape and the conductors to alleviate twisting of the conductors. Breakaway swivels are available that release at a specified tension to avoid damage to the conductors if excess force is applied or if the conductors get hung up during the pull.

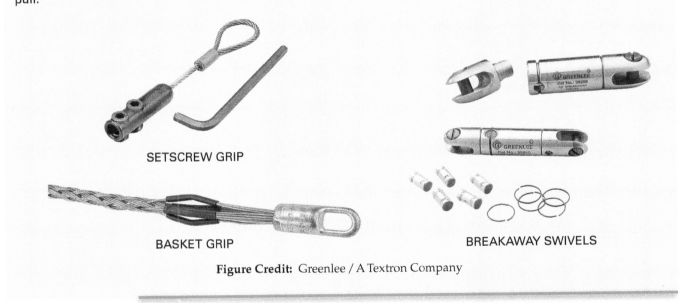

Figure Credit: Greenlee / A Textron Company

Table 1 Example Pulling Forces for Various Wraps

Number of Wraps	Operator Force (Lbs)	Pulling Force (Lbs)
1	10	21
2	10	48
3	10	106
4	10	233
5	10	512
6	10	1,127
7	10	2,478

1.6.1 Pulling Safety

Adhere to the following precautions when using power cable-pulling equipment:

> **NOTE**
> The strain placed on the wrapped cable during the pull may weaken this part of the cable. Be sure to discard the wrapped cable after making the pull.

> **CAUTION**
> Make absolutely certain that all communications equipment is in working order before the pull. Place personnel at strategic points with operable communications equipment to stop and start the pull as conditions warrant. Anyone involved with the pull has the authority to stop the pull at the first sign of danger to personnel or equipment.

- Read and understand all instructions and warnings before using the tool.
- Use compatible equipment; that is, use the properly rated cable puller for the job, along with the proper rope and accessories.
- Always be prepared for the unexpected. Make sure your footing and body position are such that you will not lose your balance. Keep out of the direct line of force.
- Make sure all cable-pulling systems, accessories, and rope have the proper rating for the pull.
- Inspect tools, rope, and accessories before using; replace damaged, missing, or worn parts.
- Personally inspect the cable-pulling setup, rope, and accessories before beginning the cable pull. Make sure that all equipment is properly and securely rigged.
- Make sure all electrical connections are properly grounded and adequate for the load.
- Use cable-pulling equipment only in uncluttered areas.
- Remind all workers that anyone can stop the pull if it seems unsafe.

1.6.2 Types of Cable Pullers

There are several types of cable pullers on the market. Most, however, operate on the same principle. The self-contained hand-crank wire puller in *Figure 17* (A) is designed to pull up to 1,500 lbs (6.7 kN) with only 30 lbs (0.13 kN) of handle force. It is used on projects where only a few cable runs need to be pulled.

Power cable pullers are available in various sizes and pulling capacities, from lightweight units that can be set up and used by a single person to heavy-duty units for high-force cable pulling. *Figure 17* (B) shows two typical units.

> **WARNING!**
> Make absolutely certain that all cable-pulling equipment is anchored properly. Follow the manufacturer's recommendations for the type of puller being used.

Be Sure to Check the Pulling Rope Rating

Many pulling ropes available today are made of synthetic materials designed for pulling by hand or with a winch-type puller. However, if you are using certain synthetic pulling ropes on friction-type capstan power pullers, make sure the rope is rated by the manufacturer for this use so that it will be able to withstand the heat generated by any extended slippage of the rope on the capstan. During high-force pulling operations, melting damage and possible failure of unrated synthetic rope can occur quickly during periods of capstan slippage.

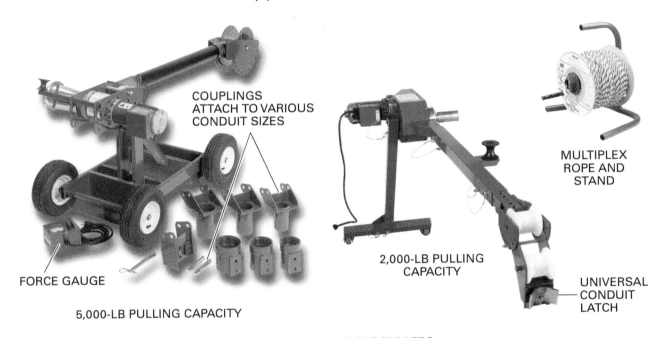

Figure 17 Cable pullers.

Guiding and Lubricating Conductors

When guiding conductors into a conduit/raceway during a pull, the conductors may tend to twist, overlap, or become crossed during the pulling operation, especially if fed from boxes instead of reels. Excessive twisting, overlaps, or crossovers can cause binding of the conductors in conduit/raceway turns, create bunching obstructions in the conduit/raceway, and contribute to insulation burns. Operators at the feeding end of the pull must attempt to keep the conductors as straight as possible during the pulling operation, and lubricants should be applied liberally during the pull to allow the conductors to slide past each other and the sides of the run.

Figure Credit: Greenlee / A Textron Company

Figure 18 shows the basic setup for a down pull using the portable puller shown in *Figure 17*. To set up for the pull, first adjust the elbow and boom to the correct angle using the attached pins. The elbow attaches to the conduit using the locking knob. This unit has a universal conduit latch so it will attach to all conduit sizes without having to change couplings.

Figure 19 shows a setup for an up pull. The setup is similar to the one described for the down pull except the elbow is attached to the bottom conduit rather than the top conduit.

1.6.3 Cable-Pulling Instruments

There are several instruments used in conjunction with cable-pulling operations. Because details of operation vary with the manufacturer, these instruments will be discussed only briefly in this module. Study the operation manuals for all instruments before using them. The following are some examples of cable-pulling instruments:

- *Cable length meter* – Cable length meters (*Figure 20*) are available for direct reading in lengths up to 20,000' (6,096 m) or more. Most are calibrated for different wire sizes whereas the sizes are selected with a selector switch on the instrument. Controls may also be set for either copper or aluminum conductors. These instruments are ideal for determining the exact length of conductors on reels before making a pull.

- *Circuit tester/wire sorter* – This instrument is used to trace conductors on de-energized circuits. See *Figure 21*. One lead of the transmitter is attached to ground while the other lead is attached to the wire being traced. The receiver is then taken to the opposite end of the circuit, where it will show a strong signal on the traced conductor. That wire is marked and other wires are traced in the same way.

- *Dynamometer (force gauge)* – This type of meter is designed to read dynamic friction (pulling force) during a cable pull. Many are designed for a specific cable-pulling tool and are shipped as an integral or optional part of the cable puller. For example, one common cable puller electronically displays both actual speed and actual force while running. It automatically shuts down when maximum preset force is reached to prevent damage to the cable being pulled. Others are portable units for use with any type of cable-pulling equipment. Such instruments are invaluable for avoiding damage to conductors during high-force cable pulls. When the instrument indicates that the maximum pulling force has been reached, the pull can be stopped before damage occurs to the cable or conductors.

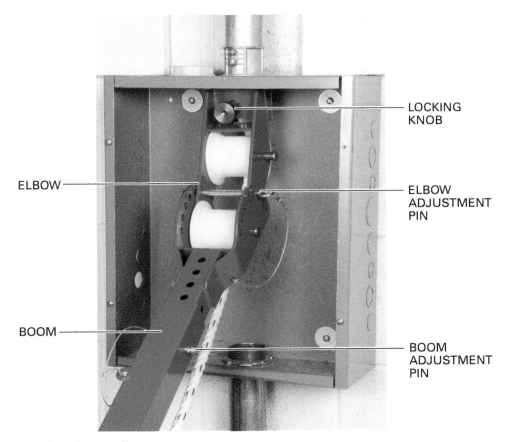

Figure 18 Puller setup for a down pull.

Case History

Cable Pulling

At a large commercial job site recently, workers began a complex pull shortly before the end of the day. At 5 PM, they left for the day, having completed only a portion of the pulling operation. The next day, they resumed work promptly at 8 AM, but the pulling lubricant had already dried in the conduit. As a result of the excess friction on the pull, the rope broke, delaying the job and causing extensive rework.

The Bottom Line: Don't stop a pull in the middle of a job.

Figure 19 Puller setup for an up pull.

Figure 20 Cable length meter.

Figure 21 Circuit tester.

Lubrication Systems

Cable-pulling lubrication systems are available to reduce mess and simplify the lubrication process. The unit shown here consists of a portable lubrication pump and applicators that attach to various conduit sizes.

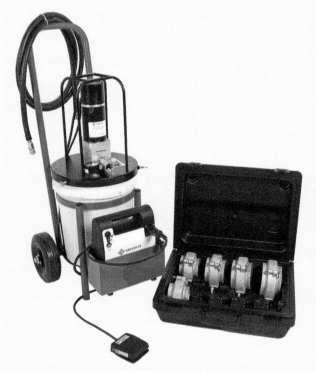
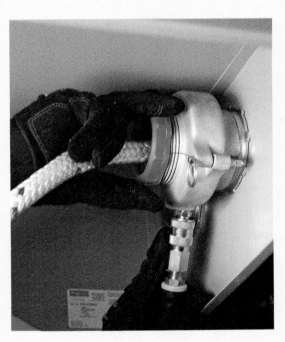

(A) LUBRICATION SYSTEM

(B) APPLICATOR ATTACHED TO CONDUIT SYSTEM

Figure Credit: Greenlee / A Textron Company

1.0.0 Section Review

1. When ordering cable for a conductor installation, it is best to _____.
    a. estimate the length required to avoid delays
    b. order three times the anticipated length to account for terminations
    c. wait until the raceways have been installed so they can be measured
    d. use a standard formula based on the square footage of the installation

2. A 48" reel should be moved using a _____.
    a. crane
    b. scissor lift
    c. dolly
    d. bucket truck

3. Test mandrels should be sized at _____.
    a. 20% of the conduit size
    b. 40% of the conduit size
    c. 80% of the conduit size
    d. 100% of the conduit size

4. A conduit piston must be _____.
    a. matched to the size of the conduit
    b. discarded after use
    c. attached to the conductors
    d. twice as wide as the conduit

5. Long pulls of large conductors are most likely to require the use of _____.
    a. taped connections
    b. basket grips
    c. setscrew grips
    d. heat-shrink grips

6. Three wraps on a capstan will provide _____.
    a. 10 pounds of pulling force
    b. 21 pounds of pulling force
    c. 48 pounds of pulling force
    d. 106 pounds of pulling force

Section Two

2.0.0 HIGH-FORCE CABLE PULLING

Objective

Set up for high-force cable pulling.
a. Set up the feeding end.
b. Support conductors.
c. Pull cable in cable trays.

Trade Terms

Conductor support: The act of providing support in vertical conduit runs to support the cables or conductors. The *NEC*® gives several methods by which cables may be supported, including wedges in the tops of conduits and supports to change the direction of cable in pull boxes.

Sheave: A pulley-like device used in cable pulls in both conduit and cable tray systems.

High-force cable pulling is not done every day. It is actually a small part of any electrical raceway installation. Workers may take days, weeks, or even months to install the complete raceway system; the cable-pulling operation typically takes less time.

There are several items to consider during high-force cable pulling:

- The design of the raceway system must be studied by consulting the working drawings and by examining the installed system to ensure that it is properly installed for the type of conductors that will be pulled through the system. Items to consider include conduit sizes, number and size of conductors, number of bends, sufficient pull boxes, and adequate supports.
- The conductors and pulling rope must be matched to the proper pulling equipment.
- A decision must be made as to the best end of the raceway for pulling. The reel setup must also be carefully placed. In general, conductors that are to be installed downward should be fed off the top of the reel; where conductors are to be fed upward, the best method is to feed from the bottom of the reel. This minimizes sharp kinks or bends in the conductors.
- The appropriate pulling equipment must be used. The equipment must be of the proper capacity for the job. Space consideration for the equipment is also important, as is the particular type of mounting.
- Having enough workers is critical. Never be caught short. This is where experienced workers earn their pay. In high-force wire pulling, experience is the best teacher.
- Safety must be foremost in everyone's mind.

When planning the cable pull, make sure that enough cable is on hand for the run. Cable may be verified while still on the reel by using a cable-length meter. Of course, the conduit run length should have already been checked, as discussed previously. Do not proceed further until everyone is assured that enough cable is on hand for all bends, sweeps in junction boxes, and other needs. If the cable falls short, take steps to correct the situation before continuing.

A feeding sheave for pulling equipment will provide a smooth guide for the cable. Also, have sufficient lubricant on hand and use it both before and during the pull.

Select and install proper cable grips on the cable ends. Gripping must be adequate to handle the imposed force.

When pulling cables in a horizontal run, the worker simply has to reduce the amount of operator pull in order to slow down the cable pull. Releasing all operator pull stops the pulling force entirely. However, when pulling vertical runs, the rope must be tied off after stopping the pull to keep the cable from reversing in the raceway.

The entire operation should be supervised from start to finish. Therefore, communications equipment must be utilized on both ends of the pull at all times.

During the pull, make sure that the pulling rope remains free and is not wrapped around any part of the body.

A typical cable-pulling setup is shown in *Figure 22*, including cable reels, reel stands, a conduit system, and other accessories necessary for the pull.

Think About It

Communications

Which type of communications equipment would most likely be used during cable-pulling operations?

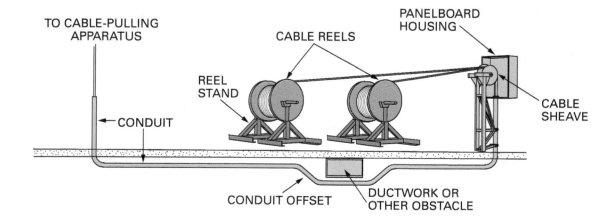

Figure 22 Typical cable-pulling setup.

2.1.0 Feeding End

When setting up the feeding end, the proper rope must be selected based on the expected force of the pull. Choose a rope that has at least four times the strength of the required pulling force; that is, if the estimated pulling force is 1,200 lbs (5.3 kN), the rope should be rated for no less than 4,800 lbs (21.4 kN). Check the rope carefully for wear or damage before the pull. Remember that a rope is only as strong as its weakest point.

The feed-in setup should unreel the cable along its natural curvature, as shown in *Figure 23 (A)*, as opposed to a reverse S curvature, as shown in *Figure 23 (B)*. Feed-in setups are shown in *Figure 24* for manhole, underfloor duct, and overhead cable tray. Note the use of auxiliary equipment in some of these drawings, that is, cable reels, guide-in tubes, and sheaves.

Single sheaves, such as those shown in *Figure 25 (A)*, may be used only for guiding cables. Multiple blocks should be arranged to hold the cable-bending radius wherever the cable is deflected. For pulling around bends, use a conveyor sheave assembly of the appropriate radius series, such as the one shown in *Figure 25 (B)*.

Sheaves and pulleys must be positioned to ensure the effective curvature is smooth and deflected evenly at each pulley. Never allow a polygon curvature, as shown in *Figure 26*.

CAUTION: Use the radius of the surface over which the cable is bent, not the outside flange diameter of the pulley. For example, a 10" cable sheave typically has an inside (bending) radius of 3".

Monitored Cable Pullers

Many power cable pullers include force gauges with an automatic over-tension shutoff device. These units prevent you from accidentally exceeding the maximum recommended pulling tension for the conductor(s) being pulled.

Figure Credit: Greenlee / A Textron Company

In general, single or multi-cable rollers are placed in the bottom of trays to protect the cable as it is pulled along. Sheaves are placed at each change of direction—either horizontally or vertically. The bottom rollers may be secured to the tray bottoms except at vertical changes in direction. Extra support is necessary at these locations to prevent damage or movement of the tray system.

> **NOTE:** If single cables are to be installed, always place them on the outside of a bend to allow room on the inside of the bend for pulling other cables.

Sheaves must be supported in the opposite direction of the pull. For example, all right-angle conveyor sheaves should be supported at two locations, as shown in *Figure 28*, to compensate for the pull of the cable.

> **NOTE:** Power cable pulls should not be stopped unless absolutely necessary. However, anyone associated with the pull—upon evidence of danger to either the cable or the workers—may stop a cable-pulling operation. Communication is the most important factor in these cases.

> **CAUTION:** Workers feeding a cable pull must carefully inspect the cable as it is paid off the cable reel. Any visible defects in the cable at the feeding end warrants stopping the pull.

> **WARNING!** At the first sign of any type of malfunctioning equipment, broken sheaves, or other events that could present a danger to either the workers or the cable, the pull should be stopped. Make certain that all communications equipment is in proper order before starting a pull.

Case History
Pulling Cable in a Tray

At a clothing manufacturing plant in New York, an electrician attempted to pull cable in a tray already partly filled with energized cables. The electrician used plenty of cable-pulling lubricant but did not use sheaves and rollers to isolate the new cables from the existing ones. During the pull, the new cables wedged against the live ones, yanking one hot wire apart at a splice. The wire grounded itself on the tray and arced for 30' down the tray, fatally shocking the electrician.

The Bottom Line: Use sheaves and rollers when pulling cables, especially when pulling them between existing live cables.

Big Pulls

For very long pulls, some contractors use an auxiliary winch. Power winches may exceed their preset tension limits and abort the pull, so a second winch is installed in a high-tension area of the pull in a straight section of tray. The second winch is used to pull a loop of slack cable and reduce the tension on the main winch, which now pulls only the length of slack cable between itself and the auxiliary winch.

2.0.0 Section Review

1. If the estimated pulling force is 800 pounds, the pulling rope should be rated for no less than _____.
    a. 800 pounds
    b. 1,600 pounds
    c. 2,400 pounds
    d. 3,200 pounds

2. The requirements for intermediate supports can be found in _____.
    a. *NEC Table 310.16*
    b. *NEC Chapter 9, Table 8*
    c. *NEC Table 402.3*
    d. *NEC Table 300.19(A)*

3. Interlocked armor cable should have a bend radius of no less than _____.
    a. twice its diameter
    b. three times its diameter
    c. five times its diameter
    d. eight times its diameter

SECTION THREE

3.0.0 CABLE LIMITATIONS WHEN PULLING

Objective

Identify cable limitations when pulling.
  a. Calculate the allowable tension on pulling devices.
  b. Calculate the allowable tension on conductors.
  c. Calculate the sidewall loading.

Performance Tasks

1. Prepare multiple conductors for pulling in a raceway system.
2. Prepare multiple conductors for pulling using a wire-pulling basket.

Trade Term

Tensile strength: The pulling force required to break a rope, wire, or structural beam.

Consideration must be given to the physical limitations of a cable as it is being pulled into position. Pulling subjects cable to extreme stress and, if done improperly, can displace a cable's components. Thus, it is important that the following guidelines be observed:

- While reels are in storage—either before or after a pull—the conductor ends must be sealed to prevent moisture from entering or creeping into the cable ends.
- The minimum ambient working temperatures for cable-pulling operations depend on the cable jackets. In cold weather, cable reels should be stored in a warm area overnight so that the cable jackets will be at the proper temperature for pulling the next day.
- Calculate and stay within the cable's maximum pulling tension, maximum sidewall load, and minimum bending radii.
- Ensure that the raceway joints are aligned and that the wiring space is sufficient.
- Train the cable to avoid dragging on the edge of the raceway; also avoid laying or dragging cable on the ground.
- If using a basket grip, secure it to the cable with steel strapping. After the cable is in place, cut well beyond the covered area and discard that portion of the cable.
- Ensure that the elongation of the pull rope minimizes jerking.
- Pull with a capstan and no faster than 40' (12.2 m) per minute.
- Do not stop a pull unless absolutely necessary.
- Never pull the middle of the cable.
- Seal the ends with appropriate putty or silicone caulking and overwrap with tape until the conductors are terminated.

The maximum pulling tension should not exceed the smaller of the following values:

- Allowable tension on pulling device
- Allowable tension on conductor
- Allowable sidewall load

> **NOTE:** Metric values for pulling tension are listed in newtons (N) or kilonewtons (kN) and can be found on the manufacturer data sheets for the conductor or pulling device in use.

3.1.0 Allowable Tension on Pulling Devices

Do not exceed the working load stated by the manufacturer of the pulling devices (pulling eyes, ropes, anchors, basket grips, or other equipment). If catalog information is not available, work at 10% of the rated braking tensile strength.

The allowable tension with a basket grip must not exceed the lbs/cmil value (as shown in *Table 3*) or 1,000 lbs (4.4 kN), whichever is smaller. Exceptions to this rule, however, do occur, but seldom will this figure rise to over 1,250 lbs (5.6 kN).

3.2.0 Allowable Tension on Conductors

The metallic phase conductors are the strength members of the cables and should bear all of the pulling force. Never use shielding drain wires or braids for pulling. *Table 3* provides the allowable pulling tensions of various types of conductors. Any wire or conductor will break if stretched beyond its tensile strength. Never exceed the listed value of a conductor.

Reduce the maximum pulling tension by 20% to 40% if several conductors are being pulled simultaneously because the tension is not always evenly distributed among the conductors.

> **CAUTION**
> When smaller conductors are pulled with large conductors, the smaller conductors may be damaged.

The maximum tension for a specific type of cable can be found using manufacturer data. In general, the maximum tension for a single-conductor cable should not exceed 6,000 lbs (26.7 kN). The maximum tension for two or more conductors should not exceed 10,000 lbs (44.5 kN).

The maximum stress for leaded cables must not exceed 1,500 pounds per square inch (psi) or 10.3 megapascals (MPa) of lead sheath area when pulled with a basket grip. The maximum tension must not exceed 1,000 psi (6.9 MPa) of insulation area for non-leaded cables when pulled with a basket grip.

For a straight horizontal raceway section, the pulling tension is equal to the length of the duct run multiplied by the weight of the cable and the coefficient of friction, which will vary depending on the type and amount of lubrication used. Therefore, the equation is as follows:

$$T = L \times w \times f$$

Where:

T = total pulling tension
L = length of raceway run in feet or meters
w = weight of cable in lbs per foot or kg/meter
f = coefficient of friction

For ducts having curved sections, the following equation applies:

$$T_{OUT} = T_{IN}\ e^{fa}$$

Where:

T_{OUT} = tension of bend
T_{IN} = tension into bend
f = coefficient of friction
e = Napierian logarithm base 2.718
a = angle of bends in radians

To aid in solving the above equation, values of e^{fa} for specific angles of bend and coefficients of friction are listed in *Table 4*. For more precise values, tables are available from cable manufacturers.

3.3.0 Sidewall Loading

Before calculating cable-pulling tension, sidewall loading or sidewall bearing pressure must be considered. The sidewall load is the radial force exerted on a cable being pulled around a conduit bend or sheave. Excessive sidewall loading can crush a cable and is, therefore, one of the most restrictive factors in installations having bends or high tensions. *Figure 29* shows a section taken across a conduit run in a 90° bend. Note that pulling tension is exerted parallel with the walls of the conduit. However, because of the 90° bend, pressure is also exerted downward against the wall of the conduit. Once again, this is known as the sidewall load. Sidewall loading is reduced by increasing the radius of bends.

Table 3 Physical Limitations of Cable

Material	Cable Type	Temper	Lbs/Cmil
Copper	All	Soft	0.008
Aluminum	Power	Hard	0.008
Aluminum	Power	¾ Hard	0.006
Aluminum	Power	AWM	0.005
Aluminum	URD (solid)	Soft (½ hard)	0.003
All	Thermocouple	—	0.008

Table 4 Angle of Bend vs. Coefficients of Friction

Angle of Bend (degrees/radians)	Values of e^{fa} for Coefficients of Friction		
	f = 0.75	f = 0.50	f = 0.35
15/0.2618	1.22	1.14	1.10
30/0.5236	1.48	1.30	1.20
45/0.7854	1.80	1.48	1.32
60/1.0472	2.19	1.68	1.44
90/1.5708	3.25	2.20	1.73

Installation Aids

Other tools available to simplify the pulling of cables into cable tray systems include triple pulleys, bull wheels, and both wide and narrow rollers. Be sure to position these pulleys and rollers at the proper locations to prevent damage to the cables during the installation and also to help the installation proceed as quickly as possible. Various rollers are shown here.

SNAP-IN SPINDLE AND CABLE ROLLER

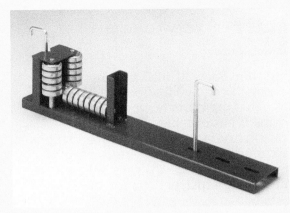

CABLE ROLLER WITH ADJUSTABLE BRACKETS

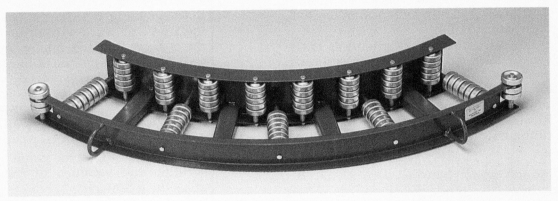

ELBOW ROLLER

Figure Credit: Greenlee / A Textron Company

Software Programs

Software programs are now available to take the math out of calculating pulling tensions. These programs typically calculate pulling tension and sidewall pressures based on conductor size using their own data on friction coefficients. These programs also determine conduit fill, conductor configuration, jam ratios, and the amount of lubricant required for the pull.

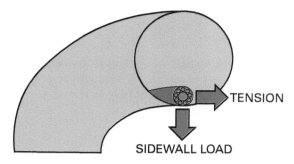

Figure 29 Sidewall loading.

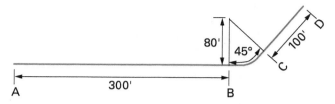

Figure 30 Sample conduit run.

In general, the sidewall load on any raceway run should not exceed 500 lbs/foot (744 kg/m) of bend radius. This pressure, however, must be reduced on some types of cables. For example, *Table 5* shows one manufacturer's recommendations for the maximum sidewall pressures permitted for various types of cables. Always refer to the cable manufacturer's instructions for the type of cable being used.

For example, *Figure 30* shows a typical raceway system containing three 500 kcmil lead sheath copper conductors. Note the straight 300' run from A to B, a 45° kick, and then another 100' straight run from C to D.

To find the calculated pulling tension from D to A, refer to cable data in the manufacturer's catalogs. Suppose this cable weighs 8 pounds per foot and has a 0.141" lead sheath. The outside diameter of the three-conductor cable assembly is 3". Use 0.5 as the coefficient of friction and calculate the pulling tension from A to B (*Figure 30*) as follows:

Step 1 Find the tension between points A and B.

$$\text{Tension at B} = T_1 = L_1 \times w \times f$$

L_1 is the length between points A and B when w is the weight of the cable per foot, and f is the coefficient of friction, which is 0.5. Substituting the known values in the equation gives:

$$T_1 = 300 \times 8 \times 0.5 = 1{,}200 \text{ pounds}$$

Step 2 Find the tension between points B and C.

$$\text{Tension at C} = T_1 e^{fa}$$

Because the distance between B and C involves an angle, refer to *Table 4* of this module. Looking in the left column of the table, we see that 45° equals 0.7854 radians; this figure multiplied by the coefficient of friction (0.5) equals 0.3927. Radians of angles may also be found with electronic pocket calculators if they have scientific functions. Follow the instructions in the manual accompanying the calculator. The exact key strokes will vary with the brand of calculator, but most require pressing the degree key, entering the numeral for degrees, then pressing the convert key, and finally pressing the radian or RAD key. The radians of the entered angle will be displayed.

Once again, refer to *Table 4*. Find the 45° angle of bend in the left column; read across the row to the column under 0.50—the coefficient of friction figure. The number is 1.48, the value of e^{fa}. Substituting these values in the equation gives the following:

$$A = 45° = 0.7854 \text{ radians}$$
$$fa = 0.7854 \times 0.5 = 0.3927$$
$$e^{fa} = 1.48$$
$$1{,}200 \times 1.48 = 1{,}776 \text{ pounds}$$

Therefore, the tension at C is 1,776 pounds.

Table 5 Maximum Sidewall Pressures for Various Types of Cable

Cable Type	Sidewall Pressure in Pounds/Foot of Bend Radius
600V nonshielded control	300
600V and a kV nonshielded EP power	500
5kV and 15kV EP power	500
25kV and 35kV power	300
Interlocked armored cable (all voltages)	300

Step 3 Find the tension from points C to D.

Tension from C to D = T_2 = $L_2 \times w \times f$

$T_2 = 100 \times 8 \times 0.5 = 400$ pounds

Step 4 Find the total pulling tension by adding the figures obtained previously.

$T = T_2 + T_1 e^{fa} = 400 + 1{,}776 = 2{,}176$ pounds

The maximum pulling force using a basket grip for this size cable should not exceed 1,900 pounds. Therefore, if the pull is made from point A to point D, a pulling eye will have to be used because the total pulling tension exceeds 1,900 pounds. However, if the pull is reversed—pulling from point D to point A—the total pulling tension will be reduced because the distance from point D to the 45° angle (point C) is $\frac{1}{3}$ the distance from point A to the 45° angle (point B).

Tension at C = 400 pounds
Tension at B = 400 × 1.48 = 592 pounds
Total tension at A = 1,200 + 592 = 1,792 pounds

Therefore, if the cable is pulled from point D to point A, either a pulling eye or basket grip may be used.

> **NOTE:** A lower tension is obtained by feeding the pull from the end nearest the bend.

Complex Installations

Complex cable-pulling installations are best handled by highly experienced contractors. This is one small portion of an award-winning installation that took over 70,000 hours to complete.

Case History

Cable Pulling

In June 2006, an electrician was tagging instrumentation cable on an elevated work platform. At the same time, a crew was pulling an assembly in an adjacent cable tray but failed to install caution signs or barrier controls. The inexperienced crew did not monitor the tension on the pulling assembly, and it failed during the pull, whipping back along the cable tray and picking up a loose cable roller, which propelled into the victim at high speed. The cable amputated the electrician's arm, broke ribs and a femur, and ruptured several internal organs. The electrician died from the injuries.

The Bottom Line: Only qualified individuals may conduct a cable pull. All pulling equipment must be tested and certified for the expected load, and the area must be secured with caution/barrier controls during the pull.

3.0.0 Section Review

1. If catalog information is unavailable for a pulling device, do not exceed _____.
    a. 10% of the rated braking tensile strength
    b. 20% of the rated braking tensile strength
    c. 30% of the rated braking tensile strength
    d. 40% of the rated braking tensile strength

2. The allowable tension for soft copper cable is _____.
    a. 0.003 lbs/cmil
    b. 0.005 lbs/cmil
    c. 0.006 lbs/cmil
    d. 0.008 lbs/cmil

3. The maximum sidewall pressure for 5kV power cable is _____.
    a. 100 lbs per foot of bend radius
    b. 200 lbs per foot of bend radius
    c. 300 lbs per foot of bend radius
    d. 500 lbs per foot of bend radius

Review Questions

1. Which of the following should be avoided when transporting or storing cable reels?
   a. Store reels in an upright position.
   b. Store reels laid flat on the side.
   c. Lift heavy reels using a crane or forklift.
   d. When carrying reels using a forklift, make sure the forks do not touch the cable.

2. Which of the following is *true* regarding conductor installations?
   a. All cables use the same type of pulling grip.
   b. Setscrew grips are easily tightened by hand.
   c. Setscrew grips are rarely used on long pulls.
   d. Strip the insulation from no-lube cables before using in a basket grip.

3. Where is the best place to seek information about whether a wire-pulling lubricant is compatible with a particular type of cable insulation?
   a. The lubricant manufacturer
   b. The cable manufacturer
   c. Both the cable and lubricant manufacturers
   d. The *NEC®*

4. What is the pulling force with five wraps on the capstan of a typical cable puller and an operator force of 10 lbs?
   a. 48 lbs
   b. 106 lbs
   c. 233 lbs
   d. 512 lbs

5. Who is allowed to stop the pull at the first sign of danger to the cable, raceway system, or personnel?
   a. Only the project supervisor
   b. Only workers on the feeding end of the run
   c. Only workers on the pulling end of the run
   d. Anyone involved with the pull

6. A dynamometer is used during a pull to _____.
   a. measure dynamic friction
   b. trace conductors on de-energized circuits
   c. determine the length of a conductor
   d. determine the ampacity of a conductor

7. How should the feed-in setup be handled during a cable installation?
   a. The cable should unreel with a reverse S curvature.
   b. It makes no difference.
   c. Only the pulling end is important when unreeling the cable.
   d. The cable should unreel along its natural curvature.

8. When pulling cable around bends, use a _____.
   a. single sheave
   b. polygon setup
   c. conveyor sheave assembly
   d. single block

9. If the estimated pulling force is 1,000 pounds, the pulling rope should be rated for no less than _____.
   a. 1,000 pounds
   b. 2,000 pounds
   c. 3,000 pounds
   d. 4,000 pounds

10. Braking systems are most likely to be used in long _____.
    a. curved horizontal pulls
    b. straight horizontal pulls
    c. vertical pulls
    d. cable tray

11. Which of the following best describes the location of sheave supports for pulling cable in cable trays?
    a. They should be supported in the opposite direction of the pull.
    b. They should be supported in the same direction as the pull.
    c. They should be supported only by the tray assembly itself.
    d. No support is necessary when sheaves are used in cable trays.

12. Which of the following pulling methods should be used when the pulling tension exceeds 1,000 pounds? 2.3.0
    a. A basket grip by itself
    b. A pulling eye
    c. Conductors bent around a hook and taped
    d. Conductors bent around a snake hook and left untapped

13. After some length of conductor has been used from a cable reel, what should be done before placing the reel in storage? 3.0.0
    a. A stress test should be made to see if the cable has been damaged.
    b. An insulation test should be made using a megger.
    c. The ends of the conductors should be sealed.
    d. The conductor jacket should be warmed before storing.

14. Which of the following precautions should be taken when making a cable pull in extremely cold weather? 3.0.0
    a. The cable should be stored in a warm area overnight.
    b. The cable should be stored outside overnight.
    c. Antifreeze should be mixed with the pulling lubricant.
    d. Electric warming blankets should be used during the pull.

15. The maximum sidewall pressure for 25kV power cable is _____. 3.0.0
    a. 100 lbs per foot of bend radius
    b. 200 lbs per foot of bend radius
    c. 300 lbs per foot of bend radius
    d. 500 lbs per foot of bend radius

Supplemental Exercises

1. Before installing conductors in a long conduit run, it is a good idea to use _____ to check for hidden obstructions.
   a. a go/no-go mandrel
   b. ¼" fish tape
   c. a flashlight
   d. a blow gun system

2. Where several wires are to be pulled together, the wires should be _____.
   a. staggered
   b. cut flush
   c. braided
   d. twisted

3. When installing large cables that require more pulling force, select a rope that is rated for at least ___4___ times the strength of the required pulling force.

4. True or False? If commercial wire lubricant is not available, any suitable petroleum-based product may be substituted.

5. When pulling vertical runs, be sure to ___tie-off___ the rope after stopping the pull to keep the cable from reversing in the raceway.

6. True or False? Runaways are most likely to occur in horizontal raceway systems.

7. True or False? When pulling long runs of cable, stopping on a regular basis will allow the cable to cool and will avoid stressing the cable.

8. True or False? Cable length meters may be used to determine the length of conductors on reels before making a pull.

9. When applying the capstan theory to a power puller, each ___additional wrap___ more than doubles the pulling force.

10. Which type of pulling device uses a flexible steel mesh grip on the ends of cable and attaches to the fish tape or pull rope?
    a. Mandrel
    b. Setscrew grip
    c. Basket grip
    d. Pulling eyes

Trade Terms Introduced in This Module

Basket grip: A flexible steel mesh grip that is used on the ends of cable and conductors for attaching the pulling rope. The more force exerted on the pull, the tighter the grip wraps around the cable.

Cable grip: A device used to secure ends of cables to a pulling rope during cable pulls.

Capstan: The turning drum of a cable puller on which the rope is wrapped and pulled. An increase in the number of wraps increases the pulling force of the cable puller.

Clevis: A device used in cable pulls to facilitate connecting the pulling rope to the cable grip.

Conductor support: The act of providing support in vertical conduit runs to support the cables or conductors. The *NEC*® gives several methods in which cables may be supported, including wedges in the tops of conduits and supports to change the direction of cable in pull boxes.

Conduit piston: A cylinder of foam rubber that fits inside the conduit and is then propelled by compressed air or vacuumed through the conduit run to pull a line, rope, or measuring tape. Also called a *mouse*.

Fish line: Light cord used in conjunction with vacuum/blower power fishing systems that attaches to the conduit piston to be pushed or pulled through the conduit. Once through, a pulling rope is attached to one end and pulled back through the conduit for use in pulling conductors.

Fish tape: A flat iron wire or fiber cord used to pull conductors or a pulling rope through conduit.

Setscrew grip: A cable grip, usually with built-in clevis, in which the cable ends are inserted in holes and secured with one or more setscrews.

Sheave: A pulley-like device used in cable pulls in both conduit and cable tray systems.

Soap: Slang for wire-pulling lubricant.

Tensile strength: The pulling force required to break a rope, wire, or structural beam.

Additional Resources

This module presents thorough resources for task training. The following resource material is suggested for further study.

Cable Installation Manual, Latest Edition. New York: Cablec Corp.
National Electrical Code® Handbook, Latest Edition. Quincy, MA: National Fire Protection Association.

Figure Credits

Greenlee / A Textron Company, Module Opener, Figures 2, 3, 8, 10, 12C, 13, 14A, 16–22, 25

Tim Dean, Figure 28

Data from *NEC® Table 300.19(A)*, Table 2. Reprinted with permission from NFPA 70-2020, *National Electrical Code®*, Copyright © 2019, National Fire Protection Association, Quincy, MA. This reprinted material is not the complete and official position of the NFPA on the referenced subject, which is represented only by the standard in its entirety which may be obtained through the NFPA website at **www.nfpa.org**.

Section Review Answer Key

Section 1.0.0

Answer	Section Reference	Objective
1. c	1.1.0	1a
2. a	1.2.0	1b
3. c	1.3.0	1c
4. a	1.4.0	1d
5. c	1.5.0	1e
6. d	1.6.0	1f

Section 2.0.0

Answer	Section Reference	Objective
1. d	2.1.0	2a
2. d	2.2.0	2b
3. c	2.3.0	2c

Section 3.0.0

Answer	Section Reference	Objective
1. a	3.1.0	3a
2. d	3.2.0; *Table 3*	3b
3. d	3.3.0; *Table 5*	3c

NCCER CURRICULA — USER UPDATE

NCCER makes every effort to keep its textbooks up-to-date and free of technical errors. We appreciate your help in this process. If you find an error, a typographical mistake, or an inaccuracy in NCCER's curricula, please fill out this form (or a photocopy), or complete the online form at **www.nccer.org/olf**. Be sure to include the exact module ID number, page number, a detailed description, and your recommended correction. Your input will be brought to the attention of the Authoring Team. Thank you for your assistance.

Instructors – If you have an idea for improving this textbook, or have found that additional materials were necessary to teach this module effectively, please let us know so that we may present your suggestions to the Authoring Team.

NCCER Product Development and Revision
13614 Progress Blvd., Alachua, FL 32615

Email: curriculum@nccer.org
Online: www.nccer.org/olf

❑ Trainee Guide ❑ Lesson Plans ❑ Exam ❑ PowerPoints Other _____

Craft / Level: _____ Copyright Date: _____

Module ID Number / Title: _____

Section Number(s): _____

Description: _____

_____

_____

_____

_____

Recommended Correction: _____

_____

_____

_____

_____

Your Name: _____

Address: _____

_____

Email: _____ Phone: _____

This page is intentionally left blank.

Cable Tray

Overview

Cable trays are the usual means of supporting cable systems in industrial applications. This module covers various types of cable tray, supports, and associated fittings. It also explains how to determine the loads on a cable tray and calculate fill per *NEC*® requirements.

Module 26207-20

Trainees with successful module completions may be eligible for credentialing through the NCCER Registry. To learn more, go to **www.nccer.org** or contact us at 1.888.622.3720. Our website, **www.nccer.org**, has information on the latest product releases and training.

Your feedback is welcome. You may email your comments to **curriculum@nccer.org**, send general comments and inquiries to **info@nccer.org**, or fill in the User Update form at the back of this module.

This information is general in nature and intended for training purposes only. Actual performance of activities described in this manual requires compliance with all applicable operating, service, maintenance, and safety procedures under the direction of qualified personnel. References in this manual to patented or proprietary devices do not constitute a recommendation of their use.

Copyright © 2020 by NCCER, Alachua, FL 32615, and published by Pearson, New York, NY 10013. All rights reserved. Printed in the United States of America. This publication is protected by Copyright, and permission should be obtained from NCCER prior to any prohibited reproduction, storage in a retrieval system, or transmission in any form or by any means, electronic, mechanical, photocopying, recording, or likewise. To obtain permission(s) to use material from this work, please submit a written request to NCCER Product Development, 13614 Progress Blvd., Alachua, FL 32615.

26207-20 V10.0

From *Electrical, Trainee Guide*. NCCER.
Copyright © 2020 by NCCER. Published by Pearson. All rights reserved.

26207-20
CABLE TRAY

Objectives

When you have completed this module, you will be able to do the following:

1. Identify cable tray components.
   a. Select cable tray fittings.
   b. Identify cable tray supports.
2. Calculate the load on a cable tray.
   a. Determine the load on supports.
   b. Identify types of failure under load.
   c. Identify installation requirements for cable tray.
3. Determine cable tray fill.
   a. Determine the number of conductors allowed in cable tray operating at 2,000V or less.
   b. Identify derating factors for cable tray conductors.

Performance Tasks

Under the supervision of the instructor, you should be able to do the following:

1. Generate a list of materials for a cable tray layout. List all the components required, including the fasteners required to complete the system.
2. Join two straight, ladder-type cable tray sections together.

Trade Terms

Barrier strip
Cable pulley
Cross
Direct rod suspension
Dropout
Dropout plate
Elbow

Expansion joints
Fittings
Ladder tray
Pipe racks
Swivel plates
Tee
Trapeze mounting

Tray cover
Trough
Unistrut®
Wall mounting
Wye

Industry Recognized Credentials

If you are training through an NCCER-accredited sponsor, you may be eligible for credentials from NCCER's Registry. The ID number for this module is 26207-20. Note that this module may have been used in other NCCER curricula and may apply to other level completions. Contact NCCER's Registry at 888.622.3720 or go to www.nccer.org for more information.

> **NOTE**
> NFPA 70®, *National Electrical Code*® and *NEC*® are registered trademarks of the National Fire Protection Association, Quincy, MA.

Contents

1.0.0 Cable Tray Components 1
   1.1.0 Cable Tray Fittings 2
      1.1.1 Cable Entry and Exit 3
      1.1.2 Cable Tray Covers 7
      1.1.3 Cable Supports in Vertical Trays 7
      1.1.4 Cable Edge Protection 7
      1.1.5 Splice Plates 7
      1.1.6 Barrier Strips 9
   1.2.0 Cable Tray Supports 9
      1.2.1 Direct Rod Suspension 11
      1.2.2 Wall Mounting 11
      1.2.3 Trapeze Mounting 11
      1.2.4 Center Hung Support 12
      1.2.5 Pipe Rack Mounting 12
      1.2.6 Monorail Systems 12
2.0.0 Cable Tray Loading 14
   2.1.0 Determining the Load on Supports 14
   2.2.0 Failure Under Load 14
   2.3.0 Installation Requirements 16
      2.3.1 Cable Tray Drawings 16
      2.3.2 Safety Precautions for Cable Tray Installation 19
3.0.0 Cable Tray Fill 22
   3.1.0 Number of Conductors Allowed in Cable Tray (2,000V or Less) ... 22
      3.1.1 All Conductors Size 4/0 or Larger 22
      3.1.2 All Conductors Smaller Than 4/0 22
      3.1.3 Combination Cables 23
      3.1.4 Solid Bottom Tray 24
      3.1.5 Single-Conductor Cables 25
   3.2.0 Ampacity of Cable Tray Conductors 25

Figures

Figure 1	Typical cable tray system	2
Figure 2	Cross section of cable tray comparing usable dimensions to overall dimensions	3
Figure 3	Cables rest on the bottom of the tray and are held in place by the longitudinal side rails	4
Figure 4	Two applications of cable tray channel	5
Figure 5	Devices used to facilitate a cable pull in a tray	7
Figure 6	Several ways in which cables may exit from a cable tray	8
Figure 7	A dropout plate provides a curved surface for the cable to follow as it leaves the tray	9
Figure 8	Typical application of supports in a vertical run	9
Figure 9	Fabricating a cable tray offset with swivel plates	10
Figure 10	Expansion joint and splice plates	10
Figure 11	Alternate ways to hang cable tray	11
Figure 12	Channel support	12
Figure 13	Center rail cable tray	12
Figure 14	Cables packed closely together can impair efficiency	14
Figure 15	Calculating tray support	15
Figure 16	Bending of loaded tray	15
Figure 17	Load of cable creates bending moments along the span	16
Figure 18	*NEC®* regulations governing the use of cable tray	17
Figure 19	*NEC®* regulations governing cable tray construction and installation	18
Figure 20	*NEC®* regulations governing cable tray grounding	19
Figure 21	Sample floor plan of a cable tray system	20

This page is intentionally left blank.

Section One

1.0.0 Cable Tray Components

Objective

Identify cable tray components.
a. Select cable tray fittings.
b. Identify cable tray supports.

Trade Terms

Barrier strip: A metal strip constructed to divide a section of cable tray so that certain kinds of cable may be separated from each other.

Cable pulley: A device used to facilitate pulling conductor in cable tray where the tray changes direction. Several types are available (single, triple, etc.) to accommodate almost all pulling situations.

Cross: A four-way section of cable tray used when the tray assembly must branch off in four different directions.

Direct rod suspension: A method used to support cable tray by means of threaded rods and hanger clamps. One end of the threaded rod is secured to an overhead structure, while the other end is connected to hanger clamps that are attached to the cable tray side rails.

Dropout: Cable leaving the tray assembly and travelling directly downward; that is, the cable is not routed into a conduit or channel.

Dropout plate: A metal plate used at the end of a cable tray section to ensure a greater cable bending radius as the cable leaves the tray assembly.

Elbow: A section of cable tray used to change the direction of the tray assembly a full quarter turn (90°). Both vertical and horizontal elbows are common.

Expansion joints: Plates used at intervals along a straight run of cable tray to allow space for thermal expansion or contraction of the tray.

Fittings: Devices used to assemble and/or change the direction of cable tray systems.

Ladder tray: A type of cable tray that consists of two parallel channels connected by rungs, similar in appearance to the common straight ladder.

Pipe racks: Structural frames used to support the piping that interconnects equipment in outdoor industrial facilities.

Swivel plates: Devices used to make vertical offsets in cable tray.

Tee: A section of cable tray that branches off the main section in two other directions.

Trapeze mounting: A method of supporting cable tray using metal channel, such as Unistrut®, Kindorf®, etc., supported by two threaded rods, and giving the appearance of a swing or trapeze.

Tray cover: A flat piece of metal, fiberglass, or plastic designed to provide a solid covering that is needed in some locations where conductors in the tray system may be damaged.

Trough: A type of cable tray consisting of two parallel channels (side rails) having a corrugated, ventilated bottom or a corrugated, solid bottom.

Unistrut®: A brand of metal channel used as the bottom bracket for hanging cable trays. Double Unistrut® adds strength and stability to the trays and also provides a means of securing future runs of conduit.

Wall mounting: A method of supporting cable tray systems using supports secured directly to the wall.

Wye: A section of cable tray that branches off the main section in one direction.

NEC Section 392.2 defines a cable tray system as a unit or assembly of units or sections and associated fittings forming a structural system used to securely fasten or support cables and raceways. *NEC Article 392* covers cable tray installations, along with the types of conductors to be used in various systems. Whenever a question arises concerning cable tray installations, this is the *NEC®* article to use.

Cable trays are the usual means of supporting cable systems in industrial applications. The trays themselves are usually made up into a system of assembled, interconnected sections and associated fittings, all of which are made of metal or noncombustible units. The finished system forms into a continuous rigid assembly for supporting and carrying single, multiconductor, or other electrical cables and raceways from their origin to their point of termination, frequently over considerable distances.

Cable tray is fabricated from both aluminum and steel. Some manufacturers provide an aluminum cable tray that is coated with PVC for installation in caustic environments. Nonmetallic trays are also available; this type of tray is ideally suited for use in corrosive areas and in areas requiring voltage isolation. Cable tray is available in various forms, including ladder, trough, center rail, and solid bottom, and can be supported by either side mounts or center mounts.

Ladder tray, as the name implies, consists of two parallel channels connected by rungs, similar in appearance to a conventional straight or extension ladder. Trough types consist of two parallel channels (side rails) having a corrugated, ventilated bottom. The solid bottom cable tray is similar to the trough. All of these types are shown in *Figure 1*. Ladder, trough, and solid bottom trays are completely interchangeable; that is, all three types can be used in the same run when needed.

Cable tray is manufactured in 12' and 24' lengths. (Common metric dimensions for cable tray are 3 m and 6 m. However, dimensions may vary depending on supplier. Refer to the manufacturer's catalog for the tray in use.) Cable tray is available in widths from 6" to 36" (150 mm to 900 mm), and depths up to 6" (150 mm).

1.1.0 Cable Tray Fittings

Cable tray sections are interconnected using various types of fittings. Fittings are also used to provide a means of changing the direction or dimension of the cable tray system. Some of the more common fittings include:

- Horizontal or vertical tee
- Wye
- Horizontal or vertical bend
- Horizontal cross
- Reducer
- Barrier strip
- Cover
- Splice plate
- Box connector

The area of a cable tray cross section that is usable for cables is defined by width (W) × depth (D), as shown in *Figure 2*. The overall dimensions of a cable tray, however, are greater than W times D because of the side flanges and seams. Therefore, overall dimensions vary according to the tray design. Cables rest on the bottom of the tray and are held within the tray area by two longitudinal side rails, as shown in *Figure 3*.

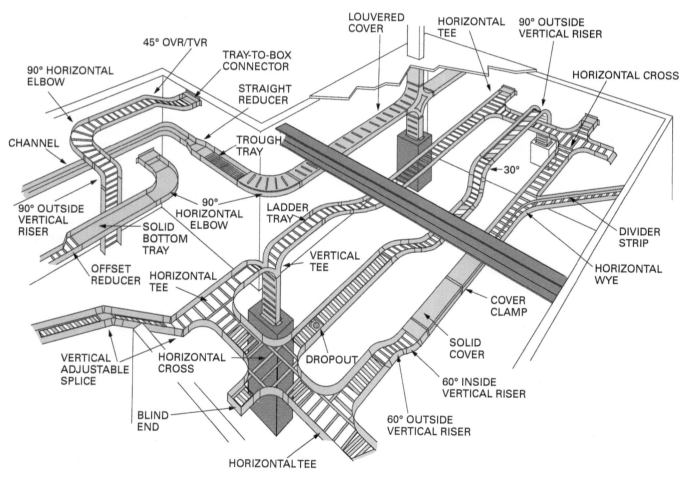

Figure 1 Typical cable tray system.

Cable Tray Installation

Cable trays are widely used in many commercial and industrial installations.

A channel is used to carry one or more cables from the main tray system to the vicinity of the cable termination (*Figure 4*). Conduit is then used to finish the run from the channel to the actual termination.

Certain *NEC®* regulations and National Electrical Manufacturers Association (NEMA) standards should be followed when designing or installing cable tray. Consequently, practically all projects of any great size will have detailed drawings and specifications for the workers to follow. Shop drawings may also be provided.

1.1.1 Cable Entry and Exit

Cables are placed in the tray either by being pulled along the tray or by being laid in over the side. *Figure 5* (*A*) shows a cable pulley being used to facilitate a cable pull in a tray, while *Figure 5* (*B*) shows the use of a cable tray roller that is clamped to the side of the tray.

Several different ways in which cables may exit from a cable tray are shown in *Figure 6*. While all of these methods are *NEC®*-approved and endorsed by most cable tray manufacturers, the engineering specifications on some projects may prohibit the use of some of these methods. Most notable are the dropout between rungs method and the dropout from the end of the tray method. The cable radius may be too short with either of these methods. Also, if a dropout plate is not used, the cable is not protected.

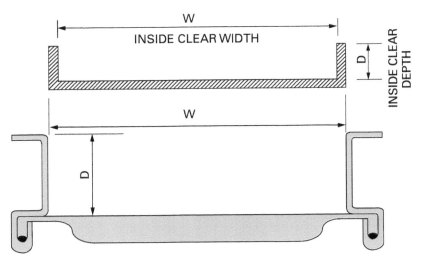

Figure 2 Cross section of cable tray comparing usable dimensions to overall dimensions.

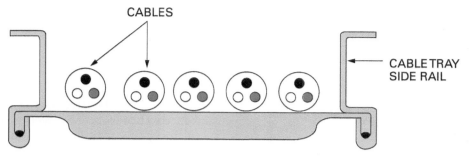

Figure 3 Cables rest on the bottom of the tray and are held in place by the longitudinal side rails.

Special-Application Cable Management Systems

This lightweight, flexible nonmetallic cable management system is used to route wiring to the large motors in a manufacturing facility.

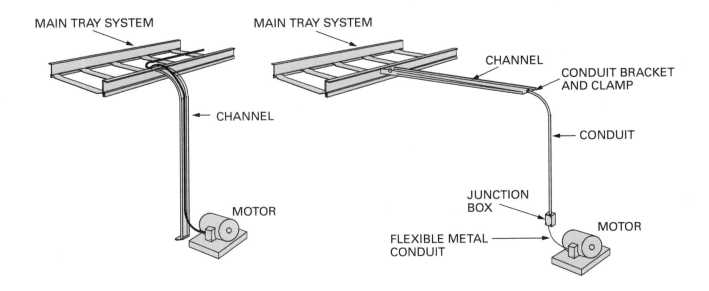

Figure 4 Two applications of cable tray channel.

Ladder Tray

Ladder tray is currently used for a majority of today's installations. It offers numerous advantages:

- Ladder tray has greater air circulation than other trays, which helps to dissipate heat.
- The ladder rungs are convenient for tying down conductors.
- Ladder tray does not collect moisture.
- Conductors can be dropped through the bottom of the tray at any point.

Figure Credit: Tim Dean

Basket Tray

Basket tray offers many of the same advantages as ladder tray. It is lightweight and easy to install using special clips that attach to threaded rods.

Basket tray can be easily cut using bolt cutters or a basket tray cutter, such as that shown here.

One additional advantage of basket tray is that it can be cut, bent, and connected to create tees, crosses, and elbows without the need for separate fittings.

Figure Credit: Tim Dean

Figure Credit: Greenlee / A Textron Company

Figure Credit: Jim Mitchem

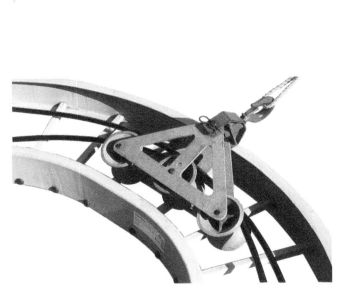
(A) CABLE PULLEY

(B) ROLLER GUIDE

Figure 5 Devices used to facilitate a cable pull in a tray.

Although *NEC Section 392.100(B)* specifically requires that cable trays have smooth edges to ensure that cable will not be damaged, accidents do occur. For example, a tool might be dropped on a cable tray rung during the installation, which may cause a burr or other sharp edge on the rung. Then, after the cable is installed and the system is in use, vibrating machinery may cause this burr to cut into the cable insulation, resulting in a ground fault and possible power outage. Always review the project specifications carefully and/or check with the project supervisor before using either of these methods.

A dropout plate provides a curved surface for the cable to follow as it passes from the tray, as shown in *Figure 7* (A). Without a dropout plate, cables can be bent sharply, causing damage to the insulation, as shown in *Figure 7* (B).

1.1.2 Cable Tray Covers

A cable tray cover is used primarily for two reasons:

- To protect the insulation of the cables against damage that might be caused if an object were to fall into the tray. Prime hazards are tools, discarded cigarettes, and weld splatter.
- To protect certain types of cable insulation against the damaging effects of direct sunlight.

When maximum protection is desired, solid covers should be used. However, if accumulation of heat from the cables is expected, caution should be used. Ventilated covers should be used if some protection of the cable is desired and provisions must be made to allow the escape of heat developed by the cables.

1.1.3 Cable Supports in Vertical Trays

A cable hanger elbow is used to suspend cables in long vertical runs. Care should be taken to ensure that the weight of the suspended length of cable does not exceed the cable manufacturer's recommendation for the maximum allowable tension in the cable.

In short vertical runs, the weight of cables can be supported either by the outside vertical riser elbow or by the vertical straight section when the cables are tied to it. *Figure 8* shows a typical application of a cable hanger elbow in a vertical run.

1.1.4 Cable Edge Protection

The bottom of the solid bottom tray might be convex, concave, or flat. When two pieces of tray are butted together, the bottoms may be out of alignment. An alignment strip, also known as an H bar, is placed between the tray bottoms.

1.1.5 Splice Plates

There are several types of splice plates available, including vertical, horizontal, and expansion plates.

Vertical adjustment splice plates – Vertical adjustment splice plates are used to change the elevation in a run of cable tray. They should not

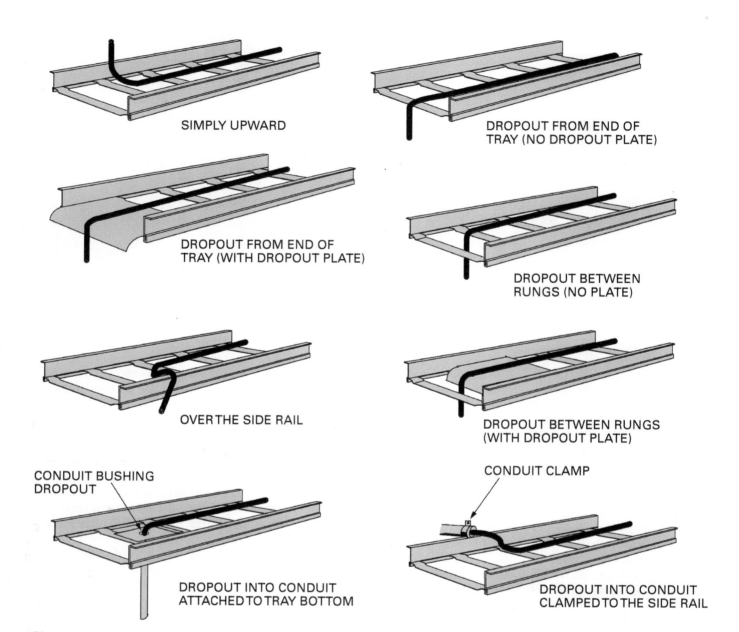

Figure 6 Several ways in which cables may exit from a cable tray.

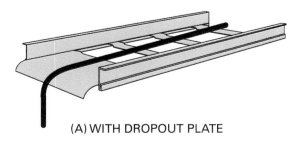
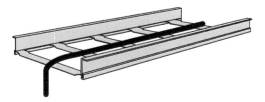

Figure 7 A dropout plate provides a curved surface for the cable to follow as it leaves the tray.

Figure 8 Typical application of supports in a vertical run.

be used where it is important to maintain a cable bending radius. Vertical adjustment splice plates are useful when the change in elevation is so slight or the angle so unusual that it would not be possible to install a standard outside vertical riser elbow and an inside vertical riser elbow in the space available. In general, four swivel plates are used to build an offset in a cable tray system. Once the proper angles have been calculated, proceed as follows:

Step 1 Bolt four swivel plates together at the proper angles, using the inner holes as the center or pivot hole (*Figure 9*).

Step 2 Using a flat surface such as a bench or concrete deck, space two swivel plates at the proper center-to-center distance apart (refer to *Figure 9*).

Step 3 Measure and cut the amount of tray needed to complete the offset.

Horizontal adjustment splice plates – These plates are sometimes used in place of horizontal elbows to change the direction in a run of cable tray. They are used primarily where there is insufficient space or an unusual angle that prevents the use of a standard elbow.

Expansion splice plates – These plates are used at intervals along a straight run of cable tray to allow space for thermal expansion or contraction of the tray to occur, or where offsets or expansion joints occur in the supporting structure. To enable the expansion joint to function properly, the cable tray must be allowed to slide freely on its supports. Any cable tray hold-down device used in an installation subject to expansion or contraction must give clearance to the tray. An expansion joint and splice plates are shown in *Figure 10*.

1.1.6 Barrier Strips

Barrier (divider) strips are used to separate certain types of cable as a result of the nature of the installation, types of circuits used, type of equipment used, local codes, or the *NEC®*. Some reasons for using divider strips are to:

- Separate or isolate electrical circuits
- Separate or isolate cables of different voltages
- Separate cable or wire runs from each other to prevent fire or ground fault damage from spreading to other cables or wires in the same tray
- Promote neatness in the arrangement of the cables
- Warn electricians of the difference between cables on each side of the divider strip

Barrier strip cable protectors are used to cover any raw metal edge over which a cable is to pass. Their purpose is to protect the cable insulation against damage.

1.2.0 Cable Tray Supports

Proper supports for cable tray installations are very important in obtaining a good overall layout. Cable is usually supported in one or more of the following ways:

- Direct rod suspension
- Wall mounting

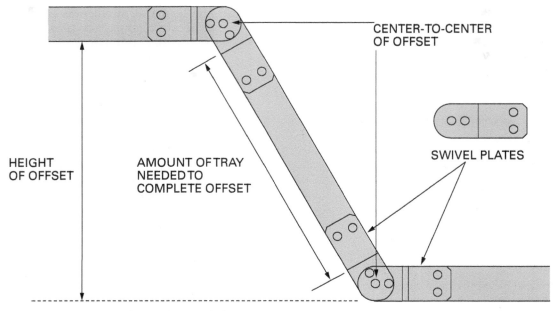

Figure 9 Fabricating a cable tray offset with swivel plates.

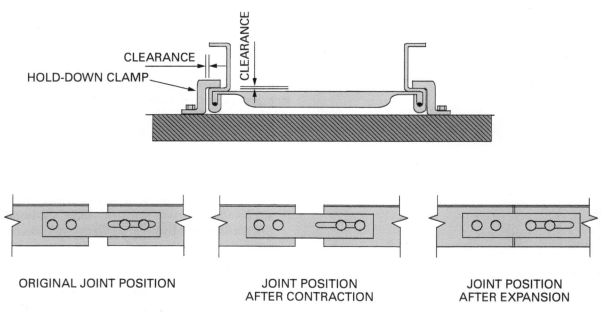

Figure 10 Expansion joint and splice plates.

Contraction and Expansion of Cable Tray

Structural expansion joints do not normally provide a solution to the thermal contraction and expansion of cable trays, because the materials of the structures and trays are different. For example, for a 100°F variance in temperature, a steel tray will require an expansion joint every 128' (39 m) and an aluminum tray will require a joint every 65' (20 m). You must determine the gap setting for the joint according to the temperature at the time of installation. Tables for setting up the gap distances are available from cable tray manufacturers.

- Trapeze mounting
- Center hung support
- Pipe rack mounting
- Monorail systems

1.2.1 Direct Rod Suspension

The direct rod suspension method of supporting cable tray uses threaded rods and hanger clamps. The threaded rod is connected to the ceiling or other overhead structure and is connected to the hanger clamps that are attached to the cable tray side rails, as shown in *Figure 11 (A)*.

1.2.2 Wall Mounting

Wall mounting is accomplished by supporting the tray with structural members attached to the wall, as shown in *Figure 11 (B)*. This method of support is often used in tunnels (mining operations) and other underground or sheltered installations where large numbers of conductors interconnect equipment that is separated by long distances. When using this or any other method of supporting cable tray, always examine the structure to which the hangers are attached, and make absolutely certain that the structure is of adequate strength to support the tray system.

1.2.3 Trapeze Mounting

Trapeze mounting of cable tray, shown in *Figure 11 (C)*, is similar to direct rod suspension mounting. The difference is in the method of attaching the cable tray to the threaded rods. When trapeze mounting is used, a structural member—usually a steel channel or Unistrut®—is connected to the vertical supports to provide an appearance

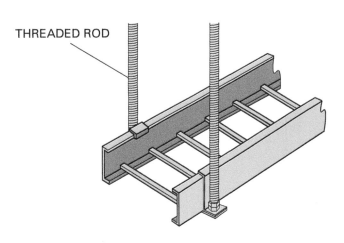

(A) DIRECT ROD SUSPENSION

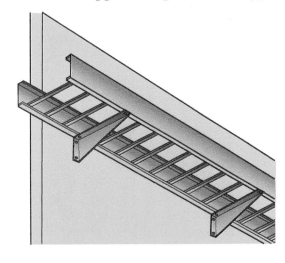

(B) WALL MOUNTING

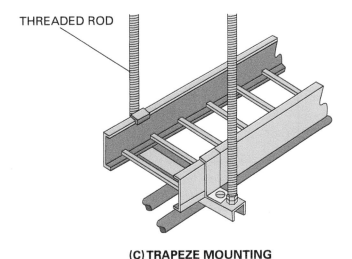

(C) TRAPEZE MOUNTING

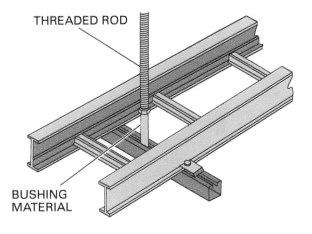

(D) CENTER HUNG SUPPORT

Figure 11 Alternate ways to hang cable tray.

similar to a swing or trapeze (*Figure 12*). The cable tray is mounted to the structural member using bolts, anchor clips, or J-clamps. The underside of the channel or Unistrut® may also be used to support conduit.

1.2.4 Center Hung Support

A method that is similar to trapeze mounting is a center hung tray support, as shown in *Figure 11 (D)*. In this case, only one rod is used and it is centered between the cable tray side rails. A bushing sleeve, such as a short piece of small-diameter PVC, may be used over the center rod to protect the conductors.

1.2.5 Pipe Rack Mounting

Pipe racks are structural frames used to support the piping that interconnects equipment in outdoor industrial facilities. Usually, some space on the rack is reserved for conduit and cable tray. Pipe rack mounting of cable tray is often seen in petrochemical plants where power distribution and instrumentation wiring is routed over a large area and for long distances.

1.2.6 Monorail Systems

Center rail or monorail cable tray systems (*Figure 13*) are light, easy to install, and provide open sides for ready access when changing or adding cables. Center rail cable tray is used in light-duty applications such as sound, telephone, and other communications systems. In addition to its weight limitations, it cannot be used with dividers so its use is restricted to systems where dividers are not necessary.

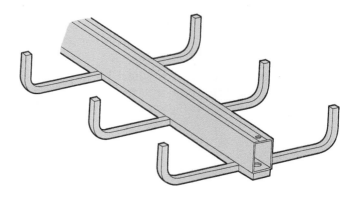

Figure 13 Center rail cable tray.

Figure 12 Channel support.

NEC Section 392.60, Grounding and Bonding

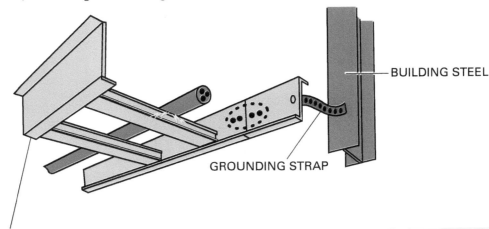

Steel or aluminum cable tray systems are permitted to be used as equipment grounding conductors if they meet all of the requirements in *NEC Section 392.60(B)*.

Proper grounding lessens hazards due to ground faults. Therefore, the *NEC®* requires all metal cable trays to be grounded as required for conductor enclosures in accordance with *NEC Section 250.96*.

Per *NEC Section 392.60(B)(2)*, the minimum cross-sectional area of cable tray must conform to the requirements in *NEC Table 392.60(B)*.

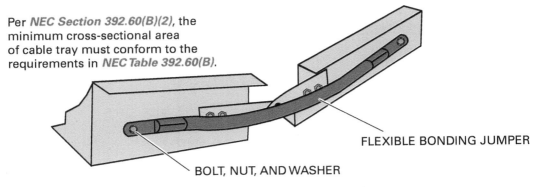

Where supervised by qualified personnel, grounded metal cable trays may also be used as an equipment grounding conductor. *NEC Section 392.60(A)*.

Cable tray sections and fittings must be bonded in accordance with *NEC Section 250.96*.

Figure 20 NEC® regulations governing cable tray grounding.

were provided with the floor plan drawing in *Figure 21*, it would provide a clearer picture of the system and leave little doubt as to how the cable tray system is to be installed. Even new workers in the trade would be able to see how the system should be installed, and this would take some of the load off experienced workers and give them more time to accomplish other tasks.

In actual practice, however, consulting engineering firms seldom furnish the isometric drawing; they merely show the layout in plan view. Consequently, plan views involving the construction details of cable tray systems must be studied carefully during the planning stage—before the work is begun.

> **NOTE:** In some areas, cable tray components are ordered in metric units. A metric conversion chart is included at the back of this module.

2.3.2 Safety Precautions for Cable Tray Installation

Installing cable tray means working at heights above the floor. Consequently, workers must take the necessary precautions. In general, workers installing cable tray will use ladders, scaffolds, or lifts, work from the tray assembly itself, or a combination of all these.

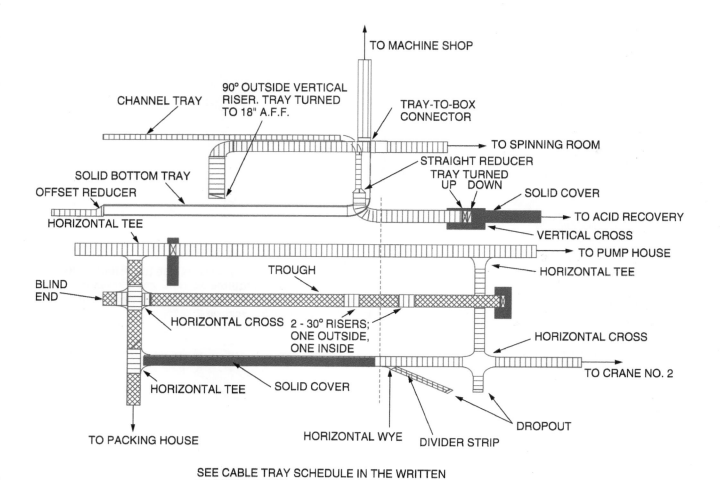

Figure 21 Sample floor plan of a cable tray system.

Keep the tray assembly uncluttered during installation. Tools and tray fittings are ideal obstacles for tripping workers. They may also fall off the tray and injure workers below. To help prevent the latter, set up barriers beneath the section of tray assembly. If you are working on the ground, never cross or move these barriers until the work above is complete.

Learn to secure your lifeline properly, and also make sure your full body harness is a proper fit. A harness that is too large may slip, resulting in a serious injury or even death.

2.0.0 Section Review

1. The expected weight of an installed cable tray system (including cable) is 400 lbs. Each support must be able to carry _____.
    a. 300 lbs
    b. 400 lbs
    c. 500 lbs
    d. 800 lbs

2. When inspecting a long expanse of cable tray, you discover that the tray is buckling. This is a type of _____.
    a. transverse failure
    b. rung failure
    c. side rail failure
    d. splice failure

3. Cable tray must be bonded in accordance with _____.
    a. *NEC Section 250.96*
    b. *NEC Section 300.21*
    c. *NEC Section 300.96*
    d. *NEC Section 392.60*

SECTION THREE

3.0.0 CABLE TRAY FILL

Objective

Determine cable tray fill.
a. Determine the number of conductors allowed in cable tray operating at 2,000V or less.
b. Identify derating factors for cable tray conductors.

NEC Section 392.20 covers the general installation requirements for all conductors used in cable tray systems; that is, splicing, securing, and running conductors in parallel. For example, cable splices are permitted in cable trays provided they are made and insulated by NEC®-approved methods. Furthermore, any splices must be readily accessible and must not project above the side rails of the tray.

In most horizontal runs, the cables may be laid in the tray without securing them in place. However, on vertical runs or any runs other than horizontal, the cables must be secured to transverse members of the cable tray.

Cables may enter and leave a cable tray system in a number of different ways, as discussed previously. In general, no junction box is required where such cables are installed in bushed conduit or tubing. Where conduit or tubing is used, it must be secured to the tray with the proper fittings. Further precautions must be taken to ensure that the cable is not bent sharply as it enters or leaves the conduit or tubing.

3.1.0 Number of Conductors Allowed in Cable Tray (2,000V or Less)

Where single-conductor cables comprising each phase, neutral, or grounded conductor of a circuit are connected in parallel as permitted in *NEC Section 310.10(G)*, the conductors must be installed in groups consisting of not more than one conductor per phase, neutral, or grounded conductor to prevent a current imbalance in the paralleled conductors due to inductive reactance. This also prevents excessive movement due to fault current magnetic forces.

The number of multiconductor cables rated at 2,000V or less that are permitted in a single cable tray must not exceed the requirements of *NEC Section 392.22(A)*. This section applies to both copper and aluminum conductors.

3.1.1 All Conductors Size 4/0 or Larger

Per *NEC Section 392.22(A)(1)(a)*, where all of the cables installed in ladder or ventilated trough tray are 4/0 or larger, the sum of the diameters of all cables shall not exceed the cable tray width, and the cables must be installed in a single layer. For example, if a cable tray installation is to contain three 4/0 multiconductor cables (1.5" in diameter), two 250 kcmil multiconductor cables (1.85"), and two 350 kcmil multiconductor cables (2.5"), the minimum width of the cable tray is determined as follows:

$$3(1.5) + 2(1.85) + 2(2.5) = 13.2"$$

The closest standard cable tray size that meets or exceeds 13.2" is 18" (450 mm). Therefore, this is the size to use.

3.1.2 All Conductors Smaller Than 4/0

Per *NEC Section 392.22(A)(1)(b)*, where all of the cables are smaller than 4/0, the sum of the cross-sectional area of all cables smaller than 4/0 must not exceed the maximum allowable cable fill area as specified in Column 1 of *NEC Table 392.22(A)*; this gives the appropriate cable tray width. To use this table, however, you must have the manufacturer's data for the cables being used. This will give the cross-sectional area of the cables.

The steps involved in determining the size of cable tray for multiconductors smaller than 4/0 AWG are as follows:

Step 1 Calculate the total cross-sectional area of all cables used in the tray. Obtain the area of each from the manufacturer's data.

Step 2 Look in Column 1 of *NEC Table 392.22(A)* and find the smallest number that is at least as large as the calculated number.

Step 3 Look at the number to the left of the row selected in Step 2. This is the minimum width of cable tray that may be used.

For example, determine the minimum cable tray width required for the following multiple conductor cables—all less than 4/0 AWG:

- Four at 1.5" diameter
- Five at 1.75" diameter
- Three at 2.15" diameter

Step 1 Determine the cross-sectional area of the cables from the equation:

$$A = \frac{\pi \times D^2}{4}$$

Where:
$A = area$
$D = diameter$

- The area of a 1.5"-diameter cable is:

$$\frac{(3.14159)(1.5^2)}{4} = 1.7671 \text{ in}^2$$

- The area of the four 1.5" cables is:

$$4 \times 1.7671 \text{ in}^2 = 7.0684 \text{ in}^2$$

- The area of a 1.75"-diameter cable is:

$$\frac{(3.14159)(1.75^2)}{4} = 2.4053 \text{ in}^2$$

- The area of the five 1.75" cables is:

$$5 \times 2.4053 \text{ in}^2 = 12.0265 \text{ in}^2$$

- The area of a 2.15" cable is:

$$\frac{(3.14159)(2.15^2)}{4} = 3.6305 \text{ in}^2$$

- Therefore, the total area of the three 2.15" cables is:

$$3 \times 3.6305 \text{ in}^2 = 10.8915 \text{ in}^2$$

- The total cross-sectional area is found by adding the above three totals to obtain:

$$7.0684 + 12.0265 + 10.8915 = 29.9864 \text{ in}^2$$

Step 2 Look in Column 1 of *NEC Table 392.22(A)* and find the smallest number that is at least as large as 29.9864 in². The number is 35 in² (22,500 mm²).

Step 3 Look to the left of 35 in² (22,500 mm²) and you will see the inside tray width of 30" (750 mm). Therefore, the minimum tray width that can be used for the given group of conductors is 30" (750 mm).

3.1.3 Combination Cables

Per *NEC Section 392.22(A)(1)(c)*, where 4/0 or larger cables are installed in the same ladder or ventilated trough cable tray with cables smaller than 4/0, the sum of the cross-sectional area of all cables smaller than 4/0 must not exceed the maximum allowable fill area from Column 2 of *NEC Table 392.22(A)* for the appropriate cable tray width. The 4/0 and larger cables must be installed in a single layer, and no other cables can be placed on them.

To determine the tray size for a combination of cables as discussed in the above paragraph, proceed as follows:

Step 1 Repeat the steps from the procedure used previously to determine the minimum tray width required for the multiconductor cables having conductors sized 4/0 and larger.

Step 2 Repeat the steps from the procedure used previously to determine the cross-sectional area of all multiconductor cables having conductors smaller than 4/0 AWG.

Step 3 Multiply the result of Step 1 by the constant 1.2 and add this product to the result of Step 2. Call this sum A. Search Column 2 of *NEC Table 392.22(A)* for the smallest number that is at least as large as A. Look to the left in that row to determine the minimum size cable tray required.

To illustrate these steps, assume that you need to find the minimum cable tray width of two multiconductor cables, each with a diameter of 2.54" (conductors size 4/0 or larger); three cables with a diameter of 3.30" (conductors size 4/0 or larger), plus eight cables with a diameter of 1.92" (conductors less than 4/0).

Step 1 The sum of all the diameters of cable having conductors 4/0 or larger is:

$$2(2.54) + 3(3.30) = 14.98"$$

Step 2 The sum of the cross-sectional areas of all cables having conductors smaller than 4/0 is:

$$\frac{(8)(3.14159)(1.92^2)}{4} = 23.1623 \text{ in}^2$$

Step 3 Multiply the result of Step 1 by 1.2 and add this product to the result of Step 2:

$$(1.2 \times 14.98) + 23.1623 = 41.1383 \text{ in}^2$$

Refer to Column 2 of *NEC Table 392.22(A)* and find the closest number to 41.1383. It is 42 in² (27,000 mm²). The tray width that corresponds to 42 in² (27,000 mm²) is 36" (900 mm). Therefore, select a cable tray width of 36" (900 mm).

Low-Voltage Cable

Increasingly, cable trays are being used to carry many low-voltage conductors in communication centers. In a large commercial building, thousands of telecommunications cables are distributed in bundles from the equipment room to the telecommunications closets on each floor. Cable trays offer a convenient means of running these cables through the building as well as a much easier method of allowing for system expansion. Instead of having to access conduit buried within the building walls, a new communication cable can simply be added to the cable tray system, which is normally accessible through a dropped ceiling.

3.1.4 Solid Bottom Tray

Per *NEC Section 392.22(A)(3)*, where solid bottom cable trays contain multiconductor power or lighting cables, or any mixture of multiconductor power, lighting, control, and signal cables, the maximum number of cables must conform to the following:

- Where all of the cables are 4/0 or larger, the sum of the diameters of all cables must not exceed 90% of the cable tray width, and the cables must be installed in a single layer.
- Where all of the cables are smaller than 4/0, the sum of the cross-sectional areas of all cables must not exceed the maximum allowable cable fill area in Column 3 of *NEC Table 392.22(A)* for the appropriate cable tray width.
- Where 4/0 or larger cables are installed in the same cable tray with cables smaller than 4/0, the sum of the cross-sectional areas of all of the smaller cables must not exceed the maximum allowable fill area resulting from the computation in Column 4, *NEC Table 392.22(A)* for the appropriate cable tray width. The 4/0 and larger cables must be installed in a single layer, and no other cables can be placed on them.

What's wrong with this picture?

Figure Credit: Tim Ely

Per *NEC Section 392.22(A)(4)*, where a solid bottom cable tray with a usable inside depth of 6" (150 mm) or less contains multiconductor control and/or signal cables only, the sum of the cross-sectional areas of all cables at any cross section must not exceed 40% of the interior cross-sectional area of the cable tray. A depth of 6" (150 mm) shall be used to compute the allowable interior cross-sectional area of any cable tray that has a usable inside depth of more than 6" (150 mm).

In a previous example, it was determined that the minimum tray size for multiconductor cables with all conductors size 4/0 or larger was 18" (450 mm). The sum of all cable diameters for this example was 13.2". To see if an 18" (450 mm) solid bottom tray can be used, multiply the tray width by 0.90 (90%):

$$18" \times 0.9 = 16.2"$$

Therefore, 16.2" is the minimum width allowed for solid bottom tray. Since we are using 18" (450 mm) tray, this meets the requirements of *NEC Section 392.22(A)(3)(a)*.

When dealing with solid bottom trays and using *NEC Table 392.22(A)*, use Columns 3 and 4 instead of Columns 1 and 2, as used for ladder and trough-type cable trays.

3.1.5 Single-Conductor Cables

Calculating cable tray widths for single-conductor cables (2,000V or under) is similar to the calculations used for multiconductor cables, with the following exceptions:

- Conductors that are 1,000 kcmil and larger are treated the same as multiconductor cables having conductors size 4/0 or larger.
- Conductors that are smaller than 1,000 kcmil are treated the same as multiconductor cables having conductors smaller than size 4/0.

NEC Section 392.22(B) covers the details of installing single-conductor cables with rated voltages of less than 2,000V in cable tray systems.

3.2.0 Ampacity of Cable Tray Conductors

NEC Section 392.80 gives the requirements for cables used in tray systems with rated voltages of 2,000 volts or less. Cables in cable tray use the same ampacity tables as other cable, with additional derating applied for the following conditions:

- *Cable tray construction* – Cable tray can be open or covered, and solid or ventilated (e.g., ladder tray). Covered, unventilated tray is subject to ampacity adjustment.
- *Type(s) of cable* – Cable is available as either single-conductor or multiconductor. Cable with more than three conductors is subject to ampacity adjustment.
- *Ampacity* – For the purposes of derating, cable is separated into two groups: systems carrying less than 2,000V and systems carrying over 2,001V. Systems carrying 2,001V and above are subject to derating per *NEC Section 392.80(B)*.
- *Number of cables and cable configuration in the tray* – Conductors installed in a single layer and those arranged in triangular or square configurations must maintain the free airspace requirements of *NEC Section 392.80(A)(2)(d)* and are subject to the ampacity adjustments listed in *NEC Section 311.60(B) and (C)*.

Think About It

Putting It All Together

If possible, examine the conduit system at your workplace or school. Did the installers make use of cable tray? If not, is the system large enough to warrant the use of it?

3.0.0 Section Review

1. A cable tray contains eight 4/0 cables, and each cable is 1.5" in diameter. The minimum width of the cable tray is _____.
    a. 6"
    b. 8"
    c. 10"
    d. 12"

2. Which of the following is least likely to result in ampacity derating of conductors in cable tray?
    a. Number of cables
    b. Tight cable configuration
    c. Covered tray
    d. Open tray

Review Questions

1. Cable tray is covered in ____.
   a. NEC Article 300
   b. NEC Article 392
   c. NEC Article 517
   d. NEC Article 550

2. Areas requiring voltage isolation are likely to use ____.
   a. stainless steel cable tray
   b. coated aluminum cable tray
   c. nonmetallic cable tray
   d. basket tray

3. When a cable tray section branches off from a main section in two 90° turns, the section is called a(n) ____.
   a. wye
   b. tee
   c. divider strip
   d. ell

4. A section of cable tray that makes a single horizontal 90° turn is known as a(n) ____.
   a. tee
   b. wye
   c. elbow
   d. cross

5. Which of the following best describes the type of fitting that will be used with a solid cover cable tray? ____
   a. An Ericson coupling
   b. Vertical splices
   c. A horizontal cross
   d. A cover clamp

6. Which of the following branches in four different directions? ____
   a. A horizontal wye
   b. A horizontal cross
   c. An inside vertical riser
   d. A dropout

7. When cable exits downward from a tray without a dropout plate, which of the following could occur? ____
   a. Current surges
   b. Reduced amperage
   c. Voltage surges
   d. Insulation damage

8. When a cable tray system must be protected from dropping objects that may damage the cable, a ____.
   a. solid bottom tray is used
   b. trough-type tray is used
   c. tray cover is used
   d. blind end is used

9. The main purpose of vertical adjustment splice plates is to ____.
   a. increase the strength of a cable tray section
   b. support the cable tray system
   c. change the elevation in a run of cable tray
   d. secure the cable within the tray

10. A barrier strip is a ____.
    a. division strip installed in a raceway to separate certain types of cables
    b. strip used to provide the tray with added support
    c. glass strip used as a cable tray insulator
    d. metal strip used solely for grounding purposes

11. Which of the following best describes how closely packed cables will be affected when energized? ____
    a. Their efficiency will be increased.
    b. Their efficiency will be decreased.
    c. No change will be encountered.
    d. The operating temperature of each cable will be lower.

12. Cable in vertical trays should be supported 1.1.3 .
    a. per the manufacturer's recommendations
    b. every 6" or 8"
    c. every 12"
    d. every 24"

13. Which of the following is a violation of *NEC Section 300.8*? 2.3.0 F19
    a. Placing an instrumentation air line in the same tray with electrical conductors
    b. Spacing tray rungs 9" or less apart
    c. Running single conductors in trays that are larger than size 1/0
    d. Using nonmetallic tray systems in corrosive areas

14. When conductors 1/0 through 4/0 are run in ladder-type trays, the maximum rung spacing must not exceed 2.3.0 F19 .
    a. 9"   9"
    b. 8"
    c. 7"
    d. 6"

15. Cable tray ampacity information can be found in 3.2.0 .
    a. *NEC Section 250.96*
    b. *NEC Section 300.96*
    c. *NEC Section 392.60*
    d. *NEC Section 392.80*

Supplemental Exercises

1. Cable supports should be capable of supporting __1.25__ times the full weight of the cable and tray.
2. Where solid bottom cable trays contain multiconductor power or lighting cables, and all cables are larger than 4/0, the sum of the diameters of all cables must not exceed __90%__ of the cable tray width.
3. The two types of cable tray failure under load are __long__ and __traverse__.
4. Cable tray is commonly manufactured in __12"__ and __24"__ lengths.
5. List the five reasons for using divider strip in cable:
   __isolation, seperation   (1.1-6)__
6. Locate splices within __1/4"__ of the span's distance from the nearest support.
7. True or False? Cable tray may require an ampacity adjustment based on construction (open vs. covered).
8. List six types of cable tray supports:
   __pg. 9-11__
9. True or False? Long runs of instrumentation wiring in petrochemical plants are often supported by pipe rack mounting.
10. True or False? Water piping may be run in cable tray with electrical conductors.

Trade Terms Introduced in This Module

Barrier strip: A metal strip constructed to divide a section of cable tray so that certain kinds of cable may be separated from each other.

Cable pulley: A device used to facilitate pulling conductor in cable tray where the tray changes direction. Several types are available (single, triple, etc.) to accommodate almost all pulling situations.

Cross: A four-way section of cable tray used when the tray assembly must branch off in four different directions.

Direct rod suspension: A method used to support cable tray by means of threaded rods and hanger clamps. One end of the threaded rod is secured to an overhead structure, while the other end is connected to hanger clamps that are attached to the cable tray side rails.

Dropout: Cable leaving the tray assembly and travelling directly downward; that is, the cable is not routed into a conduit or channel.

Dropout plate: A metal plate used at the end of a cable tray section to ensure a greater cable bending radius as the cable leaves the tray assembly.

Elbow: A section of cable tray used to change the direction of the tray assembly a full quarter turn (90°). Both vertical and horizontal elbows are common.

Expansion joints: Plates used at intervals along a straight run of cable tray to allow space for thermal expansion or contraction of the tray.

Fittings: Devices used to assemble and/or change the direction of cable tray systems.

Ladder tray: A type of cable tray that consists of two parallel channels connected by rungs, similar in appearance to the common straight ladder.

Pipe racks: Structural frames used to support the piping that interconnects equipment in outdoor industrial facilities.

Swivel plates: Devices used to make vertical offsets in cable tray.

Tee: A section of cable tray that branches off the main section in two other directions.

Trapeze mounting: A method of supporting cable tray using metal channel, such as Unistrut®, Kindorf®, etc., supported by two threaded rods, and giving the appearance of a swing or trapeze.

Tray cover: A flat piece of metal, fiberglass, or plastic designed to provide a solid covering that is needed in some locations where conductors in the tray system may be damaged.

Trough: A type of cable tray consisting of two parallel channels (side rails) having a corrugated, ventilated bottom or a corrugated, solid bottom.

Unistrut®: A brand of metal channel used as the bottom bracket for hanging cable trays. Double Unistrut® adds strength and stability to the trays and also provides a means of securing future runs of conduit.

Wall mounting: A method of supporting cable tray systems using supports secured directly to the wall.

Wye: A section of cable tray that branches off the main section in one direction.

Appendix

Metric Conversion Chart

METRIC CONVERSION CHART

INCHES Fractional	Decimal	METRIC mm	INCHES Fractional	Decimal	METRIC mm	INCHES Fractional	Decimal	METRIC mm
.	0.0039	0.1000	.	0.5512	14.0000	.	1.8898	48.0000
.	0.0079	0.2000	9/16	0.5625	14.2875	.	1.9291	49.0000
.	0.0118	0.3000	.	0.5709	14.5000	.	1.9685	50.0000
1/64	0.0156	0.3969	37/64	0.5781	14.6844	2	2.0000	50.8000
.	0.0157	0.4000	.	0.5906	15.0000	.	2.0079	51.0000
.	0.0197	0.5000	19/32	0.5938	15.0813	.	2.0472	52.0000
.	0.0236	0.6000	39/64	0.6094	15.4781	.	2.0866	53.0000
.	0.0276	0.7000	.	0.6102	15.5000	.	2.1260	54.0000
1/32	0.0313	0.7938	5/8	0.6250	15.8750	.	2.1654	55.0000
.	0.0315	0.8000	.	0.6299	16.0000	.	2.2047	56.0000
.	0.0354	0.9000	41/64	0.6406	16.2719	.	2.2441	57.0000
.	0.0394	1.0000	.	0.6496	16.5000	2 1/4	2.2500	57.1500
.	0.0433	1.1000	21/32	0.6563	16.6688	.	2.2835	58.0000
3/64	0.0469	1.1906	.	0.6693	17.0000	.	2.3228	59.0000
.	0.0472	1.2000	43/64	0.6719	17.0656	.	2.3622	60.0000
.	0.0512	1.3000	11/16	0.6875	17.4625	.	2.4016	61.0000
.	0.0551	1.4000	.	0.6890	17.5000	.	2.4409	62.0000
.	0.0591	1.5000	45/64	0.7031	17.8594	.	2.4803	63.0000
1/16	0.0625	1.5875	.	0.7087	18.0000	2 1/2	2.5000	63.5000
.	0.0630	1.6000	23/32	0.7188	18.2563	.	2.5197	64.0000
.	0.0669	1.7000	.	0.7283	18.5000	.	2.5591	65.0000
.	0.0709	1.8000	47/64	0.7344	18.6531	.	2.5984	66.0000
.	0.0748	1.9000	.	0.7480	19.0000	.	2.6378	67.0000
5/64	0.0781	1.9844	3/4	0.7500	19.0500	.	2.6772	68.0000
.	0.0787	2.0000	49/64	0.7656	19.4469	.	2.7165	69.0000
.	0.0827	2.1000	.	0.7677	19.5000	2 3/4	2.7500	69.8500
.	0.0866	2.2000	25/32	0.7813	19.8438	.	2.7559	70.0000
.	0.0906	2.3000	.	0.7874	20.0000	.	2.7953	71.0000
3/32	0.0938	2.3813	51/64	0.7969	20.2406	.	2.8346	72.0000
.	0.0945	2.4000	.	0.8071	20.5000	.	2.8740	73.0000
.	0.0984	2.5000	13/16	0.8125	20.6375	.	2.9134	74.0000
7/64	0.1094	2.7781	.	0.8268	21.0000	.	2.9528	75.0000
.	0.1181	3.0000	53/64	0.8281	21.0344	.	2.9921	76.0000
1/8	0.1250	3.1750	27/32	0.8438	21.4313	3	3.0000	76.2000
.	0.1378	3.5000	.	0.8465	21.5000	.	3.0315	77.0000
9/64	0.1406	3.5719	55/64	0.8594	21.8281	.	3.0709	78.0000
5/32	0.1563	3.9688	.	0.8661	22.0000	.	3.1102	79.0000
.	0.1575	4.0000	7/8	0.8750	22.2250	.	3.1496	80.0000
11/64	0.1719	4.3656	.	.8858	22.5000	.	3.1890	81.0000
.	0.1772	4.5000	.	.89063	22.6219	.	3.2283	82.0000
3/16	0.1875	4.7625	57/64	.9055	23.0000	.	3.2677	83.0000
.	0.1969	5.0000	29/32	.90625	23.0188	.	3.3071	84.0000
13/64	0.2031	5.1594	59/64	.92188	23.4156	.	3.3465	85.0000
.	0.2165	5.5000	.	.9252	23.5000	.	3.3858	86.0000
7/32	0.2188	5.5563	15/16	.93750	23.8125	.	3.4252	87.0000
15/64	0.2344	5.9531	.	.9449	24.0000	.	3.4646	88.0000
.	0.2362	6.0000	61/64	.95313	24.2094	3 1/2	3.5000	88.9000
1/4	0.2500	6.3500	.	.9646	24.5000	.	3.5039	89.0000
.	0.2559	6.5000	31/32	.96875	24.6063	.	3.5433	90.0000
17/64	0.2656	6.7469	.	.9843	25.0000	.	3.5827	91.0000
.	0.2756	7.0000	63/64	.98438	25.0031	.	3.6220	92.0000
9/32	0.2813	7.1438	1	1.000	25.40	.	3.6614	93.0000
.	0.2953	7.5000	.	1.0039	25.5000	.	3.7008	94.0000
19/64	0.2969	7.5406	.	1.0236	26.0000	.	3.7402	95.0000
5/16	0.3125	7.9375	.	1.0433	26.5000	.	3.7795	96.0000
.	0.3150	8.0000	.	1.0630	27.0000	.	3.8189	97.0000
21/64	0.3281	8.3344	.	1.0827	27.5000	.	3.8583	98.0000
.	0.3346	8.5000	.	1.1024	28.0000	.	3.8976	99.0000
11/32	0.3438	8.7313	.	1.1220	28.5000	.	3.9370	100.0000
.	0.3543	9.0000	.	1.1417	29.0000	4	4.0000	101.6000
23/64	0.3594	9.1281	.	1.1614	29.5000	.	4.3307	110.0000
.	0.3740	9.5000	.	1.1811	30.0000	4 1/2	4.5000	114.3000
3/8	0.3750	9.5250	.	1.2205	31.0000	.	4.7244	120.0000
25/64	0.3906	9.9219	1 1/4	1.2500	31.7500	5	5.0000	127.0000
.	0.3937	10.0000	.	1.2598	32.0000	.	5.1181	130.0000
13/32	0.4063	10.3188	.	1.2992	33.0000	.	5.5118	140.0000
.	0.4134	10.5000	.	1.3386	34.0000	.	5.9055	150.0000
27/64	0.4219	10.7156	.	1.3780	35.0000	6	6.0000	152.4000
.	0.4331	11.0000	.	1.4173	36.0000	.	6.2992	160.0000
7/16	0.4375	11.1125	.	1.4567	37.0000	.	6.6929	170.0000
.	0.4528	11.5000	.	1.4961	38.0000	.	7.0866	180.0000
29/64	0.4531	11.5094	1 1/2	1.5000	38.1000	.	7.4803	190.0000
15/32	0.4688	11.9063	.	1.5354	39.0000	.	7.8740	200.0000
.	0.4724	12.0000	.	1.5748	40.0000	8	8.0000	203.2000
31/64	0.4844	12.3031	.	1.6142	41.0000	.	9.8425	250.0000
.	0.4921	12.5000	.	1.6535	42.0000	10	10.0000	254.0000
1/2	0.5000	12.7000	.	1.6929	43.0000	20	20.0000	508.0000
.	0.5118	13.0000	.	1.7323	44.0000	30	30.0000	762.0000
33/64	0.5156	13.0969	1 3/4	1.7500	44.4500	40	40.0000	1016.000
17/32	0.5313	13.4938	.	1.7717	45.0000	60	60.0000	1524.000
.	0.5315	13.5000	.	1.8110	46.0000	80	80.0000	2032.000
35/64	0.5469	13.8906	.	1.8504	47.0000	100	100.0000	2540.000

TO CONVERT TO MILLIMETERS, MULTIPLY INCHES × 25.4*
TO CONVERT TO INCHES, MULTIPLY MILLIMETERS × 0.03937*
*FOR SLIGHTLY GREATER ACCURACY WHEN CONVERTING TO INCHES, DIVIDE MILLIMETERS BY 25.4

Additional Resources

This module presents thorough resources for task training. The following resource material is suggested for further study.

Cooper Industries Cable Tray Manual (pdf). **www.cooperindustries.com**.
Metal Cable Tray Systems, NEMA VE 1-2017. **www.nema.org**.
National Electrical Code® Handbook, Latest Edition. Quincy, MA: National Fire Protection Association.
Paralleled Phase Conductors in Cable Trays Provide Copper Savings, Cable Tray Institute Bulletin. **www.cabletrays.org**.
Safely Installing, Maintaining and Inspecting Cable Trays, Safety and Health Information Bulletin. **www.osha.org**.

Figure Credits

Greenlee / A Textron Company, Module Opener, Figure 5B
John Traister, Figures 1–3, 18
Tim Dean, Figures 5A, 8, 12

Section Review Answer Key

SECTION 1.0.0

Answer	Section Reference	Objective
1. b	1.1.5	1a
2. a	1.2.6	1b

SECTION 2.0.0

Answer	Section Reference	Objective
1. c	2.1.0	2a
2. c	2.2.0	2b
3. a	2.3.0; *Figure 20*	2c

SECTION 3.0.0

Answer	Section Reference	Objective
1. d	3.1.1	3a
2. d	3.2.0	3b

This page is intentionally left blank.

26208-20
CONDUCTOR TERMINATIONS AND SPLICES

Objectives

When you have completed this module, you will be able to do the following:

1. Strip and train conductors.
   a. Strip small conductors.
   b. Strip large conductors.
   c. Bend cable and train conductors.
2. Make wire connections.
   a. Install various types of connectors.
   b. Make aluminum connections.
   c. Install control and signal cables.
3. Reinsulate electrical connections.
   a. Tape electrical connections.
   b. Install heat-shrink insulators.
   c. Use motor connection kits.

Performance Tasks

Under the supervision of the instructor, you should be able to do the following:

1. Terminate conductors using selected crimp-type and mechanical-type terminals and connectors.
2. Terminate conductors on a terminal strip.
3. Insulate selected types of wire splices and/or install a motor connection kit.

Trade Terms

AL-CU
Amperage capacity
Connection
Connector
Drain wire
Grooming

Insulating tape
Lug
Mechanical advantage (MA)
Pressure connector
Reducing connector
Shielding

Splice
Strand
Terminal
Termination

Industry Recognized Credentials

If you are training through an NCCER-accredited sponsor, you may be eligible for credentials from NCCER's Registry. The ID number for this module is 26208-20. Note that this module may have been used in other NCCER curricula and may apply to other level completions. Contact NCCER's Registry at 888.622.3720 or go to **www.nccer.org** for more information.

> **NOTE:** NFPA 70®, *National Electrical Code*® and *NEC*® are registered trademarks of the National Fire Protection Association, Quincy, MA.

Contents

1.0.0 Stripping and Training Conductors 1
   1.1.0 Stripping Small Conductors 2
   1.2.0 Stripping Large Conductors 4
   1.3.0 Bending Cable and Training Conductors 6
      1.3.1 Minimum Bending Radius 7
      1.3.2 Short Circuit Bracing 10
2.0.0 Making Wire Connections 12
   2.1.0 Installing Connectors 14
      2.1.1 Installing Compression Connectors 14
      2.1.2 Mechanical Terminals and Connectors 16
      2.1.3 Specialized Cable Connectors 17
   2.2.0 Aluminum Connections 18
   2.3.0 Control and Signal Cables 19
      2.3.1 Crimp Connectors for Screw Terminals 19
      2.3.2 Crimping Procedure 21
      2.3.3 Termination Inspection 22
      2.3.4 Terminal Block Connections 23
3.0.0 Reinsulating Electrical Connections 26
   3.1.0 Taping 26
   3.2.0 Heat-Shrink Insulators 27
   3.3.0 Motor Connection Kits 28

Figures and Tables

Figure 1   Wire stripper/crimper 2
Figure 2   Wire strippers 3
Figure 3   Chisel point on a conductor 3
Figure 4   Cable and wire stripping tools 3
Figure 5   Improper and Proper stripping length 4
Figure 6   Ratchet-type cable cutter 5
Figure 7   Heavy-duty cable stripper 5
Figure 8   Types of cable stripping 6
Figure 9   Round cable slitting and ringing tool 7
Figure 10 Terminal bend radius 7
Figure 11 Ratchet bender 8
Figure 12 Hydraulic bender 8
Figure 13 Bending space at terminals is measured in a straight line 8
Figure 14 Conductors entering an enclosure
           opposite the conductor terminals 9

Figure 15	Incoming feeders connected to horizontal bus	9
Figure 16	MCC fed directly from a transformer secondary	10
Figure 17	Crimp-on wire lugs	12
Figure 18	Various mechanical compression connectors	13
Figure 19	Mechanical strength versus electrical performance of a crimped connector	14
Figure 20	Hand crimpers	14
Figure 21	Leveraged crimping tool	14
Figure 22	Crimping tools used to crimp large connectors	15
Figure 23	Battery-operated crimping tool	15
Figure 24	Corded crimping tool	16
Figure 25	Universal crimping tool	16
Figure 26	Multiple crimps	17
Figure 27	Weatherproof connector used with Type MC cable	18
Figure 28	Fire alarm and instrumentation cable	19
Figure 29	Basic crimp connector structure	20
Figure 30	Standard tongue styles of crimped connectors	20
Figure 31	Labeled and terminated control cable	21
Figure 32	Indent position	22
Figure 33	Crimp centering	22
Figure 34	Conductor positioning	23
Figure 35	Terminal blocks	24
Figure 36	Routing cabling	24
Figure 37	Typical method for taping motor lug connections	27
Figure 38	Typical method of taping a split-bolt connector	28
Figure 39	Method of installing heat-shrink insulators	29
Figure 40	Stub and in-line splice connections	30
Figure 41	Motor connection kits installed on splices	30
Table 1	Dimensions of Common Wire Sizes	2
Table 2	Minimum Wire Bending Space for Conductors Not Entering or Leaving Opposite Wall [Data from *NEC Table 312.6(A)*]	8
Table 3	Minimum Wire Bending Space for Conductors Entering or Leaving Opposite Wall [Data from *NEC Table 312.6(B)*]	9
Table 4	Recommended Tightening Torques for Various Bolt Sizes	17
Table 5	Typical Color Codes	20
Table 6	Tubing Selector Guide	29

This page is intentionally left blank.

Section One

1.0.0 Stripping and Training Conductors

Objective

Strip and train conductors.
a. Strip small conductors.
b. Strip large conductors.
c. Bend cable and train conductors.

Performance Tasks

1. Terminate conductors using selected crimp-type and mechanical-type terminals and connectors.
2. Terminate conductors on a terminal strip.

Trade Terms

Connection: That part of a circuit that has negligible impedance and joins components or devices.

Connector: A device used to physically and electrically connect two or more conductors.

Lug: A device for terminating a conductor to facilitate the mechanical connection.

Pressure connector: A connector applied using pressure to form a cold weld between the conductor and the connector.

Splice: The electrical and mechanical connection between two pieces of cable.

Strand: A group of wires, usually stranded or braided.

Terminal: A device used for connecting cables.

Termination: The connection of a cable.

Anyone involved with electrical systems of any type must be familiar with the wire connector and splicing, as they are both necessary to make the numerous electrical joints required during the course of an electrical installation. A properly made splice and connection should last as long as the insulation on the wire itself, while a poorly made connection will always be a source of trouble; that is, the joints will overheat under load and eventually fail with the potential for starting a fire. The majority of failures occur at terminations.

The basic requirements for a good electrical connection include the following:

- It must be mechanically and electrically secure.
- It must be insulated as well as or better than the existing insulation on the conductors.
- These characteristics should last as long as the conductor is in service.

> **NOTE**
> Every splice is a point of potential failure. Therefore, splicing should not be used as a solution to providing shorter conductor pulls. In fact, some applications prohibit the use of splices in critical areas.

There are many different types of electrical joints, and the selection of the proper type for a given application often depends on how and where the splice or connection is used. Electrical joints are normally made with a solderless pressure connector or lug to save time.

Before any connection or splice can be made, the ends of the conductors must be properly cleaned and stripped. To ensure a low-resistance connection and avoid contaminating the termination, clean the areas of the cable where it is to be cut and stripped. Remove any pulling compound, dirt, oil, grease, or water to avoid contaminating the exposed conductor.

Stripping is the removal of insulation from the end of the conductor or at the location of the splice. Conductors should only be stripped using the appropriate stripping tool. This will help to prevent cuts and nicks in the wire, which can reduce the conductor area and weaken the conductor.

Poorly stripped conductors can result in nicks, scrapes, or burnishes. Any of these can lead to a stress concentration in the damaged area. Heat, rapid temperature changes, mechanical vibration, and oscillatory motion can aggravate the damage,

Conductor Terminations and Splices

Poor electrical connections are responsible for a large percentage of equipment burnouts and fires. Many of these failures are a direct result of improper terminations, poor workmanship, and the use of improper splicing devices.

causing faults in the circuitry or even total failure. A lost strand is a problem in a splice terminal or crimp-type terminal, whereas an exposed strand might be a safety hazard.

Faulty stripping can pierce, scuff, or split the insulation. This can cause changes in dielectric strength and lower the conductor's resistance to moisture and abrasion. Insulation particles often get trapped in solder and crimp joints. These form the basis for a defective termination. A variety of factors determine how precisely a conductor can be stripped, including the wire size and type of insulation.

It is a common misconception that a certain gauge of stranded conductor has the same diameter as a solid conductor. This is a very important consideration in selecting the proper blades for strippers. *Table 1* shows the nominal sizes referenced for the different wire gauges.

To eliminate nicking, cutting, and fraying, wires should only be stripped using the appropriate stripping tool. The specific tool used depends on the size and type of wire being stripped.

1.1.0 Stripping Small Conductors

There are many kinds of wire strippers available. *Figure 1* shows a common type of wire stripper for small conductors. It can be used to cut, strip, and crimp wires from No. 22 through No. 10 AWG. To use this tool, insert the conductor into the proper size knife groove, then squeeze the tool handles. The tool cuts the conductor insulation, allowing the conductor to be easily removed without crushing its stripped end. The length of the strip is regulated by the amount of wire extending beyond the blades when it is inserted in the knife groove.

Figure 2 shows production-grade stripping tools. The tool in *Figure 2 (A)* can be used to strip conductors from No. 20 to No. 10 AWG, whereas the tool in *Figure 2 (B)* strips wires from No. 18 to No. 6 AWG. Note that this tool has front entry jaws for use in tight spaces.

When stripping small conductors such as control and signal wiring, a scissor-action type of cutting tool is preferable to cutting tools with jaws that butt against each other. These tools have a tendency to produce a flattened chisel end on the conductor, especially when the cutting edges become dull, as shown in *Figure 3*. This makes it difficult to insert the conductor into the barrel of a terminal.

Observe the following points when stripping conductors:

Table 1 Dimensions of Common Wire Sizes

Size (AWG/kcmil)	Area (Circular Mils)	Overall Diameter in Inches	
		Solid	Stranded
18	1,620	0.040	0.046
16	2,580	0.051	0.058
14	4,130	0.064	0.073
12	6,530	0.081	0.092
10	10,380	0.102	0.116
8	16,510	0.128	0.146
6	26,240	—	0.184
4	41,740	—	0.232
3	52,620	—	0.260
2	66,360	—	0.292
1	83,690	—	0.332
1/0	105,600	—	0.373
2/0	133,100	—	0.419
3/0	167,800	—	0.470
4/0	211,600	—	0.528
250	—	—	0.575
300	—	—	0.630
350	—	—	0.681
400	—	—	0.728
500	—	—	0.813
600	—	—	0.893
700	—	—	0.964
750	—	—	0.998
800	—	—	1.030
900	—	—	1.090
1,000	—	—	1.15
1,250	—	—	1.29
1,500	—	—	1.41
1,750	—	—	1.52
2,000	—	—	1.63

Figure 1 Wire stripper/crimper.

- Remove the cable jacket using strippers with an adjustable blade or a die designed for the particular wire size (*Figure 4*). The terminal manufacturer will recommend a stripping length. Be careful to avoid nicking or stretching the

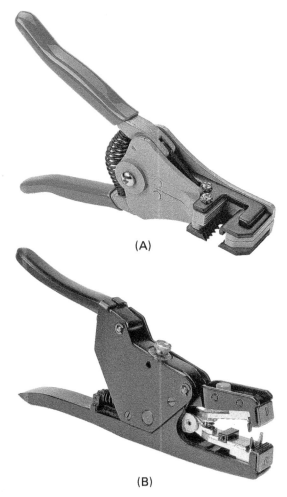

Figure 2 Wire strippers.

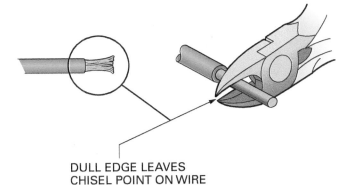

Figure 3 Chisel point on a conductor.

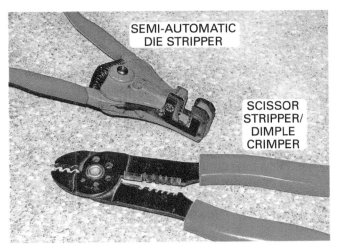

Figure 4 Cable and wire stripping tools.

conductor or insulation in a multi-conductor cable. If this type of damage occurs, it is likely that the wrong groove of the stripping tool was used, the tool was improperly set or damaged, the wrong type of stripping blade was used, or the tool was used incorrectly.

- Cut the insulation so that no frayed pieces or threads extend past the point of cutoff. Frayed pieces or threads of insulation indicate the use of an improper tool or dull cutter blades.
- If possible, do not twist, spread, or disturb the wire strands from their original position in the cable. Retwisting or tightening the twist of the strands will eventually result in damage.
- Terminate stripped wire as soon as possible. The exposed strands will invariably become bent and spread, making termination difficult. A minimum amount of handling and storage after stripping will result in better terminations.

Cable Strippers

Open-blade (knife) stripping can damage the conductors and is prohibited on many job sites. Adjustable strippers such as the tool shown here are used instead.

Figure Credit: Greenlee / A Textron Company

The last image in *Figure 5 (C)* shows the positioning of the wire in the crimp barrel when stripped to the proper length. The conductor insulation must be in the belled mouth of the terminal. This relieves stress on the strands or wire and increases the strength of the connection. Allowing conductor strands to protrude out of the inspection hole more than $\frac{1}{32}$" (1 mm) will interfere with the terminal screw. Cutting the strands too short will reduce the contact surface area.

1.2.0 Stripping Large Conductors

Larger conductors such as power cable can be cut using a ratchet-type cable cutter. The cable cutter shown in *Figure 6* can be used to cut cables up to 1,000 kcmil. Battery-operated cable cutters are also available. Heavy-duty strippers are used to strip large power cables. *Figure 7* shows a heavy-duty stripper used to strip power cables with outside diameters ranging from 1/0 through 1,000 kcmil. Strippers can be used to remove insulation from the end of a cable or to make window cuts (*Figure 8*). Window cuts are often used on shielded cable to connect earth bonds. All stripping tools should be operated according to the manufacturer's instructions. The procedures for using the tool shown in *Figure 7* to strip insulation from the end of a cable and to make a window cut are described here.

> **WARNING!** Keep fingers away from the blade when using any stripping tool and replace the blade when it becomes dull. A dull blade is dangerous because you are more likely to apply undue pressure, causing the tool to slip.

To strip insulation from the end of a cable, proceed as follows:

Step 1 Loosen the locking knob to open the tool to the maximum position. Place the cable in the V-groove and close the tool firmly around the cable. Tighten the locking knob.

Step 2 Turn the cap assembly until the blade reaches the required depth.

Stripping Cables

When using any bladed-type cable stripper, make sure to replace the blade whenever it becomes dull. Remember, a dull blade is more dangerous than a sharp one because you are more likely to apply excessive pressure when using a dull blade, causing it to slip.

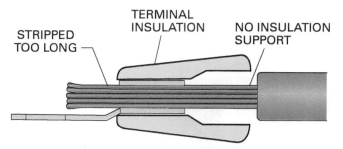

(A) STRIPPING THAT IS TOO LONG WILL INTERFERE WITH THE TERMINAL SCREW

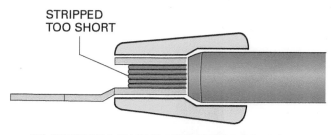

(B) STRIPPING THAT IS TOO SHORT DOES NOT PROVIDE ENOUGH CONTACT SURFACE

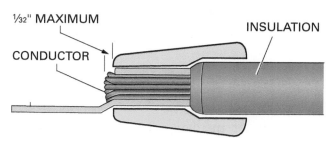

(C) PROPER STRIPPING LENGTH WITH INSULATION INSIDE THE TERMINAL

Figure 5 Improper and Proper stripping length.

> **CAUTION** Do not allow the blade to contact the conductor because damage to the conductor and/or the blade can result.

Step 3 Rotate the tool around the cable, advancing to the required strip length.

Step 4 Rotate the tool in the reverse direction to produce a square end cut.

Step 5 Loosen the locking knob to release the tool and remove it from the cable.

Step 6 Peel off the insulation.

(A) RATCHET-TYPE CABLE CUTTER

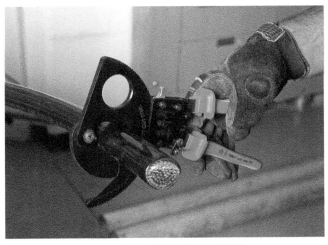

(B) RATCHET CUTTER IN USE

Figure 6 Ratchet-type cable cutter.

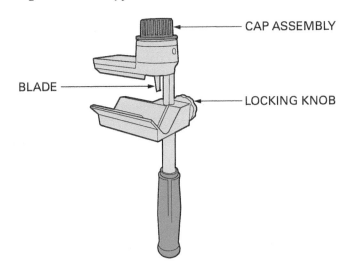

Figure 7 Heavy-duty cable stripper.

Cutting Large Power Cable

When cutting large power cables before stripping, you can make the job easier by using a battery-powered cable cutter such as the one shown here. Electric and hydraulic cable cutters are also available.

Figure Credit: Greenlee / A Textron Company

To make a window cut, proceed as follows:

Step 1 With the tool opened to the maximum position, place the cable in the V-groove and close the tool firmly around the cable. Tighten the locking knob.

Step 2 Turn the cap assembly until the blade reaches the required depth.

Step 3 Rotate the tool to produce the first square cut.

Step 4 Rotate the tool in the reverse direction to cut the required window strip length.

Step 5 Rotate the tool in the original direction to produce the second square cut.

Step 6 Loosen the locking knob assembly to release the tool and remove it from the cable.

Step 7 Peel off the insulation.

Figure 9 shows a round cable slitting and ringing tool (also known as a *cable stripping tool*) that can be used to strip single- or multi-conductor cables. This tool can be used to cut around the cable (square cut) or slit the length of the cable jacket (longitudinal cut) for easy removal. The tool blade is adjustable to accommodate different jacket thicknesses.

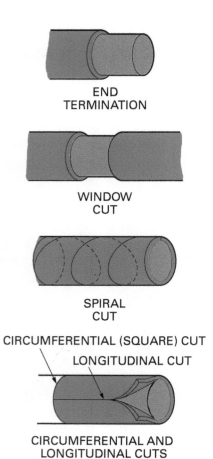

Figure 8 Types of cable stripping.

(A) CABLE STRIPPING TOOL

(B) CABLE STRIPPING TOOL IN USE

Figure 9 Round cable slitting and ringing tool.

1.3.0 Bending Cable and Training Conductors

Training is the positioning of cable so that it is not under tension. Bending is the positioning of cable that is under tension. When installing cable or any large conductors, the object is to limit the tension so that the cable's physical and electrical characteristics are maintained for the expected service life. Training conductors, rather than bending them, also reduces the tension on lugs and connectors, extending their service life considerably. Cable/conductor routing and inspection considerations include the following:

- *Wire bends* – For all wire and cable routing, use a minimum bend radius of three times the outside diameter of the wire or cable unless otherwise specified by the cable manufacturer.
- *Neatness of routing* – Provide adequate clearance for movement of mechanical parts. Allow enough slack in the cable for servicing plugs and terminal boards. Support all cables to prevent strain on the conductors or terminals. Strain can cause breaks in the terminations and render the system inoperable. Ensure that cables do not cross maintenance openings or otherwise interfere with normal operation and replacement of components.
- *Protection of wires or cable* – Provide clearance between the conductors and any heat-radiating components to prevent heat-related deterioration of the insulation. Protect all wires against abrasion. Do not route wire over sharp screws, lugs, terminals, or openings. Use grommets, chase nipples, or similar means to protect the insulation when passing wire through a metal partition.
- *Service loops* – Allow sufficient slack at the termination of each wire to permit one repair (such as cutting off a terminal and re-terminating it).
- *Jumper wires* – Bare jumper wires are not permitted. Use only insulated wire for jumpers.
- *Terminal bending limits* – Do not bend terminals more than 30 degrees above or below the termination point, as shown in *Figure 10*.

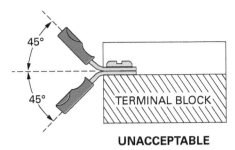
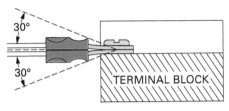

Figure 10 Terminal bend radius.

1.3.1 Minimum Bending Radius

All bends made in cable must comply with the *NEC®*. The minimum bending radius is determined by the cable diameter and, in some instances, by the construction of the cable. For example, *NEC Section 330.24* states that bends in Type MC cable must be made so that the cable is not damaged, and the radius of the curve of the inner edge of any bend shall not be less than the following:

- *Smooth sheath* – Ten times the external diameter of the metallic sheath for cable not more than 3/4" (19 mm) in external diameter, twelve times the external diameter of the metallic sheath for cable more than 3/4" (19 mm) but not more than 1 1/2" (38 mm) in external diameter, and fifteen times the external diameter of the metallic sheath for cable more than 1 1/2" (38 mm) in external diameter.
- *Interlocked-type armor or corrugated sheath* – Seven times the external diameter of the metal sheath.
- *Shielded conductors* – Twelve times the overall diameter of the individual conductors or seven times the overall diameter of the multiconductor cable, whichever is greater.

Two common types of cable bending tools are the ratchet bender and the hydraulic bender. The ratchet cable bender in *Figure 11* bends 600V copper or aluminum conductors up to 500 kcmil, and the hydraulic bender in *Figure 12* is designed for cables from 350 kcmil through 1,000 kcmil. In addition, the hydraulic bender is capable of one-shot bends up to 90° and automatically unloading the cable when the bend is finished. Either type simplifies and speeds cable installation.

Depth Gauges

This cable stripper attaches to a battery-operated drill and is equipped with a depth gauge to make precise cutbacks on a variety of cables.

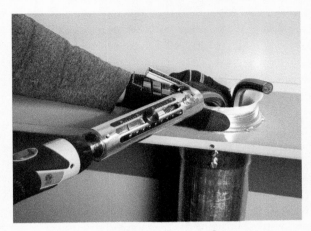

Figure Credit: Greenlee / A Textron Company

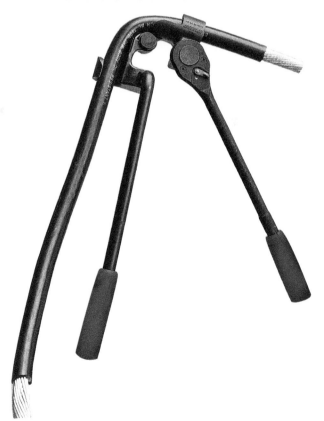

Figure 11 Ratchet bender.

Figure 12 Hydraulic bender.

Conductors at terminals or conductors entering or leaving cabinets, cutout boxes, and other enclosures must comply with certain *NEC®* requirements, many of which are covered in *NEC Article 312*. The bending radii for various sizes of conductors that do not enter or leave an enclosure through the wall opposite its terminal are shown in *Table 2*. When using this table, the bending space at terminals must be measured in a straight line from the end of the lug or wire connector (in the direction that the wire leaves the terminal) to the wall, barrier, or obstruction, as shown in *Figure 13*.

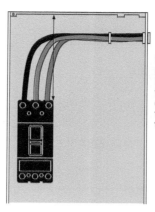

When using *NEC Table 312.6(A)*, bending space at terminals must be measured in a straight line from the end of the lug or wire connector (in the direction that the wire leaves the terminals) to the wall, barrier, or obstruction.

Figure 13 Bending space at terminals is measured in a straight line.

> **NOTE:** Remember that cable offsets are bends.

An unshielded cable can tolerate a sharper bend than a shielded cable. This is especially true of cables having helical metal tapes, which, when bent too sharply, can separate or buckle and cut into the insulation. This causes increased electrical stress. The problem is compounded by the fact that most tapes are under jackets that conceal such damage. Damaged cable may initially pass acceptance testing, but it often fails prematurely at the shield/insulation interface.

When conductors enter or leave an enclosure through the wall opposite its terminals (*Figure 14*), *NEC Table 312.6(B)* applies (*Table 3*). When using this table, the bending space at terminals must be measured in a straight line from the end of the lug or wire connector in a direction perpendicular to the enclosure wall. For removable and lay-in wire terminals intended for only one wire, the bending space in the table may be reduced by the number of inches shown in parentheses.

Figure 15 shows feeders from a 480V switchboard terminating in the main lugs only (MLO) incoming line compartment of an MCC. This connection is made directly to the horizontal bus system distributing power to the vertical bus in each vertical section of the MCC.

Figure 16 shows a circuit interrupter and fuses serving as the main disconnect and overcurrent protection for an MCC fed directly from a transformer secondary. It also illustrates the *NEC®* requirements for conductors entering enclosures and the use of enclosures for routing or tapping conductors. The load side of the overcurrent protective device is connected to the MCC busbar

Table 2 Minimum Wire Bending Space for Conductors Not Entering or Leaving Opposite Wall [Data from *NEC Table 312.6(A)*]

AWG or Circular-Mil Size of Wire	Wires Per Terminal				
	1	2	3	4	5
14–10	Not Specified	—	—	—	—
8–6	1½	—	—	—	—
4–3	2	—	—	—	—
2	2½	—	—	—	—
1	3	—	—	—	—
1/0 – 2/0	3½	5	7	—	—
3/0 – 4/0	4	6	8	—	—
250 kcmil	4½	6	8	10	—
300–350 kcmil	5	8	10	12	—
400–500 kcmil	6	8	10	12	14
600–700 kcmil	8	10	12	14	16
750–900 kcmil	8	12	14	16	18
1,000–1,250 kcmil	10	—	—	—	—
1,500–2,000 kcmil	12	—	—	—	—

Reprinted with permission from NFPA 70®-2020, *National Electrical Code®*, Copyright © 2019, National Fire Protection Association, Quincy, MA. This reprinted material is not the complete and official position of the NFPA on the referenced subject, which is represented only by the standard in its entirety which may be obtained through the NFPA website at *www.nfpa.org*

Table 3 Minimum Wire Bending Space for Conductors Entering or Leaving Opposite Wall [Data from *NEC Table 312.6(B)*]

AWG or Circular-Mil Size of Wire	Wires per Terminal			
	1	2	3	4 or More
14-10	Not Specified	—	—	—
8	1½	—	—	—
6	2	—	—	—
4	3	—	—	—
3	3	—	—	—
2	3½	—	—	—
1	4½	—	—	—
1/0	5½	5½	7	—
2/0	6	6	7½	—
3/0	6½ (½)	6½ (½)	8	—
4/0	7 (1)	7½ (1½)	8½ (½)	—
250	8½ (2)	8½ (2)	9 (1)	10
300	10 (3)	10 (2)	11 (1)	12
350	12 (3)	12 (3)	13 (3)	14 (2)
400	13 (3)	13 (3)	14 (3)	15 (3)
500	14 (3)	14 (3)	15 (3)	16 (3)
600	15 (3)	16 (3)	18 (3)	19 (3)
700	16 (3)	18 (3)	20 (3)	22 (3)
750	17 (3)	19 (3)	22 (3)	24 (3)
800	18	20	22	24
900	19	22	24	24
1,000	20	—	—	—
1,250	22	—	—	—
1,500-2,000	24	—	—	—

Reprinted with permission from NFPA 70®-2020, *National Electrical Code®*, Copyright © 2019, National Fire Protection Association, Quincy, MA. This reprinted material is not the complete and official position of the NFPA on the referenced subject, which is represented only by the standard in its entirety which may be obtained through the NFPA website at *www.nfpa.org*

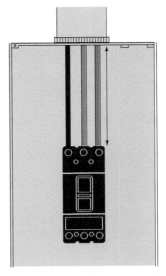

Bending space at terminals must be measured in a straight line from the end of the lug or wire connector in a direction perpendicular to the enclosure wall. Use the values in *NEC Table 312.6(B)*.

Figure 14 Conductors entering an enclosure opposite the conductor terminals.

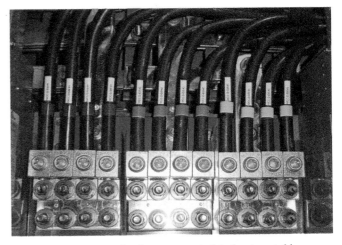

Figure 15 Incoming feeders connected to horizontal bus.

system. Where the main disconnect is rated at 400A or less, the load connection may be made with stab connections to vertical bus sections connecting to the main horizontal bus.

1.3.2 Short Circuit Bracing

All incoming lines to either incoming line lugs or main disconnects must be braced to withstand the mechanical force created by a high fault current. If the cables are not anchored sufficiently or the lugs are not tightened correctly, the connections become the weakest part of a panelboard or motor control center when a fault develops. In most cases, each incoming line compartment is equipped with a two-piece spreader bar located at a certain distance from the conduit entry. This spreader bar should be used along with appropriate lacing material to tie cables together where they can be bundled and to hold them apart where they must be separated. In other words, the incoming line cables should first be positioned and then anchored in place. Manufacturers of electrical panelboards and motor control centers normally furnish detailed information on recommended methods of short circuit bracing; follow this information exactly.

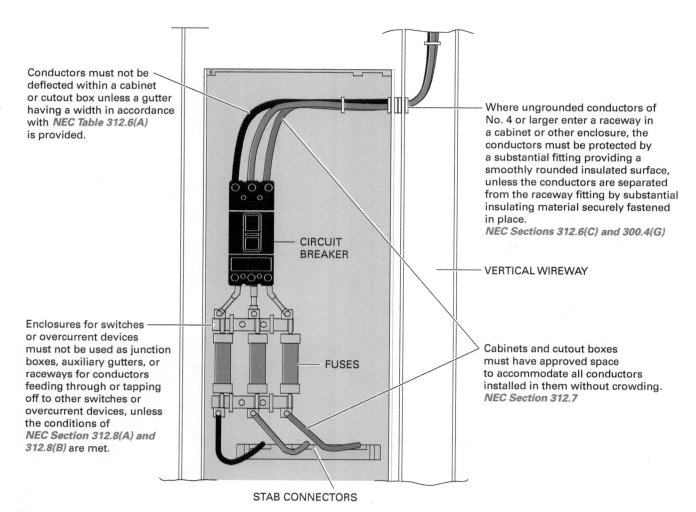

Figure 16 MCC fed directly from a transformer secondary.

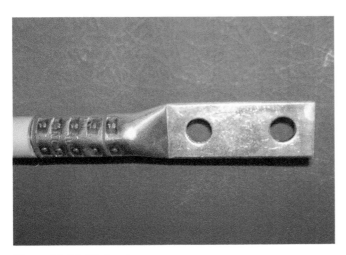

Figure 26 Multiple crimps.

Table 4 Recommended Tightening Torques for Various Bolt Sizes

Steel Hardware		Aluminum Hardware	
Bolt Size	Recommended Torque (Inch-Pounds)	Bolt Size	Recommended Torque (Inch-Pounds)
1/4-20	80	1/2-13	300
5/16-18	180	5/8-11	480
3/8-16	240	3/4-10	650
1/2-13	480	—	—
5/8-11	660	—	—
3/4-10	1,900	—	—

2.1.3 Specialized Cable Connectors

A wide variety of cables require specially designed connectors, commonly called *terminators*, to secure them to equipment enclosures. Normally, these specialized connectors are supplied with installation instructions. This section will introduce you to one type of specialized connector designed for use with metal-clad (Type MC) cable. Its construction and installation are typical of many specialized connectors.

Type MC cable is a factory assembly of one or more insulated circuit conductors with or without optical fiber members. It is enclosed in a metallic sheath of interlocking tape or a smooth or corrugated tube. Throughout the industry, there is increasing use of Type MC cable installed in trays and on racks instead of non-armored cable in conduit. At the appropriate locations, the cables are routed from the trays or racks, then along the structure to the various items of equipment. *NEC Article 330* governs the installation of Type MC cable. There are several types of connectors that can be used with Type MC cable. All fittings must be listed per *NEC Section 330.6*. The specific connector used is determined by the size and type of cable and the application.

As an example, the procedure for installing one manufacturer's weatherproof connector designed for use with Type MC cable is explained here:

Step 1 Select the correct connector size. This is normally done by comparing the physical dimensions of the cable to a cross-reference table given in the manufacturer's product literature and/or installation instructions.

Step 2 Strip back the jacket and armor of the cable as needed to meet equipment requirements. Expose the cable armoring further by stripping the cable jacket for a specified distance (L), as shown in *Figure 27*. This distance will be found in the manufacturer's installation instructions.

Step 3 Make sure that the jacket seal and retaining spring are in their uncompressed state. If necessary, loosen the connector body and the compression nut. It is not necessary to separate the connector parts.

Tightening Torques

Many terminations and types of equipment are marked with tightening torque information. For items of equipment, torque information is often marked on the equipment and/or listed in the manufacturer's installation instructions. When specific torque requirements are given, always follow the manufacturer's recommendations per *NEC Section 110.14(D)*.

For cases in which no tightening requirements are given, guidelines for tightening screw and bolt-type mechanical compression connectors can be found in the *National Electrical Code® Handbook* comments pertaining to *NEC Section 110.14*. Other good sources of tightening information are manufacturers' catalogs and product literature.

Step 7 Tighten the outer compression nut to form a seal on the cable jacket. Normally, this is hand-tight plus one full turn.

Step 8 If appropriate, terminate the individual wires contained in the cable using compression or mechanical connectors.

2.2.0 Aluminum Connections

Aluminum has certain properties that are different from copper, and special precautions must be taken to ensure reliable connections. These properties are: cold flow (aluminum does not spring back in the same way as copper and has a tendency to deform), coefficient of thermal expansion, susceptibility to galvanic corrosion, and the formation of oxide film on the surface.

Because of the thermal expansion and cold flow of aluminum, standard copper connectors cannot be safely used on aluminum wire. Most manufacturers design their aluminum connectors with greater contact area to counteract this problem. Tongues and barrels of all aluminum connectors are larger or deeper than comparable copper connectors.

The electrolytic action between aluminum and copper can be controlled by plating the aluminum with a neutral metal (usually tin). The plating prevents electrolysis from taking place, and the joint remains tight. Connectors should also be tin-plated and prefilled with an oxide-inhibiting compound.

The insulating aluminum oxide film must be removed or penetrated before a reliable aluminum joint can be made. Aluminum connectors are designed to bite through this film as they are applied to conductors. The conductor should also be wire brushed and coated with a joint compound to ensure a reliable joint.

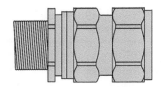

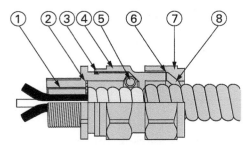

1. ENTRY COMPONENT
2. END STOP
3. O-RING
4. CONNECTOR BODY
5. RETAINING SPRING
6. WASHER
7. JACKET SEAL
8. COMPRESSION NUT

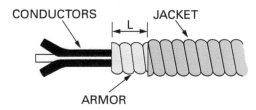

Figure 27 Weatherproof connector used with Type MC cable.

Step 4 Screw the connector body into the equipment if it has a threaded entry, or secure it with a locknut if it has an unthreaded entry.

Step 5 Pass the cable through the connector until the armor makes contact with the end stop. If it is not possible for the insulated wires to pass through the end stop, remove the end stop so that the wires can move past it and the armor can make contact with the integral end stop within the entry component.

Step 6 Tighten the connector body to compress the retaining spring and secure the armor. Normally, this is hand-tight plus one and a half full turns.

> **CAUTION**
> Never use connectors designed strictly for use on copper conductors on aluminum conductors. Connectors listed for use on both metals will be marked AL-CU. All connectors must be applied and installed in the manner for which they are listed and labeled.

NEC Section 110.14 prohibits conductors made of dissimilar metals (copper and aluminum, copper and copper-clad aluminum, or aluminum and copper-clad aluminum) from being intermixed in a terminal or splicing connector unless the device is identified for the purpose. As a general rule:

- Connectors marked with only the wire size should only be used with copper conductors.
- Connectors marked with AL and the wire size should only be used with aluminum conductors.
- Connectors marked with AL-CU and the wire size may be safely used with either copper or aluminum.

2.3.0 Control and Signal Cables

For the most part, electricians install all the associated power cables and many of the signal cables for the instrumentation systems in industrial facilities. It is necessary to select the proper type and rating of control and signal cables to meet *NEC®* requirements, local codes, and equipment manufacturer requirements for the system being installed. With the exception of optical fiber or communication cables that enter a structure and are terminated in an enclosure within 50' (15 m) of the entrance, the *NEC®* requires that control and signal cables be listed and marked with an appropriate classification code. Control and signal cables are type-classified, rated, and listed for use in various areas of a structure in accordance with the following:

- *NEC Article 725, Remote Control, Signaling, and Power-Limited Circuits (Class 1, 2, and 3 Circuits)*
- *NEC Article 727, Instrumentation Tray Cable*
- *NEC Article 760, Fire Alarm Systems*
- *NEC Article 770, Optical Fiber Cables*
- *NEC Article 800, Communications Circuits*

Fire alarm, instrumentation, and control cable carry low-level signals that require less current than power cable. Conductor sizes range from No. 12 to No. 24 AWG, depending on the circuit requirements. Conductors can be tinned or bare, solid or stranded, and twisted or parallel. Shielding, if used, is usually an aluminum-coated Mylar® film with a drain wire (ground wire) or multiple drain wires inside a jacket. The jacket and conductor insulation are rated for the applicable usage. *Figure 28* shows typical fire alarm and instrumentation cable.

To be effective, the shielding drain wire(s) must be grounded. The installation loop diagrams or instructions for the system equipment will indicate how and where the shields are connected and grounded. Typically, only one end is grounded, and the drain wire at the other end is isolated by folding it back and taping over it. This prevents the electrical noise collected by the shield from recirculating through the system and interfering with the control signals.

2.3.1 Crimp Connectors for Screw Terminals

Low-voltage control circuits typically use compression-type crimp connectors. Compression-type connectors for connecting conductors to screw terminals for low-voltage circuits include those in which hand tools indent or crimp tube-like sleeves that hold a conductor. Proper crimping action changes the size and shape of the connector and deforms the conductor strands enough to provide good electrical conductivity and mechanical strength.

Figure 29 shows the basic structure of a crimp connector. The crimp barrel receives the wire and is crimped to secure it in place. The V's or dimples inside the barrel improve the wire-to-terminal conductivity and also increase the termination tensile strength. Most crimp connectors have nylon or vinyl insulation covering the barrel

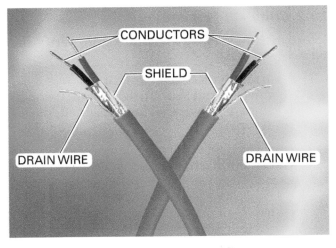

(A) FIRE ALARM CABLE

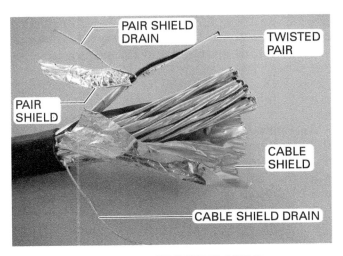

(B) INSTRUMENTATION CABLE

Figure 28 Fire alarm and instrumentation cable.

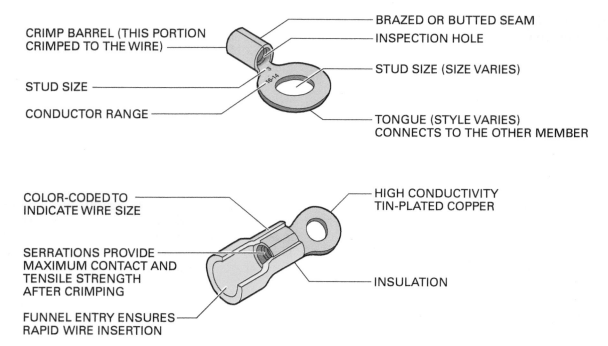

Figure 29 Basic crimp connector structure.

to reduce the possibility of shorting to adjacent terminals. The insulation is color coded according to the connector's wire range to reduce the problem of wire-to-connector mismatch. An inspection hole is provided at the end of the barrel to allow visual inspection of the wire position. For the smaller wire sizes, a sleeve is crimped over the conductor insulation in the process of crimping the barrel, providing strain relief for the conductor.

The barrel is connected to the terminal tongue, which physically connects the wire to the termination point, such as a terminal screw. Information about the connector size and conductor range is usually stamped on the tongue by the manufacturer. Tongue styles vary depending on termination requirements. *Figure 30* shows standard tongue styles. The styles most frequently used are the ring tongue and flanged or locking fork. These types are preferred because the terminals will not slip off the terminal screw as the screw is tightened. They are also compatible with most vendor-supplied termination points.

Most manufacturers color code the barrel insulation to provide quick identification for installation and as an aid to inspection. Different colors, or a combination of colors, have special meanings. Although manufacturers vary, common or standard colors have been accepted. *Table 5* lists typical color codes.

Color combinations are sometimes varied to indicate the class or grade rating of an individual lug or splice. Some manufacturers use a clear plastic or other suitable insulation on the crimp barrel with a colored line to indicate wire size range.

Cable/conductor termination first involves organizing the cables/conductors by destination and mounting or installing the appropriately sized termination panels or devices. Then the cables/conductors must be formed, supported, and dressed to length including appropriate slack. When this has been accomplished, the cables/conductors are properly labeled and then terminated to an appropriate connection device (*Figure 31*).

2.3.2 Crimping Procedure

Before making a crimped connection, verify the following items:

- The size and type of wire are correct.
- The connector and wire materials are compatible (compare cable and lug designations for material type).

> NOTE: Conductors of dissimilar metals should not be intermixed in a terminal lug or splice connector unless the device is listed and labeled for this purpose.

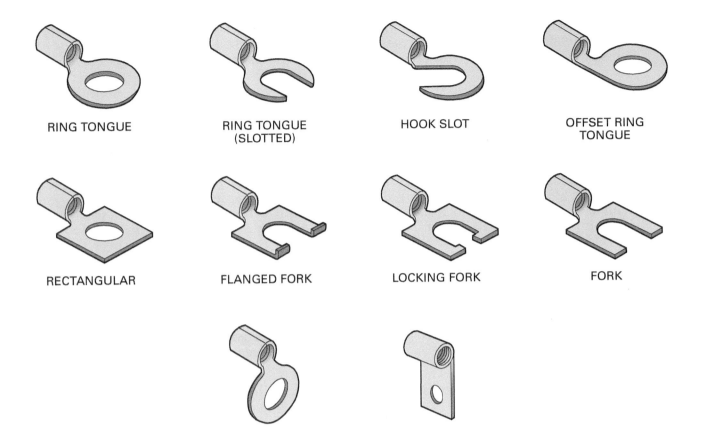

Figure 30 Standard tongue styles of crimped connectors.

- The correct crimp tool and die for the selected terminal and conductor are being used.
- The tool is in proper working order.

The installation of a compression terminal should be as follows:

Step 1 Select the proper terminal for the wire size being terminated, as specified by the terminal manufacturer. This ensures the proper fit of the lug and that the amperage capacity of the lug equals that of the conductor.

Step 2 Verify that the terminal stud size matches the terminal screw size.

Step 3 Select the proper crimping tool and die for the terminal being used. This minimizes the possibility of overcrimping or undercrimping the terminal.

Step 4 Train the wire strands if the strands are fanned. Grooming the conductor provides a proper fit in the crimp barrel.

Figure 31 Labeled and terminated control cable.

Table 5 Typical Color Codes

AWG Wire Size	Color Code
22-16	Red
16-14	Blue
12-10	Yellow

Step 5 Insert the terminal in the proper die nest. Some insulated terminals have increased wire ranges that overlap with other size terminals, such as a terminal covering wire sizes No. 14 through 18 AWG. When crimping this type of terminal, use the middle wire size that the terminal will accept to determine the correct nest to use. For example, when crimping a No. 18 AWG wire in a 14–18 terminal, use the nest marked 16–14 on the tool. Overcrimping will occur if the terminal is crimped in the nest marked 18–22. When crimping an uninsulated terminal, the top of the crimp barrel should be positioned facing the indenter. This will produce the best crimp and facilitate visual inspection of the crimp after connection to the terminal point. When crimping an insulated terminal, insert the terminal in the color-coded die corresponding to the insulation color.

Step 6 Close the crimp tool handles slightly to secure the terminal while inserting the conductor.

Step 7 Insert the stripped wire into the barrel of the terminal until the insulation butts firmly against the forward stop. Make sure that the insulation is not actually in the barrel. The strands of the conductor should be clearly visible through the inspection opening of the terminal. Make sure all strands of the conductor are inserted into the barrel of the terminal.

Step 8 Complete the crimp by closing the handles until the mechanical cycle has been completed and the ratchet releases.

Step 9 Remove the termination from the tool and examine it for proper crimping.

2.3.3 Termination Inspection

The final step of the crimping procedure is inspecting the termination to ensure that it is electrically and mechanically sound. It must be tight with no loose strands or uninsulated conductor showing. Inspect uninsulated terminals for correct positioning, centering, and size of the crimp indent. Ensure that the terminal wire size stamped on each terminal matches the conductor that was crimped.

Acceptable and unacceptable positioning of a crimp indent is illustrated in *Figure 32*. The crimp barrel is designed to provide the best mechanical strength when the indent is properly placed on the top (seam) of the barrel. An indent on the side can split the terminal seam, thereby reducing both the electrical and mechanical qualities of the termination.

The centering of a crimp indent is very important. A poor connection will result if a crimp is placed over either the belled mouth or the inspection hole of the terminal, as shown in *Figure 33*. A crimp over the belled mouth will compress the conductor insulation and result in poor or no electrical continuity. Crimping over the inspection hole reduces both electrical continuity and the holding capability of the terminal lug.

Crimping with an incorrect die changes the electrical and mechanical qualities of the termination. The crimp barrel should not be excessively distorted or show any cracks, breaks, or other damage to the base metal.

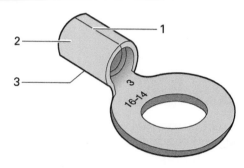

1 – ACCEPTABLE – INDENT ON SEAM (TOP)
2 – UNACCEPTABLE – INDENT ON SIDE
3 – UNACCEPTABLE – INDENT ON BOTTOM

Figure 32 Indent position.

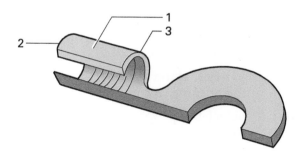

1 – ACCEPTABLE – CENTERED OVER SERRATIONS
2 – UNACCEPTABLE – OVER BELLED MOUTH
3 – UNACCEPTABLE – OVER INSPECTION HOLE

Figure 33 Crimp centering.

The position of the conductor is also very important. Tongue terminals should have the end of the conductor flush with and extending beyond the crimp barrel no more than $\frac{1}{32}$" (1 mm) to ensure proper crimp-to-wire contact area, as shown in *Figure 34*. Conductors extending more than $\frac{1}{32}$" (1 mm) past the inspection hole may interfere with the terminal screw. If a conductor is too short and does not reach the inspection hole, the connection will not provide enough contact surface area, which increases current density and may cause the lug to slip off the conductor.

Terminal lugs with a belled mouth must have the conductor insulation butted against the tapered edge of the crimp barrel, as shown in *Figure 34*. Terminal lugs without a belled mouth should not have any exposed conductor. The conductor insulation is butted to the terminal. Any exposed conductor reduces the overall strength of the connection.

Insulated terminals should be inspected in the same manner as uninsulated terminals: check for proper positioning, centering, and type of crimp. In addition, inspect the terminal insulation for breaks, cracks, holes, or any other damage. Any damage to the insulation is unacceptable. As with uninsulated terminals, inspect insulated terminals to ensure the proper terminal size for each individual conductor. Terminal conductor and stud sizes are usually stamped on the terminal tongue. They can also be checked against the manufacturer's color code.

2.3.4 Terminal Block Connections

A variety of terminal blocks are available, including clamp-type, spring-loaded, and screw-type terminal blocks (*Figure 35*). The following are three common types of terminal blocks:

- *Spring-type terminal block* – To install stripped wires in a spring-loaded terminal block, release the spring contact using a flat-blade screwdriver and insert the wire into the hole on top of the block. After the wire is inserted, remove the screwdriver. The wire is clamped in place by the spring contact.
- *Clamp-type terminal block* – To install stripped wires in a clamp-type terminal block, insert the wires into the boxed terminal and tighten the screw to clamp the wire in place.
- *Screw-type terminal block* – To install stripped wires in a common screw-type terminal block, curve the wires around the screw or fit them with a terminal lug that is inserted under the screw. After the wire or terminal lug is under the head of the screw, tighten the screw to secure the wire.

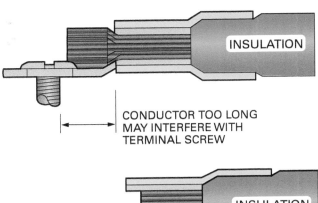

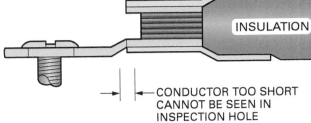

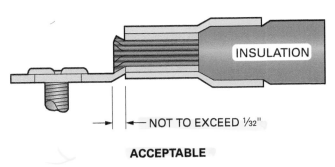

Figure 34 Conductor positioning.

> **NOTE**
> When installing wires in a common screw-type terminal block, only two terminals are permitted at any one terminal point. Also, only one flanged fork terminal may be used. When two types of terminals are located at one point, the bottom terminal should be installed upside down. This will provide easier installation and a neater appearance.

Take care to strip the cable jacket to a point as close as possible to the first termination of the cable but not to interfere with other terminations originating from that cable, as shown in *Figure 36*. Place the cable identification tag at that point on the cable jacket and make sure it can be read easily.

When multiple cables are installed, tie them neatly to a support without blocking access to the lower terminal blocks or interfering with the connection or disconnection of other wires.

Individual or multi-conductor cables should be routed parallel or at right angles to the frame or wireways. Take care to ensure that wires do not come into contact with sharp edges. Position shielded and coaxial cables on the outer perimeter of a cable bundle whenever possible, and keep wire crossover to a minimum.

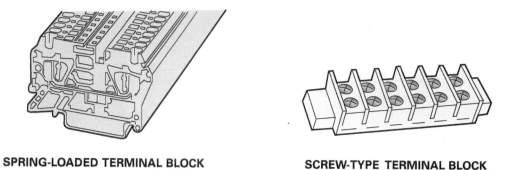

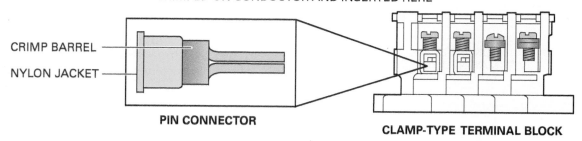

Figure 35 Terminal blocks.

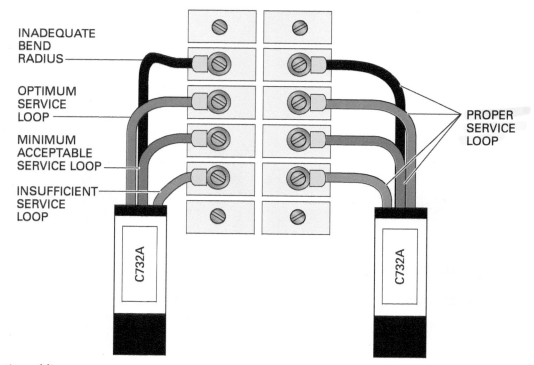

Figure 36 Routing cabling.

2.0.0 Section Review

1. Pliers are a type of _____.
   a. high-MA tool
   b. adjustable-MA tool
   c. constant-MA tool
   d. proportional-MA tool

2. Aluminum connectors are designed with greater contact area to counteract _____.
   a. thermal expansion
   b. electrolysis
   c. aluminum oxide filming
   d. galvanic corrosion

3. A common color code for a 10 AWG crimp barrel is _____.
   a. black
   b. yellow
   c. red
   d. blue

Section Three

3.0.0 Reinsulating Electrical Connections

Objective

Reinsulate electrical connections.
  a. Tape electrical connections.
  b. Install heat-shrink insulators.
  c. Use motor connection kits.

Performance Task

3. Insulate selected types of wire splices and/or install a motor connection kit.

Trade Term

Insulating tape: Adhesive tape that has been manufactured from a nonconductive material and is used for covering wire joints and exposed parts.

After making splices or other terminations, the electrical connection must be reinsulated to provide the same degree of insulation as the original conductor. Common methods of reinsulating include taping, heat-shrink insulators, and motor connection kits.

3.1.0 Taping

When it is not practical to protect a spliced joint by some other means, electrical tape may be used to insulate the joint. Joints must be taped carefully to provide the same quality of insulation over the splice as over the rest of the conductors.

Several electrical insulating tape varieties are available for use in specific applications. Some common types include vinyl plastic tape, linerless rubber tape, high-temperature silicone rubber tape, and glass cloth tape. Electrical tapes made of vinyl plastic are widely used as primary insulation on joints made with thermoplastic-insulated wires. They are used for splices up to 600V and for fixture and wire splices up to 1,000V. Depending on the product, they are made for indoor use, outdoor use, or both.

Linerless rubber splicing tape provides for a tight, void-free, moisture-resistant insulation without loss of electrical characteristics. It is typically used as primary insulation with all solid dielectric cables through 70kV. Other applications include jacketing on high-voltage splices and terminals, moisture-sealing electrical connections, busbar insulations, and end sealing high-voltage cables.

High-temperature silicone rubber tapes are used as a protective overwrap for terminating high-voltage cables. Glass cloth electrical tapes provide a heat-stable insulation for hot-spot applications such as furnace and oven controls, motor leads, and switches. They are also used to reinforce insulation where heavy loads cause high heat and breakdown of insulation, such as in motor control exciter feeds. All-metal braid tapes are also available. These are used to continue electrostatic shielding across a splice. When taping a splice, begin by selecting the correct tape for the job. Always follow the tape manufacturer's recommendations.

A general procedure describing one method of taping a splice or joint, such as encountered when connecting motor lugs, is shown in *Figure 37*. A method for taping a split-bolt connector is shown in *Figure 38*. Before taping, make sure the joined lugs are securely fastened together with the appropriate hardware. Use pieces of a suitable filler tape or putty to fill any voids around the lugs and hardware and eliminate any sharp edges. This also helps to provide a smooth, even surface that makes taping easier.

> **NOTE**
>
> For all splices and joints in which it is likely that the tape will have to be removed at some future date to perform work on the joint, apply an upside down (that is, adhesive side up) wrap of tape to the joint before applying the final layers of insulating tape to the joint in the usual manner. This will keep the area free of tape residue and facilitate the removal of the tape later on, if necessary.

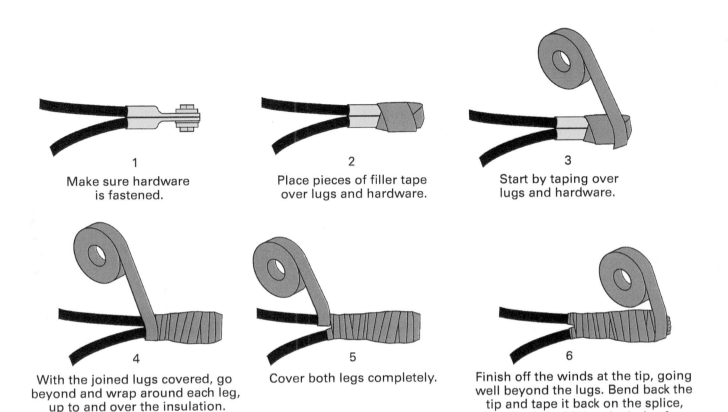

Figure 37 Typical method for taping motor lug connections.

3.2.0 Heat-Shrink Insulators

Heat-shrink insulators for small connectors provide skintight insulation protection and are fast and easy to use. They are designed to slip over wires, taper pins, connectors, terminals, and splices. When heat is applied, the insulation becomes semi-rigid and provides positive strain relief at the flex point of the conductor. A vaporproof band seals and protects the conductor against abrasion, chemicals, dust, gasoline, oil, and moisture. Extreme temperatures, both hot and cold, will not affect the performance of these insulators. The source of heat can be any number of types, but most manufacturers of these insulators also produce a heat gun especially designed for use on heat-shrink insulators. It is similar to a conventional hair dryer, as shown in *Figure 39*.

In general, a heat-shrink insulator may be thought of as tubing with a memory. After it is initially manufactured, it is heated and expanded to a predetermined diameter and then cooled. Upon a second application of heat, the tubing compound "remembers" its original size and shrinks to that smaller diameter. This property enables it to conform to the contours of any object. Various types of heat-shrink tubing are available, some of which include the following:

- *PVC* – This is a general-purpose, economical tubing widely used in the electronics industry. It provides good electrical and mechanical protection and resists cracking and splitting.
- *Polyolefin* – Polyolefin tubing has a wide range of uses for wire bundling, harnessing, strain relief, and other applications in which cables and components require additional insulation. It is flame-retardant, flexible, and comes in a variety of colors.
- *Double wall* – This tubing is designed for outstanding protective characteristics. It is a semi-rigid tubing with an inner wall that melts and an outer wall that shrinks to conform to the melted area.

(A) SPLIT-BOLT CONNECTOR

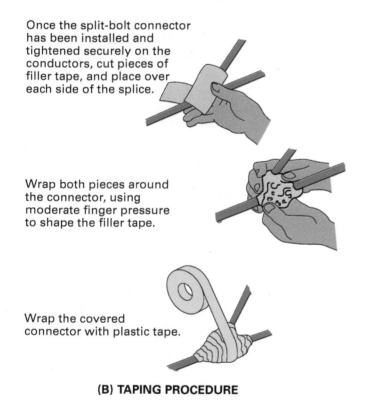

Once the split-bolt connector has been installed and tightened securely on the conductors, cut pieces of filler tape, and place over each side of the splice.

Wrap both pieces around the connector, using moderate finger pressure to shape the filler tape.

Wrap the covered connector with plastic tape.

(B) TAPING PROCEDURE

Figure 38 Typical method of taping a split-bolt connector.

- *Teflon®* – This type is considered by many users to be the best overall heat-shrink tubing—physically, electrically, and chemically. Its high temperature rating of 250°C allows Teflon® tape to resist brittleness from extended exposure to heat and to not support combustion.
- *Neoprene* – This is a highly durable and flexible tubing that provides superior protection against abrasion.
- *Kynar®* – This is a thin-wall, semi-rigid tubing with outstanding resistance to abrasion. This transparent tubing enables easy inspection of components.

Tubing is available in a wide variety of colors and configurations. The manufacturer's tubing selector guide can help in the selection of the best tubing for any given application. A typical tubing selector guide appears in *Table 6*.

3.3.0 Motor Connection Kits

Motor connection kits are available to insulate bolted splice connections, such as those in motor terminal boxes. These kits eliminate the need for taping and the use of filler tape or putty. To aid in joint reentry during rework, the insulator strips off easily, leaving a clean bolt area and thus eliminating the need to remove old tape and putty. Motor connection kits are available for use with stub (butt splice) connections (*Figure 40*) where there is insufficient room to make in-line connections. They are also made to insulate in-line splice connections where space permits. These insulating kits use a high-voltage mastic that seals the splice against moisture, dirt, and other contaminants.

One type of motor connection kit insulator is heat-shrinkable. It installs easily using heat from a propane torch to shrink the insulator in a manner similar to that of heat-shrink tubing (*Figure 41*). Another type of kit used for insulating stub connections comes in the form of an elastomeric insulating cap that is cold-applied by rolling it over the stub splice. Always follow the kit manufacturer's recommendations when selecting a kit to use for a particular application. Typically, the kit is selected based on the size of the motor feeder cable.

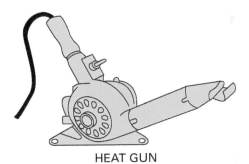
HEAT GUN

SLIP INSULATOR OVER OBJECT TO BE INSULATED, THEN APPLY HEAT FOR A FEW SECONDS

WHEN FINISHED, IT PROVIDES PERMANENT INSULATION PROTECTION

Figure 39 Method of installing heat-shrink insulators.

Electrical Tape

Most non-electricians think of electrical tape as only the simple black vinyl tape found in nearly every home toolbox. Electrical tape actually comes in a wide range of colors to be used for labeling various conductors when making terminations.

Table 6 Tubing Selector Guide

Type	Material	Temp. Range (°C)	Shrink Ratio	Max. Long. Shrinkage (%)	Tensile Strength (psi)	Colors	Dielectric Stength (V/mil)
Nonshrinkable	PVC	+ 105	-	-	2,700	White, red, clear, black	800
Shrinkable	PVC	- 35 to + 105	2:1	10	2,700	Clear, black	750
Nonshrinkable	Teflon®	- 65 to + 260	-	-	2,700	Clear	1,400
Shrinkable	Flexible polyolefin	- 55 to + 135	2:1	5	2,500	Black, white, red, yellow, blue, clear	1,300
Nonshrinkable	Teflon®	- 65 to + 260	-	-	7,500	Clear	1,400
Shrinkable	Polyolefin double wall	- 55 to 110	6:1	5	2,500	Black	1,100
Shrinkable	Kynar®	- 55 to 175	2:1	10	8,000	Clear	1,500
Shrinkable	Teflon®	+ 250	1.2:2	10	6,000	Clear	1,500
Shrinkable	Teflon®	+ 250	1½:1	10	6,000	Clear	1,500
Shrinkable	Neoprene	+ 120	2:1	10	1,500	Black	300

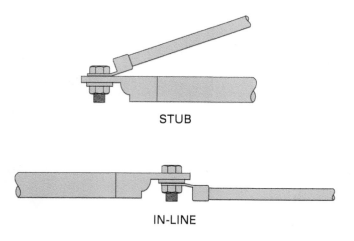

Figure 40 Stub and in-line splice connections.

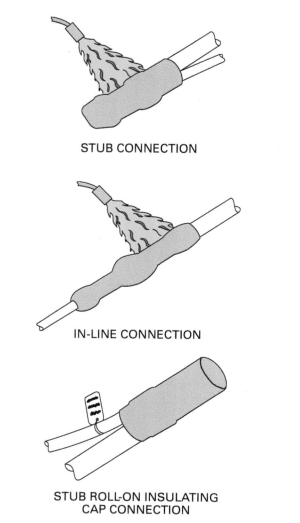

Figure 41 Motor connection kits installed on splices.

Think About It

Putting It All Together

Examine the wiring connections in your home or workplace. Are they properly made? Can you determine if any connections have the potential for failure?

3.0.0 Section Review

1. The type of electrical tape used to provide a protective overwrap for high-voltage cable is _____.
   a. linerless rubber splicing tape
   b. high-temperature silicone rubber tape
   c. vinyl plastic tape
   d. glass cloth electrical tape

2. A transparent, semi-rigid heat-shrink insulator that provides good abrasion resistance is _____.
   a. Neoprene
   b. Teflon®
   c. Kynar®
   d. PVC

3. Motor connection kits are selected based on the size of the _____.
   a. motor feeder cable
   b. overcurrent protection
   c. motor terminals
   d. motor full-load amperage

Review Questions

1. Which of the following is *true* with regard to splices?
   a. A poorly made splice is acceptable as long as it provides a path for current flow.
   b. Splices should provide a high-resistance connection.
   c. Electrical joints are normally made with soldered connections.
   d. Faulty stripping can change the dielectric strength of the conductors.

2. Which of the following best describes the term *stripping* as it applies to conductor splicing?
   a. Removal of the insulation from conductors
   b. Removal of the packing material from the carton
   c. Removal of any excess strands of wire from a splice
   d. Removal of the pulling ring from lead sheath conductors

3. What is the diameter, in inches, of 2,000 kcmil wire?
   a. 0.89
   b. 0.99
   c. 1.29
   d. 1.63

4. Which of the following is most likely to be used to connect an earth bond?
   a. End terminations
   b. Window cuts
   c. Spiral cuts
   d. Indent cuts

5. The minimum wire bending space for a single 3/0 conductor *not* entering or leaving the wall opposite its terminal is ____.
   a. 2"
   b. 3"
   c. 4"
   d. 5"

6. The minimum wire bending space for two 1/0 conductors *not* entering or leaving the wall opposite the terminal is ____.
   a. 2"
   b. 3"
   c. 4"
   d. 5"

7. The minimum wire bending space for three 3/0 conductors entering or leaving the wall opposite the terminal is ____.
   a. 7"
   b. 8"
   c. 9"
   d. 10"

8. Short circuit bracing is used to withstand the mechanical force created by ____.
   a. high fault current
   b. vibration during operation
   c. rapid temperature changes
   d. temporary overloads

9. Which of the following best describes the purpose of a reducing connector?
   a. To temporarily connect a circuit
   b. To join two different size conductors
   c. To join conductors of the same size
   d. To make a 90° bend in parallel conductors

10. Information about connector size and conductor range is ____.
    a. stamped on the connector tongue
    b. color coded
    c. printed on the insulation
    d. stamped on the connector barrel

11. When using a crimping tool, apply ____.
    a. the maximum crimping force possible
    b. a lower mechanical force to achieve maximum electrical performance
    c. the highest force needed for the maximum mechanical strength
    d. an amount of force between optimum mechanical and electrical performance

12. Using a crimp die that is too large will ____.
    a. destroy the crimp barrel
    b. weaken the conductor
    c. provide a stronger mechanical connection
    d. provide poor electrical performance

13. The most frequently used tongue styles are ____.
    a. ringed tongue and locking fork
    b. slotted and flanged fork
    c. hook shot and rectangular
    d. fork and flag

14. The crimp indent is properly placed on the ____.
    a. right side of the barrel
    b. bottom of the barrel
    c. seam of the barrel
    d. left side of the barrel

15. Which of the following types of electrical tape is best for taping most joints for voltages up to 600V?
    a. High-temperature silicone rubber
    b. Rubber
    c. Glass cloth
    d. Vinyl plastic

Supplemental Exercises

1. Heavy-duty strippers are used to strip power cables with outside diameters ranging from __1/0__ to __1,000 kcm__.
2. Properly made splices and connections should last __w/ conductor__
3. Aluminum has certain properties that are different than copper. These properties are: _____, _____, _____, and _____.
4. The electrolytic action between aluminum and copper can be controlled by __plating__.
5. Connectors marked AL should be used only with __Aluminum__ wire, whereas connectors marked AL-CU can be used with either wire.
6. True or False? Conductor nicks, scrapes, or burnishes can result in total failure over time.
7. Nearly all cable installations require the cable to be bent at terminations and other points along the cable route (these bends must comply with the NEC®). Minimum bending radii are determined by __cable dia.__.
8. The bending radii for various sizes of conductors entering or leaving cabinets or cutout boxes must comply with certain NEC® requirements, many of which are covered in __NEC 312__.
9. All incoming lines to either incoming line lugs or main disconnects must be braced to withstand the mechanical force created by a high fault current. Detailed information on the recommended methods of short circuit bracing is normally furnished by __manufacter__
10. All wire joints not protected by some other means should be taped carefully to __insulate__.

Trade Terms Introduced in This Module

AL-CU: An abbreviation for aluminum and copper, commonly marked on terminals, lugs, and other electrical connectors to indicate that the device is suitable for use with either aluminum conductors or copper conductors.

Amperage capacity: The maximum amount of current that a lug can safely handle at its rated voltage.

Connection: That part of a circuit that has negligible impedance and joins components or devices.

Connector: A device used to physically and electrically connect two or more conductors.

Drain wire: A wire that is attached to a coaxial connector to allow a path to ground from the outer shield.

Grooming: The act of separating the braid in a coaxial conductor.

Insulating tape: Adhesive tape that has been manufactured from a nonconductive material and is used for covering wire joints and exposed parts.

Lug: A device for terminating a conductor to facilitate the mechanical connection.

Mechanical advantage (MA): The force factor of a crimping tool that is multiplied by the hand force to give the total force.

Pressure connector: A connector applied using pressure to form a cold weld between the conductor and the connector.

Reducing connector: A connector used to join two different size conductors.

Shielding: The metal covering of a cable that reduces the effects of electromagnetic noise.

Splice: The electrical and mechanical connection between two pieces of cable.

Strand: A group of wires, usually stranded or braided.

Terminal: A device used for connecting cables.

Termination: The connection of a cable.

Additional Resources

This module presents thorough resources for task training. The following resource material is suggested for further study.

ANSI C119.6-2018, *American National Standard for Electric Connectors—Non-Sealed, Multiport Connector Systems Rated 600V or Less for Aluminum and Copper Conductors*. **www.ansi.org**.

National Electrical Code® Handbook, Latest Edition. Quincy, MA: National Fire Protection Association.

Figure Credits

Greenlee / A Textron Company, Module Opener, Figures 1, 2, 6, 9, 11, 12, 20–25

Jim Mitchem, Figures 15, 26, 31

Belden Inc., Figure 28A

Data from *NEC® Tables 312.6(A) and 312.6(A)*, Tables 2 and 3. Reprinted with permission from NFPA 70-2020, *National Electrical Code®*, Copyright © 2019, National Fire Protection Association, Quincy, MA. This reprinted material is not the complete and official position of the NFPA on the referenced subject, which is represented only by the standard in its entirety which may be obtained through the NFPA website at **www.nfpa.org**.

Section Review Answer Key

Section 1.0.0

Answer	Section Reference	Objective
1. a	1.1.0; Figure 5	1a
2. c	1.2.0	1b
3. a	1.3.0	1c

Section 2.0.0

Answer	Section Reference	Objective
1. c	2.1.1	2a
2. a	2.2.0	2b
3. b	2.3.1; Table 5	2c

Section 3.0.0

Answer	Section Reference	Objective
1. b	3.1.0	3a
2. c	3.2.0	3b
3. a	3.3.0	3c

This page is intentionally left blank.

26209-20
GROUNDING AND BONDING

Objectives

When you have completed this module, you will be able to do the following:

1. Identify grounding requirements and applications.
    a. Identify the purpose of grounding and bonding.
    b. Identify the grounding requirements for various systems.
2. Identify service grounding methods.
    a. Size and install a grounding electrode conductor.
    b. Select other electrodes.
3. Size and select equipment grounding.
    a. Size an equipment grounding conductor.
    b. Ground an enclosure.
4. Bond service equipment.
    a. Size the main bonding jumper.
    b. Bond multiple service disconnects.
    c. Bond enclosures and equipment.
5. Ground and bond separately derived systems.
    a. Ground separately derived systems.
    b. Install grounding at more than one building.
6. Test for effective grounds.
    a. Measure earth resistance using the fall-of-potential method.
    b. Complete a three-point test.

Performance Tasks

Under the supervision of the instructor, you should be able to do the following:

1. Size the minimum required grounding electrode conductor for a 200A service fed by 3/0 copper.
2. Using the proper fittings, connect one end of a No. 4 AWG bare copper grounding wire to a length of ¾" (MD 21) galvanized water pipe and the other end to the correct terminal in a main panelboard.
3. Install two lengths of Type NM cable in a switch box using Type NM cable clamps:
    - Strip the ends of the cable to conform with *National Electrical Code®* requirements.
    - Secure the cable in the switch box and tighten the cable clamps.
    - Connect and secure the equipment grounding conductors according to *NEC®* requirements and secure to the switch box with either a ground clip or a grounding screw.
4. Size the minimum required equipment grounding conductor in each conduit for a 400A feeder gap using two parallel runs of 3/0 copper.
5. Size the minimum required bonding jumper for a copper water pipe near a separately derived system (transformer) where the secondary conductors are 500 kcmil copper.

Trade Terms

- Auxiliary electrodes
- Bonding
- Effective ground fault path
- Equipment bonding jumper
- Equipment grounding conductor (EGC)
- Ground
- Ground current
- Ground grid
- Ground mat
- Ground resistance
- Ground rod
- Grounded
- Grounded conductor
- Grounding clip
- Grounding conductor
- Grounding connections
- Grounding electrode
- Grounding electrode conductor (GEC)
- Main bonding jumper
- Neutral conductor
- Resistivity
- Separately derived system
- Short circuit
- Step voltage
- Supplemental electrode
- Supply-side bonding jumper
- System grounding
- Touch voltage
- Ungrounded conductors

Industry Recognized Credentials

If you are training through an NCCER-accredited sponsor, you may be eligible for credentials from NCCER's Registry. The ID number for this module is 26209-20. Note that this module may have been used in other NCCER curricula and may apply to other level completions. Contact NCCER's Registry at 888.622.3720 or go to **www.nccer.org** for more information.

> **NOTE**
>
> NFPA 70®, *National Electrical Code*® and *NEC*® are registered trademarks of the National Fire Protection Association, Quincy, MA.

Contents

1.0.0 Grounding Requirements 1
  1.1.0 Grounding and Bonding 2
    1.1.1 Short Circuit 3
    1.1.2 Ground Fault 3
  1.2.0 Types of Grounding Systems 4
    1.2.1 Grounding Single-Phase Systems 4
    1.2.2 Grounding Systems Less Than 50V 5
    1.2.3 Grounding 50V to 1,000V Systems 6
    1.2.4 Grounding AC Systems Over 1kV 7
    1.2.5 Circuits That Cannot Be Grounded 8
2.0.0 Service Grounding Methods 9
  2.1.0 Grounding Electrode Conductor (GEC) 12
    2.1.1 Sizing the Grounding Electrode Conductor 12
    2.1.2 Installing the Grounding Electrode Conductor 13
  2.2.0 Other Electrodes 14
3.0.0 Equipment Grounding 17
  3.1.0 Equipment Grounding Conductor (EGC) 17
    3.1.1 Grounded and Ungrounded Systems 17
    3.1.2 Sizing an EGC 17
  3.2.0 Grounding an Enclosure 18
    3.2.1 Grounding Outlet Boxes and Devices 19
    3.2.2 Types of Equipment Grounding Conductors 20
4.0.0 Bonding Service Equipment 23
  4.1.0 Main Bonding Jumper 23
  4.2.0 Bonding Multiple Service Disconnects 24
  4.3.0 Bonding Enclosures and Equipment 24
    4.3.1 Bonding Services Over 250 Volts 24
    4.3.2 Bonding Multiple Raceway Systems 24
    4.3.3 Installing the Equipment Bonding Jumper 25
    4.3.4 Creating an Effective Grounding Path 26
5.0.0 Separately Derived Systems 28
  5.1.0 Grounding Separately Derived Systems 28
  5.2.0 Grounding at More Than One Building 31
6.0.0 Testing for Effective Grounds 32
  6.1.0 Measuring Earth Resistance 34
    6.1.1 Current Supply Circuit 35
    6.1.2 Voltmeter Circuit 35
    6.1.3 Earth Electrode Resistance 37
  6.2.0 Three-Point Test 38

Figures and Tables

Figure 1	Panelboard showing grounded and grounding conductors	2
Figure 2	Short circuit	3
Figure 3	Ground fault	4
Figure 4	Pole-mounted transformer	5
Figure 5	Wiring diagram of a 7,200V to 120/240V single-phase transformer connection	5
Figure 6	Typical service entrance and related service equipment	6
Figure 7	Systems less than 50V that must be grounded	6
Figure 8	Systems that must be grounded	7
Figure 9	Circuits that shall not be grounded	8
Figure 10	NEC®-approved grounding electrodes	10
Figure 11	Floor plan of the grounding system for an industrial building	11
Figure 12	Grounding requirements for non-industrial buildings	11
Figure 13	Requirements for ground rods	12
Figure 14	Grounding electrode connection for grounded system	12
Figure 15	Methods of splicing grounding conductors	14
Figure 16	Protection of grounding electrode conductors	15
Figure 17	Other electrodes	15
Figure 18	Equipment grounding summary	18
Figure 19	Ground faults	18
Figure 20	Installation requirements for flexible metal conduit	19
Figure 21	Installation requirements for liquidtight flexible metal conduit	19
Figure 22	Grounding clip	20

Figure 23 Grounding receptacles with different wiring methods 21
Figure 24 Main bonding jumper for parallel runs 23
Figure 25 Bonding for circuits over 250 volts 25
Figure 26 Bonding multiple raceways 25
Figure 27 Individual bonding jumpers 25
Figure 28 Bonding jumper... 26
Figure 29 Grounded service conductor run to service................................ 26
Figure 30 Transformer-supplied separately derived system 28
Figure 31 Separately derived system grounding
 and bonding locations 29
Figure 32 Generator-type separately derived system 30
Figure 33 Grounding neutral at second building 30
Figure 34 Earth ground resistance tester 32
Figure 35 Poorly grounded system.. 34
Figure 36 Fall-of-potential method of testing 35
Figure 37 Three-point testing using a ground tester 36
Figure 38 Plotted curve showing insufficient electrode spacing................ 39
Figure 39 Plotted curve showing adequate electrode spacing................... 40
Figure 40 Auxiliary electrode distance/radii chart 41
Figure 41 Typical grounding resistance curve
 to be recorded and retained 42

Table 1 Bonding of Multiple Service Disconnecting Means 24
Table 2 Number of Conductors Between Structures 31

This page is intentionally left blank.

SECTION ONE

1.0.0 GROUNDING REQUIREMENTS

Objective

Identify grounding requirements and applications.
a. Identify the purpose of grounding and bonding.
b. Identify the grounding requirements for various systems.

Trade Terms

Bonding: Connected to establish electrical continuity and conductivity.

Equipment bonding jumper: The connection between two or more portions of the equipment grounding conductor.

Equipment grounding conductor (EGC): The conductive path(s) that provide a ground-fault current path and connect normally noncurrent-carrying metal parts of equipment together and to the system grounded conductor, or to the grounding electrode conductor, or both.

Ground: The earth or a conducting connection to the earth.

Ground rod: A metal rod or pipe used as a grounding electrode.

Grounded: Connected to ground or to a conductive body that extends the ground connection.

Grounded conductor: A system or circuit conductor that is intentionally grounded.

Grounding connections: Connections used to establish a ground; they consist of a grounding conductor, a grounding electrode, and the earth surrounding the electrode.

Grounding electrode: A conducting object through which a direct connection to earth is established.

Grounding electrode conductor (GEC): A conductor used to connect the system grounded conductor or the equipment to a grounding electrode or to a point on the grounding electrode system.

Neutral conductor: The conductor connected to the neutral point of a system that is intended to carry current under normal conditions.

Short circuit: An often unintended low-resistance path through which current flows around, rather than through, a component or circuit.

System grounding: Intentional connection of one of the circuit conductors of an electrical system to ground potential.

Ungrounded conductors: Conductors in an electrical system that are not intentionally grounded.

The grounding system is a major part of the electrical system. Its purpose is to protect life and equipment against the various electrical faults that can occur. It is sometimes possible for higher-than-normal voltages to appear at certain points in an electrical system or in the electrical equipment connected to the system.

Proper grounding ensures that the electrical charges that cause these high voltages are channeled to earth or ground before damaging equipment or causing danger to human life. Therefore, a circuit is grounded to limit the voltage on the circuit and improve overall operation of the electrical system and continuity of service.

When we refer to ground, we are talking about being connected to the earth. If a conductor is connected to the earth or to some conducting body that extends the ground connection, such as a driven ground rod (electrode) or metal cold-water pipe, the conductor is said to be grounded. The neutral conductor in a three- or four-wire service, for example, is intentionally grounded and therefore becomes a grounded conductor (*Figure 1*). However, a wire used to connect this neutral conductor to a grounding electrode or electrodes is referred to as a grounding electrode conductor (GEC). Note the difference in the two meanings; one is grounded, while the other is grounding.

NEC Article 100, Informational Note No. 1 under the definition for *Grounding Conductor, Equipment*, recognizes that the equipment grounding conductor (EGC) also performs bonding.

This module is designed to explain what a ground is and to teach proper grounding and bonding techniques. *NEC®* regulations and the testing of grounding systems will be thoroughly covered.

Upon completion of this module, you should have a solid foundation in the principles of grounding and bonding, one of the most important aspects of an electrical system for the protection of life and property.

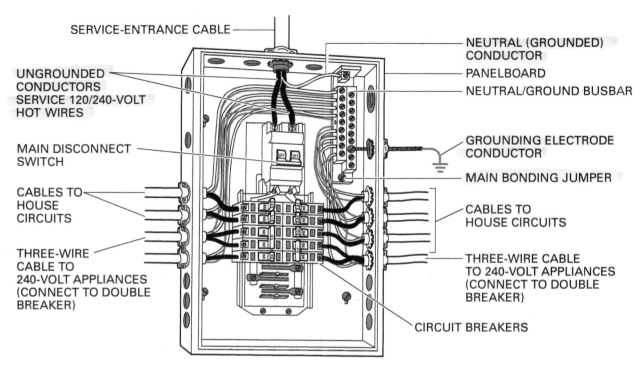

Figure 1 Panelboard showing grounded and grounding conductors.

1.1.0 Grounding and Bonding

Systems are solidly grounded to limit the voltage to ground during normal operation and to prevent excessive voltages due to lightning and line surges. They are also grounded to stabilize the voltage to ground during normal operation. Conductive materials enclosing electrical conductors or equipment, or that form a part of the equipment, are grounded to limit the voltage to ground on these materials.

The conductive materials are bonded (connected together) to establish electrical continuity and conductivity. This facilitates the operation of overcurrent devices under ground fault conditions.

NEC Article 250 is the primary governing article for the proper use and installation of grounding and bonding. This article covers general requirements for grounding and bonding of electrical installations. It presents the following specific requirements:

- Systems, circuits, and equipment that are required, permitted, or not permitted to be grounded
- Circuit conductor to be grounded on grounded systems
- Location of grounding connections
- Types and sizes of grounding and bonding conductors and electrodes
- Methods of grounding and bonding
- Conditions under which guards, isolation, or insulation may be substituted for grounding

NEC Section 250.4(A)(5) requires that the path to ground from circuits, equipment, and metal enclosures shall:

- Have sufficiently low impedance to facilitate the operation of the protective devices in the circuit or ground detector for high-impedance grounded systems
- Be capable of safely carrying the maximum ground fault likely to be imposed on it
- Not rely on the earth as an effective ground fault current path

What's wrong with this picture?

Figure Credit: Mike Powers

Neutral Conductor Size

The grounded neutral conductor is permitted to be smaller than the hot ungrounded phase conductors only when the neutral conductor is properly and adequately sized to carry the maximum load imbalance as computed according to *NEC Sections 220.61 and 230.42*.

It must be pointed out that per *NEC Section 250.24(C)(1)* the grounded conductor must not be smaller than the size referenced in *NEC Table 250.102(C)(1)*. The purpose of the grounded conductor is not only to carry the unbalanced load, but also to provide a low-impedance path back to the source (utility) to complete a circuit and allow enough current to flow to open the overcurrent devices under fault conditions. This is important when you have three-phase services with little or no neutral loads to use as a basis for sizing the grounded conductor.

NEC Section 230.42 (service-entrance conductors) references *NEC Article 220* for minimum rating and size computation requirements. *NEC Section 220.61* states that the feeder neutral load must be the maximum imbalance of the load, which is the maximum net computed load between the neutral and any one ungrounded conductor.

In summary, both the neutral current, as determined in *NEC Articles 220 and 230*, and the minimum grounding electrode conductor size, as determined in *NEC Article 250*, must be sized per *NEC Tables 250.66 and 250.102(C)(1)*.

Connections must be made as permitted in *NEC Section 250.8*.

Permanent, reliable, and continuous grounding systems are vital to the safety of electrical systems. Intermittent connections are likely to be unpredictable and may result in hazardous situations. Since current does not flow through grounding conductors in most applications, it is extremely important to have good connections that will withstand the test of time. Equipment will keep running, but when it malfunctions the enclosure may become electrified instead of clearing the fault connections.

The minimum size grounding electrode conductor, equipment bonding jumper, and equipment grounding conductor are necessary to ensure that the proper capacity to protect the system is in place. Methods to determine the proper size of these conductors will be discussed later in this module.

A low-impedance conductor is necessary because the higher the impedance, the more opposition there is to current flow. The current should flow with the least amount of opposition to ensure that personnel and equipment are protected when a fault does occur.

1.1.1 Short Circuit

Short circuits and ground faults are commonly misunderstood. Both faults typically stem from a failure of insulation resistance. It is important to have a common language for better understanding of these conditions.

A short circuit is a conducting connection, whether intentional or accidental, between any of the conductors of an electrical system, whether it is from line-to-line or from line-to-neutral (grounded). See *Figure 2*.

The failure might occur from one phase conductor to another phase conductor or from one phase conductor to the grounded conductor or neutral. The maximum value of fault current is dependent on the available capacity the system can deliver to the point of fault. The maximum value of short circuit current from line-to-neutral will vary depending upon the distance from the source to the fault. The available short circuit current is further limited by the impedance of the arc where one is established, plus the impedance of the conductors to the point of short circuit.

1.1.2 Ground Fault

A ground fault, as defined in *NEC Article 100*, is an unintentional, electrically conducting connection between an ungrounded conductor of an electrical circuit and the normally noncurrent-

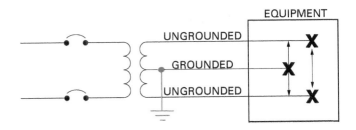

Figure 2 Short circuit.

carrying conductors, metallic enclosures, metallic raceways, metallic equipment, or the earth. While not specified in the *NEC®*, an unintentional electrically conducting connection between a grounded conductor of an electrical circuit and the normally noncurrent-carrying conductors, metallic enclosures, metallic raceways, metallic equipment, or earth would also be considered a ground fault. While the first instance may result in a large amount of fault current to flow in a properly installed grounded system, the latter may result in parallel paths being formed between the grounded conductor and the grounding system. These parallel paths would result in an unwanted current flow on the grounding system (*Figure 3*).

1.2.0 Types of Grounding Systems

The two general classifications of protective grounding are **system grounding** and equipment grounding.

System grounding and bonding relates to the service-entrance equipment and its interrelated and bonded components. That is, one system conductor is grounded to limit voltages due to lightning, line surges, or unintentional contact with higher voltage lines, as well as to stabilize the voltage to ground during normal operation.

The equipment grounding conductor is the conductive path used to connect the noncurrent-carrying metal parts of equipment, raceways, and other enclosures to the system grounded conductor, the grounding electrode conductor, or both. Equipment grounding conductors are bonded to the system grounded conductor to provide a low impedance path for fault current that will facilitate the operation of overcurrent devices under ground fault conditions. *NEC Article 250* covers general requirements for grounding and bonding. *NEC Sections 250.20 and 250.21* provide specific grounding requirements for AC systems.

1.2.1 Grounding Single-Phase Systems

To better understand a complete grounding system, we will examine a conventional residential or small commercial system beginning at the power company's high-voltage lines and transformer, as shown in *Figure 4*. The pole-mounted transformer is fed with a two-wire 7,200V system that is transformed and stepped down to a three-wire, 120/240V, single-phase electric service suitable for residential use. A wiring diagram of the transformer connections is shown in *Figure 5*. Note that the voltage between Leg A and Leg B is 240V.

However, by connecting a third wire (neutral) on the secondary winding of the transformer between the other two, the 240V splits in half, giving 120V between either Leg A or Leg B and the neutral conductor. Consequently, 240V is available for household appliances such as ranges, hot water heaters, clothes dryers, and the like, while 120V is available for lights, small appliances, televisions, and similar appliances.

Referring again to the diagram in *Figure 5*, conductors A and B are **ungrounded conductors**, while the neutral is a grounded conductor. If only 240V loads were connected, the neutral or grounded conductor would carry no current. However, since 120V loads are present, the neutral will carry the unbalanced load and become a current-carrying conductor. For example, if Leg A carries 60A and Leg B carries 50A, the neutral conductor would carry only 10A (60 − 50 = 10). This is why the *NEC®* sometimes allows the neutral conductor in an electric service to be smaller than the ungrounded conductors.

The typical pole-mounted service entrance is normally routed by a grounded (neutral) messenger cable from a point on the pole to a point on the building being served, terminating at the service drop. Service-entrance conductors are routed between the meter housing and the main service switch or panelboard. This is the point where most systems are grounded—the neutral bus in the main panelboard (*Figure 6*).

> **WARNING!** Exercise extreme caution when disconnecting a ground. Never grab a disconnected ground wire with one hand and the grounding electrode with the other. Your body will act as a conductor for any fault current; the results could be fatal.

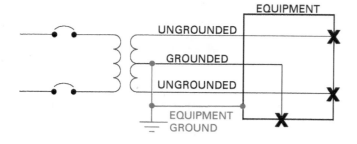

Figure 3 Ground fault.

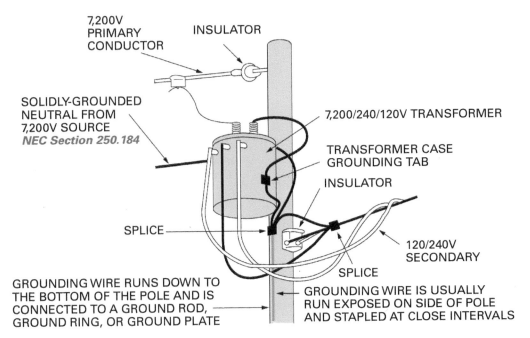

Figure 4 Pole-mounted transformer.

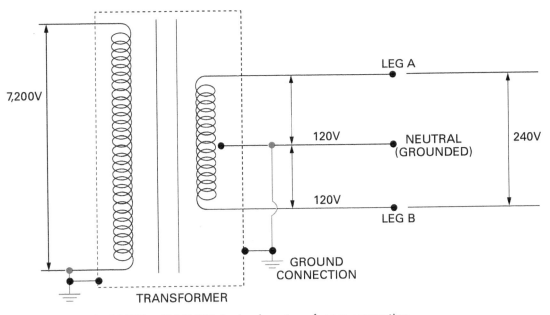

Figure 5 Wiring diagram of a 7,200V to 120/240V single-phase transformer connection.

1.2.2 Grounding Systems Less Than 50V

Grounding is required in the following situations (examples are shown in *Figure 7*):

- Where supplied by transformers if the supply voltage to the transformer exceeds 150V to ground
- Where supplied by transformers if the transformer supply system is ungrounded
- Where installed as overhead conductors outside of the building
- In other circuits and systems provided they comply with the provisions of *NEC Article 250*

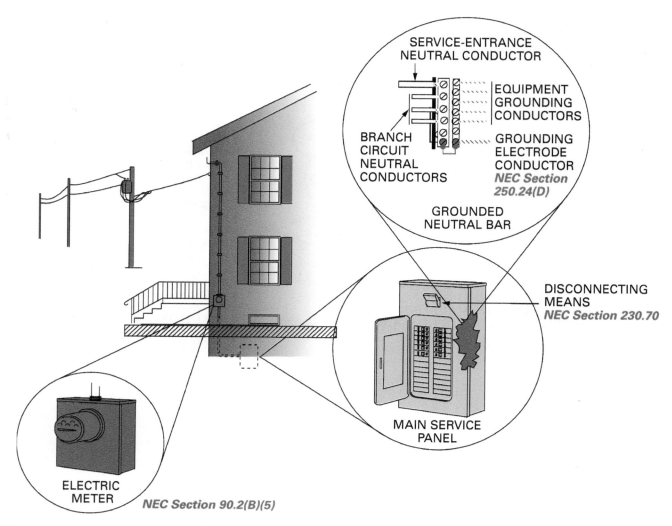

Figure 6 Typical service entrance and related service equipment.

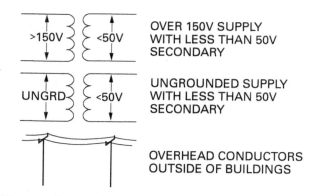

Figure 7 Systems less than 50V that must be grounded.

1.2.3 Grounding 50V to 1,000V Systems

AC systems of 50V to 1kV (1,000V) supplying premises wiring and premises wiring systems shall be grounded under any of the following conditions:

- Where the system can be grounded so that the maximum voltage to ground on the ungrounded conductors does not exceed 150V
- Where the system is three-phase, four-wire, wye-connected and the neutral is used as a circuit conductor

- Where the system is three-phase, four-wire, delta-connected and the midpoint of one phase winding is used as a circuit conductor
- In other circuits and systems provided they comply with the provisions of *NEC Article 250*

Figure 8 shows examples of 50V to 1kV grounded applications.

1.2.4 Grounding AC Systems Over 1kV

NEC Article 250, Part X discusses the grounding requirements for systems over 1kV. It also lists additional requirements for derived neutral systems, solidly grounded neutral systems, and impedance-grounded neutral systems. AC systems over 1kV must be grounded if they supply mobile or portable equipment as covered in *NEC Section 250.188*. Other AC systems over 1kV are permitted (but not required) to be grounded.

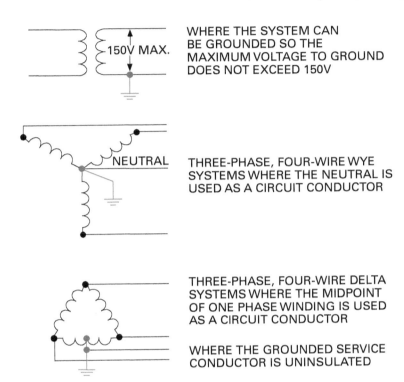

Figure 8 Systems that must be grounded.

1.2.5 Circuits That Cannot Be Grounded

According to *NEC Section 250.22*, only six types of circuits shall not be grounded. Some examples of these circuits are shown in *Figure 9*. The six types of circuits that shall not be grounded are as follows:

- Circuits for cranes that operate over combustible fibers in Class III locations, as provided in *NEC Section 503.155*
- For healthcare facilities, those isolated power circuits in hazardous (classified) locations as provided in *NEC Sections 517.61 and 517.160*
- Circuits for electrolytic cells as provided in *NEC Article 668*
- Secondary circuits of lighting systems as provided in *NEC Section 411.6(A)*
- Secondary circuits on lighting systems as provided in *NEC Section 680.23(A)(2)*
- Class 2 load-side circuits for suspended ceiling low-voltage power grid distribution systems as provided in *NEC Section 393.60(B)*

> **NOTE**
> Equipment located or used within the electrolytic cell line working zone or associated with the cell line DC power circuits are not required to comply with *NEC Article 250*.

CRANES THAT OPERATE OVER COMBUSTIBLE FIBERS IN CLASS III LOCATIONS

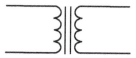
ISOLATED POWER SYSTEMS IN HEALTHCARE FACILITIES FOR HAZARDOUS INHALATION ANESTHETIZING AND WET LOCATIONS

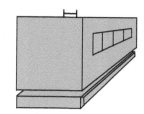
CIRCUITS FOR ELECTROLYTIC CELLS

SECONDARY CIRCUITS ON LIGHTING SYSTEMS

Figure 9 Circuits that shall not be grounded.

1.0.0 Section Review

1. Suppose a loose connection has an exposed bare conductor and it contacts the edge of a metal enclosure. Is this an example of a short circuit or a ground fault?
    a. Short circuit
    b. Ground fault

2. A single-phase system has two legs, Leg A and Leg B. If Leg A is carrying 50A and Leg B is carrying 45A, the current carried by the neutral is _1.2.1_.
    a. 5A
    b. 45A
    c. 50A
    d. 90A

Section Two

2.0.0 Service Grounding Methods

Objective

Identify service grounding methods.
a. Size and install a grounding electrode conductor.
b. Select other electrodes.

Performance Tasks

1. Size the minimum required grounding electrode conductor for a 200A service fed by 3/0 copper.
2. Using the proper fittings, connect one end of a No. 4 AWG bare copper grounding wire to a length of ¾" (MD 21) galvanized water pipe and the other end to the correct terminal in a main panelboard.

Trade Term

Supplemental electrode: An additional electrode (commonly a driven rod or pipe) required where the primary electrode is an underground metal water pipe in direct contact with the earth.

All electrical systems must be grounded and bonded in a manner prescribed by the *NEC®* to protect personnel and equipment. A grounded system must limit the voltage on the electrical system and protect it from:

- Lightning
- Voltage surges higher than that for which the circuit is designed
- An increase in maximum potential to ground due to abnormal voltage

The requirements for grounding of services are found in *NEC Section 250.24*. Methods of grounding an electric service are covered in *NEC Article 250, Part III*. In general, all of the following electrodes that are present must be bonded together to form the grounding electrode system:

- An underground metal water pipe in direct contact with the earth for no less than 10' (3 m)
- The metal in-ground support structure as described in *NEC Section 250.52(A)(2)*
- An electrode encased by at least 2" (50 mm) of concrete, located within and near the bottom of a concrete foundation or footing that is in direct contact with the earth

> **NOTE:** This electrode must be at least 20' (6 m) long and must be made of electrically conductive coated steel reinforcing bars or rods of not less than ½" (13 mm) in diameter, or consisting of at least 20' (6 m) of bare copper conductor not smaller than No. 4 AWG wire size.

- A ground ring encircling the building or structure, in direct contact with the earth, consisting of at least 20' (6 m) of bare copper not smaller than No. 2 AWG
- Rod, pipe, or plate electrodes
- Other local metal underground systems or structures

Grounding electrode systems used in industrial buildings will frequently use all of the methods shown in *Figure 10*, and the methods used will often surpass the *NEC®*, depending upon the manufacturing process and the calculated requirements made by plant engineers. *Figure 11* shows a floor plan of a typical industrial grounding system.

> **NOTE:** This ring must consist of at least 20' (6 m) of bare copper conductor not smaller than No. 2 AWG wire size (*Figure 10*).

The building in *Figure 12* has a metal underground water pipe that is in direct contact with the earth for more than 10' (3 m), so this is one valid grounding source. The building also has a metal underground gas piping system, but this may not be used as a grounding electrode [*NEC Section 250.52(B)(1)*], but it must be bonded to the electrical system if it is likely to become energized per *NEC Section 250.104(B)*.

NEC Section 250.53(D)(2) further states that the underground water pipe must be supplemented by an additional electrode of a type specified in *NEC Sections 250.52(A)(2) through 250.52(A)(8)*. Since a metal in-ground support structure, concrete-encased electrode, or ground

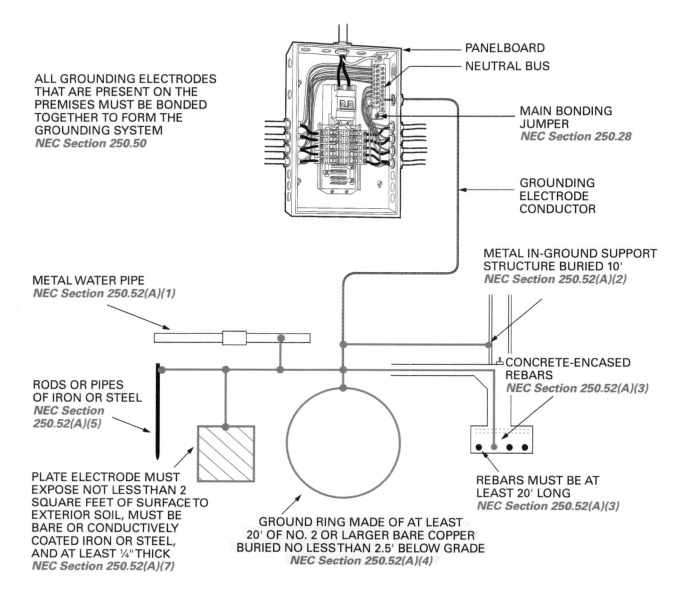

Figure 10 NEC®-approved grounding electrodes.

ring is not available in this application, *NEC Section 250.50* must be used in determining the supplemental electrode. In most cases, this **supplemental electrode** will consist of either a driven rod (*Figure 13*) or pipe electrode, the specifications for which are as follows:

- Withstand and dissipate repeated surge circuits
- Provide corrosion resistance to various soil chemistries to ensure continuous performance for the life of the equipment being protected
- Provide rugged mechanical properties for easy driving with minimum effort and rod damage

An alternative to the pipe or rod method is a plate electrode. Each plate electrode must expose not less than 2 sq ft (0.186 sq m) of surface to the surrounding earth. Plates made of iron or steel must be at least ¼" (6.4 mm) thick, while plates of nonferrous metal like copper need only be 0.06" (1.5 mm) thick.

Rod, pipe, and plate electrodes must have a resistance to ground of 25 ohms (Ω) or less. If not, they must be augmented by an additional electrode spaced not less than 6' (1.8 m) from the first. Many locations require two electrodes regardless of the resistance to ground. This is not an *NEC®* requirement, but is required by some power companies and local ordinances in some cities and counties. Always check with the local inspection authority for such rules that may go beyond the requirements of the *NEC®*.

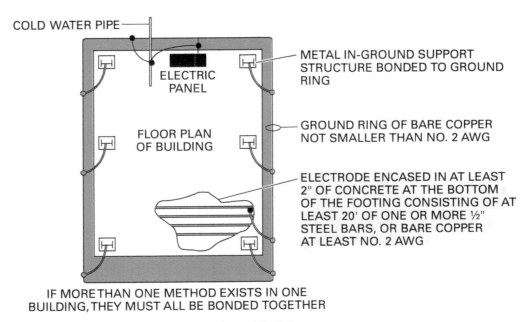

Figure 11 Floor plan of the grounding system for an industrial building.

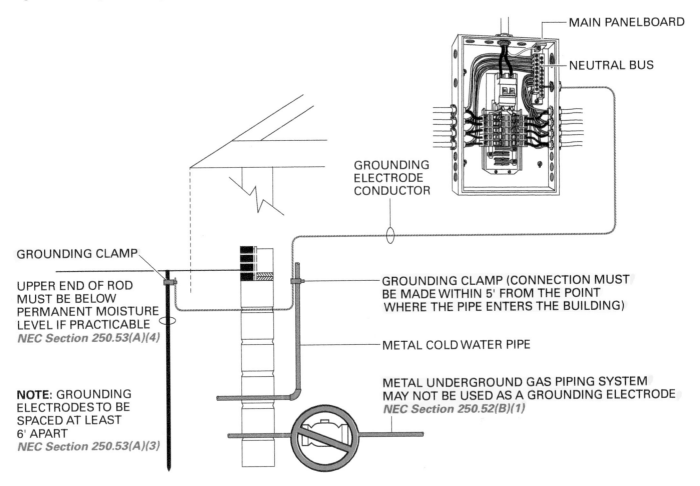

Figure 12 Grounding requirements for non-industrial buildings.

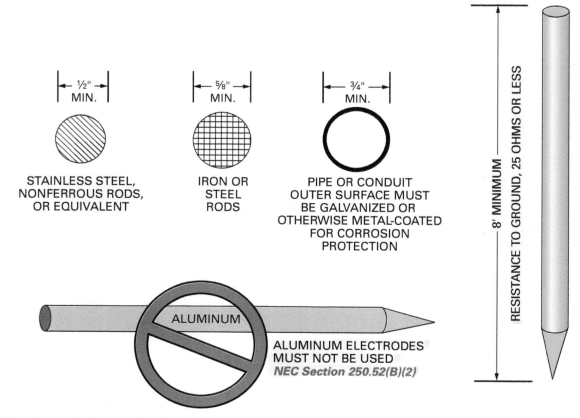

Figure 13 Requirements for ground rods.

2.1.0 Grounding Electrode Conductor (GEC)

The grounding electrode conductor is the sole connection between the grounding electrode and the grounded system conductor for a grounded system, or the sole connection between the grounding electrode and the service equipment enclosure for an ungrounded system. See *NEC Section 250.64*, which describes grounding electrode conductor installation.

A common grounding electrode conductor is required to ground both the circuit grounded conductor and the equipment grounding conductor. *Figure 14* shows the common connection for this system.

2.1.1 Sizing the Grounding Electrode Conductor

The size of grounding electrode conductors depends on service-entrance size; that is, the size of the largest service-entrance conductor or equivalent for parallel conductors. *NEC Table 250.66* gives the proper sizes of grounding electrode conductors for various sizes of electric services.

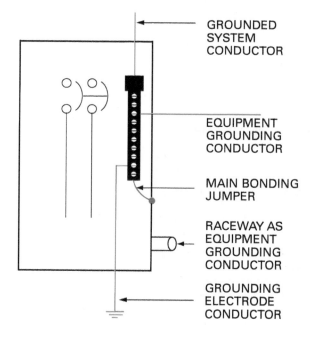

Figure 14 Grounding electrode connection for grounded system.

Around the World

The *NEC*® provides metric conversions for all listed measurements and generally lists values in whole units. In some cases, however, the code provides precise metric conversions, as with grounding electrodes. Always refer to the specific *NEC*® citation when selecting electrodes and other electrical devices and equipment.

The size of the service-entrance conductor is determined from the ampacity of that system. *NEC Table 310.16* is used to determine the size of the service conductor from its ampacity. In the case of single-phase dwelling services, *NEC Section 310.12(B)* allows an 83% adjustment.

The majority of applications use 75°C copper conductors. This information is necessary when using *NEC Table 310.16*. For a 100A commercial service application, the size of the service conductor is No. 3 AWG copper.

Now that the size of the service conductors is known, the size of the grounding electrode conductor can be determined from *NEC Table 250.66*. Using the No. 3 AWG from the 100A service example, it can be determined from the table that a No. 8 AWG copper grounding electrode conductor is required.

Now, use the following steps to determine the size of the grounding electrode conductor for a 200A, 208V, three-phase system:

Step 1 Determine the size of the service conductor from *NEC Table 310.16*. It is 3/0 AWG copper.

Step 2 Go down the left column of *NEC Table 250.66* and find 3/0.

Step 3 Go across to find the size of the grounding electrode conductor. It is a No. 4 AWG copper conductor.

NOTE: Remember, it is the total size of the service conductor that determines the size of the grounding electrode conductor.

2.1.2 Installing the Grounding Electrode Conductor

The grounding electrode conductor must be copper, aluminum, or copper-clad aluminum. The material selected must be resistant to any corrosive condition existing at the installation, or it must be suitably protected against corrosion. The conductor may be either solid or stranded, and covered or bare, but it must be in one continuous length without a splice or joint, except for the following conditions:

- Splices in busbars are permitted.
- Where a service consists of more than one single enclosure, it is permissible to connect taps to the grounding electrode conductor.
- Bonding jumper(s) from grounding electrode(s) and GECs shall be permitted to be connected to an aluminum or copper busbar not less than ¼" × 2" (6 mm × 50 mm). See *NEC Section 250.64(F)(3)*.

WARNING! Exothermic welding is a hazardous process that should be performed only by qualified personnel. Refer to your company's safety procedures and the manufacturer's recommendations before using exothermic welding equipment.

Think About It
Sizing Electrode Conductors

NEC Table 310.16 shows that the service conductor size (copper wire at 75°C) for a 150A service is 1/0 AWG. Which of the three AWG copper electrode conductor sizes listed below would be used with this service?

No. 4
No. 6
No. 8

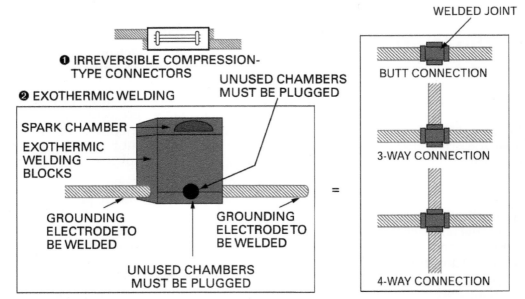

Figure 15 Methods of splicing grounding conductors.

- Per *NEC Section 250.64(C)(1)*, wire-type grounding electrode conductors may be spliced at any location by means of irreversible compression-type connectors listed for the purpose or using the exothermic welding process (*Figure 15*).

NEC Section 250.64(B) requires the grounding electrode conductor or its enclosure to be securely fastened to the surface on which it is carried.

No. 4 AWG or larger conductors require protection where exposed to physical damage. If free from exposure to physical damage, No. 6 AWG conductors may be run along the surface of the building and be securely fastened. Otherwise, the conductor must be protected by installation in rigid or intermediate metal conduit, rigid nonmetallic conduit, EMT, or cable armor. Smaller conductors shall be protected in conduit or armor. See *NEC Section 250.64(E)* for bonding requirements when using a ferrous enclosure or raceway to enclose a grounding electrode conductor. *Figure 16* shows the various protection guidelines.

2.2.0 Other Electrodes

When an electrode that meets the requirements of *NEC Section 250.50* is not present, other electrodes may be used. These may be the rod and pipe electrodes discussed earlier, other listed electrodes, plate electrodes, or other local metal underground systems such as piping and underground tanks. *NEC Sections 250.52(A)(5) through 250.52(A)(8)* provide information and requirements for these electrodes. *Figure 17* shows other electrodes.

The specific requirements for other electrodes are:

- *Local systems* – Local metallic underground systems such as piping, tanks, metal well casings, etc.

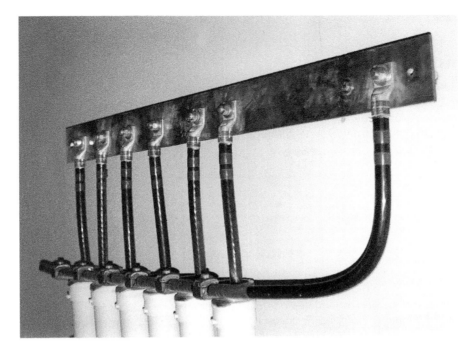

Figure 16 Grounding electrode conductors.

- *Rod and pipe electrodes* – Pipe or conduit electrodes shall be not less than 8' (2.44 m) in length nor smaller than ¾" trade size (MD 21), and if iron or steel, shall be galvanized or metal-coated for corrosion protection. Rod electrodes of stainless steel and copper- or zinc-coated steel shall be not less than 8' (2.44 m) in length and ⅝" (15.87 mm) in diameter, unless listed. Rod and pipe electrodes must be installed so that at least 8' (2.44 m) of rod/pipe is in contact with the soil. Rod and pipe electrodes must be driven vertically unless a rock obstruction is encountered. If an obstruction is encountered, the rod/pipe may be driven at not more than a 45° angle to clear the rock. The other possibility is for the rod to be buried in a trench that is at least 30" (750 mm) deep.
- *Plate electrodes* – Electrodes shall have at least 2 sq ft (0.186 sq m) of surface in contact with exterior soil. If of iron or steel, the plate shall be at least ¼" (6.4 mm) thick. If of nonferrous metal, it shall be at least 0.06" (1.5 mm) thick.

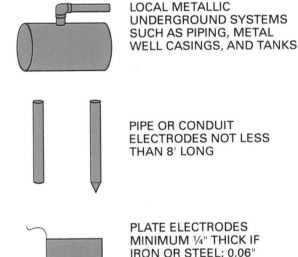

Figure 17 Other electrodes.

> **NOTE**
> Underground metal gas piping systems and structural reinforcing steel members described in *NEC Sections 680.26(B)(1) and (B)(2)* are not permitted to be used as grounding electrodes. However, this does not eliminate the requirement that these systems be bonded.

Think About It
Ground Rods

Suppose you are driving a rod electrode into the ground and you hit a rock after it has been driven about ¾ of the way in. Further effort to drive the rod deeper is of no avail. What are your options?

Exothermic Welding

Normally, the connections made by exothermic welding are permanent. They perform better than any crimped or bolted connection because the copper-to-copper or copper-to-steel bond is molecular, eliminating the risk of loosening or corrosion. Exothermic welded connections will carry more current than the conductor, resist repeated fault currents, and will not deteriorate with age.

Think About It

Plate Electrodes

Why does the *NEC®* require that electrode plates made of iron or steel be at least ¼" (6.4 mm) thick, while electrode plates made of copper (a nonferrous metal) need only be 0.06" (1.5 mm) thick?

2.0.0 Section Review

1. Assuming the use of 75°C copper conductors, a 200A commercial service would require _____.

    a. 1/0 service conductors
    b. 2/0 service conductors
    c. 3/0 service conductors
    d. 4/0 service conductors

2. A ground rod must be installed in a coastal location and can only be driven halfway before encountering a rock shelf. One alternative is to _____.

    a. bond and bury two 4' (1.2 m) lengths of ground rod
    b. allow the 4' (1.2 m) section to protrude above the ground
    c. bury the rod in a trench at least 30" (750 mm) deep
    d. bend the rod in a U shape and drive both ends into the ground

Separately Derived Systems

The requirements of *NEC Section 250.30* pertaining to the grounding of separately derived AC systems are most commonly applied to 480Y/277V transformers that transform a 480V supply to a 208Y/120V system for lighting and appliance loads. The requirements provide for a low-impedance path to ground, so that line-to-ground faults on circuits supplied by the transformer will result in sufficient current flow to operate the overcurrent devices.

For separately derived systems that are required to be grounded, an equipment grounding conductor must be supplied with the primary circuit. This will provide a low-impedance fault-current path from the transformer case to the feeder overcurrent protective enclosure. Along with the equipment grounding conductor, *NEC Section 250.30(A)(1)* requires a bonding jumper be installed for the derived system required to be grounded. This bonding jumper may be placed in either the transformer or the panel, but not in both. Installing it in both locations would create a parallel path with the grounded conductor, resulting in an unwanted current flow on the grounding conductor. Next, a grounding electrode conductor must be installed and connected at the same point. A supply-side bonding jumper is routed with the feeder conductors from the transformer to the equipment grounding bar in the panel when the bonding jumper is in the transformer. This bonding jumper is still required if the bonding is at the feeder panel, but will terminate on the neutral bar. This is illustrated in *Figure 31*.

The grounding electrode for separately derived systems must be:

- The building or structure grounding electrode system
- If located outside, in accordance with *NEC Section 250.30(C)*.

A generator-type separately derived system is shown in *Figure 32*.

The way to determine that a generator is a separately derived system is to examine the transfer switch. If the neutral and all phase conductors

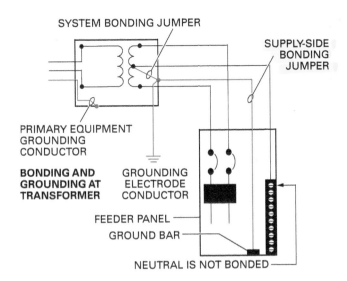
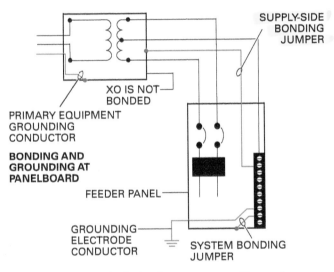

Figure 31 Separately derived system grounding and bonding locations.

are switched, then the system is separately derived. If the neutral is not switched, but solidly connected, then the system is not a separately derived system.

The system bonding jumper must be installed either at the generator or any point in between. A grounding electrode conductor must be connected at the same point as the system bonding jumper.

When the generator is not a separately derived system, the system bonding jumper must be removed from the generator and the neutral must not be grounded. Grounding is accomplished by connecting the neutral of the premises wiring to a grounding electrode.

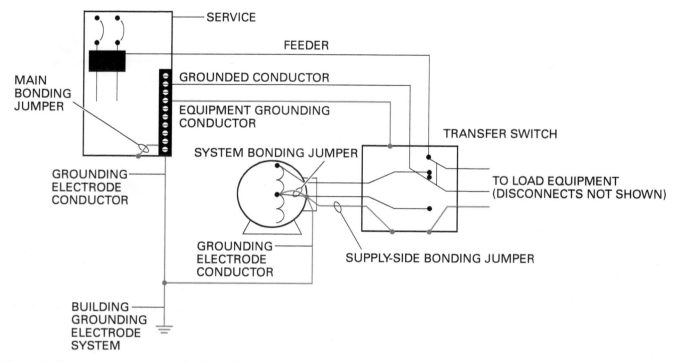

Figure 32 Generator-type separately derived system.

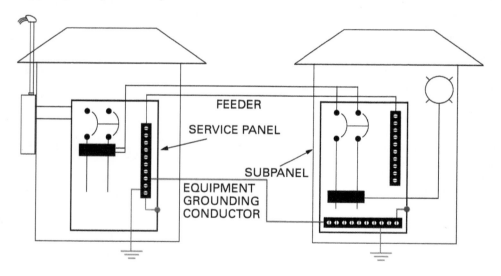

Figure 33 Grounding neutral at second building.

Table 2 Number of Conductors Between Structures

System	Ungrounded	Grounded	Equipment Ground
120V	1	1	1
120/240V	2	1	1
208/120V	3	1	1
480/277V	3	1	1

5.2.0 Grounding at More Than One Building

NEC Section 250.32 provides the method for grounding electrical systems at additional buildings on the premises (*Figure 33*).

The grounded circuit conductor and the equipment grounding conductor must both be extended to the second building. The grounded conductor is terminated on an insulated bus isolated from the metal cabinet, while the equipment grounding conductor from the main building is terminated on an equipment grounding terminal bus, directly connected to the cabinet. The equipment grounding conductor would be sized using *NEC Table 250.122* based on the overcurrent device supplying the second structure. All grounding electrodes present at the second building would connect to the grounding bus. The GEC in the second structure is sized using *NEC Table 250.66* based on the size of the feeder conductors. *Table 2* shows the number and type of conductors that must be taken from the first structure to where service is located in the second structure.

When no grounding electrodes are at the additional structures, a grounding electrode must be installed, unless the structure is supplied by a single branch circuit that includes an equipment grounding conductor. These electrodes include underground metal pipes, the metal in-ground support structure(s), or concrete-encased electrodes. Rod, pipe, or plate electrodes may also be permitted, in accordance with *NEC Section 250.52*.

By exception and in existing installations only, the *NEC®* allows a method whereby no equipment ground is extended to the second building per *NEC Section 250.32(B)(1), Exception 1*. In this case, the grounded conductor is extended to and terminated on a terminal bus that is bonded to the equipment, similar to the connection made at the grounded conductor at the service equipment. The grounded system will be connected to all grounding electrodes at the second building. The grounded conductor is sized using *NEC Sections 220.61 and 250.122*. This method may only be used when an equipment grounding conductor is not run with the supply to the second building, there are no continuous metallic paths bonded to the grounding system in both buildings, and ground fault protection has not been installed on the common electrical service in the first building. These additional provisions prevent the possibility of parallel current paths between grounded and grounding conductors, as well as nuisance tripping of a ground fault protective device.

5.0.0 Section Review

1. Which of the following is considered a separately derived system?
   a. A generator supplying power where the neutral is connected to the utility system
   b. An emergency standby system with an isolated neutral
   c. A transformer with a direct connection between the primary and the secondary
   d. A transformer with switched phase conductors but a connected neutral

2. When grounding at more than one building, the GEC at the second building is sized using ___5.2.0___.
   a. *NEC Table 250.66*
   b. *NEC Table 250.102(C)(1)*
   c. *NEC Table 250.122*
   d. *NEC Table 310.16*

Section Six

6.0.0 Testing for Effective Grounds

Objective

Test for effective grounds.
a. Measure earth resistance using the fall-of-potential method.
b. Complete a three-point test.

Trade Terms

Auxiliary electrodes: Metallic electrodes pushed or driven into the earth to provide electrical contact for the purpose of performing measurements on grounding electrodes or ground grid systems.

Ground current: Current in the earth or grounding connection.

Ground grid: System of grounding electrodes interconnected by bare cables buried in the earth to provide lower resistance than a single grounding electrode.

Ground mat: System of bare conductors, on or below the surface of the earth, connected to a ground or ground grid to provide protection from dangerous touch voltage.

Ground resistance: The ohmic resistance between a grounding electrode and a remote or reference grounding electrode that are spaced such that their mutual resistance is essentially zero.

Resistivity: Resistance between opposite faces of a unit cube. Expressed in ohm-centimeters or ohms per cubic centimeter.

Step voltage: The potential difference between two points on the earth's surface separated by a distance of one pace, or about 3' (1 m).

Touch voltage: The potential difference between a grounded metallic structure and a point on the earth's surface equal to the normal maximum horizontal reach—approximately 3' (1 m).

> **NOTE:** An ordinary ohmmeter is not used to measure the resistance of a grounding electrode to the earth because it does not provide for adequate levels of current and voltage.

> **WARNING!** Ground testers can be hazardous to both equipment and personnel if improperly used. Always check with your supervisor before using a ground tester.

One use of the ground tester is for testing electrical systems after they are installed and before normal voltage is applied. This test is made after all the conductors, fuses or circuit breakers, panelboards, outlets, etc., are in place and connected. The current used for testing is produced by a small generator within the ground tester that generates DC power, either by turning a crank handle (also a part of the ground tester), or by using a small electric DC motor within the ground tester.

When using a crank handle ground tester, the test is made by connecting the terminals to the two points between which the test is to be made and then rapidly turning the handle on the tester. The resistance in ohms can then be read from the meter dial. Satisfactory insulation resistance values will vary under different conditions, and the charts supplied with the ground tester should be consulted for the proper value for a particular installation.

An earth ground is commonly used as an electrical conductor for system returns. Although

An earth ground resistance tester (*Figure 34*) may be used to make soil **resistivity** measurements or to measure the resistance of the installed grounding electrode system to the earth.

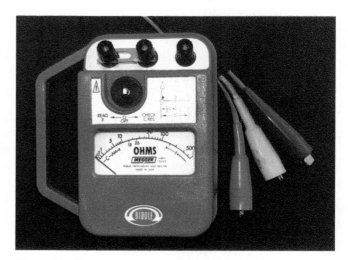

Figure 34 Earth ground resistance tester.

the resistivity of the earth is high compared to a metal conductor, its overall resistance can be quite low because of the large cross-sectional area of the electrical path.

Connections to the earth can be made using one or more grounding electrodes, a ground grid, or a ground mat. The resistance of these devices varies proportionately with the earth's resistivity, which in turn depends on the composition, compactness, temperature, and moisture content of the soil.

A good grounding system limits system-to-ground resistance to an acceptably low value. This protects personnel from a dangerously high voltage during a fault in the equipment. Furthermore, equipment damage can be limited by using this ground current to operate protective devices.

Ground testers measure the ground resistance of a grounding electrode or ground grid system. Some of the major purposes of ground testing are to verify the adequacy of a new grounding system, detect changes in an existing system, and determine the presence of hazardous step voltage and touch voltage.

In addition to personnel safety considerations, ground testing also provides information for

---

Case History

Step Voltage

In September of 1999, a single stroke of lightning is believed to have killed a herd of 56 elk on Mount Evans in Colorado. This happened because when lightning strikes the earth, it disperses in concentric rings of equal potential (equipotential zones). Because the earth has resistance, the voltage drops as it travels away from the strike point. Step voltage is the difference in potential that exists when you step between these equipotential zones. In this case, the hind legs of the elk were at a different potential than their front legs, resulting in current flow.

The Bottom Line: During a lightning strike or any other condition that creates zones of different potential, your best bet is to remain in one spot, crouched low with your feet together and your head bent forward. If you must move, hop with your feet together, moving much like a bird on a wire.

Figure Credit: National Oceanic and Atmospheric Administration (NOAA) Central Library; OAR/ERL/National Severe Storms Laboratory (NSSL). Photographer: C. Clark

equipment insulation ratings. Equipment can be damaged by an overvoltage that exceeds the rating of the insulation system. *Figure 35* depicts a poorly grounded system where the ground resistance (R_1) is 10Ω.

Assuming a power source resistance (R_2) of 40Ω, a short circuit between the 5,000V power line and the steel tower would produce 100A of short circuit current (I).

$$I = \frac{E}{R_1 + R_2} = \frac{5,000}{40 + 10} = 100\text{A of short circuit current}$$

A person touching the tower would be subjected to the voltage (E') developed across the ground resistance:

$$E = IR = 100 \times 10 = 1,000 \text{ volts between power and ground}$$

Statistics vary widely concerning what may be considered a dangerous voltage. This depends largely on body resistance and other conditions. However, to limit the touch voltage for this situation to 100V, the ground resistance for the tower would have to be less than 1Ω.

6.1.0 Measuring Earth Resistance

Several methods are used for measuring the resistance to earth of a grounding electrode; one of the most common is the fall-of-potential method (*Figure 36*).

In this method, auxiliary electrodes 1 and 2 are placed at sufficient distances from grounding electrode X. A current (I) is passed through the earth between the grounding electrode X and auxiliary current electrode 2 and is measured by the ammeter. The voltage drop (E) between

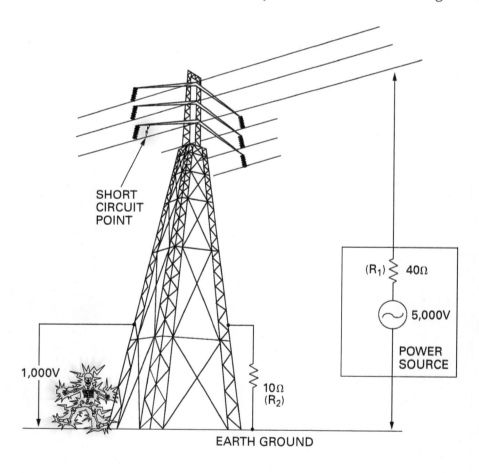

Figure 35 Poorly grounded system.

the grounding electrode X and the auxiliary potential electrode 1 is indicated on the voltmeter. Resistance (R) can therefore be calculated as follows:

$$R = E \div I$$

The following problems may arise in measuring with the simple system shown in *Figure 36*:

- Natural currents in the soil caused by electrolytic action can cause the voltmeter to read either high or low, depending on polarity.
- Induced currents in the soil, instrument, or electrical leads can cause vibration of the meter pointer, interfering with readability.
- Resistance in the auxiliary electrode and electrical leads can introduce error into the voltmeter reading.

Most ground testers use a null balance metering system. Unlike the separate voltmeter and ammeter method, this instrument provides a readout directly in ohms, thus eliminating calculation. Although the integrated systems of the ground testers are sophisticated, they still perform the basic functions for fall-of-potential testing.

6.1.1 Current Supply Circuit

As in the simple circuit, the ground tester also has a current supply circuit (*Figure 37*). This may be traced from grounding electrode X through terminal X, potentiometer R1, the secondary of power transformer T1, terminal 2, and auxiliary current electrode 2. This produces a current in the earth between electrodes X and 2.

When switch S1 is closed, battery B energizes the coil of vibrator V. Vibrator reed V1 begins oscillating, thereby producing an alternating current in the primary and secondary windings of T1. The negative battery terminal is connected alternately across first one and then the other half of the primary winding.

6.1.2 Voltmeter Circuit

The voltmeter circuit can be traced from grounding electrode X through terminal X, the T2 secondary, switch S2, resistors R2 and R3 (paralleled by the meter and V2 contacts), capacitor C, terminal 1, and auxiliary potential electrode 1. The cur-

Clamp-On Ground Testers

This clamp-on ground tester allows for checking ground grid connections from 0.10Ω to 1,200Ω without the need to disconnect the conductors.

Figure Credit: Greenlee / A Textron Company

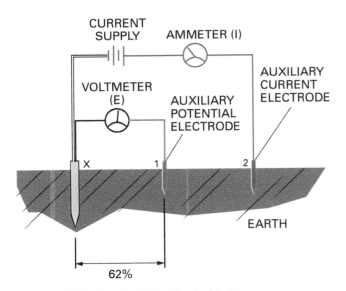

Figure 36 Fall-of-potential method of testing.

Case History

Touch Voltage

If a system is ungrounded and the metal parts become energized, a potential difference will exist between the system and ground. This is known as touch voltage. For example, a 10-year-old boy was swimming at a community pool and exited the pool to buy a snack from a vending machine. The grounding prong had been removed from the plug and the cord was pinched under the machine's metal leg, exposing the conductor and causing the metal frame of the unit to become energized. As soon as he touched the metal, he received a fatal shock.

The Bottom Line: Make sure the entire grounding system is intact for cord-and-plug connected equipment. In case of damage to the cord, the equipment grounding conductor could be damaged and deficient. When servicing any motor, check to ensure that the cord is not pinched. Never remove the grounding prong from a three-prong plug. Because of this and similar incidents, *NEC Section 422.5(A)(5)* now requires GFCI protection for cord-and-plug connected vending machines.

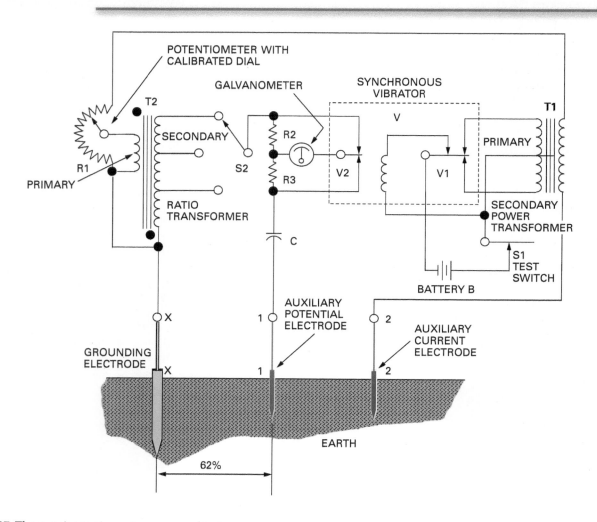

Figure 37 Three-point testing using a ground tester.

> **Think About It**
> ### Step or Touch Potential
> What are some other examples where step or touch potential may be significant enough to cause harm?

rent in the earth between grounding electrode X and auxiliary current electrode 2 creates a voltage drop due to the earth's resistance. With auxiliary potential electrode 1 placed at any distance between grounding electrode X and auxiliary current electrode 2, the voltage drop causes a current in the voltmeter circuit through balanced resistors R2 and R3. The voltage drop across these resistors causes galvanometer M to deflect from zero center scale.

Vibrator reed V2 operates at the same frequency as V1, thereby functioning as a mechanical rectifier for galvanometer M. The vibrator is tuned to operate at 97.5 hertz (Hz), a frequency unrelated to commercial power line frequencies and their harmonics. Thus, currents induced in the earth by power lines are rejected by most ground testers and have virtually no effect on their accuracy. Stray direct current in the earth is blocked out of the voltmeter circuit by capacitor C.

The current in the primary of T2 can be adjusted with potentiometer R1. Primary current in T2 induces a voltage in the secondary of T2, which is opposite in polarity to the voltage drop caused by current in the voltmeter circuit.

With R1 adjusted so the primary and opposing secondary voltages of T2 are equal, current in the voltmeter circuit is zero, and the galvanometer reads zero. The resistance of grounding electrode X can then be read on the calibrated dial of the potentiometer.

With no current in the voltmeter circuit, the lead resistance of the auxiliary potential electrode 1 has no bearing on accuracy. (With no current, there is no voltage drop in the leads.) Resistance to earth of the current electrode results only in a reduction of current, and consequently, a loss of sensitivity. Therefore, the auxiliary electrodes need only be inserted into the earth 6" to 8" (150 mm to 200 mm) to make sufficient contact. In some locations, where the soil is very dry, it may be necessary to pour water around the current electrode to lower the resistance to a practical value.

6.1.3 Earth Electrode Resistance

Current in a grounding system is primarily determined by the voltage and impedance of the electrical equipment. However, the resistance of the grounding system is very important in determining the voltage rise between the ground electrode and the earth as well as the voltage gradients that will occur in the vicinity when current is present. Three components constitute the resistance of a grounding system:

- Resistance of the conductor connecting the ground electrode
- Contact resistance between the ground electrode and the soil
- Resistance of the body of earth immediately surrounding the electrode

Resistance of the connecting conductor can be dealt with separately since this is a function of the conductor cross-sectional area and length. Contact resistance is usually negligibly small if the electrode is free from paint or grease. Therefore, the main resistance is that of the body of earth immediately surrounding the electrode. Current from a grounding electrode flows in all directions via the surrounding earth. It is as though the current flows through a series of concentric spherical shells, all of equal thickness.

The shell immediately surrounding the electrode has the smallest cross-sectional area, and therefore its resistance is highest. As the distance from the electrode is increased, each shell becomes correspondingly larger; thus, the resistance becomes smaller. Finally, a distance from the electrode is reached where additional shells do not add significantly to the total resistance. From a theoretical viewpoint, total resistance is included only when the distance is infinite. For practical purposes, only the volume that contributes the major part of the resistance need be considered. This is known as the effective resistance area and depends on electrode diameter and driven depth.

6.2.0 Three-Point Test

To measure the resistance of a grounding electrode with most ground testers, an auxiliary current electrode and an auxiliary potential electrode are required (*Figure 38* and *Figure 39*). The current electrode is placed a suitable distance from the grounding electrode under test, and the potential electrode is then placed at 62% of the current electrode distance.

> **NOTE**
>
> The 62% figure was determined by empirical data gathered by many authorities on ground resistance measurement, and in some cases, has been computed based on analysis of an equivalent hemisphere. Testing on large systems requires engineering assistance to determine proper testing procedures and electrode spacing.

If the current electrode is too near the grounding electrode, their effective resistance areas will overlap, as shown in *Figure 38*. If a series of measurements are made with the potential electrode driven at various distances in a straight line between the current electrode and grounding electrode, the readings will yield a curve as illustrated in *Figure 38*.

In *Figure 39*, a curve is plotted with the current electrode at a sufficient distance from the grounding electrode. Note that the curve is relatively flat between points B and C, which are usually considered to be at ±10' (3 m) with respect to the 62% point. Usually, a tolerance is established for the maximum allowable deviation for the second and third readings with relation to the initial reading at 62%. This tolerance is a certain ±% of the initial reading, such as ±1%, ±2%, etc.

No definite distance from the current electrode can be forecast since the optimum distance is based on the homogeneity of the earth, depth of the grounding electrode, diameter, etc. However, for a starting point for a single driven grounding electrode, the effective radius of the equivalent hemisphere can be computed. The curve of *Figure 40* can then be used for initial placement of the auxiliary electrodes. Using the practical method of moving the auxiliary potential electrode 10' (3 m) to either side of the 62% point, a curve can be plotted to determine if the current electrode spacing is adequate, as depicted by the curve flattening out between points B and C in *Figure 39*.

For example, assume that a 1" (25 mm) diameter grounding electrode is buried 10' (3.0 m) deep. The equivalent hemisphere radius is 1.7' (0.5 m). Using *Figure 40*, the current electrode would be

> **Think About It**
>
> ## Size and Depth of Ground Electrode
>
> What effect do the diameter and driven depth of a ground electrode have on its resistance?

> **Think About It**
>
> ## Poor Soil Conductivity
>
> What can be done to lower the ground electrode (rod) resistance when it is driven into soil with poor conductivity?

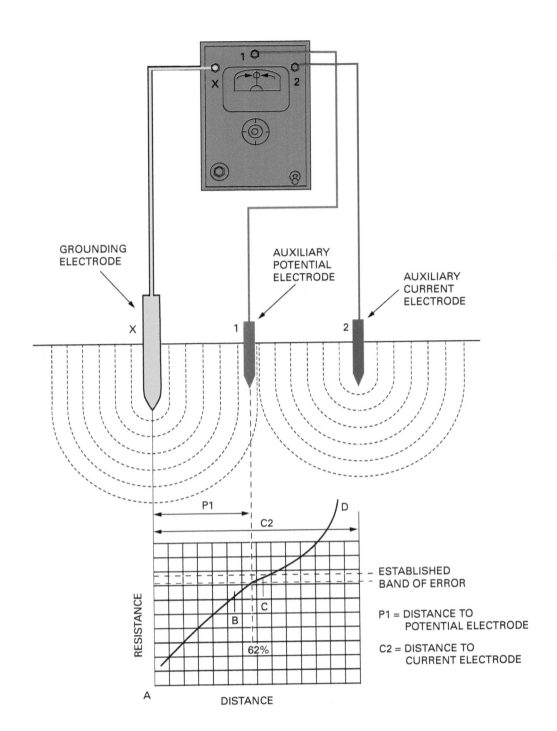

Figure 38 Plotted curve showing insufficient electrode spacing.

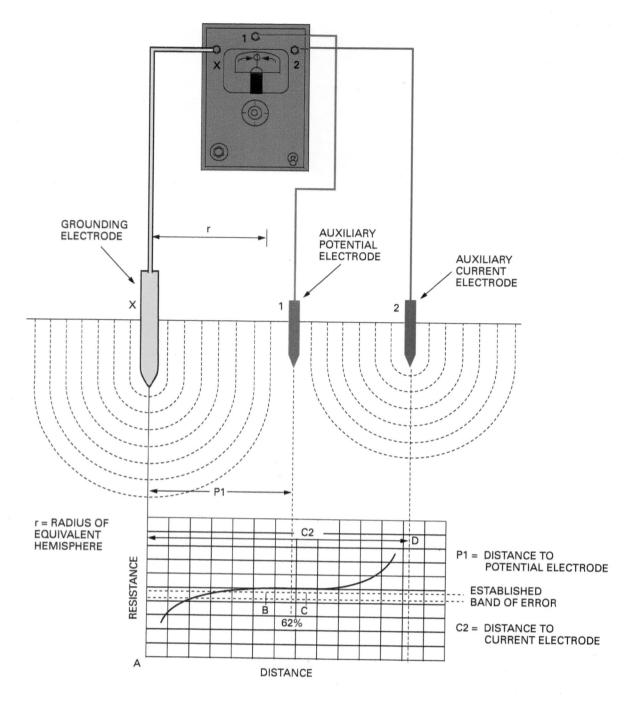

Figure 39 Plotted curve showing adequate electrode spacing.

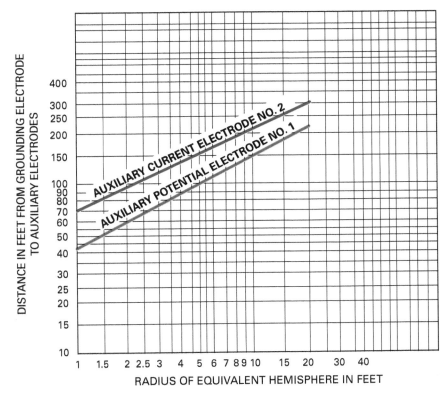

Figure 40 Auxiliary electrode distance/radii chart.

established at approximately 90' (27 m), and the potential electrode at approximately 55' (17 m) for the initial reading. The results obtained when the potential electrode is moved (to 45' [14 m] and then to 65' [20 m]) will determine if the current electrode, and consequently the potential electrode, must be spaced at a greater distance.

It is usually advisable to plot a complete curve for each season of the year (*Figure 41*). These curves should be retained for comparison purposes. Measurements at established intervals in the future need only be made at the 62% point and, if desired, 10' (3 m) on each side, providing there is no erratic deviation from the original curve. Serious deviation, other than seasonal, could mean corrosion has eaten away some of the electrode.

> **NOTE**
> Safety grounds should be applied in such a way that a zone of equal potential is formed in the work area. This equipotential zone is formed when fault current is bypassed around the work area by metallic conductors. In this situation, the worker is bypassed by the low-resistance metallic conductors of the safety ground.

Testing Station Grounds

A lethal potential can exist between the station ground and a remote ground if a system fault involving the station ground occurs while earth resistance tests are being made. Since one of the objectives of tests on a station ground is to establish the location of an effectively remote point for both current and potential electrodes, the leads to the electrodes must be treated as though a possible potential could exist between these test leads and any point on the station ground grid.

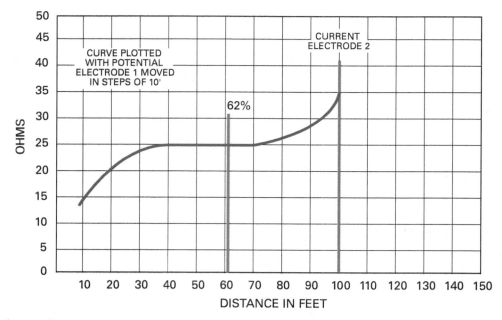
Figure 41 Typical grounding resistance curve to be recorded and retained.

Think About It
Putting It All Together

A low-impedance ground is essential to the performance of any electrical protection system. The ground must dissipate electrical transients and surges in order to minimize the chance of damage or injury. Proper grounding, which includes bonding and connections, protects personnel from the danger of shock and protects equipment and buildings from hazardous voltages. Proper grounding also contributes to the reduction of electrical noise and provides a reference for circuit conductors to stabilize their voltage to ground during normal operation. Examine the grounding system of the electrical service at your home or workplace. Is it adequate? If not, how can it be corrected to meet current *NEC*® requirements?

6.0.0 Section Review

1. If the voltage drop between a grounding electrode and an auxiliary potential electrode is 10V and a current of 10A is passed between an auxiliary current electrode and the grounding electrode, then the resistance is _____.

    $R = \frac{E}{I} = \frac{10V}{10A}$

    a. 1 ohm
    b. 10 ohms
    c. 100 ohms
    d. 1,000 ohms

2. When completing a three-point test, the potential electrode is placed at _____.

    a. 31% of the current electrode distance
    b. 45% of the current electrode distance
    c. 62% of the current electrode distance
    d. 74% of the current electrode distance

Review Questions

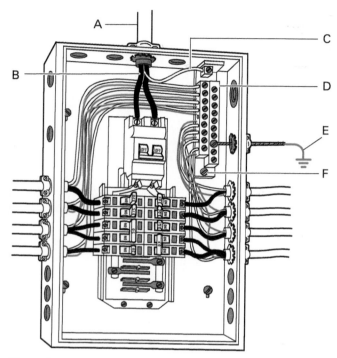

Figure RQ01

1. Item B in *Figure RQ01* represents the 1.0.0 F1.
   a. grounded conductor
   b. ungrounded conductors
   c. grounding electrode conductor
   d. main bonding jumper

2. Item E in *Figure RQ01* represents the _____.
   a. grounded conductor
   b. ungrounded conductors
   c. grounding electrode conductor
   d. main bonding jumper

3. Item C in *Figure RQ01* represents the _____.
   a. grounded conductor
   b. ungrounded conductors
   c. grounding electrode conductor
   d. main bonding jumper

4. Item F in *Figure RQ01* represents the _____.
   a. grounded conductor
   b. ungrounded conductors
   c. grounding electrode conductor
   d. main bonding jumper

5. The conducting connection, whether intentional or accidental, between any of the conductors of an electrical system is known as a 1.1.1.
   a. ground fault
   b. bonding
   c. grounded conductor
   d. short circuit

6. Which conductor(s) are grounded on a single-phase, three-wire, 120/240V system?
   a. Hot
   b. Current-carrying
   c. Neutral
   d. Black

7. Residential electrical services are grounded by connecting the grounded conductor to a grounding electrode at the 1.2.1.
   a. pole-mounted transformer
   b. service drop connection
   c. neutral bus in the main panelboard
   d. meter enclosure

8. A single-phase system has two legs, Leg A and Leg B. If Leg A is carrying 35A and Leg B is carrying 30A, the current carried by the neutral is 1.2.1.
   a. 5A
   b. 30A
   c. 35A
   d. 65A

9. A concrete-encased electrode used for service grounding must be covered with a minimum of 2.00.
   a. ½" (13 mm) of concrete
   b. 1" (25 mm) of concrete
   c. 2" (50 mm) of concrete
   d. 4" (100 mm) of concrete

10. Which of the following is an *NEC®* violation when used as a grounding electrode? 2.0.0
    a. Copper cold-water pipe
    b. Galvanized pipe
    c. An underground gas line
    d. A grounding ring

11. The maximum distance allowed by the NEC® to connect a grounding conductor to a water pipe after the pipe enters the building is _____.
    a. 5' (1.5 m)
    b. 10' (3 m)
    c. 15' (4.5 m)
    d. 20' (6 m)

12. Which of the following is used to connect a ground rod with a grounding conductor?
    a. A grounding clip
    b. A butt-and-slide connector
    c. A wire nut
    d. A grounding clamp

13. All of the following are suitable materials for ground rods, except _____.
    a. stainless steel
    b. galvanized steel
    c. iron
    d. aluminum

14. Which of the following copper AWG sizes should be used for the grounding conductor for a 200A service using 3/0 copper conductors?
    a. No. 2
    b. No. 4
    c. No. 6
    d. No. 8

15. Assuming the use of 75°C copper conductors, a 230A commercial service would require _____.
    a. 1/0 service conductors
    b. 2/0 service conductors
    c. 3/0 service conductors
    d. 4/0 service conductors

16. When a metal water pipe is used as a grounding electrode, what must be provided around the water meter?
    a. Bonding jumper
    b. Floor drain
    c. Grounding locknut
    d. Grounding bushing

17. The EGC for a flexible metal conduit connection to a motor supplied with a 15A circuit is _____.
    a. 8 AWG copper
    b. 10 AWG copper
    c. 12 AWG copper
    d. 14 AWG copper

18. The main bonding jumper for 4/0 service conductors is _____.
    a. 8 AWG copper
    b. 6 AWG copper
    c. 4 AWG copper
    d. 2 AWG copper

19. If you are bonding multiple disconnects and the size of the service-entrance conductor is 3/0, the size of the bonding jumper is _____.
    a. 4 AWG copper
    b. 2 AWG copper
    c. 1/0 copper
    d. 2/0 copper

20. Touch voltage extends to a distance of approximately _____.
    a. 1' (300 mm)
    b. 2' (600 mm)
    c. 3' (1 m)
    d. 4' (1.2 m)

Supplemental Exercises

1. A short circuit is a conducting connection, whether intentional or accidental, between any of the _Conductors_ of an electrical system.

2. True or False? The neutral or grounded conductor is a system or circuit conductor that is intentionally grounded.

3. A conductor used to connect equipment or the grounded circuit of a wiring system to a grounding electrode is called a(n) _GEC_ conductor.

4. The two types of grounding systems are system grounding and _Equ. Grou_.

5. If the loads were only 240V on a single-phase 240V system with a center-tapped transformer, the neutral conductor would carry _0_ amperes.

6. Systems less than 50V are required to be grounded if the supply voltage to the transformer exceeds _150v_ to ground.

7. Circuits that cannot be grounded include certain circuits for cranes, electrolytic cells, secondary circuits on lighting systems, and isolated power systems in _Health care_.

8. Systems are solidly grounded to limit the _V to ground_ during normal operation and to prevent excessive voltages due to lightning and line surges.

9. Grounding methods include an underground water pipe in direct contact with the earth for no less than _10ft_.

10. The size of the grounding electrode conductor is based on the size of the _____.

11. The size of the EGC is determined by the _overcurrent device_.

12. All receptacles used in residential applications must be of the _grounding_ type.

13. Electrical _____ is the key to successful clearing of ground faults.

14. The size of the bonding jumper on the supply side is based on the size of the _____ in each raceway.

15. The two main purposes of the neutral in a grounded system are to permit utilization of power at line-to-neutral voltage and _____

Trade Terms Introduced in This Module

Auxiliary electrodes: Metallic electrodes pushed or driven into the earth to provide electrical contact for the purpose of performing measurements on grounding electrodes or ground grid systems.

Bonding: Connected to establish electrical continuity and conductivity.

Effective ground fault path: An intentionally constructed, permanent, low-impedance electrically conductive path designed and intended to carry current under ground-fault conditions from the point of a ground fault on a wiring system to the electrical supply source and that facilitates the operation of the overcurrent protective device or ground-fault detectors on a high-impedance grounded system.

Equipment bonding jumper: The connection between two or more portions of the equipment grounding conductor.

Equipment grounding conductor (EGC): The conductive path(s) that provide a ground-fault current path and connect normally noncurrent-carrying metal parts of equipment together and to the system grounded conductor, or to the grounding electrode conductor, or both.

Ground: The earth or a conducting connection to the earth.

Ground current: Current in the earth or grounding connection.

Ground grid: System of grounding electrodes interconnected by bare cables buried in the earth to provide lower resistance than a single grounding electrode.

Ground mat: System of bare conductors, on or below the surface of the earth, connected to a ground or ground grid to provide protection from dangerous touch voltage.

Ground resistance: The ohmic resistance between a grounding electrode and a remote or reference grounding electrode that are spaced such that their mutual resistance is essentially zero.

Ground rod: A metal rod or pipe used as a grounding electrode.

Grounded: Connected to ground or to a conductive body that extends the ground connection.

Grounded conductor: A system or circuit conductor that is intentionally grounded.

Grounding clip: A listed spring clip used to secure a bonding conductor to an outlet box.

Grounding conductor: A conductor used to connect equipment or the grounded circuit of a wiring system to a grounding electrode or electrodes.

Grounding connections: Connections used to establish a ground; they consist of a grounding conductor, a grounding electrode, and the earth surrounding the electrode.

Grounding electrode: A conducting object through which a direct connection to earth is established.

Grounding electrode conductor (GEC): A conductor used to connect the system grounded conductor or the equipment to a grounding electrode or to a point on the grounding electrode system.

Main bonding jumper: The connection between the grounded circuit conductor and the equipment grounding conductor at the service.

Neutral conductor: The conductor connected to the neutral point of a system that is intended to carry current under normal conditions.

Resistivity: Resistance between opposite faces of a unit cube. Expressed in ohm-centimeters or ohms per cubic centimeter.

Separately derived system: A premises wiring system whose power is derived from a source of electric energy or equipment other than a service. Such systems have no direct electrical connection, including a solidly connected ground circuit conductor, to supply conductors originating in another system.

Short circuit: An often unintended low-resistance path through which current flows around, rather than through, a component or circuit.

Step voltage: he potential difference between two points on the earth's surface separated by a distance of one pace, or about 3' (1 m).

Supplemental electrode: An additional electrode (commonly a driven rod or pipe) required where the primary electrode is an underground metal water pipe in direct contact with the earth.

Supply-side bonding jumper: A conductor installed on the supply side of a service or within a service equipment enclosure(s), or for a separately derived system, that ensures the required electrical conductivity between metal parts required to be electrically connected.

System grounding: Intentional connection of one of the circuit conductors of an electrical system to ground potential.

Touch voltage: The potential difference between a grounded metallic structure and a point on the earth's surface equal to the normal maximum horizontal reach—approximately 3' (1 m).

Ungrounded conductors: Conductors in an electrical system that are not intentionally grounded.

Additional Resources

This module presents thorough resources for task training. The following resource material is suggested for further study.

National Electrical Code® Handbook, Latest Edition. Quincy, MA: National Fire Protection Association.
Ugly's Electrical Desk Reference, Latest Edition. Burlington, MA: Jones and Bartlett Learning.

Figure Credits

Greenlee / A Textron Company, Module Opener
John Traister, Figures 4, 5, 10–13, 18, 22, 23

Section Review Answer Key

Section 1.0.0

Answer	Section Reference	Objective
1. b	1.1.2	1a
2. a	1.2.1	1b

Section 2.0.0

Answer	Section Reference	Objective
1. c	2.1.1; *NEC Table 310.16*	2a
2. c	2.2.0	2b

Section 3.0.0

Answer	Section Reference	Objective
1. c	3.1.2; *NEC Table 250.122*	3a
2. d	3.2.1	3b

Section 4.0.0

Answer	Section Reference	Objective
1. c	4.1.0; *NEC Table 250.102(C)(1)*	4a
2. d	4.2.0; *Table 1*	4b
3. c	4.3.3	4c

Section 5.0.0

Answer	Section Reference	Objective
1. b	5.1.0	5a
2. a	5.2.0	5b

Section 6.0.0

Answer	Section Reference	Objective
1. a	6.1.0	6a
2. c	6.2.0	6b

Section Review Calculations

6.0.0 Section Review

Question 1

Use Ohm's law to find the resistance (R).

R = E ÷ I = 10V ÷ 10A = 1Ω

The resistance is **1 ohm**.

This page is intentionally left blank.

26210-20
CIRCUIT BREAKERS AND FUSES

Objectives

When you have completed this module, you will be able to do the following:

1. Identify the function of overcurrent protective devices.
    a. Identify types of overcurrent conditions.
    b. Identify *NEC®* requirements for overcurrent protective devices.
2. Size and select circuit breakers.
    a. Identify circuit breaker components.
    b. Identify circuit breaker types and ratings.
3. Size and select fuses.
    a. Identify fuse types and markings.
    b. Size fuses.
    c. Coordinate the operation of overcurrent protective devices.

Performance Task

Under the supervision of the instructor, you should be able to do the following:

1. Identify the following on one or more circuit breaker(s) and fuse(s):
    - Number of poles
    - Load rating
    - Voltage rating
    - Amperage interrupting rating

Trade Terms

Cartridge fuse	Frame size	Overcurrent protection
Circuit breaker	Fuse	Plug fuse
Dual-element fuse	Fuse link	Pole
Edison-base	Molded-case circuit breaker	Short circuit
Fault current	Nonrenewable fuse	

Industry Recognized Credentials

If you are training through an NCCER-accredited sponsor, you may be eligible for credentials from NCCER's Registry. The ID number for this module is 26210-20. Note that this module may have been used in other NCCER curricula and may apply to other level completions. Contact NCCER's Registry at 888.622.3720 or go to **www.nccer.org** for more information.

> **NOTE**
> NFPA 70®, *National Electrical Code®* and *NEC®* are registered trademarks of the National Fire Protection Association, Quincy, MA.

Contents

1.0.0 Overcurrent Protective Devices 1
   1.1.0 Overcurrent Conditions 1
   1.2.0 *NEC*® Regulations 2
2.0.0 Sizing and Selecting Circuit Breakers 4
   2.1.0 Circuit Breaker Components 4
   2.2.0 Circuit Breaker Types and Ratings 5
      2.2.1 Current Rating 5
      2.2.2 Interrupting Capacity Rating 6
      2.2.3 Series-Rated Systems 10
      2.2.4 Ground Fault Circuit Interrupters (GFCIs) 11
      2.2.5 Other Types of Circuit Breakers 13
3.0.0 Sizing and Selecting Fuses 15
   3.1.0 Fuse Types and Markings 15
      3.1.1 Plug Fuses 16
      3.1.2 Cartridge Fuses 17
      3.1.3 Fuse Classes and Markings 18
   3.2.0 Sizing Fuses 18
      3.2.1 Dual-Element Time Delay Fuses 19
      3.2.2 Non-Time-Delay Fuses 19
   3.3.0 Coordination of Overcurrent Protective Devices 21

Figures and Tables

Figure 1 Short circuit 1
Figure 2 Internal arrangement of a circuit breaker 5
Figure 3 Operating characteristics of a circuit breaker 5
Figure 4 Circuit breaker current rating 5
Figure 5 Normal current operation 7
Figure 6 Short circuit operation with inadequate interrupting rating ... 8
Figure 7 Short circuit operation with adequate interrupting rating 8
Figure 8 High interrupting capacity circuit breaker 9
Figure 9 Most overcurrent protective devices are labeled
         with two current ratings 11
Figure 10 GFCI protection 13
Figure 11 Standard Edison-base fuse on top;
         Type S with adapter shown on bottom 17
Figure 12 Cartridge fuses 17
Figure 13 Typical single-line schematic diagram showing
         overcurrent protective devices 20

Table 1 Molded-Case Circuit Breaker Ratings Recognized
        by Underwriters Laboratories 5

Section One

1.0.0 Overcurrent Protective Devices

Objective

Identify the function of overcurrent protective devices.
a. Identify types of overcurrent conditions.
b. Identify *NEC®* requirements for overcurrent protective devices.

Performance Task

1. Identify the following on one or more circuit breaker(s) and fuse(s):
   - Number of poles
   - Load rating
   - Voltage rating
   - Amperage interrupting rating

Trade Terms

Circuit breaker: A device that is designed to open a circuit automatically at a certain overcurrent. It can also be used to manually open and close the circuit.

Fault current: The current that exists when an unintended path is established between an ungrounded conductor and ground.

Fuse: A protective device that opens a circuit when the fusible element is severed by heating due to a fault current or overcurrent passing through it.

Overcurrent protection: De-energizing a circuit whenever the current exceeds a predetermined value; the usual devices are fuses, circuit breakers, or magnetic relays.

Pole: A portion of a device associated exclusively with one electrically separated conducting path of the main circuit or device.

Short circuit: A conducting connection, whether intentional or accidental, between any of the conductors of an electrical system, whether it is from line-to-line or from line-to-neutral (grounded).

All electrical circuits and their related components are subject to destructive overcurrents. Harsh environments, general deterioration, accidental damage, damage from natural causes, excessive expansion, and overloading of the electrical system are all factors that contribute to the occurrence of such overcurrents.

Reliable protective devices prevent or minimize costly damage to transformers, conductors, motors, equipment, and the many other components and loads that make up the complete electrical system. Therefore, reliable circuit protection is essential to avoid the severe monetary losses that can result from power blackouts and prolonged downtime. To protect electrical conductors and equipment against abnormal operating conditions and their consequences, protective devices are used in the circuits.

1.1.0 Overcurrent Conditions

An overcurrent can be an overload current, a short circuit current, or a ground fault. The overload current is an excessive current relative to normal operating current but one that is confined to the normal conductive paths provided by the conductors and other components and loads of the electrical system. A short circuit (*Figure 1*) is probably the most common cause of electrical problems. A short circuit is an undesired current path that allows the electrical current to bypass the load on the circuit. A short circuit is an unintended connection between two conductors. If an unintended connection occurs between a wire and a grounded object, such as the metal frame of a motor, then it is a ground fault.

Overloads are most often between one and six times the normal current level. Usually, they are caused by harmless temporary surge currents that occur when motors are started up or transformers are energized. Because they are of short duration, any temperature rise is trivial and has no harmful effect on the circuit components. Therefore, it is important that protective devices do not react to them.

Continuous overloads can result from motor defects (such as worn motor bearings), overloaded equipment, or too many loads on one circuit. Such

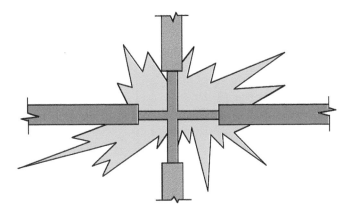

Figure 1 Short circuit.

sustained overloads are destructive and must be cut off by protective devices before they damage the electrical distribution system or affect the system loads. However, since they are of relatively low magnitude compared to short circuit currents, removal of the overload current within a few seconds will generally prevent equipment damage. A sustained overload current results in overheating of conductors and other components and will cause deterioration of insulation, which may eventually result in severe damage and short circuits if not interrupted.

The fuse and the circuit breaker are two types of automatic overload devices that are normally used in electrical circuits to prevent fires and the destruction of the circuit and its associated equipment.

The function of circuit breakers and fuses used in electrical circuits can be compared to that of a pressure relief valve used with a steam boiler. If dangerously high pressures develop within the steam boiler, the pressure relief valve opens to relieve the pressure, thus preventing damage to the boiler. Similarly, circuit breakers and fuses act as safety valves in electrical circuits by opening the circuit if a dangerous overcurrent condition exists as a result of an overload or short circuit, thus preventing damage to the circuit components.

1.2.0 NEC® Regulations

The are some critical NEC® sections with which you should become familiar. The following are not the only NEC® requirements of importance when dealing with overcurrent protection devices, but they are considered by many to be the most important:

- *NEC Section 110.3(B), Installation and Use* – Equipment that is listed, labeled, or both shall be used and installed in accordance with any instructions included in the listing or labeling.
- *NEC Section 110.9, Interrupting Rating* – Equipment intended to interrupt current at fault levels shall have an interrupting rating at nominal circuit voltage at least equal to the current that is available at the line terminals of the equipment. Equipment intended to interrupt current at other than fault levels shall have an interrupting rating at nominal circuit voltage not less than the current that must be interrupted.
- *NEC Section 110.10, Circuit Impedance, Short-Circuit Current Ratings, and Other Characteristics* – The overcurrent protection devices, the total impedance, component short-circuit current ratings, and other characteristics of the circuit to be protected shall be so selected and coordinated as to permit the circuit protective devices that are used to clear a fault to do so without extensive damage to the electrical components of the circuit. The fault shall be assumed to be either between two or more of the circuit conductors, or between any circuit conductor and the equipment grounding conductor(s) permitted in *NEC Section 250.118*.
- *NEC Section 240.1, Scope (Informational Note)* – Overcurrent protection for conductors and equipment is provided to open the circuit if the current reaches a value that will cause an excessive or dangerous temperature in conductors or conductor insulation. See also *NEC Sections 110.9 and 110.10* for requirements for interrupting ratings and protection against a fault current.
- *NEC Section 240.2, Definition of Current-Limiting Overcurrent Protective Device* – A current-limiting overcurrent protection device is a device that, when interrupting currents in its current-limiting range, reduces the current flowing in the faulted circuit to a magnitude that is substantially less than that obtainable in the same circuit if the device were replaced with a solid conductor having comparable impedance.
- *NEC Section 240.83(C), Interrupting Rating* – Every circuit breaker having an interrupting rating other than 5,000A shall have its interrupting rating shown on the circuit breaker.
- *NEC Section 240.83(E), Voltage Marking* – Circuit breakers shall be marked with a voltage rating that is not less than the nominal system voltage that is indicative of their capability to interrupt fault currents between phases or from phase to ground.
- *NEC Section 240.85, Applications* – A circuit breaker with a straight voltage rating, such as 240V or 480V, may be applied in a circuit in which the nominal voltage between any two conductors does not exceed the circuit breaker's voltage rating. A two-pole circuit breaker cannot be used for protecting a three-phase corner-grounded delta circuit unless it is marked single-phase/three-phase to indicate such suitability. (A pole is a portion of a device associated exclusively with one electrically separated conducting path of the main circuit or device.) A circuit breaker with a slash rating such as 120V/240V or 480Y/277V shall be permitted to be applied on a solidly grounded circuit where the nominal voltage of any conductor to ground does not exceed the lower of the two values of the circuit breaker's voltage rating and the nominal voltage between any

two conductors does not exceed the higher value of the circuit breaker's voltage rating.
- *NEC Section 250.4(A)(1), Electrical System Grounding* – Electrical systems that are grounded shall be connected to earth in a manner that will limit the voltages imposed by lightning, line surges, or unintentional contact with higher voltage lines and that will stabilize the voltage to earth during normal operation. System equipment and circuit conductors are solidly grounded to the system grounded conductor to facilitate overcurrent device operation in case of fault currents.

1.0.0 Section Review

1. Worn motor bearings are most likely to result in a _____.
    a. temporary overload
    b. sustained overload
    c. short circuit
    d. ground fault

2. Where in the *NEC®* would you find information on circuit breaker applications?
    a. *NEC Section 240.85*
    b. *NEC Section 250.4(A)(1)*
    c. *NEC Section 310.15*
    d. *NEC Section 340.12*

Section Two

2.0.0 Sizing and Selecting Circuit Breakers

Objective

Size and select circuit breakers.
  a. Identify circuit breaker components.
  b. Identify circuit breaker types and ratings.

Performance Task

1. Identify the following on one or more circuit breaker(s) and fuse(s):
   - Number of poles
   - Load rating
   - Voltage rating
   - Amperage interrupting rating

Trade Terms

Cartridge fuse: A fuse enclosed in an insulating tube to confine the arc when the fuse blows.

Frame size: A method used to classify circuit breakers according to given current ranges.

Fuse link: The fusible part of a cartridge fuse.

Molded-case circuit breaker: A circuit breaker enclosed in an insulating housing.

Plug fuse: A type of fuse that is held in position by a screw thread contact instead of spring clips, as is the case with a cartridge fuse.

Basically, a circuit breaker is a device used for closing and interrupting a circuit between separable contacts under both normal and abnormal conditions. This is done manually by switching the handle to the On or Off positions. However, the circuit breaker is also designed to open a circuit automatically on a predetermined overload or fault current without damage to itself or its associated equipment. As long as a circuit breaker is applied within its rating, it will automatically interrupt any fault and, therefore, must be classified as an inherently safe overcurrent protective device.

2.1.0 Circuit Breaker Components

The internal arrangement of a circuit breaker is shown in *Figure 2*, whereas its external operating characteristics are shown in *Figure 3*. Note that the handle on a circuit breaker resembles an ordinary toggle switch. On an overload, the circuit breaker opens itself, or trips. When tripped, the handle jumps to the middle position. To reset, turn the handle to the Off position and then push it as far as it will go beyond this position (Reset); finally, turn it to the On position.

A standard molded-case circuit breaker usually contains the following:

- A set of contacts
- A magnetic trip element
- A thermal trip element
- Line and load terminals
- Bussing used to connect these individual parts
- An enclosing housing of insulating material

The circuit breaker handle manually opens and closes the contacts and resets the automatic trip units after an interruption. Some circuit breakers also contain a manually operated push-to-trip testing mechanism.

Circuit breakers are classified into ampere groupings that match specific physical dimensions. These numbers, commonly referred to as the *frame classification* or frame size, are terms applied to groups of molded-case circuit breakers that are physically interchangeable according to given current ranges. Each group is classified by the largest ampere rating of its range. These groups are:

- 15A–100A
- 125A–225A
- 250A–400A
- 500A–1,000A
- 1,200A–2,000A

Therefore, they are classified as 100A, 225A, 400A, 1,000A, and 2,000A frames.

2.2.0 Circuit Breaker Types and Ratings

The established voltage rating of a circuit breaker is based on its clearances or space, both through air and over surfaces between components of the electrical circuit and between the electrical components and ground. Circuit breaker voltage ratings indicate the maximum electrical system voltage on which they can be applied. Underwriters Laboratories, Inc. (UL) recognizes only the ratings listed in *Table 1* for molded-case circuit breakers.

A circuit breaker can be rated for either alternating current (AC) or direct current (DC) system applications or for both. Single-pole circuit breakers, rated at 120/240VAC or 125/250VDC, can be used singly and in pairs on three-wire circuits having a neutral connected to the midpoint of the load. Single-pole circuit breakers rated at

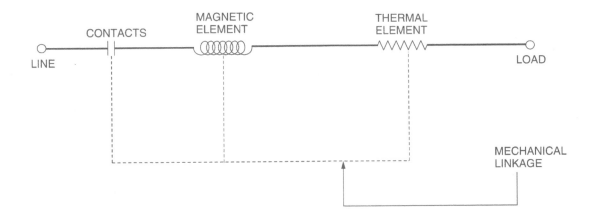

Figure 2 Internal arrangement of a circuit breaker.

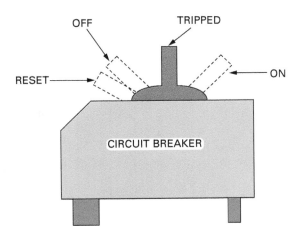

Figure 3 Operating characteristics of a circuit breaker.

120/240VAC or 125/250VDC can also be used in pairs on a two-wire circuit connected to the ungrounded conductors of a three-wire system. Two-pole circuit breakers rated at 120/240VAC or 125/250VDC can be used only on a three-wire, direct current, or single-phase, alternating current system having a grounded neutral. Circuit breaker voltage ratings must be equal to or greater than the voltage of the electrical system on which they are used.

Table 1 Molded-Case Circuit Breaker Ratings Recognized by Underwriters Laboratories

For Alternating Current	For Direct Current
120V	125V
120/240V	—
240V	250V
277V	600V
277/480V	—
480V	—
1,000V	—

Circuit breakers have two types of current ratings. The first—and the one that is used most often—is the continuous current rating. The second is the fault current interrupting capacity.

2.2.1 Current Rating

The rated continuous current of a device is the maximum current (in amperes) that it will carry continuously without exceeding the specified limits of observable temperature rise. The continuous current ratings of circuit breakers are established based on standard UL ampere ratings. As listed in *NEC Table 240.6(A)*, these range from 15A to 6,000A. The ampere rating of a circuit breaker is located on the handle of the device, as shown in *Figure 4*.

Generally, the circuit breaker current rating must be equal to or less than the load circuit conductor current-carrying capacity (ampacity).

Most overcurrent protective devices are labeled with the following current ratings:

- Normal current rating
- Interrupting rating

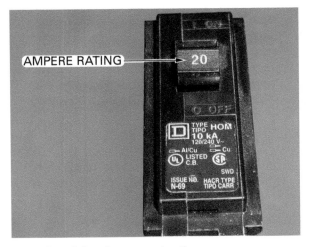

Figure 4 Circuit breaker current rating.

Tripped Circuit Breaker Indications

When tripped, the handle of a circuit breaker can position itself in one of several ways, depending on the design of the breaker. Some common handle positions (indications) of a tripped circuit breaker are:

- *Center trip* – The handle of the breaker moves to a center position on the breaker.
- *Center position flag* – The handle of the breaker moves to the center position on the breaker and displays a red flag.
- *Full off position* – The handle of the breaker moves to the full Off position on the breaker.
- *Off position outward* – A center button protrudes to display Off.

2.2.2 Interrupting Capacity Rating

The amperage interrupting capacity (AIC) of a circuit breaker is the maximum short circuit current at which the breaker will safely interrupt the circuit. The AIC is at the rated voltage and frequency. *NEC Section 110.9* states that equipment intended to interrupt current at fault levels (fuses and circuit breakers) must have an interrupting rating at nominal circuit voltage at least equal to the current that is available at the line terminals of the equipment.

Equipment intended to break current at other than fault levels must have an interrupting rating at nominal circuit voltage not less than the current that must be interrupted.

These *NEC®* statements mean that fuses and circuit breakers (and their related components) that are designed to break fault or operating currents (open the circuit) must have a rating sufficient to withstand such currents. This *NEC®* section emphasizes the difference between clearing fault level currents and clearing operating currents. Protective devices such as fuses and circuit breakers are designed to clear fault currents and, therefore, must have short circuit interrupting ratings that are sufficient for fault levels.

> NOTE: *NEC Section 240.83(C)* requires that circuit breakers having a current interrupting rating other than 5,000A have the rating clearly shown on the circuit breaker. By definition, circuit breakers that have not been marked with an interrupting rating have a rating of 5,000A.

Equipment such as contactors and safety switches have interrupting ratings for currents at levels other than fault levels. Therefore, the interrupting rating of electrical equipment is now divided into two parts:

- Current at fault (short circuit) levels
- Current at operating levels

Circuit Breakers

Circuit breakers are available in a wide variety of sizes and types to suit various applications.

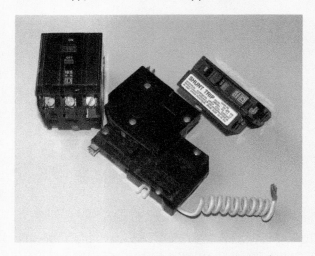

Most people are familiar with the normal current-carrying ampere rating of fuses and circuit breakers. If an overcurrent protective device is designed to open a circuit when the circuit load exceeds 20A for a given time period, as the current approaches 20A, the overcurrent protective device begins to overheat. If the current barely exceeds 20A, the circuit breaker will open normally or the fuse link on a cartridge fuse will melt after a given period of time with little, if any, arcing. If 40A of current are instantaneously applied to the circuit, the overcurrent protective device will open faster, but again with very little arcing. However, if a fault current occurs that runs the amperage up to 5,000A, an explosion effect would occur within the protective device. One simple indication of this explosion effect is demonstrated by the blackened window of a blown plug fuse.

If this fault current exceeds the interrupting rating of a fuse or circuit breaker, the protective

60°C/75°C Ratings

Some circuit breakers are suitable for use with both 60°C and 75°C rated conductors. Others can be used with only 75°C rated conductors. Be sure to check the rating of the circuit breaker selected. For example, to use the 75°C ampacity rated conductors No. 1 and smaller, you must make sure to use a circuit breaker rated for 75°C. To comply with *NEC Section 110.14(C)*, the termination at the load end of the same circuit must also be rated for 75°C.

---

device can be damaged or destroyed; such current can also cause severe damage to equipment and injure personnel. Therefore, selecting overcurrent protective devices with the proper interrupting capacity is extremely important in all electrical systems.

Consider the following analogy using a dammed stream as an example of interrupting rating (*Figure 5*). The reservoir capacity represents the available fault current in an electrical circuit; the flood gates (located downstream from the dam) represent the overcurrent protective device in the circuit rated at 10,000 gallons per minute (10,000A); and the stream of water coming through the discharge pipes in the dam represents the normal load current. *Figure 5* shows a normal flow of 100 gallons per minute (100A). Note the bridge downstream from the flood gates. This bridge represents downstream circuit components or equipment connected to the circuit.

Figure 6 shows this same diagram with a fault in the dam, creating a water short circuit that allows 50,000 gallons per minute to flow (50,000 fault circuit amperes). Such a situation destroys the flood gates because of inadequate interrupting rating. The overcurrent protective device in the circuit will also be destroyed. With the flood gates damaged, this surge of water continues downstream, wrecking the bridge. Similarly, the downstream components may not be able to withstand the let-through current in an electrical circuit.

Figure 7 shows the same situation but with adequate interrupting capacity. Note that the flood gates have adequately contained the surge of water and restricted the let-through current to an amount that can be withstood by the bridge or the components downstream.

There are several factors that must be considered when calculating the required interrupting capacity of an overcurrent protection device. *NEC Section 110.10* states that the overcurrent protective devices, the total impedance, the equipment short circuit current ratings, and other characteristics of the circuit to be protected shall be so selected and coordinated to permit the circuit protective devices used to clear a fault to do so without extensive damage to the electrical components of the circuit. This fault shall be assumed to be either between two or more of the circuit conductors, or between any circuit conductor and the equipment grounding conductor(s) permitted in *NEC Section 250.118*.

The component short circuit rating is a current rating given to conductors, switches, circuit breakers, and other electrical components, which, if exceeded by fault currents, will result in extensive damage to the component. Short circuit damage can be the result of heat generated or the

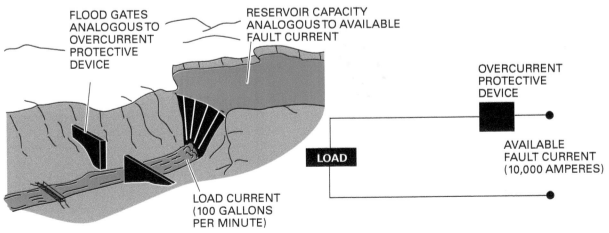

Figure 5 Normal current operation.

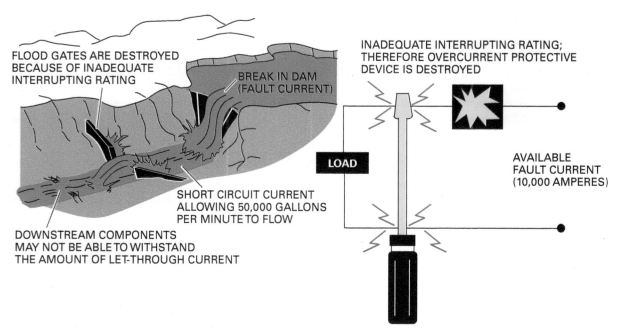

Figure 6 Short circuit operation with inadequate interrupting rating.

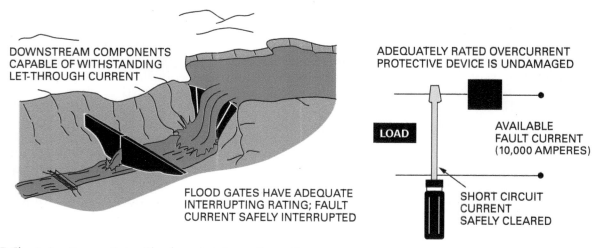

Figure 7 Short circuit operation with adequate interrupting rating.

electromechanical force of a high-intensity magnetic field. The rating is expressed in terms of time intervals and/or current values.

The intent of the *NEC®* is that the design of a system must be such that short circuit currents cannot exceed the short circuit current ratings of the components selected as part of the system. Given specific system components and the level of available short circuit currents that could occur, overcurrent protection devices (mainly fuses and/or circuit breakers) must be used that will limit the let-through current to levels within the withstand ratings of the system components.

In most large commercial and industrial installations, it is necessary to calculate the available short circuit current at various points in a system to determine if the equipment meets the requirements of *NEC Sections 110.9 and 110.10*. There are a number of methods used to determine the short circuit requirements in an electrical system. Some give approximate values; others require extensive computations and are quite exacting. A simple, yet accurate method is the point-by-point method, which will be discussed in more detail later in this module.

An overcurrent protection device must be selected with an interrupting capacity that is equal to or greater than the available short circuit current at the point at which the circuit breaker or fuse is applied in the system. The breaker interrupting capacity is based on tests to which the breaker is subjected. There are two such tests; one is set up by Underwriters Laboratories, Inc. (UL) and the other by the National Electrical

> **Think About It**
> ## Current Interrupting Rating
> When replacing a circuit breaker, is it necessary to take into account the current interrupting rating of the replacement breaker? If so, why?

Manufacturers Association (NEMA). The NEMA tests are self-certification tests, whereas the UL tests are certified by unbiased witnesses. UL tests have been limited to a maximum of 10,000A in the past, so the emphasis was placed on the NEMA tests with higher ratings. The UL tests now include the NEMA tests plus other ratings. Consequently, emphasis is now placed on the UL tests. It is important to remember that NEMA does not list devices. Two organizations that both test and list devices are UL and the Canadian Standards Association (CSA).

UL requires that molded-case circuit breakers open within a certain period of time during overcurrent conditions. For example, UL requires that a 240V molded-case circuit breaker trip at 300% of its rated continuous current within 50 seconds. If the breaker were rated at 600V, it would be required to trip within 70 seconds at 300% rated current. Currents between 7 and 15 times the rated current (depending on voltage rating and frame size) are handled as overcurrents by the thermal magnetic trip element. Currents above this level activate the magnetic trip element.

The interrupting capacity of a circuit breaker is based on its rated voltage. Where the circuit breaker can be used on more than one voltage, the interrupting capacity will be shown for each voltage level. For example, the LA-type circuit breaker has 42,000A symmetrical interrupting capacity at 240V, 30,000A symmetrical at 480V, and 22,000A symmetrical at 600V.

There are three main categories of interrupting capacity: standard interrupting capacity, high interrupting capacity, and current-limiting. Details about these categories are as follows:

- *Standard interrupting capacity circuit breakers* – Standard interrupting capacity circuit breakers can be identified by their black operating handles and black printed interrupting rating labels. The interrupting rating of a circuit breaker is as important in application as the voltage and current ratings and should be considered each time a breaker is applied. In residential applications, the available fault current is seldom higher than 10,000A or even near this value. The QO and A1 breakers (15A–150A) with the interrupting rating of 10,000A are used in these applications. Higher interrupting ratings are available when required.
- *High interrupting capacity circuit breakers* – If a circuit breaker requires higher interrupting capacity than the standard ratings, the FH (H for high interrupting capacity) is available (*Figure 8*). The FH-type circuit breaker has an interrupting rating of 65,000A symmetrical at 240VAC. The continuous current ratings are duplicated in these breakers (15A–100A), but the interrupting capacity has been increased to satisfy the need for greater interrupting capacity. This type of breaker is applied in installations where higher fault currents are available (such as large industrial plants). These breakers are built with a case material that will withstand the higher shocks from heat and interrupting forces. Circuit breakers with high interrupting capacity are available in the FH, KH, LH, MH, NH, and PH types. Three ampere ratings (15A, 20A, and 30A) are also available in the QH high interrupting capacity breakers. These breakers satisfy conditions of lighting circuits supplied from these high available fault current circuit breakers.
- *Current-limiting circuit breakers* – The need for current limitation is the result of the increasingly higher available fault currents associated with the growth and interconnection of modern power systems. To meet this need, current-limiting circuit breakers have been developed. This type of breaker operates extremely fast to provide downstream protection for other types of overcurrent protective devices with as little as 10,000 AIC on systems with 100,000A available fault current.

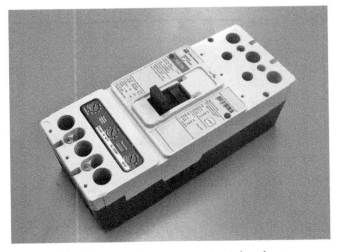

Figure 8 High interrupting capacity circuit breaker.

There are circuit breakers available for almost every need in any electrical system. For example, in circuits where the breaker must be highly sensitive to amperage changes (in an environment where the temperature changes are great), an ambient-compensating circuit breaker may be used. Its design permits it to compensate for temperature variation.

A current-limiting circuit breaker is basically a conventional, common-trip, thermomagnetic circuit breaker with an independent limiter section in series with each pole. For purposes of explanation, each limiter can be represented by a set of contacts shunted by a transformable resistor.

At currents below a threshold of about 1,000A, the current-limiting circuit breaker performs in a manner similar to conventional thermomagnetic circuit breakers. In the event of an overload or minor fault current, only the breaker section contacts open to interrupt the circuit. The limiter contacts remain closed.

For 100A frame current-limiting circuit breakers, the threshold of current limitation occurs in the range of 1,000A to 2,000A. Above this point, the limiter contacts open first—starting the limiting action within $\frac{1}{4}$ of a millisecond of fault initiation. A substantial arc voltage is generated across the limiter contacts. Rapid generation of this arc voltage is essential to drastically limit the peak let-through current in the faulted circuit.

The rapid rising equivalent resistance of the arc across the limiter contacts forces the current to transfer to the alternate path provided by the limiter resistor. This special resistor has a high positive temperature coefficient of resistance. As current is transferred to it, the temperature of the resistor increases rapidly with a consequent rapid rise in resistance. The resistor dissipates and limits let-through energy and increases the short circuit power factor of the circuit nearly to unity, resulting in an easier, more reliable, transient-free interruption.

During the extremely short period of operation of the limiter section, the breaker section tripping mechanism is activated and its contacts begin to open. After the arc between the limiter contacts has been extinguished and all the current is flowing through the limiter resistor, the breaker contacts interrupt the already drastically limited fault current. Because the fault current and system voltage are nearly in phase at this point, complete interruption of the circuit takes place near the first voltage zero of the AC wave.

When a current-limiting circuit breaker trips to clear a faulted circuit, it can be reset in the same manner as a conventional molded-case circuit breaker. This is true whether tripping was due to low-level overload currents or major faults involving the breaker's current-limiting mode of operation. The limiter contacts are automatically closed after circuit interruption. Breaker contacts are reclosed by moving the handle to the extreme Off position and then to On. There are no fusible elements in this type of circuit breaker.

Regular maintenance of circuit breakers and their enclosures is necessary to obtain the best service and performance.

Because every circuit breaker failure represents a potential hazard to other equipment in the system, it is difficult to calculate the risks involved in prolonging maintenance inspections. One of the chief causes of circuit breaker failure is high heat caused by loose connections at the load side of the breaker. Another cause of circuit breaker failure is defective loads; that is, electrical equipment that cycles too frequently, which in turn causes the circuit breaker to overheat. In the majority of circuit breaker failures, heat is the most likely cause.

2.2.3 Series-Rated Systems

NEC Sections 110.22 and 240.86(A) require special marking for UL-recognized, series-rated systems (*Figure 9*). For example, *NEC Section 110.22(B)* states that where equipment enclosures for circuit breakers or fuses applied in compliance with the series combination ratings selected under engineering supervision in accordance with *NEC Section 240.86(A)* shall be legibly marked in the field as directed by the engineer to indicate the equipment has been

---

Thermal, Magnetic, and Thermomagnetic Circuit Breakers

Thermal circuit breakers contain a spring-loaded bimetal element that pops the circuit breaker open when the element becomes overheated by excessive current. Magnetic circuit breakers use a magnetic field induced by current flow through a coil of wire to open its contacts and break the circuit. Some circuit breakers, called thermomagnetic breakers, operate on both principles. They are often used in motor protection applications.

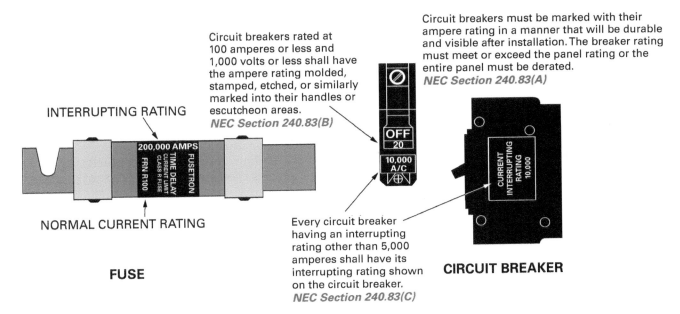

Figure 9 Most overcurrent protective devices are labeled with two current ratings.

applied with a series combination rating. The marking shall be readily visible and state the following:

```
CAUTION — ENGINEERED SERIES
       COMBINATION SYSTEM
RATED _____ AMPERES. IDENTIFIED
REPLACEMENT COMPONENTS REQUIRED.
```

Complete coverage of engineered electrical distribution systems giving complete details of these *NEC®* requirements is presented in NCCER's *Electrical Level Three* and will not be covered in depth in this module.

2.2.4 Ground Fault Circuit Interrupters (GFCIs)

Fault current protection is extremely important to anyone who works with and uses electrically operated equipment. A fault current exists when an unintended path is established between an ungrounded conductor and ground. This situation can occur not only from worn or defective electrical equipment, but also from accidental misuse of equipment that is in good working order.

Will a conventional overcurrent device (fuse or circuit breaker) detect a fault current and open the circuit before irreparable harm is done? Before answering this question, a study of the effects of current on the human body is in order.

The hand-to-hand body resistance of an adult lies between 1,000Ω and 4,000Ω, depending on moisture, muscular structure, and voltage. The average value is 2,100Ω at 240VAC and 2,000Ω at 120VAC.

Using Ohm's law, the current resulting from the above-listed average hand-to-hand resistance values is 120 milliamperes (mA) at 240VAC and 60mA at 120VAC. The effects of 60Hz alternating current on a normal healthy adult are as follows (note that the current is in milliamperes):

- *More than 5mA* – Generally painful shock.
- *More than 15mA* – Sufficient to cause freezing to the circuit for 50% of the population.
- *More than 30mA* – Breathing difficulty (possible suffocation).
- *50mA to 100mA* – Possible ventricular fibrillation or very rapid uncoordinated contractions

Series-Rated Circuit Breakers

The *NEC®* recognizes series-rated circuit breakers and requires that the end-use equipment be marked with the series combination rating. Such a series rating of devices is specific to individual manufacturers. For this reason, it is especially important to consult listing information or product literature for details of how such a combination rating was derived, especially for retrofit installations.

of the ventricles of the heart resulting in loss of synchronization between the heartbeat and pulse beat. When ventricular fibrillation occurs, it usually continues and death will happen within a few minutes unless special defibrillation equipment is used.

- *100mA to 200mA* – Certain ventricular fibrillation.
- *Over 200mA* – Severe burns and muscle contractions. The heart is more likely to stop than fibrillate.

A conventional overcurrent device will not open a circuit without causing harm. The current that would flow from a defective electric drill through the metal housing and through the human body to ground would be 60mA, calculated using the above value for 120VAC. The 60mA current is less than one-half of 1% of the rating of 15A circuit breaker or fuse, and yet it approaches the current level, which may produce ventricular fibrillation. Obviously, the standard circuit breaker or fuse will not open the circuit under such low levels of current flow.

Ground fault circuit interrupters (GFCIs) are Class A protective devices built in accordance with *UL Standard No. 943* for ground fault circuit interrupters. UL defines a Class A device as one that will trip when a fault current to ground is 6mA or more. Also, Class A devices must not trip below 4mA.

Class A GFCIs provide a self-contained means of testing the fault current circuitry, as required by UL. To test, simply push the test button, and the device will respond with a trip indication. UL requires that the current generated by the test circuit shall not exceed 9mA. Also, UL requires the device to be functional at 85% of the rated voltage.

Knowing these facts, OSHA and other codes require the use of fault current circuit interrupters to be used on certain circuits, namely, all construction sites where temporary service is in use for electric power tools. Other required uses include:

- Circuits for electrically operated pool covers
- Power or lighting circuits for swimming pools, fountains, and similar locations
- Receptacles in both commercial and residential garages
- Receptacles in residential bathrooms
- Receptacles installed outdoors in public spaces
- Receptacles installed in residential crawl spaces or unfinished basements
- All kitchen countertop receptacles or any receptacles mounted within 6' (1.8 m) of a wet bar sink
- Receptacles installed in bathhouses
- Receptacles installed in bathrooms of commercial or industrial buildings
- Receptacles installed on the roof of any building
- Dockside receptacles
- Branch circuits derived from autotransformers

There are two types of GFCI receptacles. The feed-through type will provide GFCI protection to the remainder of the outlets in a circuit, similar to a GFCI circuit breaker. The second, a nonfeed-through type, offers protection only at the point of installation. Both types are available in various styles.

When installed properly, GFCIs continuously monitor the current in the two conductor wires of the circuit—both the grounded and ungrounded conductors. The current in these two conductors should always be equal. If the GFCI senses a difference between them of more than 6mA, it assumes that the difference is due to fault current and automatically trips the circuit. Power is interrupted within $\frac{1}{40}$ of a second or less, which is fast enough to prevent injury to anyone in normal health. Note that the ungrounded and grounded conductors are constantly being monitored. If at any time the sensor detects a difference in current flow between these conductors, this difference will be amplified. See *Figure 10*. The amplified signal will then cause a circuit breaker to open to de-energize the circuit.

Listed HACR Circuit Breakers

HACR circuit breakers are UL listed for use in group motor applications such as heating, air conditioning, and refrigeration (HACR) applications. An HACR breaker has a built-in time delay that allows a higher current than its rating to momentarily flow in the circuit. This compensates for the large starting current drawn by such loads. The equipment being protected must also be marked by the manufacturer as being suitable for protection by this type of breaker. (Note: The *NEC*® now permits the use of any inverse-time circuit breaker in heating, air conditioning, and refrigeration equipment comprising group motor installations.)

Case History

GFCI Circuit Protection

Electricity to operate portable power tools at a construction site was supplied by a temporary service. A panel box on the pole had two duplex receptacles in waterproof boxes. One was a regular receptacle and the other a GFCI receptacle. The pole had not been inspected by the city and was not in compliance with code requirements. (It was not grounded.)

A worker used extension cords plugged into the regular receptacle to supply a power tool. The site was wet, humidity was high, and the worker was sweating. Reportedly, he was getting shocks whenever he used the tool. Unknown to the worker, the tool had a ground fault. As he was climbing down to the ground using a piece of metal floor truss as a makeshift ladder, he received a shock, causing him to fall from the floor truss into a puddle of water while still holding onto the tool. He was electrocuted.

The Bottom Line: This accident could have been prevented if the temporary electrical service at the construction site had complied with all local regulations and OSHA standards, including proper grounding. Also, OSHA requires that all 120V, single-phase 15A and 20A receptacles at construction sites that are not part of the permanent wiring be GFCI protected.

Another factor that contributed to this accident was the continued use of a power tool that should have been taken out of service immediately at the occurrence of the first shock. Also, the worker should have used an approved ladder instead of the floor truss to climb down to the ground.

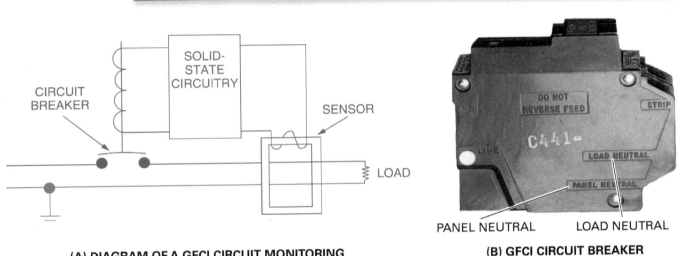

(A) DIAGRAM OF A GFCI CIRCUIT MONITORING   (B) GFCI CIRCUIT BREAKER

Figure 10 GFCI protection.

2.2.5 Other Types of Circuit Breakers

In addition to circuit breakers that protect against overloads and ground fault short circuit protection (overcurrents), some circuit breakers are designed with additional capabilities or other functions. The following are some examples of these breakers:

- *Shunt trip circuit breaker* – In addition to the normal operating devices that protect against overcurrents, this breaker has a built-in electric coil that causes it to open the breaker contacts when the coil is energized by an outside source. Typical sources may include fire suppression circuits, pushbuttons, or alarm circuits.

- *Arc fault circuit breaker* – In addition to the normal operating devices that protect against overcurrents, this breaker includes electronic circuits to monitor current flow within the breaker. If it detects a pattern of small continuous surges or spikes in that current flow, the breaker will operate and open the circuit. This pattern of surges or spikes is typical of currents in a short, high-resistance arcing circuit, such as a frayed electrical cord or a loose connection. *NEC Section 210.12(A)* requires these types of breakers to be used to protect all 120V, 15A and 20A branch circuits in dwelling unit kitchens, family rooms, dining rooms, living rooms, parlors, libraries, dens, bedrooms,

sunrooms, recreation rooms, closets, hallways, laundry areas, or similar rooms or areas by any of the means described in *NEC Sections 210.12(A)(1) through (6)*.
- *GFCI breaker* – In addition to the normal operating devices that protect against overcurrents, this breaker includes circuitry to monitor currents on both conductors in the circuit. If there is an imbalance of those currents of 6mA or more, the breaker will operate and open the circuit. This imbalance would indicate that a current path other than that intended has formed, possibly through equipment or persons, and is flowing to ground. GFCI breakers have different ranges to enable protection of people, equipment, or both.
- *Switched neutral breaker* – In addition to the normal operating devices that protect against overcurrents, this breaker disconnects the neutral or grounded conductor simultaneously with all ungrounded conductors. Breakers that switch grounded conductors are used in installations such as fuel dispensing equipment, as identified in *NEC Section 514.11*.
- *Non-automatic breaker* – This breaker has no devices to protect the circuit against overcurrent. It is used as a means of manually disconnecting circuits by operating the handle. Its use would be similar to a non-fused disconnect switch. It is also known as a molded-case switch.

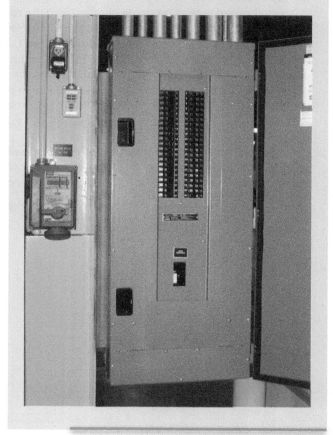

Think About It
Installing Circuit Breakers

In addition to using the correct size circuit breaker (trip current and interrupt current), which other factors should be taken into consideration when selecting and installing a circuit breaker in a panel?

2.0.0 Section Review

1. A 250A fuse has a frame size of _____.
    a. 250A
    b. 300A
    c. 350A
    d. 400A

2. At 300% of its rated current, a circuit breaker rated at 600V must trip within _____.
    a. 7 milliseconds
    b. 0.7 seconds
    c. 7 seconds
    d. 70 seconds

Section Three

3.0.0 Sizing and Selecting Fuses

Objective

Size and select fuses.
a. Identify fuse types and markings.
b. Size fuses.
c. Coordinate the operation of overcurrent protective devices.

Performance Task

1. Identify the following on one or more circuit breaker(s) and fuse(s):
   - Number of poles
   - Load rating
   - Voltage rating
   - Amperage interrupting rating

Trade Terms

Dual-element fuse: A fuse having two fuse characteristics; the usual combination is an overcurrent limit and a time delay before activation.

Edison-base: The standard screw base used for ordinary lamps and Edison-base plug fuses.

Nonrenewable fuse: A fuse that must be replaced after it interrupts a circuit.

A fuse is the simplest device for opening an electric circuit when excessive current flows due to an overload or such fault conditions as grounds and short circuits. A fusible link or links encapsulated in a tube and connected to contact terminals comprise the fundamental elements of the basic fuse. The electrical resistance of the link is so low that it simply acts as a conductor, and every fuse is intended to be connected in series with each phase conductor so that current flowing through the conductor or to any load must also pass through the fuse. The continuous current rating of the fuse in amps establishes the maximum amount of current the fuse will carry without opening. When the circuit current flow exceeds this value, an internal element (link) in the fuse melts due to the heat of the current flow and opens the circuit.

3.1.0 Fuse Types and Markings

Fuses are manufactured in a wide variety of types and sizes with different current ratings, different abilities to interrupt fault currents, various speeds of operation (either quick-opening or time delay types), different internal and external constructions, and voltage ratings for both low-voltage (600V and below) and medium-voltage (over 600V) circuits. The following are details of these ratings:

- *Voltage rating* – Most low-voltage power distribution fuses have 250V or 600V ratings (other ratings are 125V and 300V). The voltage rating of a fuse must be at least equal to the circuit voltage. It can be higher but never lower. For instance, a 600V fuse can be used in a 208V circuit. The voltage rating of a fuse is a function of or depends upon its ability to open a circuit under an overcurrent condition. Specifically, the voltage rating determines the ability of the fuse to suppress the internal arcing that occurs after a fuse link melts and an arc is produced. If a fuse is used with a voltage rating lower than the circuit voltage, arc suppression will be impaired and, under some fault current conditions, the fuse may not safely clear the overcurrent.

- *Ampere rating* – Every fuse has a specific ampere rating. In selecting the ampacity of a fuse, consideration must be given to the type of load and code requirements. The ampere rating of a fuse should normally not exceed the current-carrying capacity of the circuit. For instance, if a conductor is rated to carry 20A, a 20A fuse is the largest that should be used in the conductor circuit. However, there are some specific circumstances under which the ampere rating is permitted to be greater than the current-carrying capacity of the circuit. A typical example is the motor circuit; a dual-element fuse is generally permitted to be sized up to 175%, and a non-time-delay fuse up to 300% of the motor full-load amperes. Generally, the ampere rating of a fuse and switch combination should be selected at 125% of the load current. There are exceptions, such as when the fuse-switch combination is approved for continuous operation at 100% of its rating. A protective device must be able to withstand the destructive energy of a short circuit. If a fault current exceeds a level beyond the capability of the protective device, the device may actually rupture and cause severe damage. Thus, it is important when applying a fuse or circuit breaker to use one that can sustain the largest potential short circuit currents. The rating that defines the capacity of a protective device to maintain its integrity when reacting to fault currents is termed its interrupting rating. The interrupting rating of most branch circuit, molded-case circuit breakers typically used in residential service entrance boxes is 10,000A. The rating is usually expressed as 10,000 AIC. Larger, more expensive circuit breakers may have ratings of 14,000 AIC or higher. In contrast, most modern, current-limiting fuses have an interrupting capacity of 200,000A and are commonly used to protect the lower-rated circuit breakers. *NEC Section 110.9* requires equipment intended to interrupt current at fault levels to have an interrupting rating not less than the current that must be interrupted.

- *Time delay ratings* – The time delay rating of a fuse is established by standard UL tests. All fuses have an inverse time-current characteristic. That is, the fuse will open quickly on high currents and after a period of time delay on low overcurrents. Specific types of fuses are made to have specific time delays. The basic UL requirement for Class RK-1, RK-15, and J fuses that are marked time delay is that the fuse must carry a current equal to five times its continuous rating for a period of not less than ten seconds. UL has not developed time delay tests for all fuse classes. Fuses are available for use when a time delay is needed along with current limitation on high-level short circuits. In all cases, the manufacturer's literature should be consulted to determine the degree of time delay in relation to the operating characteristics of the circuit being protected.

Hurricane and Flood Damage Victims

The US Consumer Product Safety Commission has issued a warning to hurricane and flood damage victims that circuit breakers, GFCIs, and fuses that have been submerged under water must be replaced to avoid electrocutions, explosions, and fires. This also applies to panels and electrical devices in general. This is because any flood water and silt trapped inside the devices may prevent them from performing properly. In addition, silt, sand, and salt can also allow tracking to occur between conductors, buses, and connection points.

3.1.1 Plug Fuses

Plug fuses have a screw-shell base and are commonly used in dwellings for circuits that supply lighting, heating, and appliances. Plug fuses are supplied with standard screw bases (Edison-base) or Type S bases. An Edison-base fuse consists of a strip of fusible (capable of being melted) metal in a small porcelain or glass case, with the fuse strip or link visible through a window in the top of the fuse. The screw base corresponds to the base of a standard medium-base incandescent lamp. Edison-base fuses are permitted only as replacements in existing installations; all new work must use Type S fuses in accordance with *NEC Section 240.51(B)*, which states that plug fuses of the Edison-base type may be only used for replacements in existing installations where there is no evidence of overfusing or tampering.

The chief disadvantage of the Edison-base plug fuse is that it is made in several ratings (from 0 to 30A), all with the same size base—permitting unsafe replacement of one rating by a higher rating. Type S fuses were developed to reduce the possibility of over-fusing a circuit (inserting a fuse with a rating greater than that required by the circuit). There are 15 classifications of Type S fuses, from 0 to 30A. Each Type S fuse has a base of a different size and a matching adapter. Once an adapter is screwed into a standard Edison-base fuseholder, it locks into place and is not readily removed without destroying the fuseholder. As a result, only a Type S fuse with a size the same as that of the adapter may be inserted. Two types of plug fuses are shown in *Figure 11*; a Type S adapter is also shown.

Plug fuses are also made in time delay types that permit a longer period of overload flow before operation, such as on motor inrush current and other higher-than-normal currents. They are available in ratings up to 30A, in both Edison-base and Type S.

The principal use of plug fuses is in motor circuits, in which the starting inrush current to the motor is much higher than the running or continuous current. The time delay fuse will not open on the inrush of high starting current. If, however, the high current persists, the fuse will open the circuit just as if a short circuit or heavy overload current had developed. All Type S fuses are time delay fuses.

Plug fuses are normally permitted to be used in circuits of no more than 125V between phases, but they may also be used where the voltage between any ungrounded conductor and ground is not more than 150V. The screwshell of the fuseholder for plug fuses must be connected to the

EDISON-BASE FUSE
(MAXIMUM RATING 30 AMPERES)

TYPE S
FUSE ADAPTER

TYPE S FUSE
(MAXIMUM RATING 30 AMPERES)

Figure 11 Standard Edison-base fuse on top; Type S with adapter shown on bottom.

load side circuit conductor; the base contact is connected to the line side or conductor supply. A disconnecting means (switch) is not required on the supply side of a plug fuse.

The plug fuse is a nonrenewable fuse, or onetime fuse; that is, once it has opened the circuit because of a fault or overload, it cannot be used again or renewed. For safe and effective restoration of circuit operation, it must be replaced by a new fuse of the same rating and characteristics.

3.1.2 Cartridge Fuses

In most industrial and commercial applications, cartridge fuses are used because they have a wider range of types, sizes, and ratings than plug fuses. Older cartridge fuses were provided with a means to renew the fuse by unscrewing the end caps and replacing the links. These fuses presented safety hazards because it was easy to tamper with them to override the safety feature. Their production has been discontinued. Two types of cartridge fuses are shown in *Figure 12*.

Single-element cartridge fuses – The basic component of this fuse is the link. Depending upon the ampere rating of the fuse, the single-element fuse may have one or more links. They are electrically connected to the end blades (or ferrules) and enclosed in a tube or cartridge surrounded by an arc-quenching filler material.

Under normal operation when the fuse is operating at or near its ampere rating, it simply

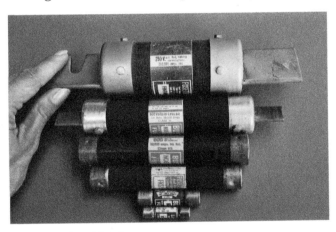

Figure 12 Cartridge fuses.

Plug Fuses

Plug fuses are used in the following applications:

- *W series* – Residential loads and nonmotor industrial circuits
- *TL series* – Normal-duty motor loads
- *T series* – When extra time delay or sensitivity to temperature is needed
- *S series* – Time delay fuses with tamper-resistant dimensions to prevent substitution of incorrect fuses

The condition of plug fuses can be checked visually by looking into the fuse window to observe the condition of the fusible link. If the element is broken and/or the window is black or discolored, the fuse is blown and must be replaced. Note that the condition of the window gives an indication of why the fuse blew. If the element is broken but the window is clear, this indicates an overload occurred; if the window is black or discolored, this indicates that a short circuit occurred.

functions as a conductor. However, if an overload current occurs and persists for more than a short interval of time, the temperature of the link eventually reaches a level that causes a restricted segment of the link to melt; as a result, a gap is formed and an electric arc is established. As the arc causes the link metal to burn back, the gap becomes progressively larger. The electrical resistance of the arc eventually reaches such a high level that the arc cannot be sustained and is extinguished; the fuse will have then completely cut off all current flow in the circuit. Suppression or quenching of the arc is accelerated by the filler material.

Single-element fuses have a very high speed of response to overcurrents. They provide excellent short circuit component protection. However, temporary harmless overloads or surge currents may cause nuisance openings unless these fuses are oversized. They are best used, therefore, in circuits that are not subject to heavy transient surge currents and the temporary overload of circuits with inductive loads such as motors, transformers, and solenoids. Because single-element fuses have a high speed of response to short circuit currents, they are particularly suited for the protection of circuit breakers with low interrupting ratings.

Dual-element cartridge fuses – Unlike single-element fuses, the dual-element fuse can be applied in circuits subject to temporary motor overload and surge currents to provide both high performance short circuit and overload protection. Oversizing in order to prevent nuisance openings is not necessary. The dual-element fuse contains two distinct types of elements. Electrically, the two elements are series connected. The fuse links, similar to those used in the single-element fuse, perform the short circuit protection function; the overload element provides protection against low-level overcurrents or overloads and will hold an overload that is five times greater than the ampere rating of the fuse for a minimum time of ten seconds.

The overload section consists of a copper heat absorber and a spring-operated trigger assembly. The heat-absorber strip is permanently connected to the first short circuit link and to the second short circuit link on the opposite end of the fuse by the S-shaped connector of the trigger assembly. The connector electronically joins the one short circuit link to the heat absorber in the overload section of the fuse. These elements are joined by a calibrated fusing alloy. An overload current causes heating of the short circuit link connected to the trigger assembly. Transfer of heat from the short circuit link to the heat absorber in the midsection of the fuse begins to raise the temperature of the heat absorber. If the overload is sustained, the temperature of the heat absorber eventually reaches a level that permits the trigger spring to fracture the calibrated fusing alloy and pull the connector free. The short circuit link is electrically disconnected from the heat absorber, the conducting path through the fuse is opened, and the overload current is interrupted. A critical aspect of the fusing alloy is that it retains its original characteristic after repeated temporary overloads without degradation.

Dual-element fuses may also be used in circuits other than motor branch circuits and feeders, such as lighting circuits and those feeding mixed lighting and power loads. The low-resistance construction of the fuses offers cooler operation of the equipment, which permits higher loading of fuses in switches and panel enclosures without heat damage and without nuisance openings from accumulated ambient heat.

3.1.3 Fuse Classes and Markings

Most fuses are used for the protection of branch circuits and feeders on systems operating at 600V or below. Cartridge fuses in this category include UL Classes H, K, G, J, R, T, CC, and L. There are also many varieties of miscellaneous fuses used for special purposes or for supplementary protection of individual types of electrical equipment. All fuses are tested and listed by Underwriters Laboratories, Inc., in accordance with established standards of construction and performance.

NEC Section 240.60(C) requires that any cartridge fuses used for branch circuit or feeder protection must be plainly marked, either by printing on the fuse barrel or by a label attached to the barrel that shows the following:

- Ampere rating
- Voltage rating
- Interrupting rating (if other than 10,000A)
- Current limiting, where applicable
- The name or trademark of the manufacturer

3.2.0 Sizing Fuses

General guidelines for sizing fuses are given here for most circuits that will be encountered on conventional systems. Some specific applications may warrant other fuse sizing; in these cases, the load characteristics and appropriate *NEC*® sections should be considered. The selections shown here are not, in all cases, the maximum or minimum ampere ratings permitted by the *NEC*®. Demand factors as permitted by the *NEC*® are not included here.

Fuse Classes

Cartridge fuses are classified by Underwriters Laboratories, Inc., according to their maximum voltage and amperage ranges. Some typical values assigned for common fuse classifications are:

Class G 480V, 1 to 60A, interrupting rating of 100,000 rms symmetrical; listed as current limiting

Class H 250/600V, 1 to 600A, interrupting rating of 10,000 rms symmetrical

Class J 600V, 1 to 600A, interrupting ratings ranging from 200,000 to 300,000 rms symmetrical; listed as current limiting

Class K 250/600V, 1/10 to 600A, interrupting ratings ranging from 50,000 to 200,000 rms symmetrical

Class L 600V, 601 to 6,000A, interrupting ratings ranging from 200,000 to 300,000 rms symmetrical; listed as current limiting

Class R 250/600V, 1/10 to 600A, interrupting rating of 200,000 rms symmetrical; listed as current limiting

Class T 300/600V, 1 to 1,200A, interrupting rating of 200,000 rms symmetrical; listed as current limiting

Class CC 600V, 1/10 to 30A, interrupting rating of 200,000 rms symmetrical; listed as current limiting

3.2.1 Dual-Element Time Delay Fuses

Application considerations for dual-element time delay fuses include the following:

- *Main service* – Each ungrounded service-entrance conductor shall have an overcurrent device in series with a rating not higher than the ampacity of the conductor, except as permitted in *NEC Section 230.90(A), Exceptions*. As specified in *NEC Section 230.91*, the service overcurrent device must be part of the service disconnecting means or located immediately adjacent to it.
- *Feeder circuit with no motor load* – The fuse size must be at least 125% of the continuous load plus 100% of the non-continuous load.
- *Feeder circuit with all motor loads* – Select the largest branch circuit overcurrent device and add 100% of the full-load current of all other motors connected on the same feeder. The *NEC®* will not permit you to size up; you must size down.
- *Feeder circuit with mixed loads* – Select the largest motor branch circuit fuse and add 100% of the full-load current of all other motors plus 125% of the continuous, non-motor load plus 100% of the non-continuous, non-motor load.
- *Branch circuit with no motor load* – The fuse size must be at least 125% of the continuous load plus 100% of the non-continuous load.

3.2.2 Non-Time-Delay Fuses

Application considerations for non-time-delay fuses include the following:

- *Main service* – Service-entrance conductors shall have a short circuit protective device in each ungrounded conductor, either on the load side of or as an integral part of the service-entrance switch. The protective device shall be capable of detecting and interrupting all values of current in excess of its trip setting or melting point, which can occur at its location. A fuse rated in continuous amperes not to exceed three times the ampacity of the conductor, or a circuit breaker with a trip setting of not more than six times the ampacity of the conductors, shall be considered as providing the required short circuit protection.
- *Feeder circuit with no motor load* – The fuse size must be at least 125% of the continuous load plus 100% of the non-continuous load.
- *Feeder circuit with all motor loads* – Select the largest branch circuit overcurrent device and add 100% of the full-load current of all other motors connected on the same feeder. The *NEC®* will not permit you to size up; you must size down.
- *Feeder circuit with mixed loads* – Select the largest motor branch circuit fuse and add 100% of the full-load current of all other motors plus 125% of the continuous, non-motor load plus 100% of the non-continuous, non-motor load.

- *Branch circuit with no motor load* – The fuse size must be at least 125% of the continuous load plus 100% of the non-continuous load.

When sizing fuses for a given application, a schematic drawing of the system will help tremendously. Such drawings do not have to be detailed; a single-line schematic will suffice, such as the one shown in *Figure 13*.

For example, assume there is one 25hp squirrel-cage induction motor (full starting voltage, nameplate current 31.6A, service factor 1.15, Code letter F), and two 30hp wound rotor induction motors (nameplate primary current 36.4A, nameplate secondary current 65A, 40°C rise), on a 460V, three-phase, 60Hz electrical supply. The feeder ampacity, motor overload protection, branch circuit short circuit and fault current protection, and feeder protection are determined as follows:

- *Feeder ampacity* – The full-load current value used to determine the ampacity of conductors for the 25hp motor is 34A [*NEC Section 430.6(A) and Table 430.250*]. A full-load current of 34A × 1.25 = 42.5A (*NEC Section 430.22*). The full-load current value used to determine the ampacity of primary conductors for each 30hp motor is 40A [*NEC Section 430.6(A) and Table 430.250*]. A full-load primary current of 40A × 1.25 = 50A (*NEC Section 430.22*). A full-load secondary current of 65A × 1.25 = 81.25A [*NEC Section 430.23(A)*]. The feeder ampacity will be (40A × 1.25) + 40A + 34A = 124A (*NEC Section 430.24*).
- *Overload* – Where protected by a separate overload device, the 25hp motor with a nameplate current of 31.6A must have overload protection of not over 39.5A [*NEC Sections 430.6(A) and 430.32(A)(1)*]. Where protected by a separate

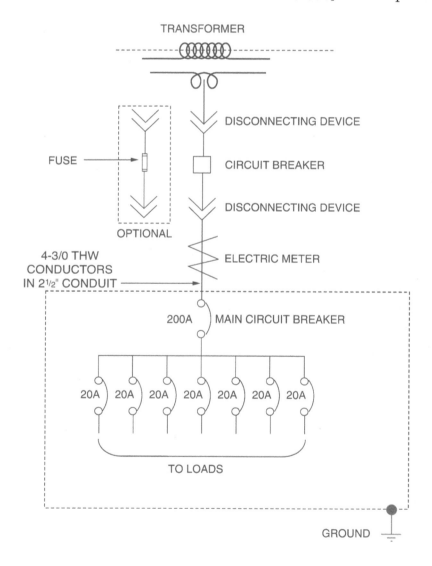

Figure 13 Typical single-line schematic diagram showing overcurrent protective devices.

overload device, the 30hp motor with a nameplate current of 36.4A must have overload protection of not over 45.5A [*NEC Section 430.6(A)*]. If the overload protection is not sufficient to start the motor or carry the load, it may be increased according to *NEC Section 430.32(C)*. For a motor marked thermally protected, overload protection is provided by the thermal protector [*NEC Sections 430.7(A)(13) and 430.32(A)(2)*].

- *Branch circuit short circuit and fault current* – The branch circuit of the 25hp motor must have short circuit and fault current protection of not over 300% for a non-time-delay fuse (*NEC Table 430.52*) or 3.00 × 34A = 102A. The next smallest standard size fuse is 100A. The fuse size may be increased to 110A [*NEC Section 430.52(C)(1), Exception No. 1 and NEC Section 240.6*]. Where the maximum value of branch circuit short circuit and fault current protection is not sufficient to start the motor, the value for a non-time-delay fuse may be increased to 400% [*NEC Section 430.52(C)(1), Exception No. 2(1)*]. If a time delay fuse is to be used, see *NEC Section 430.52(C)(1), Exception No. 2(2)*.

- *Feeder circuit* – The maximum rating of the feeder short circuit and fault current protection device is based on the sum of the largest branch circuit protective device (110A fuse) plus the sum of the full-load currents of the other motors, or 110A + 40A + 40A = 190A. The nearest standard fuse that does not exceed this value is 175A [*NEC Section 430.62(A)*].

3.3.0 Coordination of Overcurrent Protective Devices

Coordination is the name given to the time-current relationship among a number of overcurrent devices connected in series, such as fuses in a main feeder, subfeeder, and branch circuits. Safety is the prime consideration in the operation of fuses; however, coordination of the characteristics of fuses is also an important factor in large and complex electrical systems. Every fuse must be properly rated for continuous current and overloads and for the maximum short circuit current the electrical system could feed into a fault on the load side of the fuse, but this is not enough. It might still be possible for a fault on a feeder to open the main service fuse before the feeder fuse opens. Alternately, a branch circuit fault might open the feeder fuse before the branch circuit device opens. Such applications are said to be uncoordinated or nonselective—the fuse closest to the fault is not faster than the one farthest from the fault. When a fault on a feeder opens the main service fuse instead of the feeder fuse, the entire electrical system is taken out of service instead of just the one faulted feeder. Effective coordination minimizes the extent of electrical outage when a fault occurs. It therefore minimizes loss of production, interruption of critical continuous processes, and loss of vital facilities.

Selective coordination is the selection of overcurrent devices with time/current characteristics that ensure the clearing of a fault or short circuit by the device nearest the fault on the line side of the fault. A fault on a branch circuit is cleared by the branch circuit device. The subfeeder, feeder, and main service overcurrent devices will not operate. Alternately, a fault on a feeder is opened by the feeder fuse without opening any other fuse on the supply side of the feeder. With selective coordination, only the faulted part of the system is taken out of service, which represents the condition of minimum outage.

With proper selective coordination, every device is rated for the maximum fault current it might be called upon to open. Coordination is achieved by studying the curve of current versus the time required for the operation of each device. The selection is then made so that the device nearest any load is faster than all devices closer to the supply, and each device going back to the service entrance is faster than all devices closer to the supply. The main service fuses must have the longest opening time for any branch or feeder fault.

Overcurrent protection offers one of the greatest means of protection against electrical faults to both equipment and personnel. However, there are certain precautionary measures that must be observed at all times:

- Make certain the switch or circuit breaker is open before making any inspections or changes to either the overcurrent device or the circuit it is protecting.

Current-Limiting Fuses

Current-limiting fuses have a much faster response to short circuit currents and are typically used to provide protection to mains, feeders, circuit breakers, and other components that lack an adequate interrupting rating.

- Make certain that all components are compatible with each other and with the system on which they are used; that is, manufacturer, type, normal current ratings, interrupting capacity, voltage, and other factors.
- Use the proper tools for the job.
- When an overcurrent protection device opens a circuit, determine the reason before changing fuses or resetting the circuit breaker. A severe fault may not damage the circuit or equipment the first time, but repeated hits with high inrush current may do so the second or third time.
- Make sure that adjustable circuit breakers are calibrated correctly (refer to the manufacturer's instructions for calibration details). An improperly calibrated circuit breaker is considered by many to be the most dangerous of circuit breaker faults.
- Use the test breaker to trip devices one phase at a time.
- Never substitute a fuse or circuit breaker that is rated higher than the circuit conductors.
- When working on circuits protected by overcurrent protective devices, lock out and tag the device so that it cannot be inadvertently closed.

Think About It

Cartridge Fuse Removal

After the On-Off handle of a general-purpose disconnect switch is set to Off and the door has been opened, is there any danger involved in the removal or installation of the cartridge fuses?

Think About It

Putting It All Together

Examine the electrical service in your home. Does it use fuses or circuit breakers? What are the listed amperages? Are there any double breakers? Which devices do they serve? Does your furnace or HVAC system use a separate fused disconnect? If the fuses/breakers aren't labeled, take some time to label them now. Can you visualize the internal circuits of your home based on the areas that are linked to the same breaker?

3.0.0 Section Review

1. An acceptable fuse combination is a _____.
   a. 30A fuse used to protect a non-motor 20A circuit
   b. 10A fuse used to protect a non-motor 20A circuit
   c. 30A non-time-delay fuse used to protect a 15A motor
   d. 45A dual-element fuse used to protect a 15A motor

2. When sizing a non-time-delay fuse for a branch circuit with no motor load, the fuse size must be at least _____.
   a. 125% of the continuous load plus 100% of the non-continuous load
   b. 150% of the continuous load
   c. 100% of the continuous load plus 100% of the non-continuous load
   d. 200% of the non-continuous load

3. Selective coordination is used in complex systems in order to _____.
   a. shut down the whole system if a fault occurs in a feeder
   b. ensure that the device nearest any load is faster than all devices closer to the supply
   c. allow branch circuit faults to trip both feeder and service overcurrent devices
   d. ensure that the main service fuses have the shortest opening time for any branch or feeder fault

Review Questions

1. Every circuit breaker must have its interrupting rating shown on the circuit breaker, unless it is rated for _____.
   a. 1,000A
   b. 5,000A
   c. 10,000A
   d. 65,000A

2. When a common molded-case circuit breaker trips, in what position will the operating handle be?
   a. Halfway between the On and Off positions
   b. In the On position
   c. In the Off position
   d. In the Reset position

3. What is the highest voltage rating for circuit breakers used on DC systems that UL recognizes?
   a. 240V
   b. 277V
   c. 480V
   d. 600V

4. Where is the ampere rating normally marked on circuit breakers?
   a. On the side of the molded case
   b. On the handle
   c. Beneath the terminal lug
   d. On the bottom of the molded case

5. Which of the following is the most important consideration when working with overcurrent protection devices?
   a. The brand name
   b. The type of project
   c. The correct interrupting capacity
   d. The age of the device

6. What is normally the highest short circuit current in residential applications?
   a. 10,000A
   b. 30,000A
   c. 50,000A
   d. 100,000A

7. Which of the following best describes where the *NEC®* allows the use of Edison-base fuses?
   a. On new installations only
   b. On circuits rated over 40A
   c. On circuits rated under 40A
   d. For replacements in existing installations

8. What must be used in conjunction with Type S plug fuses when they are installed in a fuse holder?
   a. A Type S adapter
   b. A copper penny
   c. A circuit breaker connected in series
   d. A setscrew holding device

9. On renewable cartridge fuses, which part of the fuse is normally renewed?
   a. Barrel
   b. Link
   c. Ferrule
   d. Ferrule ring

10. Which of the following best describes a dual-element fuse?
    a. A fuse having an overcurrent limit and a time delay before activation
    b. A fuse with two attachment points
    c. A fuse having two ferrules at each end
    d. A fuse with two barrels

Supplemental Exercises

1. A standard circuit breaker contains _____.
   a. a set of contacts
   b. a thermal element
   c. a magnetic element
   d. all of the above
2. The smallest standard circuit breaker size is _____.
   a. 5A
   b. 10A
   c. 15A
   d. 20A
3. The amperage interrupting capacity (AIC) is the _____.
4. A current-limiting circuit breaker provides downstream protection for _____.
5. A ground fault circuit interrupter will trip when a fault current _____.
   a. is 1A or more
   b. is 2mA
   c. is 6mA or more
   d. reaches the rated ampacity of the circuit
6. A dual-element fuse can be applied in circuits to provide both short circuit and _____ protection.
7. An overcurrent condition can be defined as _____.
   a. an overload current
   b. a short circuit current
   c. either an overload or a short circuit current
8. Each ungrounded service-entrance conductor shall have a(n) _____ in series with a rating not higher than the ampacity of the conductor.
9. Coordination is the name given to the _____.
   a. sizing of all overcurrent devices from the service to branch circuit
   b. sizing of panelboards to hold the proper circuit breakers
   c. sizing of S-type fuses
   d. time-current relationship of overcurrent devices connected in series with each other
10. True or False? Edison-base fuses are outdated and should not be used in new installations.

Trade Terms Introduced in This Module

Cartridge fuse: A fuse enclosed in an insulating tube to confine the arc when the fuse blows.

Circuit breaker: A device that is designed to open a circuit automatically at a certain overcurrent. It can also be used to manually open and close the circuit.

Dual-element fuse: A fuse having two fuse characteristics; the usual combination is an overcurrent limit and a time delay before activation.

Edison-base: The standard screw base used for ordinary lamps and Edison-base plug fuses.

Fault current: The current that exists when an unintended path is established between an ungrounded conductor and ground.

Frame size: A method used to classify circuit breakers according to given current ranges.

Fuse: A protective device that opens a circuit when the fusible element is severed by heating due to a fault current or overcurrent passing through it.

Fuse link: The fusible part of a cartridge fuse.

Molded-case circuit breaker: A circuit breaker enclosed in an insulating housing.

Nonrenewable fuse: A fuse that must be replaced after it interrupts a circuit.

Overcurrent protection: De-energizing a circuit whenever the current exceeds a predetermined value; the usual devices are fuses, circuit breakers, or magnetic relays.

Plug fuse: A type of fuse that is held in position by a screw thread contact instead of spring clips, as is the case with a cartridge fuse.

Pole: That portion of a device associated exclusively with one electrically separated conducting path of the main circuit or device.

Short circuit: A conducting connection, whether intentional or accidental, between any of the conductors of an electrical system, whether it is from line-to-line or from line-to-neutral (grounded).

Additional Resources

This module presents thorough resources for task training. The following reference material is recommended for further study.

Feeder Circuit Protection (PDF). Available from Cooper Industries at **www.cooperindustries.com**.
National Electrical Code® Handbook. Latest Edition. Quincy, MA: National Fire Protection Association.

Figure Credits

Greenlee / A Textron Company, Module Opener
John Traister, Figures 2, 3, 11
Tim Dean, Figure 4
Courtesy of Cooper Bussmann, division of Cooper Industries plc, Figures 5–7
Tim Ely, Figure 8

Section Review Answer Key

SECTION 1.0.0

Answer	Section Reference	Objective
1. b	1.1.0	1a
2. a	1.2.0	1b

SECTION 2.0.0

Answer	Section Reference	Objective
1. d	2.1.0	2a
2. d	2.2.2	2b

SECTION 3.0.0

Answer	Section Reference	Objective
1. c	3.1.0	3a
2. a	3.2.2	3b
3. b	3.3.0	3c

NCCER CURRICULA — USER UPDATE

NCCER makes every effort to keep its textbooks up-to-date and free of technical errors. We appreciate your help in this process. If you find an error, a typographical mistake, or an inaccuracy in NCCER's curricula, please fill out this form (or a photocopy), or complete the online form at **www.nccer.org/olf**. Be sure to include the exact module ID number, page number, a detailed description, and your recommended correction. Your input will be brought to the attention of the Authoring Team. Thank you for your assistance.

Instructors – If you have an idea for improving this textbook, or have found that additional materials were necessary to teach this module effectively, please let us know so that we may present your suggestions to the Authoring Team.

NCCER Product Development and Revision
13614 Progress Blvd., Alachua, FL 32615

Email: curriculum@nccer.org
Online: www.nccer.org/olf

❏ Trainee Guide   ❏ Lesson Plans   ❏ Exam   ❏ PowerPoints   Other _____

Craft / Level: _____   Copyright Date: _____

Module ID Number / Title: _____

Section Number(s): _____

Description: _____

_____

_____

_____

_____

Recommended Correction: _____

_____

_____

_____

_____

Your Name: _____

Address: _____

_____

Email: _____   Phone: _____

This page is intentionally left blank.

Control Systems and Fundamental Concepts

Overview

Contactors are used to make or break the circuit to high-current, non-motor loads, or are used in motor circuits if overload protection is separately provided. Meanwhile, a relay is an electromagnetic device whose contacts are used in the control circuits of magnetic starters, contactors, solenoids, timers, and other relays. This module describes the operating principles of contactors and relays, including both mechanical and solid-state devices. It also explains how to select and install relays and troubleshoot control circuits.

Module 26211-20

Trainees with successful module completions may be eligible for credentialing through the NCCER Registry. To learn more, go to **www.nccer.org** or contact us at 1.888.622.3720. Our website, **www.nccer.org**, has information on the latest product releases and training.

Your feedback is welcome. You may email your comments to **curriculum@nccer.org**, send general comments and inquiries to **info@nccer.org**, or fill in the User Update form at the back of this module.

This information is general in nature and intended for training purposes only. Actual performance of activities described in this manual requires compliance with all applicable operating, service, maintenance, and safety procedures under the direction of qualified personnel. References in this manual to patented or proprietary devices do not constitute a recommendation of their use.

Copyright © 2020 by NCCER, Alachua, FL 32615, and published by Pearson Education, Inc. or its affiliates, 221 River Street, Hoboken, NJ 07030. All rights reserved. Printed in the United States of America. This publication is protected by Copyright, and permission should be obtained from NCCER prior to any prohibited reproduction, storage in a retrieval system, or transmission in any form or by any means, electronic, mechanical, photocopying, recording, or likewise. To obtain permission(s) to use material from this work, please submit a written request to NCCER Product Development, 13614 Progress Blvd., Alachua, FL 32615.

26211-20 V10.0

From *Electrical, Trainee Guide*. NCCER.
Copyright © 2020 by NCCER. Published by Pearson. All rights reserved.

26211-20
CONTROL SYSTEMS AND FUNDAMENTAL CONCEPTS

Objectives

When you have completed this module, you will be able to do the following:

1. Identify magnetic and mechanically held contactors.
   a. Select lighting contactors.
   b. Make forward and reverse motor contactor connections.
   c. Select mechanically held contactors.
2. Select and troubleshoot relays.
   a. Select control relays.
   b. Select timers and timing relays.
   c. Select solid-state relays.
   d. Select overload relays.
   e. Troubleshoot relays.
3. Install low-voltage remote control switching systems.
   a. Identify remote control switching system components and operating characteristics.
   b. Plan and install a remote control switching system.

Performance Task

Under the supervision of the instructor, you should be able to do the following:

1. Mount and connect a 120V lighting contactor with a three-wire pushbutton control.

Trade Terms

Ambient temperature
Break
Instantaneous
Latch coil
Make
Mechanically held relay

Normally closed (N.C.) contacts
Normally open (N.O.) contacts
Shading coil
Solid-state relay
Unlatch coil

Industry Recognized Credentials

If you are training through an NCCER-accredited sponsor, you may be eligible for credentials from NCCER's Registry. The ID number for this module is 26211-20. Note that this module may have been used in other NCCER curricula and may apply to other level completions. Contact NCCER's Registry at 888.622.3720 or go to **www.nccer.org** for more information.

> **NOTE:** NFPA 70®, *National Electrical Code*® and *NEC*® are registered trademarks of the National Fire Protection Association, Quincy, MA.

Contents

1.0.0 Magnetic and Mechanically Held Contactors 1
  1.1.0 Lighting Contactors 2
  1.2.0 Forward and Reverse Motor Contactor Connections 6
  1.3.0 Mechanically Held Contactors 7
2.0.0 Relays 9
  2.1.0 Control Relays 9
  2.2.0 Timers and Timing Relays 11
  2.3.0 Solid-State Relays 11
  2.4.0 Overload Relays 14
    2.4.1 Bimetallic Overload Relays 15
    2.4.2 Fuses vs. Overload Relays 15
    2.4.3 Overload Relay Operation 16
    2.4.4 Overload Relay Classifications 16
    2.4.5 Relay Selection 16
  2.5.0 Troubleshooting Relays 17
3.0.0 Low-Voltage Remote Control Switching Systems 21
  3.1.0 System Components and Operating Characteristics 22
    3.1.1 Power Supplies 24
    3.1.2 Switches 24
    3.1.3 Master Sequencer 24
    3.1.4 Operating Characteristics 24
  3.2.0 Planning a Remote Control Switching System 25
    3.2.1 Enclosures 29
    3.2.2 Switch Points 29

Figures and Tables

Figure 1   A magnetic contactor 2
Figure 2   Contactor operating on the solenoid principle 2
Figure 3   Pushbutton control circuit with a magnetic relay 6
Figure 4   Emergency lighting circuit using a relay for operation 6
Figure 5   Several lighting circuits controlled by one contactor 7
Figure 6   Three-pole reversing starting diagram 7
Figure 7   Relay used to amplify contact capacity 10
Figure 8   Relay used for voltage amplification 10
Figure 9   Relay used to multiply the switching functions of a pilot device 11
Figure 10   Plug-in relays 12
Figure 11   Wiring diagram for a plug-in relay 13
Figure 12   Basic solid-state relay 13
Figure 13   Solid-state relay circuits 14
Figure 14   Simplified bimetallic overload relay 15
Figure 15   Melting alloy thermal unit 17
Figure 16   Basic circuit of a remote control switching system 21
Figure 17   Several types of remote control circuits and related components 22
Figure 18   Relay operation 23
Figure 19   Relay installed in an outlet box 24
Figure 20   Control transformer with power-supply wiring diagram 24
Figure 21   Switch operation 25
Figure 22   Master sequencer connections 26
Figure 23   Single-switch control 26
Figure 24   Schematic wiring diagram of single switches controlling each circuit 27
Figure 25   Diagram of multipoint switching of a single circuit 27
Figure 26   Switching circuits being controlled from one location 27
Figure 27   Recommended symbols for low-voltage drawings 28
Figure 28   Floor plan of a residence using low-voltage switching 30
Figure 29   Schematic wiring diagram of lighting circuits 31

Table 1A   Electrical Ratings for AC Magnetic Contactors and Starters (1 of 2) 4
Table 1B   Electrical Ratings for AC Magnetic Contactors and Starters (2 of 2) 5
Table 2A   Contactor and Relay Troubleshooting Chart (1 of 2) 18
Table 2B   Contactor and Relay Troubleshooting Chart (2 of 2) 19

This page is intentionally left blank.

SECTION ONE

1.0.0 MAGNETIC AND MECHANICALLY HELD CONTACTORS

Objective

Identify magnetic and mechanically held contactors.
a. Select lighting contactors.
b. Make forward and reverse motor contactor connections.
c. Select mechanically held contactors.

Trade Terms

Break: The opening of closed contacts to interrupt an electrical circuit.

Latch coil: The coil of a latching relay that, when energized, causes the relay to assume the switched or latched position.

Make: The closing of open contacts to establish an electrical circuit.

Mechanically held relay: A latching relay in which the switched or latched position is maintained by the interference of a mechanical part that moves to a non-interfering position when the unlatch coil is energized, allowing the relay to return to the normal or unlatched position.

Normally closed (N.C.) contacts: A combination of a stationary contact and a movable contact, which are engaged when the coil is de-energized; in the case of a latching relay, the contacts that are engaged after the unlatch coil is energized.

Normally open (N.O.) contacts: A combination of a stationary contact and a movable contact, which are disengaged when the coil is de-energized; in the case of a latching relay, the contacts that are disengaged after the unlatch coil is energized.

Shading coil: A shorted turn surrounding a portion of the pole of an AC electromagnet, producing by mutual inductance a delay in the change of the magnetic field in that part, thereby tending to prevent chatter and reduce hum.

Unlatch coil: The coil of a latching relay that, when energized, causes the relay to assume the normal or unlatched position.

A contactor is an electromagnetic apparatus that is designed to handle relatively high currents. The conventional contactor is identical in appearance, construction, and current-carrying ability to the equivalent National Electrical Manufacturers Association (NEMA) size magnetic motor starter. The magnet assembly and coil, contacts, holding circuit interlock, and other structural features are the same. The difference is that the contactor does not provide overload protection. Contactors are used to make or break the circuit to high-current, non-motor loads, or are used in motor circuits if overload protection is separately provided. A typical application of the latter is in a reversing motor starter.

The typical magnetic contactor (*Figure 1*) contains a coil of wire wound around an iron core or a laminated iron core (as shown) and an armature. The coil acts as an electromagnet, and when the proper electric current is applied, it attracts or picks up the armature. Since the movable contacts are attached via an iron bar to the armature, when the armature moves upwards, the movable contacts make contact with a set of stationary contacts to close and complete the circuit.

When power is being applied to the coil, the device is said to be energized. The motion of the armature moves the contact points together or apart, according to their arrangement. When the power is disconnected from the coil, a spring returns the armature to its original position. When no power is being applied, the device is said to be in its de-energized or normal condition. Contacts that are together when the relay is de-energized are said to be normally closed (N.C.) contacts. Contacts that are apart when the relay is de-energized are said to be normally open (N.O.) contacts.

Notice that a shading coil has been added to the iron core in *Figure 1*. Shading coils are used with AC relays to prevent contact chatter and hum. They operate in the same way as in the shaded-pole motor. In general, the shading coil is used to provide a continuous magnetic flow to the armature when the voltage of the AC waveform is zero.

The relay operates on the solenoid principle; that is, it converts electrical energy to linear motion, as shown in *Figure 2*. In this example, a coil of wire is wound around an iron core. When current flows through the coil, a magnetic field

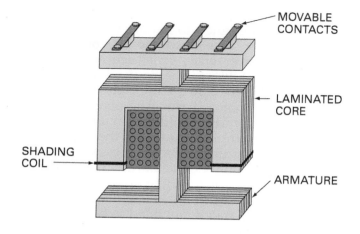

Figure 1 A magnetic contactor.

develops in the iron core. The magnetic field of the iron core attracts the movable arm, known as the armature. In its present position, the movable contact makes connection with a stationary contact. This contact is normally closed. When the armature is attracted to the iron core, the movable contact breaks the connection with the normally closed stationary contact and makes the connection with another stationary contact. This contact is normally open. In a wiring diagram, the movable contact would be the common, and the stationary contacts would be labeled N.O. and N.C.

1.1.0 Lighting Contactors

Filament-type lamps (e.g., tungsten, infrared, and quartz) have inrush currents of approximately 15 to 17 times their normal operating currents. To prevent welding of their contacts on the high initial current, standard motor control contactors must be derated if used to control this type of load.

Table 1A and *Table 1B* list electrical ratings for AC magnetic contactors and starters. As shown, a NEMA Size 1 contactor has a continuous current rating of 27A, but if it is used to switch certain lighting loads, it must be derated to 15A. The standard contactor, however, need not be derated for resistance heating or fluorescent lamp loads that do not impose as high an inrush current.

Lighting contactors differ from standard contactors in that the contacts are rated for high inrush currents. A holding circuit interlock is not normally provided because this type of contactor

> **Think About It**
> ## Coil Shading
> Why does shading the coil prevent the contacts of a contactor from chattering?

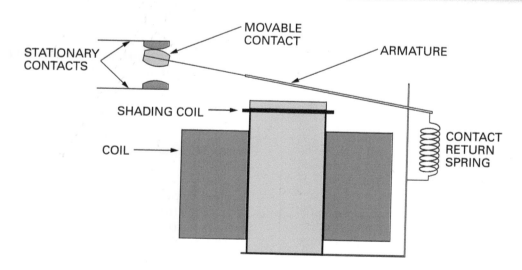

Figure 2 Contactor operating on the solenoid principle.

Troubleshooting Contactors

If the contacts of a contactor are defective, it can usually be spotted during a visual inspection. Another way to check is to place a voltmeter across the closed contacts with voltage applied. If the voltage drop is more than 5% of the rated value of the contactor, the contacts are defective.

---

is frequently controlled by a two-wire pilot device, such as a time clock or photoelectric relay.

Unlike standard contactors, lighting contactors are not horsepower-rated or categorized by NEMA size, but are designated by ampere ratings (20A, 30A, 60A, 100A, 200A, and 300A). It should be noted that lighting contactors are specialized in their application and should not be used on motor loads.

A basic circuit using a lighting contactor and serving as a control for a high-amperage lamp is shown in *Figure 3*. Its purpose is to control the lamp from two different locations. The relay coil is indicated by the standard circular symbol and the relay contacts are indicated in the diagram by two parallel lines. The contacts are drawn separated from the coil to show their function. In reality, they are next to the coil in the relay itself.

Closing the N.O. pushbutton energizes the coil of relay R1, and both sets of normally open R1 contacts close. The R1-2 contacts provide a closed circuit to light the lamp. The R1-1 contacts act as holding or seal-in contacts for the relay coil, so that the coil remains energized when the N.O. pushbutton is released.

The lamp will remain lighted until either of the N.C. pushbuttons is momentarily pressed. Pressing either of these buttons opens the circuit to the coil. The coil will drop out or become de-energized and will open the contacts of the relay. The circuit through the lamp is now open and the lamp is turned off. The lamp remains off until one of the N.O. pushbuttons is pressed again. One set of N.O. and N.C. pushbuttons (on and off) is placed at one location where remote control of the lamp is desired, and another set is placed in another convenient location. The lamp can be turned on and off from either of these locations.

Another application of a magnetic relay is its use in an emergency lighting circuit. The lamps in such a circuit should be lighted instantly after the failure of the main power supply. A simple emergency lighting system controlling one lamp is shown in *Figure 4*. The main power is supplied through the power supply terminals to the relay coil. The relay has a set of normally open contacts and a set of normally closed contacts. As long as the main power is available, it energizes the relay coil, which in turn keeps the N.O. contacts closed. The power is therefore supplied to the transformer, whose primary is connected in series with the N.O. contacts. The secondary of the transformer keeps the storage battery charged through the rectifier. Since the coil is energized, the N.C. contacts are open and the emergency lamp is not lighted.

In the event of main power failure or low voltage in the line, the relay will drop out, the N.O. contacts will open, and the N.C. contacts will close. The lamp will be connected to the battery and give light as long as the relay coil remains de-energized. Such relay and circuit arrangements are frequently used in the emergency battery packs found in many commercial, industrial, and public buildings of all types.

The following example demonstrates how a relay and one single-pole, 15A toggle switch can be used to control 60A of lighting.

Replacing Contactors

Be careful when selecting replacement contactors because they are rated for different uses. Some are rated for inductive loads; others are rated for resistive loads; and still others are rated for both types of loads.

Emergency Lighting Circuits

A typical emergency lighting circuit uses an automatic transfer switch. A potential transformer in the switch senses the absence of normal voltage and switches on the emergency generator. When it senses power from the generator, it closes the contactor to the main circuit.

Table 1A Electrical Ratings for AC Magnetic Contactors and Starters (1 of 2)

NEMA Size	Volts	Maximum Horsepower Rating—Nonplugging and Nonjogging Duty Single Phase	Maximum Horsepower Rating—Nonplugging and Nonjogging Duty Poly-Phase	Maximum Horsepower Rating—Plugging and Jogging Duty Single Phase	Maximum Horsepower Rating—Plugging and Jogging Duty Poly-Phase	Continuous Current Rating, Amperes—600 Volt Max.	Service-Limit Current Rating, Amperes	Tungsten and Infrared Lamp Load, Amperes—250 Volts Max.	Resistance Heating Loads, kW—other than Infrared Lamp Loads Single Phase	Resistance Heating Loads, kW—other than Infrared Lamp Loads Poly-Phase	KVA Rating for Switching Transformer Primaries at 50 to 60 Cycles Single Phase	KVA Rating for Switching Transformer Primaries at 50 to 60 Cycles Poly-Phase	3-Phase Rating for Switching Capacitors Kvar
00	115	⅓	—	—	—	9	11	5	—	—	—	—	—
	200	—	1½	—	—	9	11	5	—	—	—	—	—
	230	1	—	—	—	9	11	5	—	—	—	—	—
	380	—	1½	—	—	9	11	—	—	—	—	—	—
	460	—	2	—	—	9	11	—	—	—	—	—	—
	575	—	2	—	—	9	11	—	—	—	—	—	—
0	115	1	—	½	—	18	21	10	—	—	0.9	1.2	—
	200	—	3	—	1½	18	21	10	—	—	—	1.4	—
	230	2	—	1	—	18	21	10	—	—	1.4	1.7	—
	380	—	5	—	1½	18	21	—	—	—	—	2	—
	460	—	5	—	2	18	21	—	—	—	1.9	2.5	—
	575	—	5	—	2	18	21	—	—	—	1.9	2.5	—
1	115	2	—	1	—	27	32	15	3	5	1.4	1.7	—
	200	—	7½	—	3	27	32	15	—	9.1	—	3.5	—
	230	3	—	2	—	27	32	15	6	10	1.9	4.1	—
	380	—	10	—	5	27	32	—	—	16.5	—	4.3	—
	460	—	10	—	5	27	32	—	—	20	3	5.3	—
	575	—	10	—	5	27	32	15	12	25	3	5.3	—
1P	115	3	—	1½	—	36	42	24	—	—	—	—	—
	230	5	—	3	—	36	42	24	—	—	—	—	—
2	115	3	—	2	—	45	52	30	5	8.5	1	4.1	—
	200	—	10	—	7½	45	52	30	—	15.4	—	6.6	11.3
	230	7½	15	5	10	45	52	30	10	17	4.6	7.6	13
	380	—	25	—	15	45	52	—	—	28	—	9.9	21
	460	—	25	—	15	45	52	—	20	34	5.7	12	26
	575	—	25	—	15	45	52	—	25	43	5.7	12	33

Table 1B Electrical Ratings for AC Magnetic Contactors and Starters (2 of 2)

NEMA Size	Volts	Maximum Horsepower Rating—Nonplugging and Nonjogging Duty		Maximum Horsepower Rating—Plugging and Jogging Duty		Continuous Current Rating, Amperes—600 Volt Max.	Service-Limit Current Rating, Amperes	Tungsten and Infrared Lamp Load, Amperes—250 Volts Max.	Resistance Heating Loads, kW—other than Infrared Lamp Loads		KVA Rating for Switching Transformer Primaries at 50 to 60 Cycles		3-Phase Rating for Switching Capacitors
		Single Phase	Poly-Phase	Single Phase	Poly-Phase				Single Phase	Poly-Phase	Single Phase	Poly-Phase	Kvar
3	115	7½	–	–	–	90	104	60	10	17	4.6	7.6	–
	200	–	25	–	15	90	104	60	–	31	–	13	23.4
	230	15	30	–	20	90	104	60	20	34	8.6	15	27
	380	–	50	–	30	90	104	–	–	56	–	19	43.7
	460	–	50	–	30	90	104	–	40	68	14	23	53
	575	–	50	–	30	90	104	–	50	86	14	23	67
4	200	–	40	–	25	135	156	120	–	45	–	20	34
	230	–	50	–	30	135	156	120	30	52	11	23	40
	380	–	75	–	50	135	156	–	–	86.7	–	38	66
	460	–	100	–	60	135	156	–	60	105	22	46	80
	575	–	100	–	60	135	156	–	75	130	22	46	100
5	200	–	75	–	60	270	311	240	–	91	–	40	69
	230	–	100	–	75	270	311	240	60	105	28	46	80
	380	–	150	–	125	270	311	–	–	173	–	75	132
	460	–	200	–	150	270	311	–	120	210	40	91	160
	575	–	200	–	150	270	311	–	150	260	40	91	200
6	200	–	150	–	125	540	621	480	–	182	–	79	139
	230	–	200	–	150	540	621	480	120	210	57	91	160
	380	–	300	–	250	540	621	–	–	342	–	148	264
	460	–	400	–	300	540	621	–	240	415	86	180	320
	575	–	400	–	300	540	621	–	300	515	86	180	400
7	230	–	300	–	–	810	932	720	180	315	–	–	240
	460	–	600	–	–	810	932	–	360	625	–	–	480
	575	–	600	–	–	810	932	–	450	775	–	–	600
8	230	–	450	–	–	1215	1400	1080	–	–	–	–	360
	460	–	900	–	–	1215	1400	–	–	–	–	–	720
	575	–	900	–	–	1215	1400	–	–	–	–	–	900

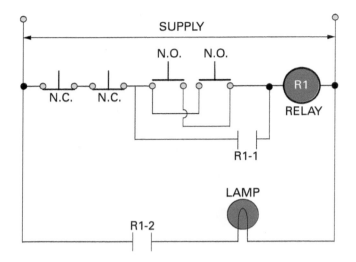

Figure 3 Pushbutton control circuit with a magnetic relay.

Figure 5 shows a diagram of a lighting contactor controlled by a single-pole switch. The contactor coil operates six sets of contacts, each of which is connected to a lighting circuit with lamps totaling 1,200W (10A). When the single-pole switch is turned to the On position (contacts closed), the coil is energized and moves the armature, which in turn closes all of the contacts simultaneously. When these contacts are closed, this completes the circuit to all of the lamps. Thus, all of the lamps will come on at the same time.

Six single-pole switches could have been used in place of the contactor, but simultaneous switching of the lamps would not be possible.

1.2.0 Forward and Reverse Motor Contactor Connections

Reversing the direction of motor shaft rotation is often required. Three-phase squirrel cage motors can be reversed by reconnecting any two of the three line connections to the motor. By interwiring two contactors, an electromagnetic method of making the reconnection can be obtained.

As seen in the power circuit in *Figure 6*, the contacts (F) of the forward contactor, when closed, connect lines L1, L2, and L3 to the motor terminals T1, T2, and T3, respectively. As long as the forward contacts are closed, mechanical and electrical interlocks prevent the reverse contactor from being energized.

When the forward contactor is de-energized, the second contactor can be picked up, closing its contacts (R), which reconnect the lines to the motor. Note that by running through the reverse contacts, L1 is connected to motor terminal T3, and L3 is connected to motor terminal T1. The motor will now run in the opposite direction.

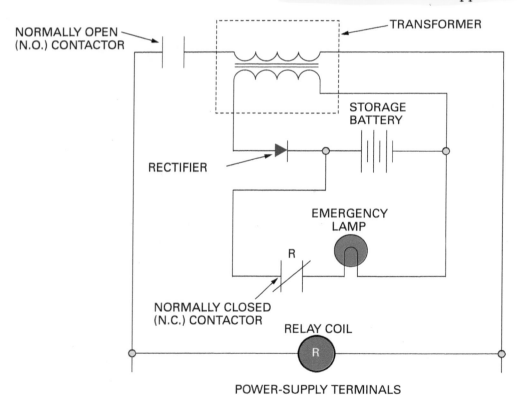

Figure 4 Emergency lighting circuit using a relay for operation.

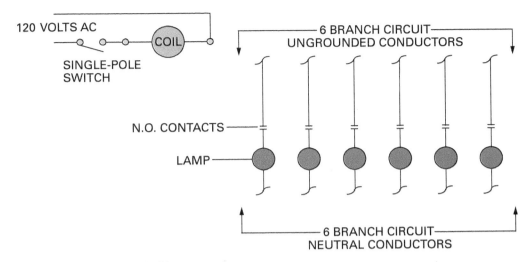

Figure 5 Several lighting circuits controlled by one contactor.

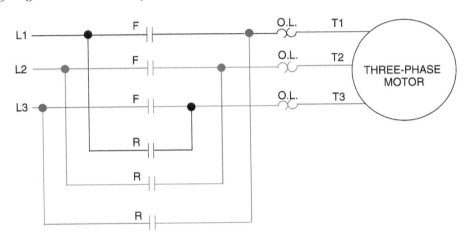

Figure 6 Three-pole reversing starting diagram.

Whether operating through either the forward or reverse contactor, the power connections are run through an overload relay assembly, which provides motor overload protection. A magnetic reversing starter, therefore, consists of a starter and a set of contactors, suitably interwired, with electrical and mechanical interlocking to prevent the coil of both units from being energized at the same time.

Manual reversing starters (employing two manual starters) are also sometimes used. As in the magnetic version, the forward and reverse switching mechanisms are mechanically interlocked, but since coils are not used in manually operated equipment, electrical interlocks are not furnished.

1.3.0 Mechanically Held Contactors

In a conventional contactor, current flow through the coil creates a magnetic pull to seal in the armature and maintain the contacts in a switched position; that is, the N.O. contacts will be held closed while the N.C. contacts will be held open. Because the contactor action is dependent on the current flow through the coil,

Relays with Many Contacts

A common example of a relay that controls many circuits is the defrost control on a large refrigeration unit, such as a walk-in cooler. When a timing relay closes the defrost circuit, one set of N.C. contacts opens to shut the fans off, another set of N.C. contacts opens to turn the compressor off, and a set of N.O. contacts closes to turn on the heaters. A set of N.C. contacts then opens to activate a solenoid that mechanically impedes the refrigerant.

the contactor is described as electrically held. As soon as the coil is de-energized, the contacts will revert to their initial position.

Versions of a **mechanically held relay** and contactor are also used in some applications. The action is accomplished using two coils and a latching mechanism. Energizing one coil (called the **latch coil**) through a momentary signal causes the contacts to switch, and a mechanical latch holds the contacts in this position even though the initiating signal is removed. The coil is then de-energized. To restore the contacts to their initial position, a second coil (called the **unlatch coil**) is momentarily energized.

Mechanically held relays and contactors are used where the hum of an electrically held device would be objectionable, such as in auditoriums, hospitals, churches, and so on.

1.0.0 Section Review

1. The maximum tungsten/infrared load amperage for a NEMA Size 0 magnetic contactor is _____.
    a. 5A
    b. 10A
    c. 18A
    d. 21A

2. A manual reversing starter uses _____.
    a. a mechanical interlock
    b. an electrical interlock
    c. two separate coils
    d. a single coil

3. Mechanically held relays would most likely be used in a _____.
    a. grocery store
    b. cafeteria
    c. sports arena
    d. hospital

Section Two

2.0.0 Relays

Objective

Select and troubleshoot relays.
a. Select control relays.
b. Select timers and timing relays.
c. Select solid-state relays.
d. Select overload relays.
e. Troubleshoot relays.

Trade Terms

Ambient temperature: The temperature of the medium (usually air) surrounding a device and into which the heat losses in the device are dissipated.

Instantaneous: Contacts that make or break a circuit immediately upon coil energization or de-energization.

Solid-state relay: A semiconductor switch having no moving parts. These devices control power to the load and are analogous to relay contacts.

Relays

The term *relay* defines a device that accepts a signal and sends it on down the line, just as a runner in a relay race passes a baton to the next runner.

The invention of the relay is credited to Samuel Morse, one of the inventors of the telegraph. Morse developed the relay to boost signal strength. An incoming signal activated an electromagnet, which closed a battery circuit, thereby transmitting the signal to the next relay.

A control relay is an electromagnetic device that is similar in operating characteristics to a contactor. The contactor, however, is generally employed to switch power circuits or relatively high-current loads. Relays are normally used in control circuits, and consequently, their lower ratings (15A maximum at 600V) reflect the reduced current levels at which they operate.

Contactors generally have from one to five poles, with most applications using normally open contacts. There is little, if any, conversion of contact operation on the job. In contrast, relays are often used in applications requiring 10 or 12 poles per device, with various combinations of normally open and normally closed contacts. In addition, some relays have convertible contacts, permitting changes to be made in the field from N.O. to N.C. operation, or vice versa, without requiring modification kits or additional components.

Relays are used in the control circuits of magnetic starters, contactors, solenoids, timers, and other devices. Relays are generally used to amplify the contact capability or multiply the switching functions of a pilot device. A tremendous variety of relays are used for lighting, motor, and HVAC control circuits. The magnetic relay is still common, although solid-state control devices are rapidly finding their way into all types of applications.

Besides controlling electrical circuits, some types of relays are also used for circuit and equipment protection. For example, thermal overload relays are intended to protect motors, controllers, and branch circuit conductors against excessive heating due to prolonged motor overcurrents up to and including locked rotor currents. These relays sense motor current by converting this current to heat in a resistance element. The heat generated is used to open a set of contacts in series with a starter coil, causing the motor to be disconnected from the line.

There are also a great many electrical protective relays installed in modern power systems, including:

- Overcurrent (time delay and **instantaneous**)
- Over-/under-voltage
- Directional overcurrent
- Percentage differential

The operating characteristics of these protective relays are all similar and are covered in this module. However, the sensing, adjustment, and actuating functions are quite different and are beyond the scope of this module. For specific instructions, always consult the manufacturer's literature.

2.1.0 Control Relays

A control relay is an electromagnetic device whose contacts are used in the control circuits of magnetic starters, solenoids, timers, and other relays. The wiring diagrams in *Figure 7* and *Figure 8* demonstrate how a relay amplifies contact capacity. *Figure 7* represents a cur-

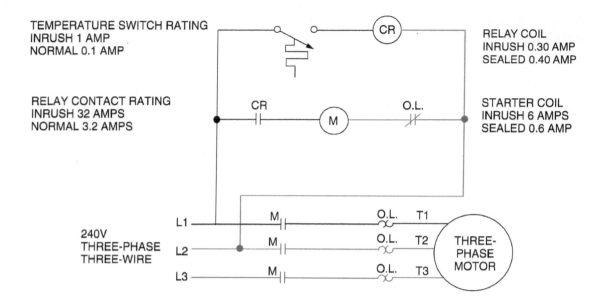

Figure 7 Relay used to amplify contact capacity.

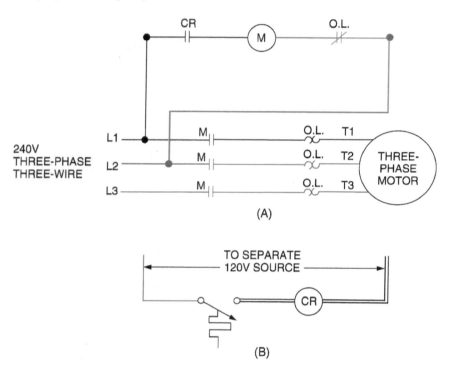

Figure 8 Relay used for voltage amplification.

rent amplification. The relay and starter coil voltages are the same, but the ampere rating of the temperature switch is too low to handle the current drawn by the starter coil (M). A relay is interposed between the temperature switch and the starter coil. The current drawn by the relay coil (CR) is within the rating of the temperature switch, and the relay contact (CR) has a rating that is adequate for the current drawn by the starter coil.

Figure 8 represents a voltage amplification. A condition may exist in which the voltage rating of the temperature switch is too low to permit its direct use in a starter control circuit operating at a higher voltage. In this application, the coil of the interposing relay and the pilot device are both wired to a low-voltage source of power that is compatible with the rating of the pilot device. The relay contact, with its higher voltage rating, is then used to control the operation of the starter.

Figure 9 represents another use of relays, which is to multiply the switching functions of a pilot device with a single or limited number of contacts. In this circuit, a single-pole pushbutton contact can, through the use of an interposing six-pole relay, control the operation of a number of different loads such as a pilot light, starter, contactor, solenoid, and timing relay.

Relays are commonly used in complex controllers to provide the logic required to set up and initiate the proper sequencing and control of a number of interrelated operations.

Relays differ in voltage ratings (up to 600V), number of contacts, contact convertibility, physical size, and attachments for accessory functions such as mechanical latching and timing.

In selecting a relay for a particular application, one of the first steps should be a determination of the control voltage at which the relay will operate. Once the voltage is known, the relays that have the necessary contact rating can be further reviewed and a selection made on the basis of the number of contacts and other characteristics required. *Figure 10* and *Figure 11* show an 8-pin plug-in relay with its base and wiring diagram.

2.2.0 Timers and Timing Relays

A pneumatic timer or timing relay is similar to a control relay, except that it contains contacts that are designed to operate at a specific time interval after the coil is energized or de-energized. A delay on energization is also referred to as an on delay. A delay on de-energization is called an off delay.

A timed function is useful in applications such as the lubricating system of a large industrial machine, in which a small oil pump must deliver lubricant to the bearings of the main motor for a set period of time before the main motor starts. Furnaces also incorporate fan on delays to avoid blowing cold air into the space when the system starts up.

In pneumatic timers, the timing is accomplished by the transfer of air through a restricted orifice. The amount of restriction is controlled by an adjustable needle valve, permitting changes to be made in the timing period.

Timing relays are often motor-driven by a clock-type mechanism, but solid-state timing circuits will also be encountered.

2.3.0 Solid-State Relays

Like a magnetic relay, a **solid-state relay** is also used in switching applications (*Figure 12*). Solid-state relays are available in different styles and ratings, including time-delay relays. One common use for a solid-state relay is the I/O track of a programmable controller, which is a controller that can be changed by reprogramming rather than requiring an extensive rewiring of the control system. This type of relay can be used to switch a 120V signal into a 5VDC signal.

Some of the advantages of solid-state relays include the following:

- No moving parts
- Resistant to shock and vibration
- Sealed against dirt and moisture
- Control input voltage isolated from relay-controlled circuit

> **NOTE:** Solid-state relays are much more sensitive to loads than equivalent mechanical devices. Always follow the manufacturer's application instructions.

The following are some typical solid-state relay control circuits (*Figure 13*):

- *Power transistors for DC loads* – In this DC circuit, a power transistor is used to connect the load to the line, as shown in *Figure 13 (A)*. This is an opto-isolated relay with a light-emitting diode (LED) connected to the input or control voltage. This is a common circuit arrangement in solid-state relays, and relays that use this coupling method are referred to

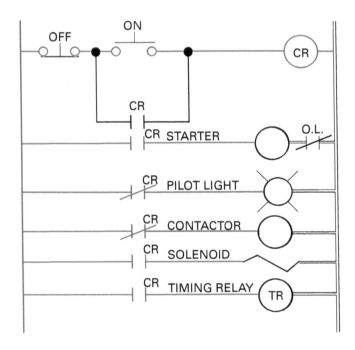

Figure 9 Relay used to multiply the switching functions of a pilot device.

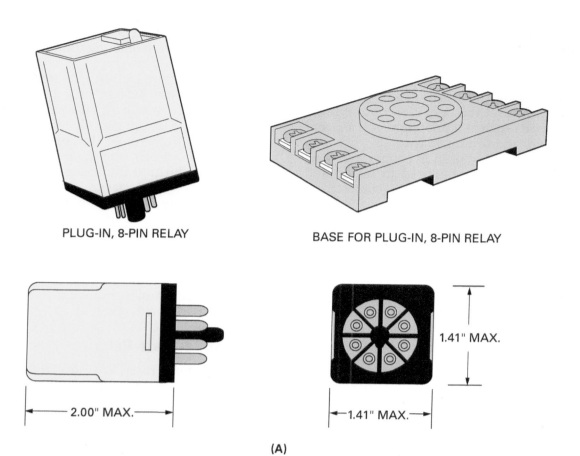

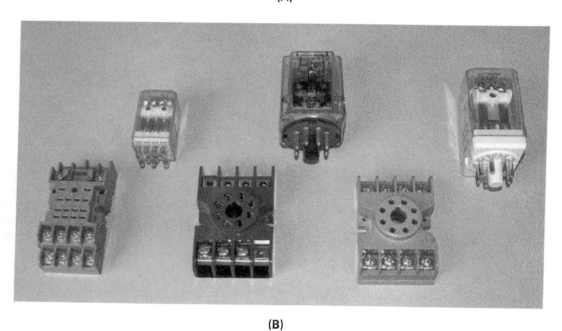

Figure 10 Plug-in relays.

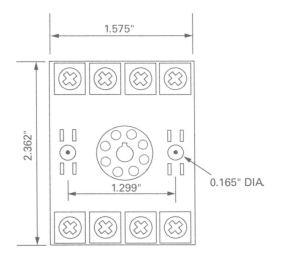
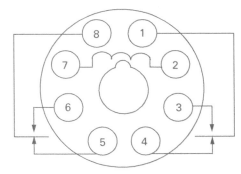

DOUBLE-POLE, DOUBLE-THROW

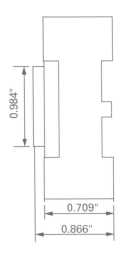

SCREW AND CAPTIVE WASHERS

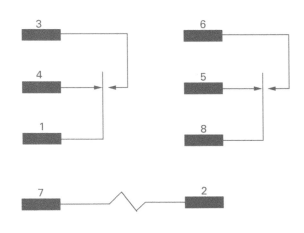

WIRING DIAGRAM

Figure 11 Wiring diagram for a plug-in relay.

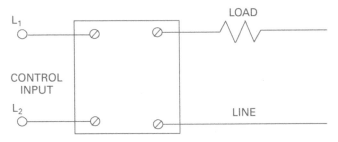

Figure 12 Basic solid-state relay.

as opto-isolated because the load side of the relay is optically isolated from the control side of the relay. Since a light beam (rather than electricity) is used to control the relay, no voltage spikes or electrical noise produced on the load side of the relay can be transmitted to the control side of the relay. When the LED is activated by the input voltage, a photo-detector connected to the base of the transistor turns the transistor on and connects the load to the line via this optical coupling.

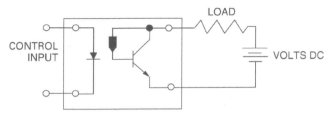

(A) POWER TRANSISTOR USED TO CONTROL A DC LOAD

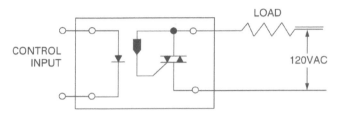

(B) TRIAC USED TO CONTROL AN AC LOAD

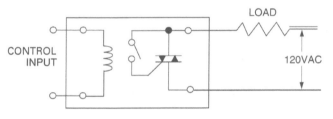

(C) REED RELAY CONTROLS THE OUTPUT

Figure 13 Solid-state relay circuits.

- *Triac-controlled AC relays* – In this AC circuit, a triac is connected to the load in the place of a power transistor. Refer to *Figure 13 (B)*. This circuit also uses an LED as the control device. When the photo-detector detects the LED, it triggers the gate of the triac and connects the load to the line.
- *Reed-controlled AC relays* – Another method of controlling solid-state relays in AC circuits is to use a small reed relay to control the output, as shown in *Figure 13 (C)*. In this circuit, small sets of reed contacts are connected to the gate of the triac and the control circuit is connected to the coil of the reed relay. When the control voltage sends a current through the coil, the magnetic field produced around the coil of the relay closes the reed contacts and activates the triac. Unlike opto-isolated relays, this type of solid-state relay uses a magnetic field to isolate the control circuit from the load circuit.

Plug-In Relays

Plug-in relays make troubleshooting and replacement of relays much easier. Instead of having to connect and disconnect wiring, a new relay can simply be snapped into place.

Solid-state relays normally have control voltages from 3V to 32V. Triac-controlled relays commonly have load voltage ratings of 120VAC to 240VAC, with current ratings between 5A and 25A.

Many solid-state relays incorporate a zero switching feature, which means that if the relay is programmed to turn off when the AC voltage is in the middle of a cycle, it will continue to conduct until the AC voltage returns to zero and turns off.

2.4.0 Overload Relays

To protect a motor and its related circuits from accidental or prolonged overloads, either the starter or the motor should be equipped with automatic devices that will open the circuit in the event of an overload. This protection may be provided by fuses, circuit breakers, or overload relays.

Overcurrent protection must be provided in the line of every motor circuit, but additional protection should be provided in the form of magnetic overload relays. These are used in both manual and automatic starters.

Timing Relays

Some of the simplest examples of timing relays are the plug-in devices that turn a household lamp on and off or the clocks that control a night-rate water heater. These relays are usually mechanical switches. A more sophisticated example of a timing relay is the relay that energizes half the windings of a split-wound, three-phase motor a microsecond after initially energizing the motor.

2.4.1 Bimetallic Overload Relays

A simplified bimetallic overload relay is shown in *Figure 14*. The load current passes through a heater element in the overload relay. The heater element is replaceable and contains a resistance. When the current exceeds a certain value, the heat radiated from the element causes a bimetallic strip to spring away, interrupting the control circuit or other mechanical action. The heater element is sized to the full load current of the motor. A small overload may have a long delay, while a larger overload will have a shorter delay.

Most overload relays used on modern motor starters are thermally operated and usually consist of two strips of metal that are welded together. Each has a different degree of thermal expansion. If this bimetallic strip is heated, it will deflect sufficiently to trip two normally closed contacts, which in turn will open the circuit connected to the holding coil of the magnetic contactor, causing the main contacts to open. An advantage of this protective device is that it provides a time delay, which prevents the circuit from being opened by momentary high starting currents and short overloads. At the same time, it protects the motor from prolonged overloading.

Conditions of motor overload may be caused by an overload on driven machinery, by a low line voltage, or by an open line in a polyphase system, which results in single-phase operation. Under any conditions of overload, a motor draws excessive current that causes overheating. Because the motor winding insulation will deteriorate when subjected to overheating, there are established limits on motor operating temperatures. To protect a motor from overheating, overload relays are employed on a motor control to limit the amount of current drawn. This is known as overload protection or running protection.

The ideal overload protection for a motor is an element with current-sensing properties that are very similar to the heating curve of the motor it protects and will act to open the motor circuit when the full-load current is exceeded. The operation of the protective device should be such that the motor is allowed to carry harmless overloads but is quickly removed from the line when an overload has persisted for too long.

2.4.2 Fuses vs. Overload Relays

Fuses are not the best choice to provide overload protection. Their basic function is to protect against short circuits (overcurrent protection). Motors draw a high inrush current when starting, and conventional single-element fuses have no way of distinguishing between this temporary and harmless inrush current and a damaging overload. Such fuses, if chosen on the basis of motor full-load current, would blow every time the motor started. On the other hand, if a fuse were chosen so that it was large enough to pass the starting or inrush current, it would not protect the motor against small, sustained overloads that might occur later.

Solid-State Relays

Several advantages of solid-state relays are longer life, higher reliability, higher speed switching, and high resistance to shock and vibration. The absence of mechanical contacts eliminates contact bounce, arcing when contacts open, and hazards from explosives and flammable gases.

Motor Protection

Some motors have a bimetal thermal device embedded in the stator winding. This device opens when the circuit overheats; when the device cools, the motor will resume operation. The electrician must determine whether the application of the motor warrants further overload protection.

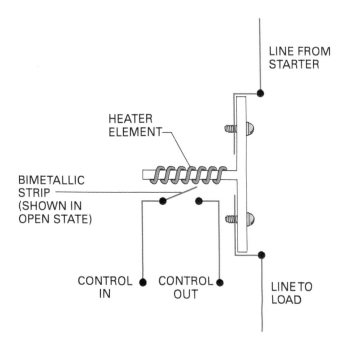

Figure 14 Simplified bimetallic overload relay.

Dual-element or time-delay fuses can provide motor overload protection but have the disadvantage of being nonrenewable and must be replaced once tripped.

The overload relay is the heart of motor protection. It has inverse trip time characteristics, permitting it to hold in during the accelerating period (when inrush current is drawn) yet providing protection on small overloads above the full-load current when the motor is running. Unlike dual-element fuses, overload relays are renewable and can withstand repeated trip and reset cycles without needing replacement. They cannot, however, take the place of overcurrent protective equipment.

2.4.3 Overload Relay Operation

An overload relay consists of a current-sensing unit connected in the line to the motor, plus a mechanism that is actuated by the sensing unit and serves to directly or indirectly break the circuit. In a manual starter, an overload trips a mechanical latch, causing the starter contacts to open and disconnect the motor from the line. In magnetic starters, an overload opens a set of contacts within the overload relay itself. These contacts are wired in series with the starter coil in the control circuit of the magnetic starter. Breaking the coil circuit causes the starter contacts to open, disconnecting the motor from the line.

2.4.4 Overload Relay Classifications

Overload relays can be classified as being either thermal or magnetic. Magnetic overload relays react only to current excesses and are not affected by temperature. As the name implies, thermal overload relays rely on the rising temperatures caused by the overload current to trip the overload mechanism. Thermal overload relays can be further subdivided into two types: melting alloy (eutectic) and bimetallic.

The melting alloy assembly of a heater element and solder pot are shown in *Figure 15*. The motor load current passes through the heater element. When it exceeds the designated capacity of the heater, it melts an alloy solder pot. The ratchet wheel will then be allowed to turn in the molten pool, and a tripping action of the starter control circuit results, stopping the motor. A cooling-off period is required to allow the solder pot to harden before the overload relay assembly may be reset and the motor service restored.

Melting alloy thermal units are interchangeable and of a one-piece construction, which ensures a constant relationship between the heater element and solder pot and allows for factory calibration, making them virtually tamperproof in the field. These important features are not possible with any other type of overload relay construction. Many types of interchangeable thermal units are available to provide precise overload protection from any full-load current.

Bimetallic overload relays have two major advantages: first, the automatic reset feature is desirable when devices are mounted in locations that are not easily accessible for manual operation; and, second, these relays can easily be adjusted to trip within a range of 85% to 115% of the nominal trip rating of the heater unit. This feature is useful when the recommended heater size might result in unnecessary tripping, while the next larger size would not give adequate protection.

The ambient temperature affects overload relays operating on the principle of heat. Ambient-compensated bimetallic overload relays were designed for one particular situation, that is, when the motor is at a constant temperature and the controller is located separately and has a varying temperature. In this case, if a standard thermal overload relay were used, it would not trip consistently at the same level of motor current if the controller temperature changed. A thermal overload relay is always affected by the surrounding temperature. To compensate for temperature variations, an ambient-compensated overload relay is applied. Its trip point is not affected by temperature and it performs consistently at the same value of current.

Melting alloy and bimetallic overload relays are designed to approximate the heat actually generated in the motor. As the motor temperature increases, so does the temperature of the thermal unit. Motor and relay heating curves show this relationship. From such graphs (obtainable from the manufacturers of overload relays), you can see that, no matter how high the current drawn, the overload relay will provide protection, yet the relay will not trip out unnecessarily.

2.4.5 Relay Selection

When selecting thermal overload relays, the following must be considered:

- Motor full-load current
- Type of motor
- Difference in ambient temperature between the motor and controller

Motors of the same horsepower and speed do not all have the same full-load current, and the motor nameplate must always be checked to

(A) FRONT

(B) BACK

Figure 15 Melting alloy thermal unit.

obtain the full-load amperes for a particular motor. Thermal unit selection tables are published based on continuous-duty motors operating under normal conditions with a 1.15 service factor. These tables are shown in manufacturers' catalogs and also appear on the inside of the door or cover of the motor controller. These values are selected to properly protect the motor and allow it to develop its full horsepower, allowing for the service factor if the ambient temperature is the same at the motor as at the controller. If the temperatures are not the same, or if the motor service factor is less than 1.15, a special procedure is required to select the proper thermal unit.

Standard overload relay contacts are closed under normal conditions and open when the relay trips. An alarm signal is sometimes required to indicate when a motor has stopped due to an overload trip. Also, with some machines, particularly those associated with continuous processing, it may be required to signal an overload condition, rather than have the motor and process stop automatically. This is done by fitting the overload relay with a set of contacts that close when the relay trips, thus completing the alarm circuit. These contacts are appropriately called alarm contacts.

A magnetic overload relay has a movable magnetic core inside a coil that carries the motor current. The flux set up inside the coil pulls the core upward. When the core rises far enough, it trips a set of contacts on the top of the relay. The movement of the core is slowed by a piston working in an oil-filled dashpot mounted below the coil. This produces an inverse-time characteristic. The effective tripping current is adjusted by moving the core on a threaded rod. The tripping time is varied by uncovering oil bypass holes in the piston. Because of the time and current adjustments, the magnetic overload relay is sometimes used to protect motors having long accelerating times or unusual duty cycles.

2.5.0 Troubleshooting Relays

Electromagnetic contactors and relays of various types are used to make and break circuits that control and protect the operation of such devices as electric motors, combustion and process controls, alarms, and annunciators. Successful operation of a relay is dependent on maintaining the proper interaction between its solenoid and its mechanical elements—springs, hinges, contacts, dashpots, and the like.

The operating coil is usually designed to operate the relay from 80% to 110% of rated voltage. At low voltage, the magnet may not be strong enough to pull in the armature against the armature spring and gravity, while at high voltages excessive coil temperatures may develop. Increasing the armature spring force or the magnet gap will require higher pull-in voltage or current values and will result in higher drop-out values.

When sufficient voltage is applied to the operating coil, the magnetic field builds up, and the armature is attracted and begins to close. The air gap is shortened, increasing the magnetic attraction and accelerating the closing action, so that the armature closes with a snap, closing normally open contacts and opening normally closed contacts. Binding at the hinges or excessive armature spring force may cause a contactor or relay that normally has snap action to make and break sluggishly. This condition is often encountered in relays that are rarely operated. Contactors and relays should be operated and observed from time to time to make sure their parts are working freely with proper clearances and spring actions.

Troubleshooting charts are provided in *Table 2A* and *Table 2B*. Knowing these symptoms and their causes should help you to quickly spot and repair contactor and relay problems.

Table 2A Contactor and Relay Troubleshooting Chart (1 of 2)

Symptoms	Probable Cause	Action or Items to Check
Failure to pull in	Either no voltage or low voltage at coil terminals	Blown fuse, open line switch, break in wiring Line voltage below normal Overload relay open or set too low Tripping toggle (non-automatic breakers) fouled Control level or start button in off position Pull-in circuit open, shorted, or grounded Contacts in protective or controlling circuit open or one of their pigtail connections broken
	Operating coil open or grounded	Inspect and test coil
	Loose or disconnected coil lead wire	Inspect and correct
	Excessive magnet gap, improper alignment	Inspect and correct
	Armature obstructed or gumming deposits between armature and pole face	Inspect and correct
	Binding caused by deformed or gummy hinge	Replace if bent; degrease if gummed up
	Excessive armature spring force	Lessen spring force
	Normally closed contacts welded together	Replace contacts
Failure of equipment to start with contactor closed	One contact not closing	Replace set of contacts
	Contacts burned	Replace
	Contact pigtail connection broken	Replace
Failure to drop out	Operating coil is energized	Contacts in controlling or protective tripping circuits closed, shorted, or shunted Tripping devices defective, such as overload tripping toggles do not strike release, undervoltage relay plunger stuck or out of adjustment, defective stop button, or defective time-delay escape mechanism (closed air vent) Current supplied over an unintentional path due to grounds, defective insulation, pencil markings, moisture, or lacquer chipped off relay base
	Residual magnetism excessive due to armature closed tightly against pole face	Adjust or replace
	Armature obstructed or gumming deposits between armature and pole face	Clear and clean
	Binding caused by deformed or gummy hinge	Replace or degrease
	Contact pressure spring or armature spring too weak or improperly adjusted	Replace or adjust
	Improper mounting position (upside down)	Mount correctly
	Normally open contacts welded together	Replace contacts

Table 2B Contactor and Relay Troubleshooting Chart (2 of 2)

Symptoms	Probable Cause	Action or Items to Check
Time-delay relays operate too fast	Air escapes too freely due to holes in bellows or open air vent	Escape mechanism faulty
	Non-magnetic shim in air gap too large or armature spring too strong	Worn dashpot plunger, dashpot oil too thin Adjust or replace
	Magnets out of adjustment	Adjust
Contacts pitted or discolored	Contacts overheated from overload	Reduce load and replace contacts
	Contacts not fitted properly	Refit
	Barriers broken from rough usage or breakers closing with too much force	Replace barriers; check for high voltage
	Wiping action of contacts on closing is insufficient	Adjust
	Excessive chatter or hum	Check causes listed below
	Exposure to weather, dripping water, salt air, or vibration	Use suitable NEMA enclosure
Excessive chatter or hum	Vibration of surrounding devices communicated to relay	Arrange differently
	Relay receiving contradictory signals	Correct
	Bouncing of controlling or protective contacts	Adjust
	Line voltage fluctuating or insufficient	Consider using buck-and-boost transformers
	Excessive coil circuit resistance	Lessen resistance
	Armature spring or contact pressure spring too strong	Lessen
	Excessive coil voltage drop on closing	Adjust
	Missing shading coil	Replace the contactor
	Improper antifreeze pin location	Locate in correct position
	Free movement of armature hindered due to deformed parts or dirt	Clean and replace appropriate parts
Excessive coil temperature	Excessive current or voltage	Reduce load or adjust taps on transformer
	Short circuit in coil	Replace coil

2.0.0 Section Review

1. A single-contact device can be used to control multiple loads using a _____.

    a. single-pole relay
    b. current amplifier relay
    c. voltage amplifier relay
    d. multiple-pole relay

2. A furnace might use an on delay to _____.

    a. save energy during periods when the space is unoccupied
    b. avoid blowing cold air into the space when the system starts up
    c. prevent the system from cycling between heating and cooling
    d. prevent thermostat tampering

3. A type of relay controlled by a magnetic field is a(n) _____.

    a. opto-isolated relay
    b. power transistor
    c. triac
    d. reed relay

4. A type of renewable overload device is a _____.

    a. time-delay fuse
    b. dual-element fuse
    c. melting alloy relay
    d. fusible link

5. The operating coil of a relay is designed to work properly at _____.

    a. 80% to 110% of the rated voltage
    b. 90% to 100% of the rated voltage
    c. 95% to 105% of the rated voltage
    d. 100% of the rated voltage only

Section Three

3.0.0 Low-Voltage Remote Control Switching Systems

Objective

Install low-voltage remote control switching systems.
a. Identify remote control switching system components and operating characteristics.
b. Plan and install a remote control switching system.

Performance Task

1. Mount and connect a 120V lighting contactor with a three-wire pushbutton control.

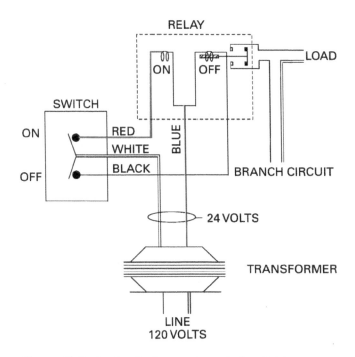

Figure 16 Basic circuit of a remote control switching system.

In applications where lighting must be controlled from several points, where there is a complexity of lighting or power circuits, or where flexibility is desirable in certain systems, low-voltage remote control relay systems have been applied. Basically, these systems use special low-voltage components operated from a transformer to switch relays, which in turn control the standard line voltage circuits. Because the control wiring does not carry the load current directly, small, lightweight cable can be used. It can be installed wherever and however convenient—placed behind moldings, stapled to woodwork, buried in shallow plaster channels, or installed in holes bored in wall studs. A basic circuit of a remote control switching system is shown in *Figure 16*. In both circuits, the relay permits positive control for on and off. It can be located near the load or installed in centrally located distribution panel boxes, depending on the application.

Because no line current flows through the control circuits and low voltage is used for all switch and relay wiring, it is possible to place the controls at a great distance from the source or load, thus offering many advantages.

Since branch circuits go directly to the loads in this type of system, no line voltage switch legs are required. This saves costly runs of larger conductor through all switches and saves installation time and costs if multipoint switches are used.

Under certain conditions, the *NEC*® allows low-voltage conductors to be run in open spaces above ceilings and through wall partitions without further protection. An outlet box is not required at the switch locations. When the electricians rough-in the wiring, they merely secure a plaster ring of the correct depth at each switch location and usually wrap the low-voltage cable around a nail driven behind the plate. The switch and its cover are then installed after the wall is finished.

Step-Down Transformers

Step-down transformers like the one shown here can provide low voltage for remote control switching circuits.

Low-voltage switching is especially useful in rewiring existing buildings, because the small cables are as easy to run as telephone wires. They are easy to hide behind baseboards or even behind quarter-round molding. The cable can be run exposed without being very noticeable because of its small size. The small, flexible wires are easily fished in partitions.

Figure 17 shows several types of remote control circuits. In the circuits shown, any number of on-off switches can be connected to provide control from many remote points. In a typical installation, outside lights could be controlled from any area within the building without the need to run full-size cable for three- or four-way switch legs through the building. By running a low-voltage line from each relay to a central point, such as an exit door, all relays can be operated from one spot using a selector switch; this allows quick, convenient control of all exterior and interior lighting and power circuits.

A motor-driven rotary switch is available that allows control of 25 (or more) separate circuits with the push of one button. Dimming can also be accomplished by using the motorized control unit together with a modular incandescent dimming system.

3.1.0 System Components and Operating Characteristics

The relay is the heart of a remote control switching system. It employs a split low-voltage coil to move the line voltage contact armature to the On (latched) position. As illustrated in *Figure 18*, the On coil moves the armature to the right when a 24VAC control signal is impressed across its leads. This is similar to a magnet attracting the

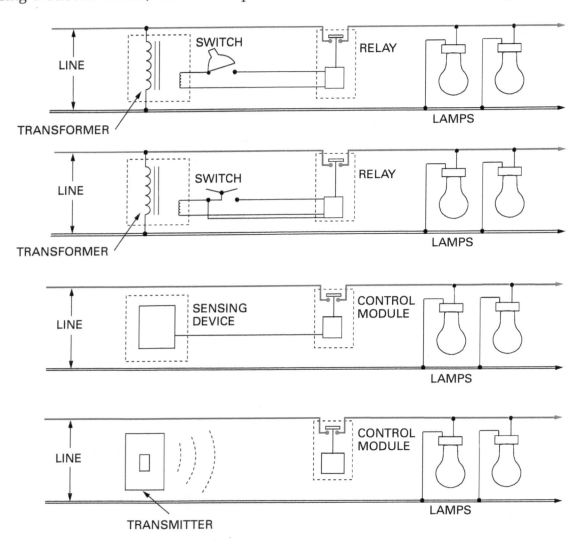

Figure 17 Several types of remote control circuits and related components.

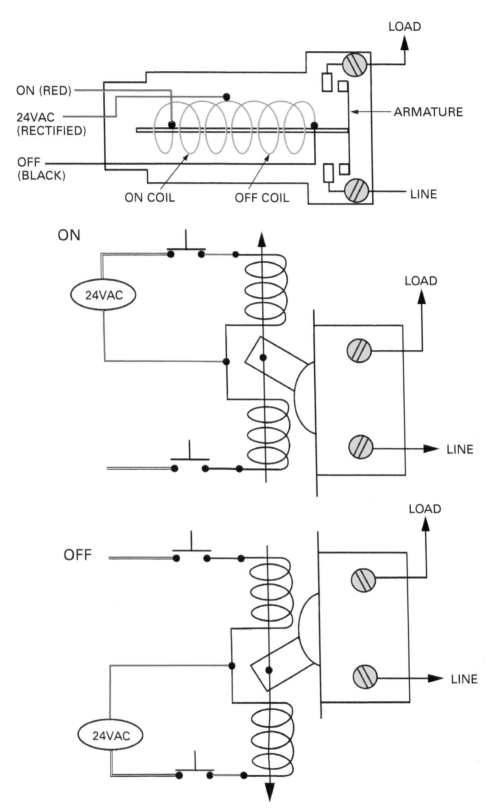

Figure 18 Relay operation.

handle of a standard single-pole switch to the On position when energized. The armature (handle) latches in the On position and will remain there until the Off coil is energized, drawing the armature into the Off position. *Figure 19* shows how a low-voltage relay is installed in an outlet box.

3.1.1 Power Supplies

Transformers designed for use on remote control switching systems supply 24VAC power. This is rectified to 24VDC to operate the relays and pilot lights. The AC input voltage to the power supply can vary by manufacturer, with the more common voltages being 120V, 208V, 240V, and 277V. The type of voltage used depends on the system. *Figure 20* shows a control transformer along with connection details.

3.1.2 Switches

The standard low-voltage switch uses a rocker or two-button configuration to provide a momentary single-pole, double-throw action. Pushing the On button completes the circuit to the On coil of the relay, shifting the contact armature to the corresponding position. When the button is released, the relay remains in that position.

The pulse operation allows any number of switches to be wired in parallel, as shown in *Figure 21*. A group of relays could also be wired for common switch control by paralleling their control leads. Those relays would operate as a group.

Pilot switches include a lamp wired between the switch common (white) and a pilot terminal. The auxiliary contact in the relay provides power to drive this lamp when the relay is energized.

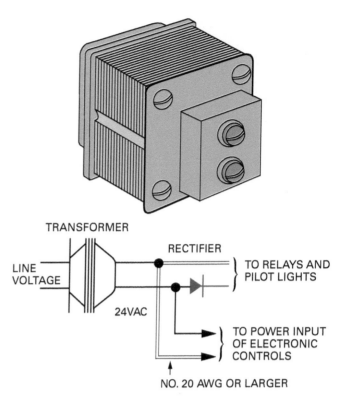

Figure 20 Control transformer with power-supply wiring diagram.

3.1.3 Master Sequencer

The master sequencer allows relays to be controlled as a group while still allowing individual switch control for each. When the master switch is turned On, the sequencer pulses each of its On relay outputs sequentially. A local switch can control an individual relay without affecting any other.

A second input channel allows time clocks, building automation system outputs, photocells, or other maintained contact devices to also control the sequencer. This provides simple automation coupled with local override of individual loads (*Figure 22*).

3.1.4 Operating Characteristics

Figure 23 shows a single-switch control of a light or a group of lights in one area. This is the basic circuit of relay control and is similar to single-pole conventional switch circuits. The schematic wiring diagram of this circuit is shown in *Figure 24*.

The addition of a time clock and/or photoelectric cell allows applications in which any number of control circuits are turned on or off at desired intervals (e.g., the dusk-to-dawn control of outdoor lighting circuits).

Multipoint switching, or the switching of a single circuit from two or more switches, is shown in *Figure 25*, along with its related wiring diagram.

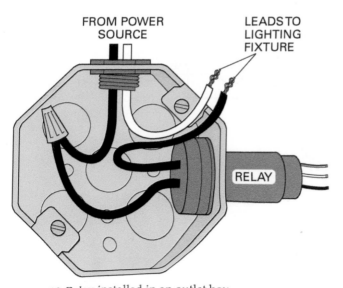

Figure 19 Relay installed in an outlet box.

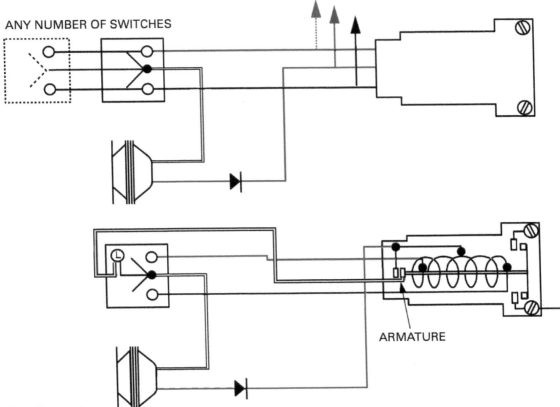

Figure 21 Switch operation.

In conventional wiring, multipoint switching requires costly three-way and four-way switches, plus the extension of switch legs and traveler wires through all switches. One of the greatest advantages of relay switching is the low cost of adding switch points.

Figure 26 shows many lighting circuits being controlled from one location. In conventional line voltage switches, this is accomplished by ganging the individual switches together. The same procedure can be employed with remote control switches. When more than six switches are required, it is usually desirable to install a master-selector switch that permits the control of individual circuits.

Master control of many individual and isolated circuits is possible. Where more than 12 circuits are to be controlled from a single location and the selection of individual circuits is not required, a motor master control unit automatically sweeps 25 circuits from either On or Off at the touch of a single master switch, which can be placed at one or more locations.

3.2.0 Planning a Remote Control Switching System

Low-voltage lighting control systems can be installed in both residential and commercial installations. The systems installed in each type of location may differ. The manufacturer's instructions and recommendations should always be consulted and adhered to, as should the blueprint specifications. The first step in the installation procedure, whether the building is residential or commercial, is to develop a plan. When laying out a remote control switching system for a building, drawing symbols such as the ones shown in *Figure 27* are normally used to identify the various components. The drawing may be simple, requiring only the following:

- The location of each low-voltage control with a dashed line drawn from the switch to the outlet to be controlled
- A schematic wiring diagram showing all components

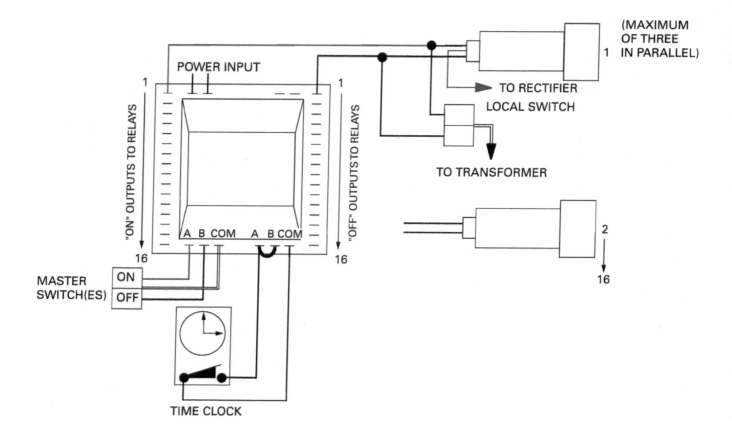

Figure 22 Master sequencer connections.

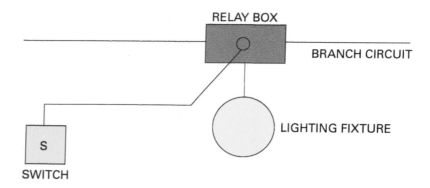

Figure 23 Single-switch control.

- The following note:

  Furnish and install complete remote control wiring system for control of lighting and other equipment as indicated on the drawings, diagrams, and schedules. System shall be complete with transformers, rectifiers, relays, switches, master-selector switches, wallplates, and wiring. All remote control wiring components shall be of the same type and installed in accordance with the recommendations of the manufacturer. Except where otherwise indicated, all remote control wiring shall be installed in accordance with NEC Article 725, Class 2.

Many times, the low-voltage lighting control system will be designed by architects, especially in new installations. However, in existing buildings, the electrician may be expected to design and install the system. If this situation occurs, work from the set of job blueprints, and use the following general procedure:

Step 1 Assign each light or receptacle a specific relay. This can be accomplished by using the letters R1 for relay 1, R2 for relay 2, and so on.

Step 2 Group together all the lights and outlets that are to be switched together.

Step 3 Determine the number of relays that can be controlled by each control module.

Step 4 Divide the number of relays by the number computed in Step 3. This is the number of control modules needed to control the relays.

Step 5 Determine the location of the switch points on the building blueprints.

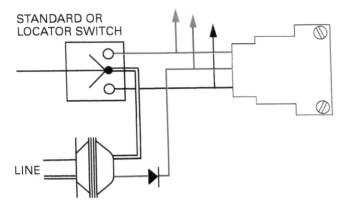

Figure 24 Schematic wiring diagram of single switches controlling each circuit.

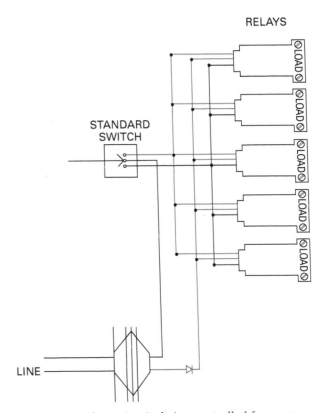

Figure 26 Switching circuits being controlled from one location.

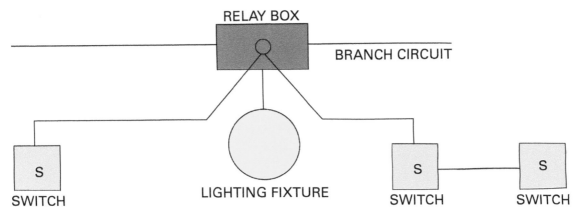

Figure 25 Diagram of multipoint switching of a single circuit.

Symbol	Description	Symbol	Description
◇	THE DIAMOND IS A WIDELY RECOGNIZED SYMBOL THAT IS USED FOR DENOTING LOW VOLTAGE.		
— - - —	Remote control low-voltage wire	$\langle S_T \rangle$	Low-voltage switch for interchangeable-type bracket
$\langle T \rangle$	Low-voltage transformer	$\langle S_X \rangle$	Extension switch receptacle
◇▶◇	Rectifier or rectifier assembly	$\langle MS \rangle_8$	Master-selector pushbutton switch; 8 position
$\langle C \rangle_A$	Components cabinet; A refers to first cabinet	$\langle MS_P \rangle_8$	Master-selector pushbutton switch with pilot lights; 8 position
$\langle R \rangle_1$	Relay, standard type, spec. grade; 1 refers to number of relay	$\langle MM_R \rangle$	Motor master; red for ON
$\langle R_P \rangle_2$	Relay with pilot switch, spec. grade; 2 refers to number of relay	$\langle MM_B \rangle$	Motor master; black for OFF
$\langle S \rangle$	Low-voltage switch, spec. grade, standard type	$\langle P_L \rangle$	Separate pilot light
$\langle S_P \rangle$	Low-voltage switch, spec. grade, with pilot light	**NOTE:** At controlled receptacles (ceiling or wall) use a standard relay (e.g., RA6, etc. which designates component cabinet A, relay #6).	

Figure 27 Recommended symbols for low-voltage drawings.

Step 6 Indicate what lights and receptacles are being controlled by what relay. This step may be accomplished by indicating the group of lights on the blueprint. However, it is easier if a switch schedule is used.

Step 7 Review the electrical plans or schedule to determine the number and size of the switching stations.

Figure 28 shows the floor plan of a building. Symbols are used to show the location of all lighting fixtures just as they are used to show the location of conventional switches. Line voltage circuits feeding the lighting outlets are indicated in the same way (by solid lines) that they are in conventional switching. All of the remote control switches, however, are indicated by the symbol S_L, instead of S, S_3, S_4, etc. Lines from these remote control switches to the lighting fixtures that they control are shown by dashed lines instead of the conventional solid line. Notice that the door switches controlling the closet lights use conventional switches since low-voltage switches of this type are uncommon.

The schematic wiring diagram in *Figure 29* shows further details of all components and the related wiring connections. This wiring diagram aids the contractor or electrician in the installation of the system and leaves little doubt as to exactly what is required.

A brief note or specification, such as described earlier, completes the design of the low-voltage remote control system. Once the layout plan has been developed and materials are acquired, the installation of the system can begin. The rough-in instructions presented here are only an example; always follow the manufacturer's recommendations and job specifications for specific requirements.

3.2.1 Enclosures

The enclosure should be placed in a location that is easily accessible but hidden from sight. If at all possible, install a unit that is prewired. This will decrease installation time and also limit the possibility of mistakes. Note that even though the panel may be prewired, the supply cables will still have to be installed. This should be done according to the manufacturer's recommendations. Since heat will be produced by the enclosed control modules, vents are placed around the enclosure to facilitate cooling. The location chosen for the enclosure should not hinder the cooling capabilities for which the panels are designed. The manufacturer may require that the high-voltage portion of the enclosure panel be mounted in the upper right-hand corner. This is due to the *NEC®* requirement that is designed to keep low-voltage conductors separated from high-voltage conductors. The panel itself should be mounted at eye level between two studs.

The power supply should be mounted on the enclosure and the power connections made. The number of power supplies needed per system will depend on the size of the system and the capabilities of the power supply. Some power supplies are capable of supplying up to 10 modules if the average number of LEDs per channel does not exceed 2 or 2.5. Be sure to check the manufacturer's specifications to ensure that the power supply will not be overloaded.

3.2.2 Switch Points

The first step when installing the switch points is to mark their location. As with ordinary switches, the switch points should be located as specified by the plans or specifications. Once the location has been determined, mount either a plaster ring or a nonmetallic box. A box is not needed in most frame construction installations.

A plaster ring or Romex® staple should then be positioned partway in the stud, behind the plaster ring or nonmetallic box. The preparation for running the low-voltage wiring is the same as for running 120V wiring. Per *NEC Section 300.4(A)*, if the edge of the hole is closer than 1¼" (32 mm) from the nearest edge of the stud, a 1/16" (1.6 mm) steel nail plate or bushing must be used to protect the cable from future damage by nails or screws.

The wires are then pulled from the switch station to the panel enclosure. They should come from the left side if the manufacturer recommends that the enclosure be installed with the high-voltage section in the upper right-hand corner. There should be approximately 10" to 12" (250 mm to 300 mm) of extra wire left at each switch location and 12" to 18" (300 mm to 450 mm) left at the panel enclosure.

Multiconductor cable is frequently used due to its simpler installation. Run one wire (low-voltage lead) from the enclosure to the farthest switching point and anchor the wire to the plaster ring. This wire will supply the low-voltage power to each of the switches. Working back toward the enclosure, loop the wire down and anchor it to each switch location. Take care to leave a sufficient amount of wire for connecting the switch. The low-voltage lead is then connected to the appropriate power supply lead at the enclosure. The process is repeated for the DC common lead. When this wiring method is used, the DC positive and common lead will have the same color throughout the system and future connection mistakes will be minimized.

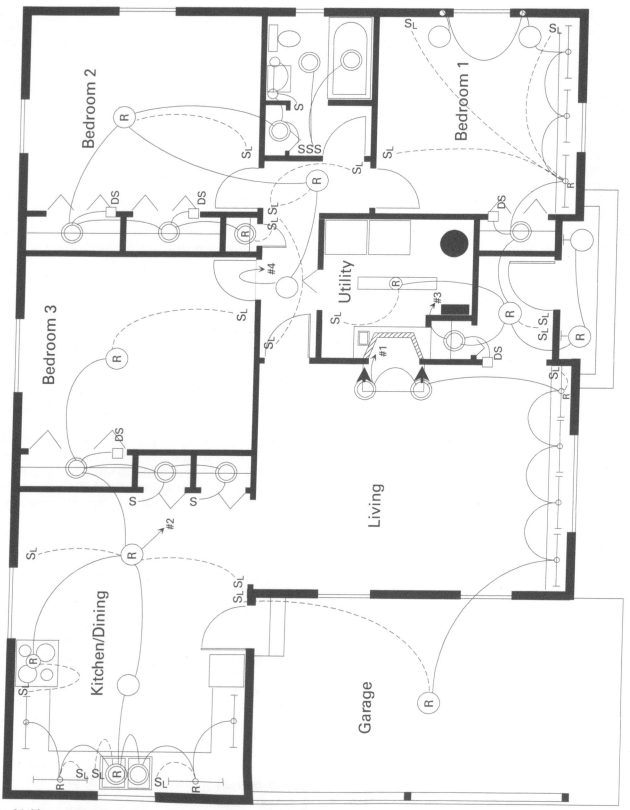

Figure 28 Floor plan of a residence using low-voltage switching.

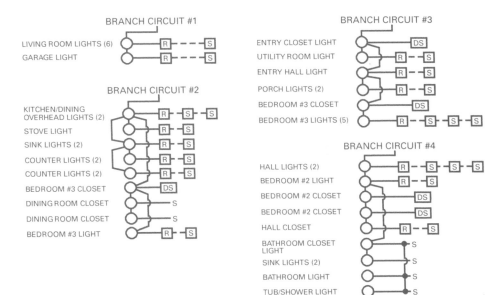

Figure 29 Schematic wiring diagram of lighting circuits.

The next step is to select a multiple-switch circuit. Pull a different colored wire from the enclosure to the farthest switching location in this circuit. Working back toward the panel enclosure as before, loop the wire to the other switch locations in the multiple-switch circuit. Thus, there are two leads going to each switch with one of the leads common to all the switches in the system. However, if the switches contain a pilot light, two additional No. 18 wires must be run to the LED. The wires are connected to the appropriate pilot light transformer located in the panel enclosure.

If possible, use different colored wires for each run. This will limit confusion in the future when connecting the individual circuits to power. Wire schedules should be used to limit hookup mistakes and provide a clearer picture of the system.

Remember that the colors for the control wiring should not be the same as the colors for the DC power and common conductors. Additional marking methods should also be used for future reference. Note that if the wires are marked using adhesive tabs or punch cards, they should be placed in a plastic bag and shoved behind the ring or securely into the plastic box. This will prevent the markings from being covered by paint or plaster when the walls are finished.

The low-voltage wiring should also be kept apart from the high-voltage wiring whenever possible. If the cables must be placed in parallel runs, they should be separated by at least 6" (150 mm), otherwise the noise generated from the high-voltage conductors could seriously interfere with the operation of the low-voltage system.

Power and finish trim – Connecting the circuits to power is now a fairly simple task. Each wire from the switch is connected to a relay with the other side of the relay connected to the common of the DC supply.

When the walls of the building are finished, the switches are connected to the conductors located at each switch point location. Wall plates are then added. A common type of wall plate used with low-voltage lighting control systems requires no screws because it can be snapped into place.

Think About It
Putting It All Together

Imagine you are designing your dream home. What systems would you like to be able to control from a central location? What types of devices would be required for this system?

3.0.0 Section Review

1. A device that allows relays to be controlled as a group while still allowing individual switch control is a _____.
   a. pilot switch
   b. master sequencer
   c. three-way switch
   d. rectifier

2. Where low-voltage cable must be run parallel with high-voltage cable, it should be separated by a minimum of _____.
   a. 4" (100 mm)
   b. 6" (150 mm)
   c. 8" (200 mm)
   d. 12" (300 mm)

Review Questions

1. The function of a shading coil in a contactor is to ____.
   a. keep ultraviolet light off the contacts
   b. prevent contact chatter and hum
   c. act as a holding circuit to keep the contactor energized when power is removed
   d. help the contacts return to their normal position when the contactor becomes de-energized

2. What is the major difference between a contactor and a motor starter?
   a. Contactors are always larger than motor starters for the same current-carrying capacity.
   b. All contactors are suitable for full-load lighting loads; starters are not.
   c. A contactor always has a disconnect while a starter does not.
   d. The starter provides overload protection while a contactor does not.

3. The approximate inrush current of filament-type lamps is how many times the normal operating current?
   a. 15 to 17
   b. 20 to 30
   c. 30 to 40
   d. 50 to 60

4. The maximum tungsten/infrared load amperage for a NEMA Size 2 magnetic contactor at 230V is ____.
   a. 11A
   b. 15A
   c. 24A
   d. 30A

5. The maximum tungsten/infrared load amperage for a NEMA Size 00 magnetic contactor at 230V is ____.
   a. 5A
   b. 9A
   c. 11A
   d. 18A

6. Which of the following *best* describes a control relay?
   a. A device designed to handle relatively high currents
   b. A fuse with two attachment points
   c. An electromagnetic device used in control circuits
   d. A fuse with two barrels

7. Which of the following *best* describes a relay that is used for motor protection?
   a. Amplifying relay
   b. Thermal overload relay
   c. Solenoid
   d. Contactor

8. The normal maximum amperage rating of control relays at 600V is ____.
   a. 15A
   b. 20A
   c. 30A
   d. 65A

9. The highest voltage rating for a magnetic relay is ____.
   a. 240V
   b. 277V
   c. 480V
   d. 600V

10. When selecting overload relays for motors, which of the following *must* be considered?
    a. The motor manufacturer
    b. The cost of the motor
    c. The length of time the motor has been in use
    d. The motor full-load current

11. Transformers for remote control switching systems supply ____.
    a. 24V
    b. 60V
    c. 120V
    d. 208V

12. On electrical drawings, the symbol used for denoting low-voltage devices is a _____.
    a. square
    b. circle
    c. triangle
    d. diamond

13. The letters RP appear in a diamond shape on a schematic diagram. This denotes _____.
    a. relay power
    b. rectifier power
    c. relay with pilot
    d. receptacle power

14. The enclosure for a low-voltage switching system in a residential installation should be _____.
    a. readily accessible and in plain sight
    b. readily accessible but hidden from sight
    c. concealed behind drywall
    d. in an attic or crawl space

15. When color coding low-voltage remote control switching systems, _____.
    a. use the same color for the power and control wiring
    b. use the same color for the common conductors and control wiring
    c. use the same color for each run
    d. use the same color for the DC positive and common lead throughout the system

Supplemental Exercises

1. What does the word *normally* refer to in regard to relay contacts? 1.0.0
   a. The placement of the contacts when the circuit is energized
   b. The placement of the contacts when the circuit is de-energized
   c. The way the contact is most of the time
   d. The way the contact is when it operates
2. Because filament-type lamps have large inrush currents, standard motor control contactors must be __derated__ if used to control this type of load. 1.1.0
3. If a solid-state relay is used to control a DC circuit, a(n) __power transistor__ is used to connect the load to the line. 2.3.0
4. Solid-state AC relays can be controlled by triacs or a(n) __reed relay__. 2.3.0
5. A(n) __shading coil__ is used to provide a continuous magnetic flow to the armature when the voltage of the AC waveform is zero. 1.2.0
6. Mechanically held relays and contactors use two coils and a(n) __latching__ mechanism to provide for quiet operation of the circuit. 1.3.0
7. True or False? Most overload relays used on modern motor starters are thermally operated. 2.4.1
8. If the equipment fails to start with the contactor closed, probable causes would be one contact not closing, burned contacts, or a broken __contact pigtail conn.__. 2.5.0
9. Unlike contactors, which are generally used to switch power circuits on relatively high-current loads, relays are normally used in low-current __control__ circuits. 2.0.0
10. Per *NEC Section 300.4(A)*, if the edge of the hole is closer than 1¼" (32 mm) from the nearest edge of the stud, the cable must be protected by a __steel plate__ or __bushing__. 3.2.2

Trade Terms Introduced in This Module

Ambient temperature: The temperature of the medium (usually air) surrounding a device and into which the heat losses in the device are dissipated.

Break: The opening of closed contacts to interrupt an electrical circuit.

Instantaneous: Contacts that make or break a circuit immediately upon coil energization or de-energization.

Latch coil: The coil of a latching relay that, when energized, causes the relay to assume the switched or latched position.

Make: The closing of open contacts to establish an electrical circuit.

Mechanically held relay: A latching relay in which the switched or latched position is maintained by the interference of a mechanical part that moves to a non-interfering position when the unlatch coil is energized, allowing the relay to return to the normal or unlatched position.

Normally closed (N.C.) contacts: A combination of a stationary contact and a movable contact, which are engaged when the coil is de-energized; in the case of a latching relay, the contacts that are engaged after the unlatch coil is energized.

Normally open (N.O.) contacts: A combination of a stationary contact and a movable contact, which are disengaged when the coil is de-energized; in the case of a latching relay, the contacts that are disengaged after the unlatch coil is energized.

Shading coil: A shorted turn surrounding a portion of the pole of an AC electromagnet, producing by mutual inductance a delay in the change of the magnetic field in that part, thereby tending to prevent chatter and reduce hum.

Solid-state relay: A semiconductor switch having no moving parts. These devices control power to the load and are analogous to relay contacts.

Unlatch coil: The coil of a latching relay that, when energized, causes the relay to assume the normal or unlatched position.

Additional Resources

This module presents thorough resources for task training. The following reference material is suggested for further study.

Lessons in Electric Circuits, Volume IV (free online textbook). **www.allaboutcircuits.com**.
National Electrical Code® Handbook, Latest Edition. Quincy, MA: National Fire Protection Association.

Figure Credits

Greenlee / A Textron Company, Module Opener
National Electrical Manufacturers Association, Tables 1A, 1B
Tim Dean, Figure 15

Section Review Answer Key

Section 1.0.0

Answer	Section Reference	Objective
1. b	1.1.0; Table 1	1a
2. a	1.2.0	1b
3. d	1.3.0	1c

Section 2.0.0

Answer	Section Reference	Objective
1. d	2.1.0	2a
2. b	2.2.0	2b
3. d	2.3.0	2c
4. c	2.4.4	2d
5. a	2.5.0	2e

Section 3.0.0

Answer	Section Reference	Objective
1. b	3.1.3	3a
2. b	3.2.2	3b

NCCER CURRICULA — USER UPDATE

NCCER makes every effort to keep its textbooks up-to-date and free of technical errors. We appreciate your help in this process. If you find an error, a typographical mistake, or an inaccuracy in NCCER's curricula, please fill out this form (or a photocopy), or complete the online form at **www.nccer.org/olf**. Be sure to include the exact module ID number, page number, a detailed description, and your recommended correction. Your input will be brought to the attention of the Authoring Team. Thank you for your assistance.

Instructors – If you have an idea for improving this textbook, or have found that additional materials were necessary to teach this module effectively, please let us know so that we may present your suggestions to the Authoring Team.

NCCER Product Development and Revision
13614 Progress Blvd., Alachua, FL 32615

Email: curriculum@nccer.org
Online: www.nccer.org/olf

❏ Trainee Guide ❏ Lesson Plans ❏ Exam ❏ PowerPoints Other _____

Craft / Level: _____ Copyright Date: _____

Module ID Number / Title: _____

Section Number(s): _____

Description: _____

_____

_____

_____

Recommended Correction: _____

_____

_____

_____

_____

Your Name: _____

Address: _____

_____

Email: _____ Phone: _____

This page is intentionally left blank.

Glossary

AL-CU: An abbreviation for aluminum and copper, commonly marked on terminals, lugs, and other electrical connectors to indicate that the device is suitable for use with either aluminum conductors or copper conductors.

Ambient temperature: The temperature of the medium (usually air) surrounding a device and into which the heat losses in the device are dissipated.

Amperage capacity: The maximum amount of current that a lug can safely handle at its rated voltage.

Approximate ram travel: The distance the ram of a hydraulic bender travels to make a bend. To simplify and speed bending operations, many benders are equipped with a scale that shows ram travel. Using a simple table (supplied with many benders), the degree of bend can easily be converted to inches of ram travel. This measurement, however, can only be approximated because of the variation in springback of the conduit being bent.

Armature: The rotating windings of a DC motor.

Auxiliary electrodes: Metallic electrodes pushed or driven into the earth to provide electrical contact for the purpose of performing measurements on grounding electrodes or ground grid systems.

Back-to-back bend: Any bend formed by two 90° bends with a straight section of conduit between the bends.

Ballast: A circuit component in fluorescent and high-intensity discharge (HID) luminaires that provides the required voltage surge at startup and then controls the subsequent flow of current through the lamp during operation.

Barrier strip: A metal strip constructed to divide a section of cable tray so that certain kinds of cable may be separated from each other.

Basket grip: A flexible steel mesh grip that is used on the ends of cable and conductors for attaching the pulling rope. The more force exerted on the pull, the tighter the grip wraps around the cable.

Bending protractor: Made for use with benders mounted on a bending table and used to measure degrees; also has a scale for 18, 20, 21, and 22 shots when using it to make a large sweep bend.

Bending shots: The number of shots needed to produce a specific bend.

Bonding: Connected to establish electrical continuity and conductivity.

Branch circuits: The circuit conductors between the final overcurrent device protecting the circuit and the outlet(s).

Break: The opening of closed contacts to interrupt an electrical circuit.

Brush: A conductor between the stationary and rotating parts of a machine. It is usually made of carbon.

Cable grip: A device used to secure ends of cables to a pulling rope during cable pulls.

Cable pulley: A device used to facilitate pulling conductor in cable tray where the tray changes direction. Several types are available (single, triple, etc.) to accommodate almost all pulling situations.

Capacitance: The storage of electricity in a capacitor; capacitance produces an opposition to voltage change. The unit of measurement for capacitance is the farad (F) or microfarad (µF).

Capstan: The turning drum of a cable puller on which the rope is wrapped and pulled. An increase in the number of wraps increases the pulling force of the cable puller.

Cartridge fuse: A fuse enclosed in an insulating tube to confine the arc when the fuse blows.

Circuit breaker: A device designed to open and close a circuit by nonautomatic means and to open the circuit automatically on a predetermined overcurrent without injury to itself when properly applied within its rating.

Circuit breaker: A device that is designed to open a circuit automatically at a certain overcurrent. It can also be used to manually open and close the circuit.

Clevis: A device used in cable pulls to facilitate connecting the pulling rope to the cable grip.

Color rendering index (CRI): A measurement of the way a light source reproduces color, with a range from 0 to 100. The higher the index number, the closer colors are to how an object appears in full sunlight or incandescent light.

Commutator: A device used on electric motors or generators to maintain a unidirectional current.

Concentric bending: The process of making 90° bends in parallel runs of conduit. This requires increasing the radius of each conduit from the inside of the bend toward the outside.

Conductor support: The act of providing support in vertical conduit runs to support the cables or conductors. The *NEC®* gives several methods in which cables may be supported, including wedges in the tops of conduits and supports to change the direction of cable in pull boxes.

Conduit: Piping designed especially for pulling electrical conductors. Types include RMC, IMC, EMT, PVC, aluminum, and other materials.

Conduit body: A separate portion of a conduit or tubing system that provides access through a removable cover (or covers) to the interior of the system at a junction of two or more sections of the system or at a terminal point of the system. Boxes such as FS and FD or larger cast or sheet metal boxes are not classified as conduit bodies.

Conduit piston: A cylinder of foam rubber that fits inside the conduit and is then propelled by compressed air or vacuumed through the conduit run to pull a line, rope, or measuring tape. Also called a mouse.

Connection: That part of a circuit that has negligible impedance and joins components or devices.

Connector: A device used to physically and electrically connect two or more conductors.

Continuous duty: Operation at a substantially constant load for an indefinitely long time.

Controller: A device that serves to govern, in some predetermined manner, the electric power delivered to the apparatus to which it is connected.

Cross: A four-way *section* of cable tray used when the *tray* assembly must branch off in four *different* directions.

Degree indicator: An instrument designed to indicate the exact degree of bend while it is being made.

Delta-connected motor: A type of motor terminal arrangement in which there are three sets of two coils connected together.

Developed length: The amount of straight pipe needed to bend a given radius. Also, the actual length of the conduit that will be bent.

Dip tolerance: The ability of an HID lamp or luminaire circuit to ride through voltage variations without the lamp extinguishing and cooling down.

Direct rod suspension: A method used to support cable tray by means of threaded rods and hanger clamps. One end of the threaded rod is secured to an overhead structure, while the other end is connected to hanger clamps that are attached to the cable tray side rails.

Drain wire: A wire that is attached to a coaxial connector to allow a path to ground from the outer shield.

Driver: The circuit components in LED lighting that take incoming power and convert it to the required stable voltage and current for the LED.

Dropout: Cable leaving the tray assembly and travelling directly downward; that is, the cable is not routed into a conduit or channel.

Dropout plate: A metal plate used at the end of a cable tray section to ensure a greater c_{ab}le bending radius as the cable leaves the tray assembly.

Dual-element fuse: A fuse having two fuse characteristics; the usual combination is an overcurrent limit and a time delay before activation.

Duty: Describes the length of operation. There are four designations for circuit duty: continuous, periodic, intermittent, and varying.

Edison-base: The standard screw base used for ordinary lamps and Edison-base plug fuses.

Effective ground fault path: An intentionally constructed, permanent, low-impedance electrically conductive path designed and intended to carry current under ground-fault conditions from the point of a ground fault on a wiring system to the electrical supply source and that facilitates the operation of the overcurrent protective device or ground-fault detectors on a high-impedance grounded system.

Efficacy: The light output of a light source divided by the total power input to that source. It is expressed in lumens per watt (LPW).

Elbow: A section of cable tray used to change the direction of the tray assembly a full quarter turn (90°). Both vertical and horizontal elbows are common; a 90° bend

Equipment: A general term including material, fittings, devices, appliances, fixtures, apparatus, and the like used as a part of, or in connection with, an electrical installation.

Equipment bonding jumper: The connection between two or more portions of the equipment grounding conductor.

Equipment grounding conductor (EGC): The conductive path(s) that provide a ground-fault current path and connect normally noncurrent-carrying metal parts of equipment together and to the system grounded conductor, or to the grounding electrode conductor, or both.

Expansion joints: Plates used at intervals along a straight run of cable tray to allow space for thermal expansion or contraction of the tray.

Explosion-proof: Designed and constructed to withstand an internal explosion without creating an external explosion or fire.

Fault current: The current that exists when an unintended path is established between an ungrounded conductor and ground.

Field poles: The stationary portion of a DC motor that produces the magnetic field.

Fish line: Light cord used in conjunction with vacuum/blower power fishing systems that attaches to the conduit piston to be pushed or pulled through the conduit. Once through, a pulling rope is attached to one end and pulled back through the conduit for use in pulling conductors.

Fish tape: A flat iron wire or fiber cord used to pull conductors or a pulling rope through conduit.

Fittings: Devices used to assemble and/or change the direction of cable tray systems.

Frame size: A method used to classify circuit breakers according to given current ranges.

Frequency: The number of cycles an alternating electric current, sound wave, or vibrating object undergoes per second.

Fuse link: The fusible part of a cartridge fuse.

Fuse: A protective device that opens a circuit when the fusible element is severed by heating due to a fault current or overcurrent passing through it.

Gain: The amount of pipe saved by bending on a radius as opposed to right angles. Because conduit bends in a radius and not at right angles, the length of conduit needed for a bend will not equal the total determined length. Gain is the difference between the right angle distances A and B and the shorter distance C—the length of conduit actually needed for the bend.

Grooming: The act of separating the braid in a coaxial conductor.

Ground: The earth or a conducting connection to the earth.

Ground current: Current in the earth or grounding connection.

Ground grid: System of grounding electrodes interconnected by bare cables buried in the earth to provide lower resistance than a single grounding electrode.

Ground mat: System of bare conductors, on or below the surface of the earth, connected to a ground or ground grid to provide protection from dangerous touch voltage.

Ground resistance: The ohmic resistance between a grounding electrode and a remote or reference grounding electrode that are spaced such that their mutual resistance is essentially zero.

Ground rod: A metal rod or pipe used as a grounding electrode.

Grounded: Connected to ground or to a conductive body that extends the ground connection.

Grounded conductor: A system or circuit conductor that is intentionally grounded.

Grounding clip: A listed spring clip used to secure a bonding conductor to an outlet box.

Grounding conductor: A conductor used to connect equipment or the grounded circuit of a wiring system to a grounding electrode or electrodes.

Grounding connections: Connections used to establish a ground; they consist of a grounding conductor, a grounding electrode, and the earth surrounding the electrode.

Grounding electrode: A conducting object through which a direct connection to earth is established.

Grounding electrode conductor (GEC): A conductor used to connect the system grounded conductor or the equipment to a grounding electrode or to a point on the grounding electrode system.

Handhole enclosure: An enclosure used with underground systems to provide access for installation and maintenance.

Hertz (Hz): A unit of frequency; one hertz equals one cycle per second.

Horsepower: The rated output capacity of the motor. It is based on breakdown torque, which is the maximum torque a motor will develop without an abrupt drop in speed.

Hours: The duty cycle of a motor. Most fractional horsepower motors are marked continuous for around-the-clock operation at the nameplate rating in the rated ambient conditions. Motors marked one-half are for 1/2-hour ratings, and those marked one are for 1-hour ratings.

Impedance: The opposition to current flow in an AC circuit; impedance includes resistance (R), capacitive reactance (XC), and inductive reactance (XL). Impedance is measured in ohms (Ω).

Incandescence: The self-emission of visible radiant energy from an object heated above 750°K, such as occurs when an electric current is passed through the filament in an incandescent lamp.

Incident light: The light emitted from a self-luminous object such as the sun, a flame, or an electric source.

Inductance: The creation of a voltage due to a time-varying current; also, the opposition to current change, causing current changes to lag behind voltage changes. The unit of measure for inductance is the henry (H).

Inside diameter (ID): The inside diameter of conduit. All electrical conduit is measured in this manner. The outside dimensions, however, will vary with the type of conduit used.

Instantaneous: Contacts that make or break a circuit immediately upon coil energization or de-energization.

Insulating tape: Adhesive tape that has been manufactured from a nonconductive material and is used for covering wire joints and exposed parts.

Intermittent duty: Operation for alternate intervals of (1) load and no load, (2) load and rest, or (3) load, no load, and rest.

Junction box: An enclosure where one or more raceways or cables enter, and in which electrical conductors can be, or are, spliced.

ick: A bend made to change the direction of the conduit. Kicks are typically less than 90°.

Ladder tray: A type of cable tray that consists of two parallel channels connected by rungs, similar in appearance to the common straight ladder.

Latch coil: The coil of a latching relay that, when energized, causes the relay to assume the switched or latched position.

Leg length: The distance from the end of the straight section of conduit to the bend, measured to the center line or to the inside or outside of the bend or rise.

Lug: A device for terminating a conductor to facilitate the mechanical connection.

Lumen (lm): The basic measurement of light. One lumen is defined as the amount of light cast upon one square foot of the inner surface of a hollow sphere with a one-foot radius with a light source of one candela at its center.

Lumen maintenance: A measure of how a lamp maintains its light output over time. It may be expressed either numerically or as a graph of light output versus time. A commonly used value is L70, the number of hours before the lamp output drops to 70% of its initial value.

Lumens per watt (LPW): A measure of the efficiency, or, more properly, the efficacy of a light source. The efficacy is calculated by taking the lumen output of a lamp and dividing by the lamp wattage. For example, a 100W lamp producing 1,750 lumens has an efficacy of 17.5 lumens per watt.

Luminaire: A complete lighting unit consisting of a light source such as a lamp or lamps, together with the parts designed to position and protect the light source and to distribute the light. It may also include components to match the incoming electricity with the needs of the light source.

Luminance: The intensity of light traveling in a given direction from any surface. It is measured in candela/meter2. The term luminance is commonly used to express brightness.

Main bonding jumper: The connection between the grounded circuit conductor and the equipment grounding conductor at the service.

Make: The closing of open contacts to establish an electrical circuit.

Mechanical advantage (MA): The force factor of a crimping tool that is multiplied by the hand force to give the total force.

Mechanically held relay: A latching relay in which the switched or latched position is maintained by the interference of a mechanical part that moves to a non-interfering position when the unlatch coil is energized, allowing the relay to return to the normal or unlatched position.

Micro (µ): Prefix designating one-millionth of a unit. For example, one microfarad is one-millionth of a farad.

Mogul: A type of conduit body with a raised cover to provide additional space for large conductors.

Molded-case circuit breaker: A circuit breaker enclosed in an insulating housing.

Neutral conductor: The conductor connected to the neutral point of a system that is intended to carry current under normal conditions.

Nonrenewable fuse: A fuse that must be replaced after it interrupts a circuit.

Normally closed (N.C.) contacts: A combination of a stationary contact and a movable contact, which are engaged when the coil is de-energized; in the case of a latching relay, the contacts that are engaged after the unlatch coil is energized.

Normally open (N.O.) contacts: A combination of a stationary contact and a movable contact, which are disengaged when the coil is de-energized; in the case of a latching relay, the contacts that are disengaged after the unlatch coil is energized.

Offset: Two equal bends made to avoid an obstruction blocking the run of the conduit.

One-shot shoe: A large bending shoe that is designed to make 90° bends in conduit.

Outside diameter (OD): The size of any piece of conduit measured on the outside diameter.

Overcurrent protection: De-energizing a circuit whenever the current exceeds a predetermined value; the usual devices are fuses, circuit breakers, or magnetic relays.

Overcurrent: Any current in excess of the rated current of equipment or the ampacity of a conductor. It may result from an overload, short circuit, or ground fault.

Overload: Operation of equipment in excess of the normal, full-load rating, or of a conductor in excess of rated ampacity, which, after a sufficient length of time, will cause damage or dangerous overheating. (A fault, such as a short circuit or ground fault, is not an overload.)

Peak voltage: The peak value of a sinusoidally varying (cyclical) voltage or current is equal to the root-mean-square (rms) value multiplied by the square root of two (1.414). AC voltages are usually expressed as rms values; that is, 120 volts, 208 volts, 240 volts, 277 volts, 480 volts, etc., are all rms values. The peak voltage, however, differs. For example, the peak value of 120 volts (rms) is actually $120 \times 1.414 = 170$ volts.

Periodic duty: Intermittent operation at a substantially constant load for a short and definitely specified time.

Pipe racks: Structural frames used to support the piping that interconnects equipment in outdoor industrial facilities.

Plug fuse: A type of fuse that is held in position by a screw thread contact instead of spring clips, as is the case with a cartridge fuse.

Pole: That portion of a device associated exclusively with one electrically separated conducting path of the main circuit or device.

Pressure connector: A connector applied using pressure to form a cold weld between the conductor and the connector.

Pull box: A sheet metal box-like enclosure used in conduit runs to facilitate the pulling of cables from point to point in long runs, or to provide for the installation of conduit support bushings needed to support the weight of long riser cables, or to provide for turns in multiple-conduit runs.

Radian: An angle at the center of a circle, subtending (opposite to) an arc of the circle that is equal in length to the radius.

Radius: The relative size of the bent portion of a pipe.

Raintight: Constructed or protected so that exposure to a beating rain will not result in the entrance of water under specified test conditions.

Reactance: The opposition to alternating current (AC) due to capacitance (XC) and/or inductance (XL).

Reducing connector: A connector used to join two different size conductors.

Reflected light: Any light that comes from a surface that is not self-illuminating.

Reflection: The bouncing back of light waves or rays by a surface.

Refraction: The bending of a light wave or ray as it passes obliquely (at an angle) from one medium to another of different density, or through layers of different density in the same medium.

Resistivity: Resistance between opposite faces of a unit cube. Expressed in ohm-centimeters or ohms per cubic centimeter.

Revolutions per minute (rpm): The approximate full-load speed at the rated power line frequency. The speed of a motor is determined by the number of poles in the winding.

Rise: The length of the bent section of conduit measured from the bottom, center line, or top of the straight section to the end of the bent section.

Root-mean-square (rms): The square root of the average of the square of the function taken throughout the period. The rms value of a sinusoidally varying voltage or current is the effective value of the voltage or current.

Rotation: For single-phase motors, the standard rotation, unless otherwise noted, is counterclockwise facing the lead or opposite shaft end. All motors can be reconnected at the terminal board for opposite rotation unless otherwise indicated.

Segment bend: Any bend formed by a series of bends of a few degrees each, rather than a single one-shot bend.

Segmented bending shoe: A smaller type of shoe designed for bending segmented bends only (always less than 15°).

Self-inductance: A magnetic field induced in the conductor carrying the current.

Separately derived system: A premises wiring system whose power is derived from a source of electric energy or equipment other than a service. Such systems have no direct electrical connection, including a solidly connected ground circuit conductor, to supply conductors originating in another system.

Setscrew grip: A cable grip, usually with built-in clevis, in which the cable ends are inserted in holes and secured with one or more setscrews.

Shading coil: A shorted turn surrounding a portion of the pole of an AC electromagnet, producing by mutual inductance a delay in the change of the magnetic field in that part, thereby tending to prevent chatter and reduce hum.

Sheave: A pulley-like device used in cable pulls in both conduit and cable tray systems.

Shielding: The metal covering of a cable that reduces the effects of electromagnetic noise.

Short circuit: A conducting connection, whether intentional or accidental, between any of the conductors of an electrical system, whether it is from line-to-line or from line-to-neutral (grounded).

Short circuit: An often unintended low-resistance path through which current flows around, rather than through, a component or circuit.

Soap: Slang for wire-pulling lubricant.

Solid-state relay: A semiconductor switch having no moving parts. These devices control power to the load and are analogous to relay contacts.

Splice: The electrical and mechanical connection between two pieces of cable.

Springback: The amount, measured in degrees, that a bent conduit tends to straighten after pressure is released on the bender ram. For example, a 90° bend, after pressure is released, will pull back about 2° to 88°.

Step voltage: he potential difference between two points on the earth's surface separated by a distance of one pace, or about 3' (1 m).

Strand: A group of wires, usually stranded or braided.

Stub-up: Another name for the rise in a section of conduit.

Supplemental electrode: An additional electrode (commonly a driven rod or pipe) required where the primary electrode is an underground metal water pipe in direct contact with the earth.

Supply-side bonding jumper: A conductor installed on the supply side of a service or within a service equipment enclosure(s), or for a separately derived system, that ensures the required electrical conductivity between metal parts required to be electrically connected.

Sweep bend: A 90° bend with a radius larger than that produced by a standard one-shot shoe.

Swivel plates: Devices used to make vertical offsets in cable tray.

Synchronous speed: The speed of the revolving field of the stator, which is dependent on the supply frequency and number of poles in a stator winding.

System grounding: Intentional connection of one of the circuit conductors of an electrical system to ground potential.

Take-up (deduct): The amount that must be subtracted from the desired stub length to make the bend come out correctly using a point of reference on the bender or bending shoe.

Tee: A section of cable tray that branches off the main section in two other directions.

Tensile strength: The pulling force required to break a rope, wire, or structural beam.

Terminal: A device used for connecting cables.

Termination: The connection of a cable.

Thermal protector: A protective device for assembly as an integral part of a motor or motor compressor that, when properly applied, protects the motor against dangerous overheating due to overload or failure to start.

Touch voltage: The potential difference between a grounded metallic structure and a point on the earth's surface equal to the normal maximum horizontal reach—approximately 3' (1 m).

Trapeze mounting: A method of supporting cable tray using metal channel, such as Unistrut®, Kindorf®, etc., supported by two threaded rods, and giving the appearance of a swing or trapeze.

Tray cover: A flat piece of metal, fiberglass, or plastic designed to provide a solid covering that is needed in some locations where conductors in the tray system may be damaged.

Troffers: Recessed luminaires installed with the opening flush with the ceiling.

Trough: A type of cable tray consisting of two parallel channels (side rails) having a corrugated, ventilated bottom or a corrugated, solid bottom.

Ungrounded conductors: Conductors in an electrical system that are not intentionally grounded.

Unistrut®: A brand of metal channel used as the bottom bracket for hanging cable trays. Double Unistrut® adds strength and stability to the trays and also provides a means of securing future runs of conduit.

Unlatch coil: The coil of a latching relay that, when energized, causes the relay to assume the normal or unlatched position.

Varying duty: Operation at varying loads and/or intervals of time.

Wall mounting: A method of supporting cable tray systems using supports secured directly to the wall.

Watertight: Constructed so that water will not enter the enclosure under specified test conditions.

Weatherproof: Constructed or protected so that exposure to the weather will not interfere with successful operation.

Wye-connected motor: A type of motor terminal arrangement in which three coils are connected together and three coils are isolated.

Wye: A section of cable tray that branches off the main section in one direction.

This page is intentionally left blank.